Applied Mathematical Sciences
Volume 95

Applied Mathematical Sciences

(continued following index)

Wolfgang Hackbusch

Iterative Solution of Large Sparse Systems of Equations

With 16 Illustrations

Springer-Verlag
New York Berlin Heidelberg London Paris
Tokyo Hong Kong Barcelona Budapest

Wolfgang Hackbusch
Institut für Informatik und Praktische Mathematik
Christian-Albrechts-Universität zu Kiel
Olshausenstr. 40
D 24098 Kiel
Germany

Library of Congress Cataloging-in-Publication Data
Hackbusch, W., 1948–
[Iterative Lösung grosser schwachbesetzter Gleichungssysteme, English]
Iterative solution of large sparse systems of equations / Wolfgang Hackbusch.
p. cm.—(Applied mathematical science; v. 95)
Includes bibliographical references and indexes.
ISBN 0-387-94064-2
1. Sparse matrices. 2. Iterative methods (Mathematics)
3. Differential equations, Partial—Numerical solutions. I. Title.
II. Series: Applied mathematical sciences (Springer-Verlag New York Inc.); v. 95.
QA1.A647 vol. 95
[QA188]
510 s—dc20
[512.9′434] 93-2098

Printed on acid-free paper.

Production managed by Henry Krell; manufacturing supervised by Jacqui Ashri.
Typeset by Asco Trade Typesetting Ltd., Hong Kong.
Printed and bound by R.R. Donnelley and Sons, Harrisonburg, VA.
Printed in the United States of America.

9 8 7 6 5 4 3 2 1

ISBN 0-387-94064-2 Springer-Verlag New York Berlin Heidelberg
ISBN 3-540-94064-2 Springer-Verlag Berlin Heidelberg New York

Preface

C. F. Gauß in a letter from Dec. 26, 1823 to Gerling:

> Ich empfehle Ihnen diesen Modus zur Nachahmung. Schwerlich werden Sie je wieder direct eliminiren, wenigstens nicht, wenn Sie mehr als 2 Unbekannte haben. Das indirecte Verfahren lässt sich halb im Schlafe ausführen, oder man kann während desselben an andere Dinge denken.

[C. F. Gauß: Werke vol. 9, Göttingen, p. 280, 1903]

What difference exists between solving large and small systems of equations? The standard methods well-known to any student of linear algebra are applicable to all systems, whether large or small. The necessary amount of work, however, increases dramatically with the size, so one has to search for algorithms that most efficiently and accurately solve systems of 1000, 10,000, or even one million equations. The choice of algorithms depends on the special properties the matrices in practice have. An important class of large systems arises from the discretisation of partial differential equations. In this case, the matrices are sparse (i.e., they contain mostly zeros) and well-suited to iterative algorithms. Because of the background in partial differential equations, this book is closely connected with the author's *Theory and Numerical Treatment of Elliptic Differential Equations*, whose English translation has also been published by Springer-Verlag.

This book grew out of a series of lectures given by the author at the Christian-Albrecht University of Kiel to students of mathematics. It tries to describe the recent state of iterative and related methods, without, however, delving into specialised areas. Even the volume's limitation to iterative techniques entails a selection: Various fast direct algorithms for special problems as well as the optimised versions of Gauß elimination, the Cholesky method, or band-width reduction are not taken into consideration.

Although special attention is devoted to the modern effective algorithms (conjugate gradients, multi-grid methods, parallel techniques), the theory of classical iterative methods should not be neglected. On the other hand, some effective algorithms are not or only marginally considered, if they are connected too closely with discretisation techniques that are not the subject of this book. A discussion of the iterative treatment of nonlinear problems or of eigenvalue problems is completely avoided. A chapter on saddle-point problems (special indefinite problems) arising in many interesting applications could not be realized because of the need to limit the size of the book.

This volume requires no basic mathematical knowledge other than courses on analysis and linear algebra. The principles of linear algebra are summarised in Chapter 2 of this book in order to provide as complete a presentation as possible and present the formulation and notation needed here.

With respect to a course of study, a selection of the given material is best suited to a full-semester course (four hours a week) in the second part of the study (between Vordiplom and Diplom). A partial selection can also be recommended for the second part of a course on numerical analysis.

The exercises cited, which may also be understood as remarks without proof, are an integral part of the presentation. Should this book be used as the basis for an academic course, they can be assigned as problems for students. However, the non-student reader should also try to test his comprehension by working on these exercises.

The discussion of the various methods is illustrated by very many numerical examples, mostly for the Poisson-model problem. To enable the interested reader to test the algorithms with other parameters, step sizes, etc., the PASCAL programs are explicitly given. A collection of the source codes required is available on disk (see [Prog] in the bibliography). The programs can also be used independently of the book for producing numerical examples for courses or seminars.

The present English version contains corrections of several misprints still in the original German edition. New publications have been added to the bibliography. Furthermore, we replaced references to German textbooks by English equivalents as much as possible.

The author would like to thank his colleagues, in particular Mr. J. Burmeister, for their help in proofreading. I am grateful for the stimulating conversations with W. Niethammer, G. Maeß, M. Dryja, G. Wittum, O. Widlund, and others. I also wish to express my gratitude to Teubner (publisher of the German version) and Springer-Verlag for their friendly cooperation on both the German and English editions of this book.

Kiel, April 1993 *W. Hackbusch*

Hints for Reading the Book:

§1: Prelude

§2: Thoughts to serve for reference. However, one should glance through §2.1.

§3: Read §§3.1–3 first. Rest ad libitum.

§4: Chapters 4.2–3 (classical iterations) are the basis of almost all other considerations. §4.4 deals with the corresponding convergence analysis. §§4.1 and 4.7 refer to the Poisson-model problem and serve as illustration. First, §§4.5 and 4.8 may be passed over.

§5: Contains the SOR analysis and may be left out during the first reading.

§6: Independent chapter. The other parts refer only very seldom to §6.

§7: Necessary preparation to §9

§§8–11: Each chapter is independent.

Contents

Notations

Numbers of formulae: Equations in a subchapter x.y are numerated by (x.y.1), (x.y.2) etc. The equation **(3.2.1)** is quoted by **(1)** in the same Section 3.2, while we write **(2.1)** in the other sections of chapter 3.

Numeration of Theorems etc.: All theorems, definitions, and lemmata etc. are enumerated together. The reference to a theorem etc., is analogous to what is said above. **Lemma 3.2.7** is cited as «**Lemma 7**» in Section 3.2, while in the other sections of chapter 3 it is denoted by «**Lemma 2.7**». However, §1 indicates Chapter 1 and never the sections §3.1 or §3.2.1.

Special Symbols, Abbreviations, and Conventions

a, b	bounds for $\sigma(M)$; cf. (4.8.4c), Theorem 7.3.8
A, A_ℓ	matrix of the linear system; cf. (1.2.5), (10.1.8a)
$A^{\alpha\beta}, A^{ij}$	block of A; cf. (2.5.2b, c)
$A_{\alpha\beta}, a_{\alpha\beta}, A_{ij}, a_{ij}$	components of matrix A
b, b_ℓ	right-hand side of the linear system; cf. (1.2.5), (10.1.8a)
blockdiag$\{\dots\}$	block-diagonal matrix; cf. (2.5.3a, b)
blocktridiag$\{\dots\}$	block-tridiagonal matrix; cf. (2.5.4)
$\mathbb{C}$	complex numbers
cond	condition, cond_2: spectral condition; cf. (2.10.7)
$D, D', D_1, \dots$	(block-)diagonal matrix
det	determinant
diag$\{\dots\}$	diagonal matrix or diagonal part; cf. (2.1.5a–c)

e^m	error $x^m - x$ of the mth iterate
E	strictly lower triangular matrix; cf. (1.4.2)
F	strictly upper triangular matrix; cf. (1.4.2)
$G(A)$	graph of a matrix A; cf. (6.2.1)
h, h_ℓ	grid size; cf. (1.2.2)
i, j, k	indices of the ordered index set $I = \{1, \ldots, n\}$
I	identity matrix
I	index set (not necessarily ordered)
I_α	subset of block indices; cf. (2.5.1)
$\mathbb{K}$	the field $\mathbb{R}$ or $\mathbb{C}$
$\mathbb{K}^I$	space of the vectors corresponding to the index set I
$\mathbb{K}^{I \times I}$	space of the matrices corresponding to the index set I
ℓ	level number in the discretisation hierarchy; cf. §10.1.2
$L, L', \hat{L}$	lower (block-)triangular matrix
log	natural logarithm
m	iteration number; cf. x^m
M, M^{xyz}	iteration matrix (of the iteration xyz); cf. §3.2.1
n, n_ℓ	dimension of the linear system; cf. §3.3, (10.1.8b)
N	number of the grid points per row or column; cf. (1.2.2)
N, N^{xyz}	matrix of the 2nd normal form (of iteration xyz); cf. (3.2.4)
$\mathbb{N}$	natural numbers $\{1, 2, 3, ..\}$
$\mathbb{N}_0$	$\mathbb{N} \cup \{0\} = \{0, 1, 2, \ldots\}$
$O(\cdot)$	Landau symbol: $f(\alpha) = O(g(\alpha))$ if $\lvert f(\alpha)\rvert \leqslant C\lvert g(\alpha)\rvert$ for the underlying limit process $\alpha \to 0$ or $\alpha \to \infty$. The notation $f(h) = 1 - O(h^\kappa)$ is more special and means that $f(h) \leqslant 1 - Ch^\kappa$ with fixed $C > 0$ for $h \to 0$.
p	prolongation; §10.1.3, (11.2.1)
Q	often: unitary matrix
r	restriction; cf. §10.1.4, (11.2.8a)
$r(A)$	numerical radius of matrix A; cf. §2.9.5
$\mathbb{R}$	real numbers
range(Φ)	range (image space) of a mapping Φ
S_ℓ	iteration matrix of the smoother $\mathcal{S}_\ell$; cf. Lemma 10.2.1
$\mathcal{S}_\ell$	smoothing iteration; cf. §10.1.1, §10.2.1
span$\{\ldots\}$	linear space spanned by $\{\ldots\}$
T_ℓ, T_r	left- and right-sided transformation; cf. (8.1.4), (8.1.11)
tridiag$\{\ldots\}$	tridiagonal matrix; cf. (2.1.6)
u_{ij}	components of the grid function $u = x$; cf. (1.2.6b)
$U, U', \hat{U}$	upper (block-)triangular matrix
W, W^{xyz}	matrix of the 3rd normal form (of iteration xyz); cf. (3.2.5)
x	vector; often solution of the equation $Ax = b$
x^*	solution of the equation $Ax = b$, if the symbol x is needed as a variable
x_ℓ, x_ℓ^*	vectors x, x^* at the level ℓ; cf. §10
x^0	starting value of the iteration

x^m	mth iterate
x^α, x^i	block of x corresponding to the index α or i; cf. (2.5.2a)
x_α, x_i	components of a vector x
$\mathbb{Z}$	integers $\{0, \pm 1, \pm 2, \ldots\}$

Greek Letters

α, β, γ	indices of the index set; cf. §2.1
γ	in §10: number of the secondary multi-grid steps for the coarse-grid equation; cf. (10.4.2d$_2$)
γ, Γ	lower and upper eigenvalue bounds of $W^{-1}A$; cf. (4.8.4c)
δ_{ij}	Kronecker symbol: $\delta_{ij} = 1$ for $i = j$, $\delta_{ij} = 0$ otherwise
ζ	often contraction number; cf. (10.3.21b), (10.5.3)
$\eta(\nu)$	zero sequence for smoothing property, cf. (10.6.4b)
$\eta_0(\nu)$	special function, defined in Lemma 10.6.1
ϑ, Θ	damping factor; cf. §§4.3.1–2
$\kappa(A)$	condition number (2.10.8)
λ, Λ	eigenvalue bounds of A; cf. Theorem 9.2.3, Theorem 4.4.8
$\lambda_{\max}(A)$	maximal eigenvalue of the matrix A, if $\sigma(A) \subset \mathbb{R}$
$\lambda_{\min}(A)$	minimal eigenvalue of the matrix A, if $\sigma(A) \subset \mathbb{R}$
ν, ν_1, ν_2	in §10: number of the smoothing steps; cf. §§10.2.1–2
$\rho(A)$	spectral radius of the matrix A; cf. Definition 2.4.13
$\rho_{m+k,m}$	convergence factors; cf. (3.2.18a, b)
$\sigma(A)$	spectrum of the matrix A; cf. (2.4.1)
ω	relaxation parameter; cf. §4.3.3.1
Ω_h	grid; cf. (1.2.3)

Symbols

$\mathbb{1}$	vector $(1, 1, \ldots, 1)^T$
A^T, A^H, A^{-H}	cf. (2.1.1/2)
Δ	Laplace operator; cf. (1.2.1a)
$\langle \cdot, \cdot \rangle$	(Euclidean) scalar product; cf. (2.2.1a–c)
$\langle \cdot, \cdot \rangle_A$	energy scalar product; cf. (2.10.5b)
$\Vert \cdot \Vert$, $\vert\vert\vert \cdot \vert\vert\vert$	norm (of vectors or matrices)
$\Vert \cdot \Vert_A$	energy norm; cf. (2.10.5a)
$\Vert \cdot \Vert_2$	Euclidean norm; cf. (2.6.2). spectral norm; cf. (2.9.4a)
$\Vert \cdot \Vert_\infty$	maximum norm; cf. (2.6.2). row sum norm; cf. (2.6.8)
$\Vert \cdot \Vert_{Y \leftarrow X}$	cf. (2.6.11): norm of a mapping (matrix) from X into Y
$\vert \cdot \vert$	absolute value, in §6 also applied to matrices
$<$, $\leqslant$, $>$, $\geqslant$	in connection with matrices, the order relation from §2.10.2 is used; only in §6 (and parts of §8.5) is the order relation of (6.1.1a, b) meant

1
Introduction

1.1 Historical Remarks Concerning Iterative Methods

Iterative methods are almost 170 years old. The first iterative method for systems of linear equations is due to Carl Friedrich Gauß (the less correct spelling is: Gauss). His least squares method led him to a system of equations that was too large for the use of direct Gauß elimination. The iterative method described in "Supplementum theoriae combinationis observationum erroribus minime obnoxiae" (1819–1822) today would be called the blockwise Gauß-Seidel method. The value that Gauß attributed to his iterative method can be seen in the excerpt from his letter of 1823 preceding the foreword of this book.

A very similar method was described by Carl Gustav Jacobi in his 1845 paper "Über eine neue Auflösungsart der bei der Methode der kleinsten Quadrate vorkommenden linearen Gleichungen" ("On a new solution method of linear equations arising from the least square method"; *Astronom. Nachr.*). In 1874 Phillip Ludwig Seidel, a student of Jacobi, wrote about "a method, to solve the equations arising from the least squares method as well as general linear equations by successive approximation" ("Über ein Verfahren, die Gleichungen, auf welche die Methode der kleinsten Quadrate führt, sowie lineare Gleichungen überhaupt, durch successive Annäherung aufzulösen"; *Münch. Abh.*).

Since the time that electronic computers became available for solving systems of equations, the number of equations has increased many orders of magnitudes and the methods mentioned above have proved to be too slow. After 100 years of stagnation in this field, Southwell [1–3] experimented with variants of the Gauß-Seidel method («relaxation») and, in 1950, Young [1] succeeded in a breakthrough. His variant of the Gauß-Seidel method leads to an important acceleration of the convergence. This so-called SOR method will be described in §1.4 as an example of an iterative method. Since then, numerous other methods even more effective have been developed. Most of them are discussed in this book.

1.2 Model Problem (Poisson Equation)

During the time of Gauß, Jacobi and Seidel, the equations of the least square method have led to a larger number of equations. Today, in particular the boundary-value problems, i.e. the partial differential equations of elliptic type, lead to a number of equations of orders between 1000 and one million. In the following we will often refer to a model problem representing the simplest nontrivial example of a boundary-value problem. It is the *Poisson equation*

$$-\Delta u(x, y) = f(x, y) \quad \text{for } (x, y) \in \Omega, \tag{1.2.1a}$$

$$u(x, y) = \varphi(x, y) \quad \text{on } \Gamma = \partial\Omega. \tag{1.2.1b}$$

Here

$$\Delta = \frac{\partial^2}{\partial x^2} + \frac{\partial^2}{\partial y^2}$$

abbreviates the *Laplace operator*. As domain Ω we choose the unit square

$$\Omega = (0, 1) \times (0, 1). \tag{1.2.1c}$$

In (1a, b) the *source term* f and the *boundary value* φ are given, while the function u is unknown.

To discretise the differential equation (1a–c) the domain Ω is covered with a grid of the step size h (cf. Figure 1a). Each grid point (x, y) must have the representation $x = ih$, $y = jh$ $(0 \leqslant i,j \leqslant N)$, where

$$h = 1/N. \tag{1.2.2}$$

By the term *grid* the set of *inner* grid points is meant:

$$\Omega_h := \{(x, y) = (ih, jh)\colon 1 \leqslant i,j \leqslant N-1\}. \tag{1.2.3}$$

We abbreviate the desired values $u(x, y) = u(ih, jh)$ by u_{ij}. An approximation for the differential equation (1a) is given by the so-called *five-point formula*

$$h^{-2}[4u_{ij} - u_{i-1,j} - u_{i+1,j} - u_{i,j-1} - u_{i,j+1}] = f_{ij} \tag{1.2.4a}$$

with $f_{ij} := f(ih, jh)$ for $1 \leqslant i, j \leqslant N-1$. The left side in (4a) coincides with $-\Delta u(ih, jh)$ up to a consistency error $O(h^2)$, when the solution u of (1a, b) is inserted (cf. Hackbusch [15]). For grid values on the boundary, i.e. for $i = 0$, $i = N, j = 0$ or $j = N$, u_{ij} is known from the boundary data (1b):

$$u_{ij} := \varphi(ih, jh) \quad \text{for } i = 0, i = N, j = 0 \text{ or } j = N. \tag{1.2.4b}$$

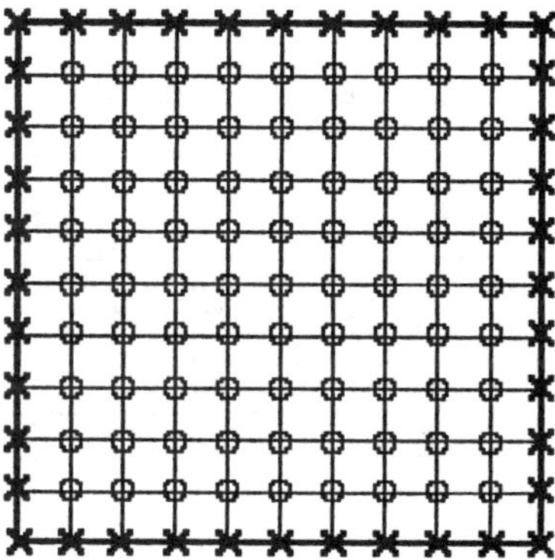

Fig. 1.2.1a Inner grid points (○) and boundary points (×)

The number of the unknown u_{ij} is $n := (N-1)^2$ and corresponds to the number of the inner grid points. In order to form the system of equations, one has to eliminate the boundary values (4b), which possibly may appear in (4a). For example for $N \geqslant 3$ the equation corresponding to the index $(i,j) = (1,1)$ reads

$$h^{-2}[4u_{11} - u_{12} - u_{21}] = g_{11} \quad \text{with } g_{11} := f(h,h) + h^{-2}[\varphi(0,h) + \varphi(h,0)].$$

In order to write the equations in the commonly used matrix formulation

$$Ax = b \tag{1.2.5}$$

with an $n \times n$ matrix A and n-dimensional vectors x and b with $n = (N-1)^2$, one is forced to represent the twofold indexed unknowns u_{ij} by a singly indexed vector x. This implies that the (inner) grid points must be enumerated in some way. Figure 1b shows the lexicographic ordering. The exact definition of the matrix A and of the right-hand side b can be seen from the following

Definition of the matrix A and of the vectors b by lexicographical ordering for the Poisson-model problem: (1.2.6a)

$N1 := N - 1$; $n := N1 * N1$; $h2 := 1/(N * N)$;
$k := 0$; $\{1 \leqslant k \leqslant n$ is the index with respect to the lex. ordering$\}$
$A := \mathbf{0}$; {all entries of A are initialised by zero}
for $j := 1$ to $N1$ do for $i := 1$ to $N1$ do
begin $k := k + 1$; $a_{kk} := 4 * h2$; $b_k := f(ih, jh)$;
 if $i > 1$ then $a_{k-1,k} := -h2$ else $b_k := b_k + h2 * \varphi(0, jh)$;
 if $i < N1$ then $a_{k+1,k} := -h2$ else $b_k := b_k + h2 * \varphi(1, jh)$;
 if $j > 1$ then $a_{k,k-N1} := -h2$ else $b_k := b_k + h2 * \varphi(ih, 0)$;
 if $j < N1$ then $a_{k,k+N1} := -h2$ else $b_k := b_k + h2 * \varphi(ih, 1)$
end;

Vice versa, the solution x of $Ax = b$ is to be interpreted as

$$x_k = u_{ij} = u(ih, jh) \quad \text{for } k = i + (j-1)*(N-1) \text{ with } 0 \leqslant i,j \leqslant N. \tag{1.2.6b}$$

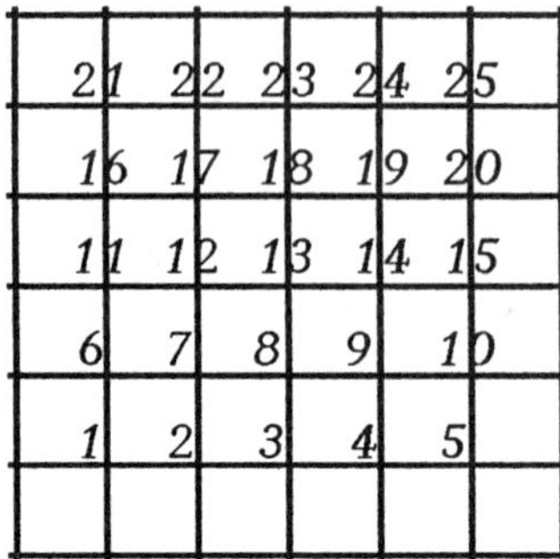

Fig. 1.2.1b Lexicographical ordering

When x is interpreted as the grid function, we shall use the notation u_{ij} or $u(x, y)$ with $x = ih$, $y = jh$.

Remark 1.2.1. The reformulation of the two-dimensionally ordered unknowns into a one-dimensionally ordered vector is rather unobvious. The reason should not be sought in the two-dimensional nature of the problem, but rather in the questionable concept of enumerating the components of a vector from 1 to n. We shall see that **the matrix A will never be needed in the presentation (2.6a).**

If, nevertheless, one wants to represent the matrix A in the usual form, A must be written as a *block matrix*. The vector x decomposes naturally into $N - 1$ blocks

$$x^{(j)} := \begin{bmatrix} x_{k+1} \\ \vdots \\ x_{k+N-1} \end{bmatrix} = \begin{bmatrix} u_{1,j} \\ \vdots \\ u_{N-1,j} \end{bmatrix} \quad \begin{array}{l} \text{with } k := (j-1)*(N-1) \\ \text{for } j = 1, \ldots, N-1, \end{array} \tag{1.2.7}$$

corresponding to the jth row in the grid Ω_h. Accordingly, A takes the form of a block-tridiagonal matrix built from $(N-1) \times (N-1)$ blocks T, which again are tridiagonal $(N-1) \times (N-1)$ matrices:

$$A = h^{-2} \begin{bmatrix} T & -I & & & & \\ -I & T & -I & & & \\ & -I & T & -I & & \\ & & \ddots & \ddots & \ddots & \\ & & & -I & T & -I \\ & & & & -I & T \end{bmatrix}, \quad T = \begin{bmatrix} 4 & -1 & & & & \\ -1 & 4 & -1 & & & \\ & -1 & 4 & -1 & & \\ & & \ddots & \ddots & \ddots & \\ & & & -1 & 4 & -1 \\ & & & & -1 & 4 \end{bmatrix}. \tag{1.2.8}$$

I is the $(N-1) \times (N-1)$ identity matrix. Unmarked matrix entries or blocks are zeros or zero blocks, respectively. The representation (8) proves

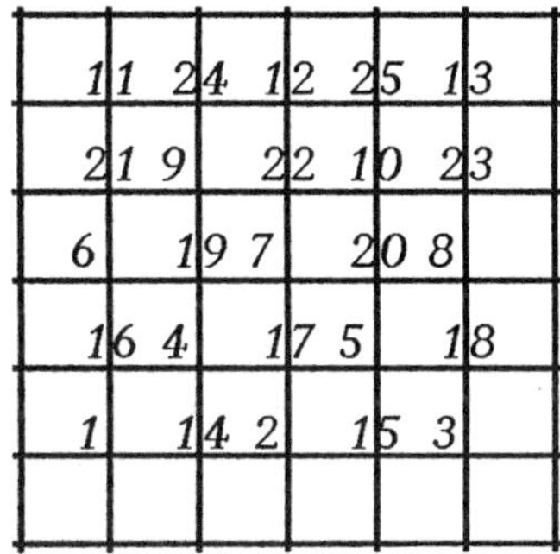

Fig. 1.2.1c Chequer-board ordering

Remark 1.2.2. For the lexicographical ordering of the unknowns, the matrix A has a block-tridiagonal structure.

Lexicographical ordering is by no means the only ordering one can think of. Another frequently used aproach is the chequer-board ordering (cf. Fig. 1c). In that case, the components u_{ij} with an even sum $i + j$ («black squares») are enumerated first and thereafter those with an odd sum $i + j$ («red squares») are numbered lexicographically. In the course of the next chapters further orderings will be mentioned. A broad collection of orderings of practical interest is given by Duff–Meurant [1].

Exercise 1.2.3. In the case of the chequer-board ordering, A decomposes into two blocks corresponding to the «red» and «black» indices. Prove that A has the block structure (9) with a rectangular submatrix B and identity matrices I_r, I_b, where the block size is given by the number of the red and black grid points:

$$A = \begin{bmatrix} D_r & B \\ B^T & D_b \end{bmatrix}, \qquad D_r = 4h^{-2}I_r, \qquad D_b = 4h^{-2}I_b. \tag{1.2.9}$$

1.3 Amount of Work for the Direct Solution of the System of Equations

Those methods are called *direct* that terminate after finitely many operations with an exact solution (up to rounding errors). The best known direct method is Gauß elimination. In the case of the model problem from §1.2, one may perform this method *without* pivoting (cf. §6.4.4). Concerning the valuation of the amount of computational work, we do not distinguish between additions, subtractions, multiplications, or divisions. Each is counted as one (arithmetical) operation. Traditionally, arithmetical operations for indices, data transfer, and similar activities are not counted.

Remark 1.3.1. In the general case the Gauß elimination for the solution of a system of n equations $Ax = b$ requires $2n^3/3 + O(n^2)$ operations. The storage amounts to $n^2 + n$.

Proof. During the ith elimination step, the ith row contains $n - i$ nonzero elements, whose multiples are to be subtracted from $n - i - 1$ matrix rows. Summation of these $2(n - i)^2 + O(n)$ operations over $1 \leqslant i \leqslant n$ yields the assertion. □

In the model case, n equals $(N - 1)^2 = h^{-2} + O(h^{-1})$ and implies

Remark 1.3.2. A naive application of the Gauß elimination to the model problem from §1.2 leads to $2N^6/3 + O(N^5) = 2h^{-6}/3 + O(h^{-5})$ operations and requires a storage of $N^4 + O(N^3) = h^{-4} + O(h^{-3})$.

Halving the grid size h yields the 64-fold amount of computational work. If we assume 1 CPU sec for the solution of grid size h, the same computation for the quartered grid size $h/4$ consumes more than 1 CPU hr!

However, the amount of work diminishes if the system matrix A is a band matrix.

Definition 1.3.3. A is a *band matrix* of *band width* w, if $a_{ij} = 0$ for all indices with $|i - j| > w$.

A band matrix has at maximum $2w$ side-diagonals besides the main diagonal. Concerning the analysis of the properties of band matrices, we refer to Berg [1].

Remark 1.3.4. The matrix A arising from the lexicographical ordering for the model problem according to (2.7) is a band matrix of band width $w = N - 1$.

The major part of the amount of work given in Remark 1 consists of unnecessary multiplications and additions by zeros. During the ith elimination step the ith row contains $w + 1$ nonzero elements. It suffices to eliminate the next w rows. This leads to $2w^2$ operations. In total one obtains

Remark 1.3.5. The amount of work for the Gauß elimination without pivoting for solving a system with an $n \times n$ matrix of band width w amounts to $2nw^2 + O(nw + w^3)$. The storage requirement reduces to $2n(w + 1)$, when only the $2w + 1$ diagonals of A and the right-hand side b are stored.

Remark 1.3.6. In the case of the model problem from §1.2, w equals $N - 1$. Therefore, the banded Gauß elimination requires $2N^4 + O(N^3) = 2h^{-4} + O(h^{-3})$ operations and a storage of $2N^3 + O(N^2)$.

In the latter version, $2w + 1$ diagonals of A are used, although the matrix A from (2.8) has only five diagonals: the main diagonal, two side-diagonals with distance 1, and two further ones with distance $N - 1$. Unfortunately, one cannot exploit this property for the Gauß elimination because of

Remark 1.3.7. The zeros in the second to $(N - 2)$th side-diagonals of the matrix A from (2.8) are completely filled during the elimination process with nonzeros (with the exception of the first block).

This occurrence is called *fill-in* and indicates a principal disadvantage of Gauß elimination when applied to *sparse* matrices. Here, we call an $n \times n$ matrix sparse, if the number of nonzero entries is by far smaller than n^2. Otherwise, the matrix is called a *full* matrix. Because of the equivalence of Gauß elimination to the *triangular or LU decomposition* (cf. Stoer [1, §4.1], Stoer-Bulirsch [1, §4.1], the same difficulties exist in the latter technique.

Remark 1.3.8. The decomposition $A = LU$ into a lower triangular matrix L and an upper triangular matrix U for the sparse matrix A from (2.8) yields factors L and U, which are full band matrices of width $w = N - 1$. The same holds for Cholesky decomposition.

There are special direct methods solving the system described in §1.2 with an amount of work between $O(n) = O(N^2)$ and $O(n \log n) = O(N^2 \log N)$. Examples are the Buneman algorithm and the method of total reduction, both described in Meis–Markowitz [1] (cf. also Buneman [1], Björstad [1], Duff–Erisman–Reid [1], Schröder–Trottenberg [1]).

1.4 Examples of Iterative Methods

For the iterative solution of a system of equations, one starts with an arbitrary *starting vector* x^0 and computes a sequence of *iterates* (*iterands*) x^m for $m = 1, 2, \ldots$:

$$x^0 \mapsto x^1 \mapsto x^2 \mapsto x^3 \mapsto \cdots \mapsto x^m \mapsto x^{m+1} \mapsto \cdots.$$

In the following x^{m+1} is dependent only on x^m, so that the mapping $x^m \mapsto x^{m+1}$ determines the iteration method. The choice of the starting value x^0 is not part of that method.

The already mentioned Gauß-Seidel iteration for solving the system (2.5): $Ax = b$ is as follows:

Gauß-Seidel iteration

$$\text{for } i := 1 \text{ to } n \text{ do } x_i^{m+1} := \left(b_i - \sum_{j=1}^{i-1} a_{ij}x_j^{m+1} - \sum_{j=i+1}^{n} a_{ij}x_j^m\right)\Big/ a_{ii} \quad (1.4.1)$$

Remark 1.4.1. (a) The Gauß-Seidel iteration (1) can be performed whenever all diagonal entries satisfy $a_{ii} \neq 0$.
(b) During the execution of the iteration, the variable x_i^m may be overwritten by the new value x_i^{m+1}.
(c) Different orderings (e.g., lexicographical or chequer-board ordering) yield different results.

Each matrix A can be uniquely decomposed into the sum

$$A = D - E - F \qquad \left\{\begin{array}{l} D \text{ diagonal matrix,} \\ E \text{ strictly lower triangular matrix} \\ F \text{ strictly upper triangular matrix} \end{array}\right\}. \quad (1.4.2)$$

Here, E is called a *lower triangular matrix*, if $E_{ij} = 0$ for $j > i$, and a *strictly lower triangular matrix*, if $E_{ij} = 0$ for $j \geqslant i$. The (*strictly*) *upper triangular matrix* is defined analogously. The system of equations $Ax = b$ is equivalent to

$$(D - E)x = b + Fx. \quad (1.4.3)$$

Replacing x by x^m on the right-hand side and by x^{m+1} on the left-hand side, one obtains the iterative description (4a) or (4b):

$$(D - E)x^{m+1} = b + Fx^m, \quad (1.4.4a)$$

$$x^{m+1} = (D - E)^{-1}(b + Fx^m). \quad (1.4.4b)$$

Exercise 1.4.2. Prove that (4a/b) and (1) are equivalent, i.e. (4a) and (4b) are the vector representations of the Gauß-Seidel iteration, while (1) is the componentwise representation.

A PASCAL procedure performing one iteration step $x^m \mapsto x^{m+1}$ in the case of a *general* matrix A could be as follows:

```
const n = ···;
type vector = array [1:n] of real; matrix = array [1:n] of vector;

procedure GaußSeidel (var x, b: vector; var A: matrix);
var i, j: integer; s: real;
begin for i := 1 to n do
      begin s := 0; for j := 1 to i - 1 do s := s + A[i,j] * x[j];        (1.4.5)
                    for j := i + 1 to n do s := s + A[i,j] * x[j];
            x[i] := (b[i] - s)/A[i,i]
end   end;
```

Instead of defining A and b for the model problem (2.4a, b) by (2.6a) and using (5), one should use the original data $f_{ij} = f[i,j]$ from (2.4a) and the boundary data $\varphi(ih, jh) = u_{ij} = u[i,j]$, which are to be stored at the boundary points of the array u. According to (2.6b), we use the variables u and f instead of x and b. The Gauß-Seidel method for the model problem then takes the following form:

```
const N = ··· {N from (2.2)};
type gridfunction = array [0:N,0:N] of real;
var u, f: gridfunction; i, j: integer;

procedure GaußSeidel (var u, f: gridfunction);
var i, j: integer; h2: real;
begin h2 := 1/(N * N);                              {h2 = h²}
for j := 1 to N − 1 do for i := 1 to N − 1 do       {lex. ordering}    (1.4.6)
  u[i,j]:=
  (h2 * f[i,j]+u[i − 1,j]+u[i + 1,j]+u[i,j − 1]+u[i,j + 1])/4
end;

begin {main program for the example f = −4, φ(x, y) = x² + y²}
  for i := 1 to N − 1 do for j := 1 to N − 1 do f[i,j] := −4;
  for i := 0 to N do
  begin
    u[i,0]:=i * i/(N * N);       u[0,i]:=u[i,0];  {definition of the
    u[i,N]:=1+i * i/(N * N); u[N,i]:=u[i,1]        boundary values}
  end;
  for i := 1 to N − 1 do for j := 1 to N − 1 do u[i,j] := 0; {start}
  for i := 1 to 300 do GaußSeidel (u,f); ...                {iteration}
end.
```

In the double loop of (6), the matrix A is explicitly represented by its nonzero entries. The indexing is based on the «natural» double indices. The lexicographical ordering of the grid points is a consequence of the arrangement of the j- and i-loop. For chequer-board ordering, the loop in (6) can be changed as follows:

```
w := 0; {«red squares»} for j := 1 to N − 1 do
begin w := −1 − w; i := w; while i ≤ N − 3 do
  begin i := i + 2; u[i,j] := (h2 * f[i,j] + u[i − 1,j]
                                                         (1.4.7)
                  + u[i + 1,j] + u[i,j − 1] + u[i,j + 1])/4
end end;
w := −1; {«black squares»} for j := 1 to N − 1 do ... {as above}
```

Since one may immediately store $h2 * f[i,j]$ instead of $f[i,j]$, one obtains

Remark 1.4.3. In the case of the model problem, the Gauß-Seidel method (independently of the ordering) requires $5n$ operations ($4n$ additions and n divisions) per iteration.

Remark 1.4.4. Algorithm (6) realises the example $f_{ij} = -4$, $\varphi(x, y) = x^2 + y^2$. The corresponding solution is $u_h(x, y) = x^2 + y^2$, i.e., $u_{ij} = (i^2 + j^2)h^2$. As a starting value, $u_{ij}^0 = 0$ is used. In the following, the system (4a) with these data will be called the *Poisson-model problem.* In the course of the next chapters, various iterative methods will be tested for this example.

Table 1 shows the error $\varepsilon_m := \max\{|u_{ij}^m - (i^2 + j^2)h^2|: 1 \leqslant i,j \leqslant N-1\}$ of the mth iterate for $h = 1/32$ and the value $u_{16,16}^m$ at the midpoint $(16h, 16h) = (\frac{1}{2}, \frac{1}{2})$, which should converge to $u(\frac{1}{2}, \frac{1}{2}) = 0.5$. The listed values indicate the convergence of the Gauß-Seidel method, but its slowness is disappointing. After 100 iterations the first decimal of $u_{16,16}^m$ is still completely wrong! The third column contains the so-called «reduction factor»: the ratio $\varepsilon_{m-1}/\varepsilon_m$ of the successive errors. The factor indicates how fast the error decreases per iteration. A comparison of the data in Table 1 demonstrates that ordering influences the results, but not the convergence speed.

The Gauß-Seidel iteration (1) is equivalent to the representation

Table 1.4.1 Results of the Gauß-Seidel iteration for $N = 32$

	lexicographical ordering			chequer-board ordering		
m	$u_{16,16}$	ε_m	$\varepsilon_{m-1}/\varepsilon_m$	$u_{16,16}$	ε_m	$\varepsilon_{m-1}/\varepsilon_m$
0	0.0	1.877		0.0	1.877	
1	−0.002	1.760	0.93756	−0.001	1.759	0.93704
2	−0.004	1.646	0.93563	−0.003	1.589	0.90323
9	−0.018	1.276		−0.017	1.202	
10	−0.019	1.246	0.97637	−0.019	1.165	0.96903
99	0.1102	0.404		0.1353	0.380	
100	0.1135	0.400	0.98989	0.1385	0.376	0.98994
199	0.3479	0.152		0.3585	0.142	
200	0.3494	0.151	0.99041	0.3598	0.140	0.99041
299	0.4421	0.058		0.4461	0.054	
300	0.4426	0.057	0.99039	0.4466	0.053	0.99039

Table 1.4.2 SOR results (lexicographical ordering, $N = 32$, $\omega = 1.821465$)

m	$u_{16,16}^m$	ε_m	$\varepsilon_{m-1}/\varepsilon_m$	m	$u_{16,16}^m$	ε_m	$\varepsilon_{m-1}/\varepsilon_m$
0	0.0	1.877					
1	−0.016	1.777	0.94677	39	0.4805	0.050	
2	−0.027	1.680	0.94512	40	0.4838	0.043	0.85661
9	−0.065	1.046		49	0.4964	0.0055	
10	−0.070	0.962	0.91970	50	0.4970	0.0049	0.88303
19	0.1111	0.399		99	0.4999996	$9.05_{10}-7$	
20	0.1486	0.365	0.91550	100	0.4999997	$7.23_{10}-7$	0.79768
29	0.4198	0.166		129	$0.5 - 1.5_{10}-9$	$3.57_{10}-9$	
30	0.4445	0.150	0.90620	130	$0.5 - 1.2_{10}-9$	$2.81_{10}-9$	0.78805

$$\text{for } i := 1 \text{ to } n \text{ do } x_i^{m+1} := x_i^m - \left(\sum_{j=1}^{i-1} a_{ij}x_j^{m+1} + \sum_{j=i}^{n} a_{ij}x_j^m - b_i\right)\Big/ a_{ii}, \tag{1.4.8}$$

which shows that x_i^{m+1} is obtained from x_i^m by subtraction of a correction. Note that, in contrast to (1), the second sum in (8) starts with $j = i$. A seemingly insignificant modification is the multiplication of this correction by a factor ω. The method obtained is called «*successive overrelaxation method*» and abbreviated *SOR*. In the general case, it is as follows.

SOR method

$$\text{for } i := 1 \text{ to } n \text{ do } x_i^{m+1} := x_i^m - \omega\left(\sum_{j=1}^{i-1} a_{ij}x_j^{m+1} + \sum_{j=i}^{n} a_{ij}x_j^m - b_i\right)\Big/ a_{ii} \tag{1.4.9}$$

In the model case, the only change in (6) or (7) is the replacement of the assignment to u by $u[i,j] := u[i,j] - \omega * (4 * u[i,j] - u[i-1,j] - u[i+1,j] - u[i,j-1] - u[i,j+1] - h2 * f[i,j])/4$. In §5.6 we will prove that $\omega = 2/[1 + \sin(\pi h)]$ (i.e., $\omega = 1.821 \ldots$ for $N = 32$) is a suitable value. Table 2 shows the error ε_m of the first 150 iterations for the same example as above. Convergence is evidently much faster than for the Gauß-Seidel method. Analysis of the above-mentioned methods and the construction of even more effective iterations are the focus of the next chapters.

2
Recapitulation of Linear Algebra

2.1 Notations for Vectors and Matrices

2.1.1 Nonordered Index Sets

According to Remark 1.2.1, the indices of the vectors are considered as *nonordered*. The (always finite) *index set* is denoted by I. A vector $b \in \mathbb{R}^I$ or $b \in \mathbb{C}^I$ is a mapping $b\colon I \to \mathbb{K}$ into $\mathbb{K} = \mathbb{R}$ (real case) or $\mathbb{K} = \mathbb{C}$ (complex case). The value of b at $\alpha \in I$ is denoted as vector component b_α. A vector composed of its components b_α is written in the form

$$b = (b_\alpha)_{\alpha \in I}.$$

If the index set is *ordered*, we will identify the indices with $1, 2, \ldots, n := \#I$ (number of elements in I) except when another notation is explicitly given. While the indices $\alpha, \beta, \gamma, \ldots$ are used for nonordered indices, we use $i, j, k, \ldots$ in the ordered case.

Remark 2.1.1. Let $n := \#I$ and $I_n = \{1, \ldots, n\}$. The ordering of the elements of I can be represented by a surjective mapping $\alpha\colon I_n \to I$. $\alpha(i) \in I$ is the ith index in I. Replacing the name $\alpha(i)$ by i, one obtains the above-mentioned identification of I with I_n.

In general, a subscript is only used for the notation of a component. Sometimes, a subscript has to be used for vectors, e.g., the first column vector of a matrix A could be written as $\boldsymbol{a}_1$. In order to avoid confusion with vector components, the vectors will be represented in boldface as in the previous example. If not defined differently, $\boldsymbol{e}_\alpha$ abbreviates the *α-unit vector* with the components $(e_\alpha)_\beta = \delta_{\alpha\beta}$. Here,

$$\delta_{\alpha\beta} = 1 \quad \text{for} \quad \alpha = \beta \quad \text{and} \quad \delta_{\alpha\beta} = 0 \quad \text{for } \alpha \neq \beta \qquad (\alpha, \beta \in I)$$

is the *Kronecker symbol.*

Square matrices are mappings of the set $I \times I$ of index pairs into $\mathbb{K}$. The set of these matrices is denoted by $\mathbb{K}^{I \times I}$. Matrices are always symbolised by uppercase letters. The matrix entry of A corresponding to the index pair $(\alpha, \beta) \in I \times I$ is written as $a_{\alpha\beta}$ or $a_{\alpha,\beta}$, and occasionally as $A_{\alpha\beta}$. In particular, the notation $(A + B)_{\alpha\beta}$, $(A^{-1})_{\alpha\beta}$, etc. is used for the components of matrix expressions. The matrix composed of the entries $a_{\alpha\beta}$ is denoted by

$$A = (a_{\alpha\beta})_{\alpha,\beta \in I}.$$

In this notation the matrix multiplication reads $(AB)_{\alpha\beta} = \sum_{\gamma \in I} a_{\alpha\gamma} b_{\gamma\beta}$. Similarly, we have $(Ax)_\alpha = \sum_{\beta \in I} a_{\alpha\beta} x_\beta$. The symbol

$$I = (\delta_{\alpha\beta})_{\alpha,\beta \in I}$$

also abbreviates the *identity matrix*, since it cannot be confused with the index set I.

In the case of rectangular (sub)matrices, the indices α and β may belong to different sets I and J: $A = (a_{\alpha\beta})_{\alpha \in I, \beta \in J}$ is a $I \times J$ matrix. The set of these matrices is $\mathbb{K}^{I \times J}$. If the set J contains only one element, the $I \times J$ matrix can be identified with an I vector (more precisely, column vector). If vice versa, I has only one element, $I \times J$ matrices can be considered J-row vectors.

For an ordered index set I, e.g., $I = \{1, \ldots, n\}$, we use the standard index notation a_{ij} or A_{ij}.

2.1.2 Notations

Let $A = (a_{\alpha\beta})_{\alpha,\beta \in I}$. A^T denotes the *transposed* matrix, $\bar{A}$ the *complex conjugate* matrix and A^H the *adjoint* (or Hermitian transposed) matrix corresponding to A:

$$A^T = (a_{\beta\alpha})_{\alpha,\beta \in I}, \qquad \bar{A} = (\bar{a}_{\alpha\beta})_{\alpha,\beta \in I}, \qquad A^H = \bar{A}^T = (\bar{a}_{\alpha\beta})_{\alpha,\beta \in I}. \quad (2.1.1)$$

The notation

$$A^{-H} := (A^H)^{-1} \qquad (2.1.2)$$

has been adopted for the inverse of the adjoint matrix. In this book the notation A^T is used for real-valued matrices only, therefore, A^T coincides with A^H. Since $(x_1, x_2, \ldots)$ indicates a row vector, $(x_1, x_2, \ldots)^T$ is used for a column vector. The definition of the notations A^T, A^H for rectangular matrices $A \in \mathbb{K}^{I \times J}$, $I \neq J$, is analogous.

Exercise 2.1.2. Prove the following rules for ^{T} and ^{H} (where $\lambda \in \mathbb{K}$):

$$(A+B)^T = A^T + B^T, \quad (AB)^T = B^T A^T, \quad (\lambda A)^T = \lambda A^T, \quad (A^{-1})^T = (A^T)^{-1}, \tag{2.1.3a}$$

$$(A+B)^H = A^H + B^H, (AB)^H = B^H A^H, (\lambda A)^H = \bar{\lambda} A^H, (A^{-1})^H = (A^H)^{-1} = A^{-H}. \tag{2.1.3b}$$

Definition 2.1.3. Let $A \in \mathbb{K}^{I \times I}$. The matrix A is called

symmetric, if $A = A^T$, (2.1.4a)

Hermitian, if $A = A^H$, (2.1.4b)

regular, if A^{-1} exists, (2.1.4c)

unitary, if $AA^H = I$ (i.e., A regular and $A^{-1} = A^H$), (2.1.4d)

normal, if $AA^H = A^H A$. (2.1.4e)

Remark 2.1.4. (a) If a matrix A is Hermitian or unitary, it is also normal.
(b) Each of the properties (4a–e) carries over from A to the adjoint A^H.
(c) Products of regular (unitary) matrices are again regular (unitary).

A *diagonal matrix* D is completely described by its diagonal entries. We write

$$D = \operatorname{diag}\{d_\alpha : \alpha \in I\} \quad \text{for } D \text{ with } D_{\alpha\alpha} = d_\alpha,\ D_{\alpha\beta} = 0\ (\alpha \neq \beta). \tag{2.1.5a}$$

If I is ordered, we may also write

$$D = \operatorname{diag}\{d_1, d_2, \ldots, d_n\}. \tag{2.1.5b}$$

For an arbitrary matrix $A \in \mathbb{K}^{I \times I}$,

$$D = \operatorname{diag}\{A\} \tag{2.1.5c}$$

denotes the *diagonal part* $\operatorname{diag}\{a_{\alpha\alpha} : \alpha \in I\}$ of A.

In the case of an ordered index set, a matrix T is called *tridiagonal* if $T_{ij} = 0$ for all $|i - j| > 1$, i.e. if T has the band width 1 (cf. Definition 1.3.3). The entries $\alpha_i = T_{i,i-1}$ define the lower side diagonal, $\beta_i = T_{ii}$ the (main) diagonal, and $\gamma_i = T_{i,i+1}$ the upper side diagonal, while all other entries of T vanish. Such a matrix is abbreviated by

$$T = \operatorname{tridiag}\{(\alpha_i, \beta_i, \gamma_i) : i \in I\}. \tag{2.1.6}$$

Note that the values α_1 and γ_n, $n = \# I$, in (6) are meaningless. Tridiag$\{A\}$ denotes the tridiagonal part of an arbitrary matrix A.

2.1.3 Star Notation

In §1.2 the index set $I = \Omega_h$ has occurred. In the following, Ω_h can be more general than in (1.2.3): It may be an arbitrary subset of the two-dimensional infinite grid $\{(x, y) = (ih, jh): i, j \in \mathbb{Z}\}$. The vector $x \in \mathbb{K}^I$ can be interpreted as a *grid function*. Since the symbol x represents the vector as well as the first component in the index $(x, y) \in \Omega_h$, we write u instead of $x \in \mathbb{K}^I$ in accordance with Eqs. (1.2.1a, b):

$$x_\alpha = u(x, y) \quad \text{for } \alpha = (x, y) \in I = \Omega_h. \tag{2.1.7a}$$

If it seems to be favourable, the argument $(x, y) = (ih, jh)$ is replaced by the indices «$_{ij}$»:

$$u(ih, jh) = u_{ij} \quad \text{for } (ih, jh) \in \Omega_h. \tag{2.1.7b}$$

The first index component x or i corresponds to the grid row (oriented from left to right), the second component y or j to the grid column (from the bottom to the top).

Mappings (matrices) defined in $\mathbb{K}^I$ with $I = \Omega_h$ can be described conveniently by means of so-called *star notation*. The *nine-point star*

$$\begin{bmatrix} a_{-1,1} & a_{01} & a_{1,1} \\ a_{-1,0} & a_{00} & a_{1,0} \\ a_{-1,-1} & a_{0,-1} & a_{1,-1} \end{bmatrix} \quad \text{or} \quad \begin{bmatrix} a^{ij}_{-1,1} & a^{ij}_{01} & a^{ij}_{1,1} \\ a^{ij}_{-1,0} & a^{ij}_{00} & a^{ij}_{1,0} \\ a^{ij}_{-1,-1} & a^{ij}_{0,-1} & a^{ij}_{1,-1} \end{bmatrix} \tag{2.1.8a}$$

represents a matrix A containing in its *rows* the nine coefficients a_{pq} $(-1 \leqslant p, q \leqslant 1)$ appearing in (8a). The component of Ax associated with the index $(ih, jh) \in \Omega_h$ is

$$\sum_{p,q=-1}^{1} a_{pq} u_{i+p,j+q} \quad \text{or} \quad \sum_{p,q=-1}^{1} a^{ij}_{pq} u_{i+p,j+q}, \text{ resp.,} \tag{2.1.8b}$$

where $u = x$ according to (7a). In the first case, the matrix entries are independent of the grid point (as, e.g., for the Poisson-model problem), whereas in the second case they depend on $(ih, jh) \in \Omega_h$. $a_{00} = a^{ij}_{00}$ is the diagonal element (corresponding to the index «ij»). For example, the element $a_{1,0}$ in (8a) at the position right at the middle is the matrix entry by which the right neighbour $u_{i+1,j}$ (corresponding to the grid point $(ih + h, jh) \in \Omega_h$) is to be multiplied in (8b); etc.

Although $(ih, jh) \in \Omega_h$, the index «$_{i+p,j+q}$» appearing in (8b), more precisely the grid point $((i + p)h, (j + q)h)$, may not belong to Ω_h. In this case, the term $a^{ij}_{pq} u_{i+p,j+q}$ in (8b) is to be ignored. The same effect is obtained by the formal definition $u_{i+p,j+q} := 0$.

The *five-point formula* of the Poisson-model problem is

$$h^{-2} \begin{bmatrix} & -1 & \\ -1 & 4 & -1 \\ & -1 & \end{bmatrix}. \tag{2.1.9}$$

Unmarked entries a_{pq} (as in the positions p, $q = \pm 1$ for (9)) are defined by zero. The stars (or stencils) may have another format than 3×3. The interpretation of the 3×5-star

$$\frac{h^{-2}}{12}\begin{bmatrix} & & -12 & & \\ 1 & -16 & 54 & -16 & 1 \\ & & -12 & & \end{bmatrix}$$

is evident. The format $\ell \times k$ should be used only for odd ℓ, k. Otherwise, the midpoint (corresponding to the diagonal entry) is not recognisable.

2.2 Systems of Linear Equations

Let $A \in \mathbb{K}^{I \times I}$ and $b \in \mathbb{K}^I$. The system of equations to be solved is

$$Ax = b, \text{ i.e. } \sum_{\beta \in I} a_{\alpha\beta} x_\beta = b_\alpha \quad \text{for all } \alpha \in I. \tag{2.2.1}$$

General conditions for the solvability of this equation are less interesting in regard to numerical treatment. Since the right-hand side b may be perturbed (by rounding errors, etc.), the correct question is, when is (1) solvable for *all* $b \in \mathbb{K}^I$? The following theorem recalls that this property is equivalent to the regularity of A.

Theorem 2.2.1. *Let $A \in \mathbb{K}^{I \times I}$. The following properties are equivalent:*
(a) *A is regular.*
(b) $\operatorname{rank}(A) = \#I$ *(number of elements in I).*
(c) $\det(A) \neq 0$.
(d) *$Ax = 0$ has only the trivial solution $x = 0$.*
(e) *$Ax = b$ is solvable for all b.*
(f) *$Ax = b$ has at most one solution.*
(g) *$Ax = b$ is uniquely solvable for all b.*

2.3 Permutation Matrices

Any surjective mapping $\pi\colon I \to I$ is called a *permutation*. The corresponding *permutation matrix* $P = P_\pi$ is defined by

$$(Px)_\alpha = x_{\pi(\alpha)} \quad \text{for all } \alpha \in I \text{ and } x \in \mathbb{K}^I. \tag{2.3.1}$$

Lemma 2.3.1. (a) *The componentwise representation of the permutation matrix P associated with π is $P_{\alpha\beta} = \delta_{\pi(\alpha),\beta} = \delta_{\alpha,\pi^{-1}(\beta)}$.*
(b) *P is real and unitary: $P^{-1} = P^T$.*

Proof. $P_{\alpha\beta} = (Pe_\beta)_\alpha$. $(PP^H)_{\alpha\beta} = \sum_k \delta_{k,\pi(\alpha)}\delta_{k,\pi(\beta)} = \delta_{\pi(\alpha),\pi(\beta)} = \delta_{\alpha\beta} = (I)_{\alpha\beta}$. □

Exercise 2.3.2. Prove that the multiplication $A \mapsto PA$ permutes the rows of the matrix A, while $A \mapsto AP$ permutes the columns.

Given $A \in \mathbb{K}^{I \times I}$, we define the π-permuted matrix A_π by

$$A_\pi = P_\pi A P_\pi^T. \tag{2.3.2}$$

Definition 2.3.3. Matrices A, $B \in \mathbb{K}^{I \times I}$ are called *p-equivalent* (permutation equivalent, abbreviated by $A \underset{\mathrm{P}}{\sim} B$) if a permutation $\pi: I \to I$ exists such that $B = A_\pi$. Each A is associated with an equivalence class $\kappa(A) := \{B: A \underset{\mathrm{P}}{\sim} B\}$.

In the following we check which properties can be attributed to matrices if we define no ordering of the indices. Let $E(A)$ denote some property of matrices A. Examples for $E(A)$ are: «A is Hermitian», «A is a diagonal matrix».

Remark 2.3.4. (a) Let the index set I be nonordered. Assume that $E(A)$ is a property that does not depend on the denomination of the indices. Since all B with $B \underset{\mathrm{P}}{\sim} A$ differ from A only with respect to the index denomination, the property $E(A)$ carries over to the whole equivalence class $\kappa(A)$.
(b) Let I be ordered. If and only if a property E is compatible with $\underset{\mathrm{P}}{\sim}$ (i.e., $E(A)$ and $E(B)$ are equivalent for matrices $A \underset{\mathrm{P}}{\sim} B$), is it possible to define E as a property of the matrix $A = (a_{\alpha\beta})_{\alpha,\beta \in I}$ with nonordered index set I.

Through Remark 4a, we see that all properties mentioned in (1.4a–e) (symmetric, Hermitian, ...) are independent of the ordering of the indices. By means of part (b), one may decide whether a property, which is usually defined for ordered-index matrices, carries over.

Example 2.3.5. (a) If $A \in \mathbb{K}^{I \times I}$ (I ordered) is a diagonal matrix, A_π is a diagonal matrix, too. Therefore, the term «diagonal matrix» is also meaningful for nonordered index sets. Its direct definition reads as follows: A is a *diagonal matrix*, if $a_{\alpha\beta} = 0$ whenever $\alpha \neq \beta$.
(b) Because of $\det(P^T) = \det(P^{-1}) = 1/\det(P)$ and $\det(A_\pi) = \det(PAP^T) = \det(P)\det(A)\det(P^T) = \det(A)$, the *determinant* is invariant hence, $\det(A)$ can also be defined for $A \in \mathbb{K}^{I \times I}$ with nonordered index sets I.

Exercise 2.3.6. Prove that the properties «tridiagonal matrix» or «upper triangular matrix» make no sense for matrices without an index ordering.

2.4 Eigenvalues and Eigenvectors

Let $A \in \mathbb{K}^{I \times I}$ ($\mathbb{K} = \mathbb{R}$ or $\mathbb{K} = \mathbb{C}$). The *spectrum* of the matrix A is defined by

$$\sigma(A) := \{\lambda \in \mathbb{C}: \det(A - \lambda I) = 0\}. \tag{2.4.1}$$

Each $\lambda \in \sigma(A)$ is called an *eigenvalue* of A. An eigenvalue has *algebraic multi-*

plicity k if it is a k-fold root of the *characteristic polynomial* $\det(A - \lambda I)$. Since $\det(A - \lambda I)$ has the degree $n = \# I$, there exist exactly n eigenvalues, when they are counted according to their algebraic multiplicity.

The properties of the determinant prove

Remark 2.4.1. $\sigma(A^T) = \sigma(A)$, $\sigma(A^H) = \sigma(\bar{A}) = \overline{\sigma(A)} := \{\bar{\lambda}: \lambda \in \sigma(A)\}$. $e \in \mathbb{K}^I$ is called an *eigenvector* of the matrix A if $e \neq 0$ and

$$Ae = \lambda e. \tag{2.4.2}$$

By Theorem 2.1c, d we conclude from (2) that λ must be an eigenvalue. Vice versa, the same theorem proves

Lemma 2.4.2. *For each $\lambda \in \sigma(A)$, there exists an eigenvector e satisfying the eigenvalue problem* (2).

Exercise 2.4.3. Let $A = (a_{ij})_{i,j \in I}$ be an upper or lower triangular matrix or diagonal matrix. Prove $\sigma(A) = \{a_{ii}: i \in I\}$.

Definition 2.4.4. Two matrices A, $B \in \mathbb{K}^{I \times I}$ are called *similar* if there is a regular matrix T such that

$$A = T^{-1}BT. \tag{2.4.3}$$

Theorem 2.4.5. (a) *The eigenvalues of similar matrices A and B coincide, including their multiplicities*: $\sigma(A) = \sigma(B)$.
(b) *If T is the similarity transformation from* (3) *and if e is an eigenvector of A, then Te is an eigenvector of B.*

Proof. (a) Part (a) is a consequence of $\det(A - \lambda I) = \det(T^{-1}(B - \lambda I)T) = \det(T^{-1})\det(B - \lambda I)\det(T) = (1/\det(T))\det(B - \lambda I)\det(T) = \det(B - \lambda I)$.
(b) $B(Te) = TT^{-1}BTe = TAe = T(\lambda e) = \lambda(Te)$. □

Theorem 2.4.6. *The products AB and BA have the same spectrum, except possibly a zero eigenvalue*:

$$\sigma(AB)\backslash\{0\} = \sigma(BA)\backslash\{0\}. \tag{2.4.4}$$

This statement is also true for rectangular matrices $A \in \mathbb{K}^{I \times J}$, $B \in \mathbb{K}^{J \times I}$.

Proof. Let the eigenvector $e \neq 0$ belong to the eigenvalue $\lambda \in \sigma(AB)\backslash\{0\}$: $ABe = \lambda e$. Since $\lambda e \neq 0$, $v := Be$ does not vanish. Multiplication by B yields $BABe = \lambda Be$, i.e., $BAv = \lambda v$ with $v \neq 0$. $\lambda \in \sigma(BA)\backslash\{0\}$ proves $\sigma(AB)\backslash\{0\} \subset \sigma(BA)\backslash\{0\}$. The reverse inclusion is analogous. □

Given a polynomial $P(\xi) = \sum_\nu a_\nu \xi^\nu$ in $\xi \in \mathbb{C}$, one can extend the domain of definition of P by

$$P(A) := \sum_\nu a_\nu A^\nu \quad \text{for arbitrary } A \in \mathbb{K}^{I \times I} \tag{2.4.5a}$$

to (square) matrices. Here, A^0 is defined as identity I. The proof of the following lemma is postponed to the end of §2.8.1.

Lemma 2.4.7. (a) *The spectra of A and $P(A)$ satisfy* $\sigma(P(A)) = P(\sigma(A)) := \{P(\lambda): \lambda \in \sigma(A)\}$.
(b) *The algebraic multiplicity of the eigenvalues $P(\lambda)$ of $P(A)$ is the sum of the multiplicities of all eigenvalues $\lambda_1, \lambda_2, \ldots, \lambda_k$ of A with $P(\lambda_j) = P(\lambda)$ $(1 \leqslant j \leqslant k)$.*
(c) *Each eigenvector of A associated with the eigenvalue λ is also an eigenvector of $P(A)$ with the eigenvalue $P(\lambda)$.*

Exercise 2.4.8. Prove the following: (a) If $\sigma(A)$ contains no zeros of $P(\xi)$, then $P(A)$ is regular.
(b) The properties «diagonal», «upper triangular matrix», «lower triangular matrix» carry over from A to $P(A)$. This statement is also true for the properties «symmetric» and «Hermitian», provided that P has real coefficients.
(c) Let A be regular. All properties mentioned in (b) carry over from A to A^{-1}.

Lemma 2.4.9. *Let $A \in \mathbb{K}^{I \times I}$ be a strictly (upper or lower) triangular matrix (cf. §1.4). Then $A^m = 0$ holds for all $m \geqslant \#I$.*

Proof. One proves by induction that A^m $(m \in \mathbb{N})$ has a vanishing main diagonal and $m - 1$ vanishing side diagonals: $(A^m)_{ij} = 0$ for $|i - j| < m$. For $m \geqslant \#I$ the inequality $|i - j| < m$ is true for all indices; hence $A^m = 0$. □

Two matrices A, B are called *commutative* (or A and B commute), if $AB = BA$.

Exercise 2.4.10. Prove: (a) If A and B are commutative, $P(A)$ and $Q(B)$ also commute for arbitrary polynomials P and Q. In particular, $P(A)$ and $Q(A)$ as well as $P(A)$ and A are commutative pairs.
(b) For regular A the matrices $P(A)$ and $P(A^{-1})$ commute.
(c) Under the assumption of (a), $P(A)$, $P(A)^{-1}$, $Q(B)$, $Q(B)^{-1}$ are pairwise commutative, as long as the inverse matrices exist.

Two polynomials P and Q define the rational function $R(\xi) := P(\xi)/Q(\xi)$. By Exercise 8a, the matrix

$$R(A) := P(A)(Q(A))^{-1} \tag{2.4.5b}$$

is defined if and only if $\sigma(A)$ contains no zero of the denominator polynomial Q. The foregoing results prove

Remark 2.4.11. Assume that $\sigma(A)$ contains no pole of the rational function R. Then the following statements hold:
(a) (5b) is equivalent to $R(A) = (Q(A))^{-1}P(A)$.

(b) Lemma 7 also holds for R instead of P. In particular, $\sigma(R(A)) = R(\sigma(A))$ is the spectrum of $R(A)$.
(c) The properties «diagonal», «upper triangular matrix», «lower triangular matrix» carry over from A to $R(A)$. The same holds for the terms «symmetric» and «Hermitian» if R has real coefficients.

Exercise 2.4.12. Assume that $\sigma(A)$ contains no pole of the rational function R and prove:
(a) The similarity $A = T^{-1}BT$ (cf. (3)) implies $R(A) = T^{-1}R(B)T$.
(b) If $D = \operatorname{diag}\{d_\alpha: \alpha \in I\}$, then $R(D) = \operatorname{diag}\{R(d_\alpha): \alpha \in I\}$.

A fundamental term for iterative methods is the following one:

Definition 2.4.13. The *spectral radius* $\varrho(A)$ of a matrix A is the largest absolute value of the eigenvalues of A:

$$\varrho(A) := \max\{|\lambda|: \lambda \in \sigma(A)\}.$$

Lemma 2.4.14. *The spectral radius satisfies the following rules*:

$$\varrho(\zeta A) = |\zeta|\varrho(A) \quad \textit{for all } \zeta \in \mathbb{C} \textit{ and } A \in \mathbb{K}^{I\times I}, \tag{2.4.6a}$$

$$\varrho(A^k) = (\varrho(A))^k \quad \textit{for all } k \in \mathbb{N}_0 \textit{ and } A \in \mathbb{K}^{I\times I}, \tag{2.4.6b}$$

$$\varrho(A) = \varrho(B) \quad \textit{for similar matrices } A, B \in \mathbb{K}^{I\times I}, \tag{2.4.6c}$$

$$\varrho(A) = \varrho(A^H) = \varrho(A^T) \quad \textit{for all } A \in \mathbb{K}^{I\times I}. \tag{2.4.6d}$$

Proof. (i) Let the maximum of $\{|\lambda|: \lambda \in \sigma(A)\}$ be attained at $\lambda' \in \sigma(A)$: $|\lambda'| = \varrho(A)$. Then $|\zeta\lambda|$ and $|\lambda^k|\,(\lambda \in \sigma(A))$ also take their maxima at $\lambda = \lambda'$, which proves (6a, b).
(ii) For similar matrices A, B, we have $\sigma(A) = \sigma(B)$ (cf. Theorem 5a). This implies (6c).
(iii) (6d) is a consequence of Remark 1. □

Exercise 2.4.15. Prove the following: (a) A diagonal or triangular matrix has the spectral radius $\varrho(A) = \max\{|a_{\alpha\alpha}|: \alpha \in I\}$.
(b) $\varrho(A) = 0$ holds for strictly triangular matrices A.

Lemma 2.4.16. $\varrho(AB) = \varrho(BA)$ *holds for all* $A \in \mathbb{K}^{I\times J}$, $B \in \mathbb{K}^{J\times I}$.

Proof. According to (4), the spectra of AB and BA differ at most by the eigenvalue 0, which is irrelevant for the definition of the spectral radius. □

Multiple products satisfy $\varrho(A_0 A_1 \cdot \ldots \cdot A_m) = \varrho(A_1 \cdot \ldots \cdot A_m A_0)$.

2.5 Block-Vectors and Block-Matrices

As demonstrated by matrix (1.2.8) of the model problems, the vectors and matrices often have a special block structure. The exact definition of *block structure* is based on a decomposition of the index set I into disjoint and non-empty subsets:

$$I = \bigcup_{\kappa \in B} I_\kappa, \qquad I_\kappa, I_\lambda (\kappa, \lambda \in B) \text{ pairwise disjoint.} \tag{2.5.1}$$

B is the index set of the blocks. The vector $x \in \mathbb{K}^I$ decomposes into the *blocks* $x^\kappa (\kappa \in B)$:

$$x^\kappa := (x_\alpha)_{\alpha \in I_\kappa}, \qquad x(x^\kappa)_{\kappa \in B}. \tag{2.5.2a}$$

Example 2.5.1. In the case of the model problem from §1.2, the grid Ω_h from (1.2.3) is the obvious index set. The grid consists of $N - 1$ «rows» $I_j = \{(ih, jh): 1 \leqslant i \leqslant N - 1\}$, $j = 1, \ldots, N - 1$. In this case, $B = \{j: 1 \leqslant j \leqslant N - 1\}$ is the block-index set.

Let $A \in \mathbb{K}^{I \times I}$. For each pair $\kappa, \lambda \in B$, the decomposition (1) of I defines a *block* (a *submatrix*)

$$A^{\kappa\lambda} := (a_{\alpha\beta})_{\alpha \in I_\kappa, \beta \in I_\lambda} \quad \text{for each } \kappa, \lambda \in B. \tag{2.5.2b}$$

In general, the blocks $A^{\kappa\lambda}$ are rectangular submatrices. The complete matrix can be built from the blocks:

$$A = (A^{\kappa\lambda})_{\kappa, \lambda \in B}. \tag{2.5.2c}$$

A block is a special submatrix. Each matrix $(a_{\alpha\beta})_{\alpha, \beta \in K}$ associated with a subset $K \subset I$ is caled a *principal submatrix*. The diagonal-blocks $A^{\kappa\kappa}$ from (2b) are special principal submatrices. The term «block» is ambiguous. It is used for the index subset I_κ, a vector block x^κ as well as a block-submatrix. There are two extreme cases of block decompositions: If B has only one element, A consists of one block that coincides with A. If $B = I$, i.e. if all subsets $I_\kappa = \{\kappa\}$ have only one element, the terms «block» and «matrix entry» coincide.

If the block-indices $B = \{1, \ldots, k\}$ are ordered, a block-matrix can be represented in the form

$$A = \begin{bmatrix} A^{11} & A^{12} & \ldots & A^{1k} \\ A^{21} & A^{22} & \ldots & A^{2k} \\ \vdots & \vdots & & \vdots \\ A^{k1} & A^{k2} & \ldots & A^{kk} \end{bmatrix}.$$

Note that, in general, only the diagonal-blocks A^{ii} must be square submatrices.

Note the difference between the terms «*diagonal-block*» and «diagonal block». The former is a block in a diagonal position, whereas the latter has diagonal form.

Example 2.5.2. In the case of the model problem, the rows can be taken as blocks according to Example 1. Then $A^{jj} = h^{-2}T$ (cf. (1.2.8)) are the diagonal-blocks and $A^{j,j-1} = A^{j,j+1} = -h^{-2}I$ are the side-diagonal blocks. All further locks are zero-blocks and therefore not represented in (1.2.8). Note that a representation by (1.2.8) is only possible if the indices are ordered.

Remark 2.5.3. Block-matrices can be interpreted twofold. First, they can be regarded as matrices that are structures by means of the index decomposition (1). Second, they can be represented as matrices of the index set B (not I) with matrix-valued (not $\mathbb{K}$-valued) entries. For example, the matrix multiplication AB can be defined directly by means of the blocks: $(AB)^{\kappa\lambda} = \sum_{\gamma \in B} A^{\kappa\gamma}B^{\gamma\lambda}$.

The second interpretation from Remark 3 allows us to generalise the terms «diagonal, tridiagonal, and triangular matrix» immediately to block-matrices: A is called a *block-diagonal matrix* with respect to an index decomposition (1) if $A^{\kappa\lambda} = 0$ (zero-block) for all $\kappa \neq \lambda$, $\kappa, \lambda \in B$. In analogy to (1.5a), we write

$$A = \text{blockdiag}\,\{D^{\kappa}: \kappa \in B\} \tag{2.5.3a}$$

for a block-diagonal matrix with $A^{\kappa\kappa} = D^{\kappa}$. For an arbitrary matrix $C \in \mathbb{K}^{I \times I}$,

$$A = \text{blockdiag}\,\{C\} := \text{blockdiag}\,\{C^{\kappa\kappa}: \kappa \in B\} \tag{2.5.3b}$$

denotes the *block-diagonal part* of C which we obtain after defining all off-diagonal blocks by zero. Different block structures B may lead to different block-diagonal parts blockdiag $\{C\}$!

Similarly, we write

$$A = \text{blocktridiag}\,\{(E^j, D^j, F^j): j \in B\} \tag{2.5.4}$$

for a *block-tridiagonal matrix* (cf. (1.6)) if B is ordered. A is an upper (lower) *block-triangular matrix* if $A^{ij} = 0$ for all $i, j \in B$ with $i > j$ $(i < j)$.

Exercise 2.5.4. Prove: (a) $(A^T)^{\kappa\lambda} = (A^{\lambda\kappa})^T$, $(A^H)^{\kappa\lambda} = (A^{\lambda\kappa})^H$.
(b) The diagonal-blocks of Hermitian matrices are again Hermitian.
(c) Let A be a block-diagonal or block-tridiagonal matrix with diagonal-blocks $A^{\kappa\kappa}$ $(\kappa \in B)$. The characteristic polynomial of A is the product of the charadteristic polynomials of $A^{\kappa\kappa}$ $(\kappa \in B)$. The spectrum and the spectral radius of A satisfy

$$\sigma(A) = \bigcup \{\sigma(A^{\kappa\kappa}): \kappa \in B\}, \tag{2.5.5a}$$

$$\begin{aligned} \varrho(A) &= \max\{|\lambda|: \lambda \text{ eigenvalue of } A^{\kappa\kappa}: \kappa \in B\} \\ &= \max\{\varrho(A^{\kappa\kappa}): \kappa \in B\}. \end{aligned} \tag{2.5.5b}$$

(d) The diagonal-blocks of block-triangular or block-diagonal matrices satisfy

$$(P(A))^{\kappa\kappa} = P(A^{\kappa\kappa}) \qquad (\kappa \in B,\ P \text{ polynomial}). \tag{2.5.5c}$$

(e) The block-diagonal structure is invariant with respect to the application of polynomials P:

$$P\,(\text{blockdiag}\,\{D^\kappa\colon \kappa \in B\}) = \text{blockdiag}\,\{P(D^\kappa)\colon \kappa \in B\}. \tag{2.5.5d}$$

2.6 Norms

2.6.1 Vector Norms

In the following let V be a finite-dimensional vector space over the field $\mathbb{K}$ that may be $\mathbb{R}$ or $\mathbb{C}$. The vector space of the previous applications has been $V = \mathbb{K}^I$. A mapping $\|\cdot\|\colon V \to [0, \infty)$ is called a *norm* (on V) if

$$\|x\| = 0 \qquad \text{only for } x = 0, \tag{2.6.1a}$$

$$\|x + y\| \leqslant \|x\| + \|y\| \qquad \text{for all } x, y \in V, \qquad \text{(triangle inequality)} \tag{2.6.1b}$$

$$\|\lambda x\| = |\lambda|\,\|x\| \qquad \text{for all } \lambda \in \mathbb{K} \text{ and } x \in V. \tag{2.6.1c}$$

Sometimes, $|||\cdot|||$ is also used as a norm symbol. Special norms are indicated by indices.

Example 2.6.1. The *maximum norm* $\|\cdot\|_\infty$ and the *Euclidean norm* $\|\cdot\|_2$ are defined as follows:

$$\|x\|_\infty := \max\{|x_\alpha|\colon \alpha \in I\}, \qquad \|x\|_2 := \left(\sum_{\alpha \in I} |x_\alpha|^2\right)^{1/2}. \tag{2.6.2}$$

Exercise 2.6.2. (a) Check the properties (1a–c) for the norms (2). Prove:
(b) Let $c > 0$. If $\|\cdot\|$ is a norm on V, then $|||x||| := c\,\|x\|$ is a norm, too.
(c) If $\|\cdot\|$ is a norm on $V = \mathbb{K}^I$ and if $A \in \mathbb{K}^{I \times I}$ is a regular matrix, then $|||x||| := \|Ax\|$ is also a norm on V.

Lemma 2.6.3. *The following «inverse triangle inequality» holds:*

$$|\,\|x\| - \|y\|\,| \leqslant \|x - y\| \quad \textit{for all } x, y \in V. \tag{2.6.3}$$

Each norm defines a topology on V. Therefore, the continuity of mappings in the normed vector space $(V, \|\cdot\|)$ is well-defined. Inequality (3) *leads to*

Remark 2.6.4. The norm $\|\cdot\|$ is a continuous (even Lipschitz continuous) mapping from $(V, \|\cdot\|)$ into $\mathbb{R}$.

2.6.2 Equivalence of All Norms

Two norms $\|\cdot\|$ and $|||\cdot|||$ on V are called *equivalent* if there is a constant C such that

$$\|x\| \leqslant C\,|||x|||,\ |||x||| \leqslant C\,\|x\| \quad \text{for all } x \in V. \tag{2.6.4}$$

Exercise 2.6.5. Prove: (a) Transitivity holds: If $(\|\cdot\|_a, \|\cdot\|_b)$ and $(\|\cdot\|_b, \|\cdot\|_c)$ are two pairs of equivalent norms, then $\|\cdot\|_a$ and $\|\cdot\|_c$ are also equivalent.
(b) The Euclidean norm and maximum norm satisfy the inequalities

$$\|x\|_\infty \leqslant \|x\|_2,\ \|x\|_2 \leqslant \sqrt{\#I}\,\|x\|_\infty \quad \text{for all } x \in V = \mathbb{K}^I. \tag{2.6.5}$$

Since in this book we always assume $\dim(V) = \dim(\mathbb{K}^I) = \#I < \infty$, the assumption of the following theorem is satisfied.

Theorem 2.6.6. *If* $\dim(V) < \infty$, *all norms on* V *are equivalent.*

Proof. (i) Let $\{e_\alpha : \alpha \in I\}$ be the basis of V. We define a reference norm by

$$\|x\| := \max\{|a_\alpha| : \alpha \in I\} \text{ with } a_\alpha \text{ from the representation } x = \sum_\alpha a_\alpha e_\alpha.$$

Let $|||\cdot|||$ be an arbitrary norm on V. Because of the transitivity (cf. Exercise 5a), it is sufficient to show the equivalence of $\|\cdot\|$ and $|||\cdot|||$.

(ii) The second part of (4), $|||x||| \leqslant c\,\|x\|$, follows from the triangle inequality with $c := \sum_\alpha |||e_\alpha|||$:

$$|||x||| = \left|\left|\left|\sum a_\alpha e_\alpha\right|\right|\right| \leqslant \sum |a_\alpha|\,|||e_\alpha||| \leqslant c \max_\alpha |a_\alpha| = c\,\|x\|.$$

(iii) The set $S := \{x \in V : \|x\| = 1\}$ is bounded in $(V, \|\cdot\|)$ (the bound is 1) and closed, because it is the inverse image of the value 1 under a continuous mapping (cf. Remark 4). Hence, by $\dim(V) < \infty$, the set S is compact. Inequality (3) and part (b) yields $\big||||x||| - |||y|||\big| \leqslant |||x - y||| \leqslant c\,\|x - y\|$, i.e., $|||\cdot|||$ is also continuous with respect to the space $(V, \|\cdot\|)$ normed by $\|\cdot\|$. Since a continuous function on a compact set attains its minimum, there is some $x_0 \in S$ with

$$|||x_0||| \leqslant |||x'||| \quad \text{for all } x' \in S.$$

(iv) Since (4) is trivial for $x = 0$, we assume $x \neq 0$. By (1a) we have $\xi := \|x\| > 0$, hence, $x' := x/\xi$ is well-defined and satisfies $\|x'\| = 1$, i.e., $x' \in S$ holds. Part (iii) yields

$$\|x\| = \xi \leqslant \xi\,|||x'|||/|||x_0||| = c'\xi\,|||x'||| = c'\,|||\xi x'||| = c'\,|||x|||$$

for $c' := 1/|||x_0|||$. Hence, (4) is proved with $C := \max\{c, c'\}$. □

Remark 2.6.7. (a) The constant C from (4) does not depend on $x \in V$ but does depend on V, more precisely on $\dim(V)$, as illustrated by the example (5).

(b) Because of inequality (4), the open sets in $(V, \|\cdot\|)$ are also open in $(V, |||\cdot|||)$. Hence, the equivalence of the norms may also be expressed as follows: In a finite-dimensional normed space $(V, \|\cdot\|)$, the choice of the norm defining the topology is arbitrary.

2.6.3 Corresponding Matrix Norms

The space $\mathbb{K}^{I\times I}$ of the $I \times I$ matrices is also a linear vector space of the dimension $(\#I)^2$; hence, one may define norms on $\mathbb{K}^{I\times I}$. These are called *matrix norms*, whereas norms on $\mathbb{K}^I$ are called *vector* norms.

Example 2.6.8. The generalisation of the Euclidean norm leads to the *Frobenius norm* $\|A\|_F := \left(\sum_{\alpha,\beta\in I} |a_{\alpha,\beta}|^2\right)^{1/2}$.

Since the matrices with their matrix multiplication form an algebra, the following subclass of matrix norms is of special interest:

Definition 2.6.9. Let $\|\cdot\|$ be a (vector) norm on $\mathbb{K}^I$. The *corresponding matrix norm* is

$$|||A||| := \sup\left\{\frac{\|Ax\|}{\|x\|} : 0 \neq x \in \mathbb{K}^I\right\}. \tag{2.6.6}$$

Exercise 2.6.10. Prove: (a) $|||\cdot|||$ from (6) is a norm on $\mathbb{K}^{I\times I}$.
(b) The supremum (6) is attained by some x, so that «sup» may be replaced by «max».
(c) Two vector norms differing only by a factor (cf. Exercise 2b) lead to identical corresponding matrix norms.
(d) $|||A|||$ is the smallest possible bound C in the inequality

$$\|Ax\| \leqslant C\|x\| \quad \text{for all } x \in \mathbb{K}^I. \tag{2.6.7}$$

In the following we will always denote the vector norm and the corresponding matrix norm by the same norm symbol, since confusion is avoided because of the disjoint domains of definition. In particular, $\|A\|_\infty$ and $\|A\|_2$ are the matrix norms corresponding to the maximum norm $\|x\|_\infty$ and the Euclidean norm $\|x\|_2$, respectively. Because of its affinity to the spectral radius (cf. §2.9.2), $\|A\|_2$ is called the *spectral norm*. The name *row-sum norm* for $\|A\|_\infty$ is explained by

Exercise 2.6.11. Prove: (a) $\|A\|_\infty$ has the representation

$$\|A\|_\infty = \max\left\{\sum_{\beta\in I} |a_{\alpha\beta}| : \alpha \in I\right\} \qquad (A \in \mathbb{K}^{I\times I}). \tag{2.6.8}$$

(b) For a diagonal matrix we have $\|D\|_\infty = \|D\|_2 = \max\{|d_\alpha|: \alpha \in I\} = \varrho(D)$.
(c) The spectral norm $\|A\|_2$ is independent of whether $K = \mathbb{R}$ or $\mathbb{K} = \mathbb{C}$ is chosen in (6). *Hint*: Use (2.9.4a).

Theorem 2.6.12. *Let* $\|\cdot\|$ *denote the vector norm on* $\mathbb{K}^I$ *as well as the corresponding matrix norm* (6) *on* $\mathbb{K}^{I\times I}$. *Then*

$$\|AB\| \leqslant \|A\|\,\|B\| \quad \textit{for all } A, B \in \mathbb{K}^{I\times I} \qquad (\textit{submultiplicativity}), \tag{2.6.9a}$$

$$\|Ax\| \leqslant \|A\|\,\|x\| \quad \textit{for all } A \in \mathbb{K}^{I\times I},\ x \in \mathbb{K}^I. \tag{2.6.9b}$$

Proof. (b) By Exercise 10d, (7) holds with $C := |||A|||$.
(a) Apply (9b) to Bx instead of x: $\|ABx\| \leqslant \|A\|\,\|Bx\|$. Application of (9b) to Bx yields $\|ABx\| \leqslant \|A\|\,\|B\|\,\|x\|$. Hence, (7) is satisfied with $C := \|A\|\,\|B\|$ and AB instead on B. Exercise 10d shows $\|AB\| \leqslant C$ □

Exercise 2.6.13. Prove: (a) $\|I\| = 1$ holds for all corresponding norms.
(b) The Frobenius norm from Example 8 does not correspond to any vector norm.
(c) Given a vector norm $\|\cdot\|$ and a regular matrix T, we may define an additional vector norm $|||\cdot|||_T$ by $|||x|||_T := \|Tx\|$ (cf. Exercise 2c). The equally denoted corresponding matrix norm satisfies

$$|||A|||_T = \|TAT^{-1}\| \quad \text{for all } A \in \mathbb{K}^{I\times I}. \tag{2.6.10}$$

Definition 9 associates each vector norm with a matrix norm. This mapping is not injective (cf. Exercise 10d). However, for any corresponding matrix norm $\|\cdot\|_M$ one can reconstruct the underlying vector norm $\|\cdot\|_V$ up to a factor as follows. Choose any $0 \neq a \in \mathbb{K}^I$. The product xa^H $(x \in \mathbb{K}^I)$ represents the matrix $(x_\alpha a_\beta)_{\alpha,\beta\in I}$. $\|x\|_a := \|xa^H\|_M$ is a vector norm differing from $\|\cdot\|_V$ only by a factor. Let $\|\cdot\|_{a,M}$ be the matrix norm corresponding to $\|\cdot\|_a$. Then an arbitrary matrix norm $\|\cdot\|_M$ corresponds to some vector norm if and only if $\|\cdot\|_M = \|\cdot\|_{a,M}$.

Let X and Y be two normed spaces with the norms $\|\cdot\|_X$ and $\|\cdot\|_Y$, and let $A: X \to Y$ be a linear mapping. Then

$$\|A\|_{Y\leftarrow X} := \sup\{\|Ax\|_Y/\|x\|_X : 0 \neq x \in X\} \tag{2.6.11}$$

denotes the corresponding matrix norm.

2.7 Scalar Product

A *scalar product* of a vector space V is a positive, symmetric sesquilinear form $\langle\cdot,\cdot\rangle: V \times V \to \mathbb{K}$ ($\mathbb{K} = \mathbb{R}$ or $\mathbb{K} = \mathbb{C}$), i.e., it satisfies

$$\langle x, x\rangle > 0 \qquad \text{for all } 0 \neq x \in V, \tag{2.7.1a}$$

$$\langle x + \lambda x', y\rangle = \langle x, y\rangle + \lambda\langle x', y\rangle \qquad \text{for all } x, x', y \in V, \lambda \in \mathbb{K}, \tag{2.7.1b}$$

$$\langle x, y\rangle = \overline{\langle y, x\rangle} \qquad \text{for all } x, y \in V. \tag{2.7.1c}$$

(1b, c) imply the semi-linearity with respect to the second argument:

$$\langle x, y + \lambda y' \rangle = \langle x, y \rangle + \bar{\lambda} \langle x, y' \rangle \quad \text{for all } x, y, y' \in V, \lambda \in \mathbb{K}. \quad (2.7.1b')$$

In the real case $\mathbb{K} = \mathbb{R}$, one may write $\langle y, x \rangle$ and λ instead of $\overline{\langle y, x \rangle}$ and $\bar{\lambda}$. The well-known properties of the scalar product are given in

Remark 2.7.1. Each scalar product induces the norm

$$\|x\| := \sqrt{\langle x, x \rangle} \quad (2.7.2)$$

on V. The *Schwarz inequality* holds:

$$|\langle x, y \rangle| \leqslant \|x\| \, \|y\| \quad \text{for all } x, y \in V. \quad (2.7.3)$$

Equality in (3) implies that x and y are linear dependent. Furthermore, the duality statement (4) holds:

$$\|x\| = \max\{|\langle x, y \rangle| / \|y\| : 0 \neq y \in V\}. \quad (2.7.4)$$

The *Euclidean scalar product* on $V = \mathbb{K}^I$ is defined by

$$\langle x, y \rangle := \sum_{\alpha \in I} x_\alpha \bar{y}_\alpha. \quad (2.7.5)$$

If not differently defined, $\langle \cdot, \cdot \rangle$ will always denote the Euclidean scalar product (5). Another representation of the Euclidean scalar product $\langle x, y \rangle$ is $y^H x$.

Remark 2.7.2. (a) The norm (2) induced by the Euclidean scalar product is the Euclidean norm $\| \cdot \|_2$ from (6.2).
(b) The Hermitian is the adjoint matrix with respect to $\langle \cdot, \cdot \rangle$:

$$\langle Ax, y \rangle = \langle x, A^H y \rangle \quad \text{for } x, y \in \mathbb{K}^I, A \in \mathbb{K}^{I \times I}. \quad (2.7.6)$$

Two vectors $x, y \in V$ with $\langle x, y \rangle = 0$ are called *orthogonal* (with respect to $\langle \cdot, \cdot \rangle$) and symbolised by $x \perp y$. x and y are *orthonormal*, if they are orthogonal and normalised, i.e., $\langle x, x \rangle = \langle y, y \rangle = 1$. A basis $\{b^\alpha : \alpha \in I\}$ is called an *orthonormal basis* if the vectors b^α are pairwise orthonormal.

Remark 2.7.3 (orthogonalisation method). If $b^1, b^2, \ldots, b^m$ are m linear independent vectors of V, then the procedure

$$w^i := b^i - \sum_{j=1}^{i-1} \langle v^j, b^i \rangle v^j, \qquad v^i := w^i / \|w^i\| \; (i = 1, 2, \ldots, m) \quad (2.7.7)$$

with $\| \cdot \|$ from (2) defines m pairwise orthonormal vectors v^i spanning the same subspace:

$$\operatorname{span}\{b^1, \ldots, b^m\} = \operatorname{span}\{v^1, \ldots, v^m\}.$$

Here, we use the notation $\operatorname{span}\{x^\alpha : \alpha \in J\} := \left\{ x = \sum_{\alpha \in J} a_\alpha x^\alpha : a_\alpha \in \mathbb{K} \right\}$.

If W is a subspace of the vector space V, then $x \in V$ is called orthogonal on W (symbolised by $x \perp W$), if $x \perp w$ for all $w \in W$. $W^\perp$ denotes the *orthogonal space* with respect to W:

$$W^\perp := \{x \in V : x \perp W\}.$$

Lemma 2.7.4. *Let* $\boldsymbol{a}_\alpha = (a_{\alpha\beta})_{\beta\in I}$ *for* $\alpha \in I$ *denote the* α*-row vector of* $A = (a_{\alpha\beta})_{\alpha,\beta\in I}$. A *is a unitary matrix if and only if* $\{\boldsymbol{a}_\alpha : \alpha \in I\}$ *represents an orthonormal basis.*

Proof. Follows from $(A^H A)_{\alpha\beta} = \boldsymbol{a}_\alpha^H \boldsymbol{a}_\beta = \langle \boldsymbol{a}_\beta, \boldsymbol{a}_\alpha \rangle = \delta_{\alpha\beta}$, i.e. $A^H A = I$. □

2.8 Normal Forms

2.8.1 Schur Normal Form

The following theorem states that all matrices are unitary-similar to an upper triangular matrix. Evidently, the upper triangular matrix could also be replaced by a lower one. To be able to define a triangular matrix, the index set I must be ordered.

Theorem 2.8.1 (Schur normal form). *For any matrix* $A \in \mathbb{K}^{I\times I}$ *there is a unitary matrix* Q *and an upper triangular matrix* U *such that*

$$A = QUQ^H. \tag{2.8.1}$$

Q describes a unitary similarity transformation of A into upper triangular form (normal form):

$$U = Q^H A Q. \tag{2.8.1'}$$

The term «normal form» does not imply that Q and U are determined uniquely.

Proof of Theorem 1 (by induction on $n := \#I$). For $n = 1$ Eq. (1) holds for $Q := I$ and $U := A$. Let the assertion be true for $n - 1$. Choose an eigenvalue $\lambda \in \sigma(A)$ and a corresponding eigenvector e (possible because of Lemma 4.2). By suitable $x^2, \ldots, x^n$ the normalised vector $x^1 := e/\|e\|_2$ can be extended to an orthonormal basis (cf. Remark 7.3). Let $X := [x^1, x^2, \ldots, x^n]$ denote the matrix with the column-vectors x^i. According to Lemma 7.4, X is a unitary matrix. Let e^1 be the first unit vector: $e_i^1 = \delta_{1i}$. The first column of $A' := X^H A X$ is $A'e^1 = X^H A X e^1 = X^H A x^1 = \lambda X^H x^1 = \lambda X^H X e^1 = \lambda e^1$, because x^1 as well as e are eigenvectors corresponding to λ. The decomposition of the index set $I = \{1, \ldots, n\}$ into $I_1 := \{1\}$ and $I_2 := \{2, \ldots, n\}$ induces the block-decomposition of A' into $A' = [\lambda e^1, \ldots] = \begin{bmatrix} \lambda & a \\ 0 & A'' \end{bmatrix}$ with an $I_2 \times I_2$-matrix A'' and an I_2-row vector a. Since $\#I_2 = n - 1$, by the induction hypothesis there is a unitary $I_2 \times I_2$-matrix Y, such that $Y^H A'' Y = U'$ is an upper triangular $I_2 \times I_2$-matrix. Lemma 7.4 shows that the $I \times I$ matrix $Y' := \begin{bmatrix} 1 & 0 \\ 0 & Y \end{bmatrix}$

augmented by one row and one column is again unitary. The product $U := Y'^H A' Y' = Y'^H X^H A X Y'$ results in

$$U := Y'^H \begin{bmatrix} \lambda & aY \\ 0 & A''Y \end{bmatrix} = \begin{bmatrix} 1 & 0 \\ 0 & Y^H \end{bmatrix} \begin{bmatrix} \lambda & aY \\ 0 & A''Y \end{bmatrix} = \begin{bmatrix} \lambda & aY \\ 0 & Y^H A'' Y \end{bmatrix} = \begin{bmatrix} \lambda & aY \\ 0 & U'' \end{bmatrix}.$$

If U'' is an upper triangular matrix, U is also. The product $Q := XY'$ is unitary (cf. Remark 1.4). Hence, (1′) and (1) are proved. □

Exercise 4.3 and Theorem 4.5 yield

Corollary 2.8.2. *The diagonal of U from* (1) *contains the eigenvalues of A:*

$$\sigma(A) = \{u_{ii} : i \in I\}.$$

Proof of Lemma 4.7. Let P be a polynomial and represent A according to (1) by QUQ^H. Because of $P(A) = QP(U)Q^H$ (cf. Exercise 4.12a), the characteristic polynomials of $P(A)$ and $P(U)$ coincide (cf. Theorem 4.5a). $P(U)$ is again an upper triangular matrix (cf. Exercise 4.11c) with the diagonal entries $(P(U))_{ii} = P(U_{ii})$ (cf. Exercise 4.12c). Since the number of identical U_{ii} corresponds to the multiplicity, the eigenvalues of A (cf. Theorem 4.5a), statements (a) and (b) of the lemma follow. Part (c) is evident. □

2.8.2 Jordan Normal Form

Analysing the kernels of $(A - \lambda I)^k$ for $\lambda \in \sigma(A)$ and $k = n, n-1, \ldots, 1$, one can construct a basis formed by principal vectors and eigenvectors generating a transformation T into the Jordan normal form (cf. Gantmacher [1, VII.§7]). The Jordan normal form is an upper triangular matrix with a more restrictive structure than U from (1). The disadvantage is that, in general, T is not unitary.

The bidiagonal $k \times k$ matrix («Jordan block»)

$$J(\lambda, k) = \left.\begin{bmatrix} \lambda & 1 & & & 0 \\ & \lambda & 1 & & \\ & & \ddots & \ddots & \\ & & & \lambda & 1 \\ 0 & & & & \lambda \end{bmatrix}\right\} k \text{ rows and columns} \tag{2.8.2}$$

has the eigenvalue λ with the algebraic multiplicity k. Since only one eigenvector exists, the geometrical multiplicity equals 1.

Theorem 2.8.3 (Jordan normal form). *For any matrix $A \in \mathbb{K}^{I \times I}$, there exists a regular matrix T transforming A into its Jordan normal form J:*

$$A = TJT^{-1} \quad \textit{or equivalently} \quad J = T^{-1}AT. \tag{2.8.3a}$$

Here, J is an upper triangular matrix with the block-diagonal structure

$$J = \text{blockdiag}\,\{J(\lambda_i, k_i): i = 1, \ldots, K\} \text{ with } k_i \geqslant 1, \sum_{i=1}^{K} k_i = n := \#I. \tag{2.8.3b}$$

The numbers λ_i run over all eiegenvalues $\sigma(A)$. The k_i corresponding to equal eigenvalues λ_i sum up to the algebraic multiplicity of λ_i. K coindices with the maximum number of linear independent eigenvectors.

Since A and J are similar, they have the same characteristic polynomial (*cf.* Theorem 4.5a). *According to Exercise* 5.4c, *the common characteristic polynomial reads*

$$\chi(\xi) = \prod_{i=1}^{K} \det(J(\lambda_i, k_i) - \xi I) = \prod_{i=1}^{K} (\lambda_i - \xi)^{k_i}. \tag{2.8.4a}$$

Since some of the λ_i in (4a) *may coincide, k_i is not necessarily the multiplicity of λ_i. We define*

$$\bar{k}(\lambda) := \textit{algebraic multiplicity of } \lambda \in \sigma(A), \tag{2.8.4b}$$

$$\underline{k}(\lambda) := \max\{k_i: \lambda_i = \lambda, 1 \leqslant i \leqslant K\} \quad \textit{for } \lambda \in \sigma(A). \tag{2.8.4c}$$

Obviously, $\bar{k}(\lambda) \geqslant \underline{k}(\lambda)$ and $\chi(\xi) = \prod_{\lambda \in \sigma(A)} (\lambda - \xi)^{\bar{k}(\lambda)}$ hold, where the product is to be taken over all different eigenvalues in $\sigma(A)$. Hence, the polynomial

$$\mu(\xi) := \prod_{\lambda \in \sigma(A)} (\lambda - \xi)^{\underline{k}(\lambda)} \tag{2.8.4d}$$

is a divisor of the characteristic polynomial $\chi(\xi)$. $\mu(\xi)$ is called the minimum function of A, because it is the polynomial of smallest degree satisfying the following requirement (5).

Theorem 2.8.4 (Cayley–Hamilton). *Let μ and χ be the minimum function and the characteristic polynomial of a matrix A, respectively. Then*

$$\mu(A) = \chi(A) = 0 \qquad \text{(0: zero matrix).} \tag{2.8.5}$$

Proof. (i) To prove $p(B) = 0$ for a polynomial p, it suffices to show $q(B) = 0$ for a divisor polynomial q.
(ii) Define $q(\xi) := (\lambda - \xi)^{\underline{k}(\lambda)}$ with $\lambda = \lambda_i$ for some $i \in \{1, \ldots, K\}$. Since $\lambda_i I - J(\lambda_i, k_i)$ is a strictly upper triangular matrix and $\underline{k}(\lambda) \geqslant k_i$ according to definition (4c), $q(J(\lambda_i, k_i))$ is the zero matrix (cf. Lemma 4.9). $q(\xi)$ is a divisor of $\mu(\xi)$; hence, $\mu(J(\lambda_i, k_i)) = 0$ follows from part (i) for all $i = 1, \ldots, K$.
(iii) Exercise 5.4e applied to the block-diagonal matrix J yields $\mu(J) = \text{blockdiag}\{\mu(J(\lambda_i, k_i)): 1 \leqslant i \leqslant K\} = \text{blockdiag}\{0: 1 \leqslant i \leqslant K\} = 0$. By Exercise 4.12a, one concludes from (3a) that $\mu(A) = T\mu(J)T^{-1} = 0$. As μ is a divisor of χ, the remaining part of assertion (5) follows from (i). □

2.8.3 Diagonalisability

If $k_i = 1$ for all $i = 1, \ldots, K$, J from (3b) becomes a diagonal matrix. In this case, (3a) describes a transformation into diagonal form.

Theorem 2.8.5 (diagonalisability). *Let $A \in \mathbb{K}^{I \times I}$. A regular matrix T transforming A into diagonal form*:

$$A = TDT^{-1}, \qquad D = \operatorname{diag}\{\lambda_\alpha : \alpha \in I\}, \tag{2.8.6}$$

exists if and only if there are $n := \#I$ linear independent eigenvectors. In this case, A is termed diagonalisable. If, furthermore, all λ_α ($\alpha \in I$) are real, A is real diagonalisable.

Proof. Assuming (6), we conclude from $AT = TD$ that the α-column vectors $e^\alpha := Te_\alpha$ (e_α: α-unit vector) of T are the (linear independent) eigenvectors of A. Vice versa, given n linear independent eigenvectors, we can build from these column vectors the matrix T satisfying $AT = TD$, i.e. (6). □

Remark 2.8.6. For similar matrices A and B, A is diagonalisable if and only if B is also.

In general, the transformation matrix T from (6) is not unitary. More precisely, the following theorem holds:

Theorem 2.8.7. *A unitary matrix Q transforming A into diagonal form*:

$$A = QDQ^H, \quad Q \text{ unitary}, \quad D = \operatorname{diag}\{\lambda_{\alpha\alpha} : \alpha \in I\}, \tag{2.8.7}$$

exists if and only if A is normal.

Proof. (i) In the case of $A = QBQ^H$ with unitary Q, A is normal if and only if B is normal, since $A^HA = (QB^HQ^H)(QBQ^H) = QB^HBQ^H$ and $AA^H = (QBQ^H)(QB^HQ^H) = QBB^HQ^H$.
(ii) In the case of (7), we can apply part (i) with $B = D$: A diagonal matrix is always normal, hence A is also.
(iii) Let A be normal and assume that QUQ^H is its Schur normal form. Following part (i), U is normal. By induction on $n := \#I$ we are proving that A normal upper triangular matrix is diagonal. For $n = 1$ both terms are identical. The $n \times n$ matrix U can be written in the block structure $U = \begin{bmatrix} \lambda & a^H \\ 0 & U' \end{bmatrix}$ with an upper triangular $(n-1) \times (n-1)$ matrix U' and an $(n-1)$ row vector a^H. The comparison of $UU^H = \begin{bmatrix} \lambda & a^H \\ 0 & U' \end{bmatrix} \begin{bmatrix} \bar\lambda & 0 \\ a & U'^H \end{bmatrix} = \begin{bmatrix} |\lambda|^2 + a^Ha & \cdots \\ \cdots & U'U'^H \end{bmatrix}$ with $U^HU = \begin{bmatrix} \bar\lambda & 0 \\ a & U'^H \end{bmatrix} \begin{bmatrix} \lambda & a^H \\ 0 & U' \end{bmatrix} = \begin{bmatrix} |\lambda|^2 & \cdots \\ \cdots & U'^HU' \end{bmatrix}$ shows $a^Ha = \langle a, a \rangle = 0$, hence $a = 0$. Furthermore, since U' is normal, it is

diagonal by the induction hypothesis. Therefore, U is also diagonal, i.e., $D := U$ satisfies (7). □

Since Hermitian matrices A are special normal matrices (cf. Remark 1.4a), they share representation (7). $A = A^H$ is equivalent to $D^H = D$. On the other hand, $D = D^H$ characterises real diagonal matrices. This proves

Theorem 2.8.8. *If and only if A is Hermitian, there is a unitary matrix Q transforming A into a real diagonal matrix D:*

$$A = QDQ^H, \quad Q \text{ unitary}, \quad D = \operatorname{diag}\{\lambda_{\alpha\alpha} : \alpha \in I\} \textit{ real}. \tag{2.8.8}$$

Not only polynomials but also general functions can be applied to diagonalisable matrices, as explained in

Remark 2.8.9. Let A be diagonalisable. For an arbitrary function $f: \sigma(A) \to \mathbb{K}$, the matrix $f(A)$ is defined by

$$f(A) := T \operatorname{diag}\{f(\lambda_\alpha) : \alpha \in I\} T^{-1} \tag{2.8.9a}$$

with T and $D = \operatorname{diag}\{\lambda_\alpha : \alpha \in I\}$ from (6). A and $f(A)$ are commutative. For $g: \sigma(A) \to \mathbb{K}$ as a second function, $f(A)$ and $g(A)$ are also commutative. Furthermore, we have for all regular $S \in \mathbb{K}^{I \times I}$

$$f(SAS^{-1}) = Sf(A)S^{-1}. \tag{2.8.9b}$$

Theorem 2.8.10. *Let A, B be normal. A, B are commutative if and only if there exists a simultaneous unitary transformation into diagonal form:*

$$Q^H A Q = \operatorname{diag}\{\lambda_\alpha : \alpha \in I\}, \qquad Q^H B Q = \operatorname{diag}\{\mu_\alpha : \alpha \in I\}. \tag{2.8.10}$$

The column vectors of Q are the joint eigenvectors of A and B.

Proof. (i) Since diagonal matrices always commute, (10) implies

$$Q^H ABQ = (Q^H AQ)(Q^H BQ) = (Q^H BQ)(Q^H AQ) = Q^H BAQ$$

and therefore $AB = BA$.
(ii) Let T be unitary with $T^H AT = D_A := \operatorname{diag}\{\lambda_\alpha : \alpha \in I\}$. $AB = BA$ implies $D_A X = X D_A$ with $X := T^H BT$. First, we suppose $\lambda_\alpha \neq \lambda_\beta$ for $\alpha \neq \beta$. From $\lambda_\alpha X_{\alpha\beta} = (D_A X)_{\alpha\beta} = (X D_A)_{\alpha\beta} = \lambda_\beta X_{\alpha\beta}$, it follows that $X_{\alpha\beta} = 0$ for $\alpha \neq \beta$. Hence, X is diagonal, i.e. $Q := T$ also transforms B into a diagonal matrix $X = T^H BT$. If, otherwise, there are multiple eigenvalues, X is a block-diagonal matrix and we can choose $S = \operatorname{blockdiag}\{S^\kappa : \kappa \in B\}$ such that S^κ is unitary and transforms the diagonal-block $X^{\kappa\kappa}$ into diagonal form. $Q := TS$ has the desired properties. □

Remark 2.8.11. Assume A, B is commutative and normal with eigenvalues λ_α, μ_α $(\alpha \in I)$. Then $aA + bB$ has the eigenvalues $a\lambda_\alpha + b\mu_\alpha$ $(\alpha \in I)$.

2.9 Correlation Between Norms and the Spectral Radius

2.9.1 Corresponding Matrix Norms as Upper Bound for the Eigenvalues

Lemma 2.9.1. *Let $\|\cdot\|$ be a corresponding matrix norm. Then*

$$|\lambda| \leqslant \|A\| \quad \textit{for all eigenvalues of the matrix } A, \tag{2.9.1a}$$

$$\varrho(A) \leqslant \|A\| \quad \textit{for all matrices } A. \tag{2.9.1b}$$

Proof. According to Lemma 4.2, for each eigenvalue λ there is an eigenvector e with $Ae = \lambda e$. (6.1c) and (6.9b) yield $|\lambda|\,\|e\| = \|\lambda e\| = \|Ae\| \leqslant \|A\|\,\|e\|$, hence, we have (1a). (1b) follows from (1a). □

2.9.2 Spectral Norm

In §2.6.3 we have defined the *spectral norm* $\|\cdot\|_2$ as the matrix norm corresponding to the Euclidean vector norm.

Lemma 2.9.2. *The Euclidean norm and spectral norm are invariant with respect to unitary transformations in the following sense. A unitary matrix $Q \in \mathbb{K}^{I\times I}$ satisfies*

$$\|Qx\|_2 = \|x\|_2 \quad \textit{for all } x \in \mathbb{K}^I, \tag{2.9.2a}$$

$$\|Q\|_2 = \|Q^H\|_2 = 1, \tag{2.9.2b}$$

$$\|QA\|_2 = \|AQ\|_2 = \|Q^HA\|_2 = \|AQ^H\|_2 = \|Q^HAQ\|_2 = \|QAQ^H\|_2 = \|A\|_2. \tag{2.9.2c}$$

Proof. (a) We have $\|Qx\|_2^2 = \langle Qx, Qx\rangle = \langle x, Q^HQx\rangle = \langle x, x\rangle = \|x\|_2^2$ thanks to (7.2), (1.4d) and (7.6).

(b) As, by Remark 1.4b, Q^H is unitary if Q is also, it suffices to prove the assertions for Q. (2b) follows from definition (6.6) by means of (2a).

(c) (6.9a) and (2b) yield $\|QA\|_2 \leqslant \|Q\|_2\|A\|_2 = \|A\|_2$. The same estimate with Q^H and QA instead of Q and A shows $\|A\|_2 = \|Q^HQA\|_2 \leqslant \|QA\|_2$, whence $\|QA\|_2 = \|A\|_2$ follows. All further statements in (2c) can be proved analogously or follow from the previous ones. □

Lemma 2.9.3. *An equivalent definition of the spectral norm is*

$$\|A\|_2 = \max\{|\langle Ax, y\rangle|/(\|x\|_2\|y\|_2): 0 \neq x, y \in \mathbb{K}^I\}. \tag{2.9.3}$$

Proof. Express $\|Ax\|_2$ in (6.6) by means of (7.4) □

(3), (7.6) and (7.1c) imply immediatey the first part of

Remark 2.9.4. $\|A^H\|_2 = \|\bar{A}\|_2 = \|A^T\|_2 = \|A\|_2$.

The term «spectral norm» is due to the fact that for normal matrices this norm coincides with the spectral radius and that even in the general case, it can be expressed by the spectral radius, as shown in the following theorem.

Theorem 2.9.5. *The spectral norm satisfies*

$$\|A\|_2 = \sqrt{\varrho(A^H A)} = \sqrt{\varrho(AA^H)} \quad \text{for all } A \in \mathbb{K}^{I\times I}, \tag{2.9.4a}$$

$$\|A\|_2 = \varrho(A) \quad \text{for all normal matrices } A \in \mathbb{K}^{I\times I}. \tag{2.9.4b}$$

(4a) *holds also for rectangular matrices* $A \in \mathbb{K}^{I\times J}$.

Proof. (i) By definition 6.9, the square $\|A\|_2^2$ is the maximum of $\|Ax\|_2^2/\|x\|_2^2 = \langle Ax, Ax\rangle/\langle x,x\rangle = \langle A^H Ax, x\rangle/\langle x,x\rangle$ over all $x \neq 0$. The Hermitian matrix $A^H A$ has the representation QDQ^H with the diagonal matrix $\mathrm{diag}\{\lambda_\alpha\}$ constructed from the eigenvalues λ_α of $A^H A$. These eigenvalues are $\geqslant 0$ (cf. Exercise 10.10a and Lemma 10.3). There is $\beta \in I$ with $\varrho(A^H A) = \lambda_\beta$. Substitution $y = Q^H x$ and (2a) yield $\|Ax\|_2/\|x\|_2 = \langle y, Dy\rangle/\langle y, y\rangle$. The latter expression is maximal for the unit vector $y = \boldsymbol{e}_\beta$ and yelds the value $\|A\|_2^2 = \varrho(A^H A)$. The second equality in (4a) follows from Lemma 4.16.
(ii) Given a normal matrix A, due to Theorem 8.6 one finds a unitary matrix Q and a diagonal matrix D such that $A = QDQ^H$. (2c) shows $\|A\|_2 = \|D\|_2$. According to Exercise 6.11b, we have $\|D\|_2 = \varrho(D)$. By the similarity of A and D, $\varrho(A) = \varrho(D)$ holds and proves (4b). □

$A^H A$ and AA^H are Hermitian matrices and therefore normal. Hence, one deduces from (4a, b) that

$$\|A\|_2^2 = \|A^H A\|_2 = \|AA^H\|_2 \quad \text{for all } A \in \mathbb{K}^{I\times J}. \tag{2.9.4c}$$

Exercise 2.9.6. Prove: (a) For all $A \in \mathbb{K}^{I\times I}$ we have $\|A\|_2 \leqslant \|A^H\|_\infty^{1/2}\|A\|_\infty^{1/2}$, $\|A\|_2 \leqslant \sqrt{n}\,\|A\|_\infty$, $\|A\|_\infty \leqslant \sqrt{n}\,\|A\|_2$ with $n := \#I$.
(b) $\|A\|_2 \leqslant \|A\|_\infty$ for normal A.
(c) $|a_{\alpha\beta}| \leqslant \|A\|_2$ for all matrix entries of A.
(d) $|a_{\alpha\beta}| \leqslant C$ for all $\alpha, \beta \in I$ implies $\|A\|_2 \leqslant nC$.

2.9.3 Matrix Norm Approximating the Spectral Radius

Lemma 2.9.7. *For each matrix* $A \in \mathbb{K}^{I\times I}$ *and any* $\varepsilon > 0$, *there exists a corresponding matrix norm* $\|\cdot\|_{A,\varepsilon}$ *with the property*

$$\varrho(A) \leqslant \|A\|_{A,\varepsilon} \leqslant \varrho(A) + \varepsilon. \tag{2.9.5}$$

Proof. Let $A = QUQ^H$ be the Schur normal form (8.1). The eigenvalues of A are the diagonal elements $\lambda_i := u_{ii}$ of U (cf. Exercise 4.3). Hence, the diagonal matrix $D := \operatorname{diag}\{\lambda_1, \ldots, \lambda_n\}$, $n := \#I$, satisfies (6a) (cf. (4b)):

$$\varrho(A) = \varrho(D) = \|D\|_2. \tag{2.9.6a}$$

We define $\xi := \min\{1, \varepsilon/[n\|A\|_2]\}$ and apply the simularity transformation with the diagonal matrix $X := \operatorname{diag}\{1, \xi, \xi^2, \ldots, \xi^{n-1}\}$ to U:

$$V := X^{-1}UX = \begin{bmatrix} \lambda_1 & \xi u_{12} & \xi^2 u_{13} & \cdots \\ & \lambda_2 & \xi u_{23} & \cdots \\ 0 & & \ddots & \end{bmatrix} = D + R, \quad R_{ij} = \begin{cases} 0 & \text{for } i \geqslant j, \\ \xi^{j-i} u_{ij} & \text{for } i < j. \end{cases}$$

According to Exercise 6c and (2c), $|R_{ij}| \leqslant \xi^{j-i}|u_{ij}| \leqslant \xi\|U\|_2 = \xi\|A\|_2 \leqslant \varepsilon/n$ holds for $i < j$ by the choice of ξ. Exercise 6d yields $\|R\|_2 \leqslant \varepsilon$. We define the vector norm $\|x\| := \|X^{-1}Q^H x\|_2$ (cf. Exercise 6.2c). The corresponding matrix norm is $\|A\| = \|X^{-1}Q^H AQX\|_2$ (cf. (6.10)). Hence, we arrive at

$$\|A\| = \|X^{-1}Q^H AQX\|_2 = \|X^{-1}UX\|_2 = \|V\|_2 \leqslant \|D\|_2 + \|R\|_2. \tag{2.9.6b}$$

(6a) and $\|R\|_2 \leqslant \varepsilon$ yield $\|A\|_{A,\varepsilon} \leqslant \varrho(A) + \varepsilon$. The first part of inequality (5) is trivial because of (1b). □

The following theorem demonstrates the asymptotical relationship between an arbitrary norm and the spectral radius.

Theorem 2.9.8. *For all $A \in \mathbb{K}^{I \times I}$ and any (even non-corresponding) matrix norm, the spectral radius is the following limit:*

$$\varrho(A) = \lim_{m \to \infty} \|A^m\|^{1/m}. \tag{2.9.7}$$

Proof. (i) Without loss of generality, we may prove (7) for a fixed, corresponding norm, since, if we assume (7) for $\|\cdot\|$ and choose another matrix norm $|||\cdot|||$, we conclude from the equivalence $\|\cdot\|/C \leqslant |||\cdot||| \leqslant C\|\cdot\|$ (cf. Theorem 6.6) that $\varrho(A) = \lim \|A^m\|^{1/m} = \lim \left(\frac{1}{C}\|A^m\|\right)^{1/m} \leqslant \underline{\lim}\, |||A^m|||^{1/m} \leqslant \overline{\lim}\, |||A^m|||^{1/m} \leqslant \lim(C\|A^m\|)^{1/m} = \lim \|A^m\|^{1/m} = \varrho(A)$, i.e. (7), holds for $|||\cdot|||$ also. In particular, this argumentation shows

$$\overline{\lim}\, |||A^m|||^{1/m} = \overline{\lim}\, \|A^m\|^{1/m} \quad \text{for arbitrary norms } |||\cdot|||,\ \|\cdot\|. \tag{2.9.7$'$}$$

(ii) First, we consider the case $\varrho(A) = 0$. The Schur normal form yields U with $\varrho(U) = 0$, i.e., U is strictly triangular (cf. Exercise 4.15a). The assertion follows from Lemma 4.9: $A^m = 0$ holds for all $m \geqslant \#I$.

(iii) Now assume $\varrho := \varrho(A) > 0$ and define $B := \frac{1}{\varrho} A$. Hence, the assertion is equivalent to

$$\lim_{m \to \infty} \|B^m\|^{1/m} = 1, \tag{2.9.7$''$}$$

because of $\varrho(B) = 1$. For the norm $\|\cdot\| = \|\cdot\|_{B,\varepsilon}$ $(\varepsilon > 0)$ from Lemma 7 inequality (5) shows $1 = \varrho(B) = \varrho(B^m)^{1/m} \leqslant \|B^m\|^{1/m} \leqslant (\|B\|^m)^{1/m} = \|B\| \leqslant \varrho(B) + \varepsilon = 1 + \varepsilon$ for all m. Hence, $\overline{\lim}\,\|B^m\|_{B,\varepsilon}^{1/m} \leqslant 1 + \varepsilon$. Since this estimate holds for all $\varepsilon > 0$ and since by (7′) the limit superior is independent of ε, (7″) follows from $1 \leqslant \underline{\lim}\,\|B^m\|^{1/m} \leqslant \overline{\lim}\,\|B^m\|^{1/m} \leqslant 1$. □

2.9.4 Geometrical Sum (Neumann's Series) for Matrices

The finite geometrical series results in

$$\left[\sum_{\nu=0}^{m-1} A^\nu\right][I - A] = I - A^m. \tag{2.9.8}$$

If 1 is not an eigenvalue of A, i.e. if $I - A$ is regular, (8) can be rewritten as

$$\sum_{\nu=0}^{m-1} A^\nu = (I - A)^{-1}(I - A^m). \tag{2.9.8′}$$

Lemma 2.9.9. *Let* $A \in \mathbb{K}^{I\times I}$. $\lim_{m\to\infty} \|A^m\| = 0$ *holds if and only if* $\varrho(A) < 1$.

Proof. (i) In the case of $\varrho := \varrho(A) < 1$, one concludes from (7), $\|A^m\|^{1/m} \to \varrho$, for any ϱ' with $\varrho < \varrho' < 1$ that $\|A^m\| < \varrho'^m \to 0$ for $m \geqslant m_0$ with sufficiently large m_0. Hence, $\varrho(A) < 1$ is sufficient for $\|A^m\| \to 0$.
(ii) In the remaining case of $\varrho := \varrho(A) \geqslant 1$, inequality (1b) shows $\|A^m\| \geqslant \varrho(A^m) = \varrho(A)^m \geqslant 1$ (cf. (4.6b)). Hence, $\varrho(A) < 1$ is also a necessary condition for $\|A^m\| \to 0$. □

Theorem 2.9.10. *If and only if* $\varrho(A) < 1$, *the geometrical sum converges and results in the value*

$$\sum_{\nu=0}^{\infty} A^\nu = (I - A)^{-1}. \tag{2.9.9}$$

Proof. (i) Assume $\varrho(A) < 1$. By Lemma 9 we can go to the limit $m \to \infty$ in (8) and obtain $(\sum_{\nu=0}^{\infty} A^\nu)(I - A) = I$, i.e. (9).
(ii) For $\varrho(A) \geqslant 1$, according to Lemma 9, the terms A^ν do not converge to zero. Hence, the geometrical sum must diverge. □

2.9.5 Numerical Radius of a Matrix

An intermediate position between the spectral radius and the spectral norm is taken by a quantity called the *numerical radius* of the matrix A:

$$r(A) := \max\{|\langle Ax, x\rangle|/\|x\|_2^2 : 0 \neq x \in \mathbb{C}^I\}. \tag{2.9.10}$$

The interesting property is that $r(A)$ can be estimated against the spectral norm (cf. (11b, d)).

Lemma 2.9.11. *The numerical radius $r(A)$ has the following properties*:

$$r(A^H) = r(A) \qquad \textit{for all } A \in \mathbb{K}^{I \times I}, \tag{2.9.11a}$$

$$\varrho(A) \leqslant r(A) \leqslant \|A\|_2 \qquad \textit{for all } A \in \mathbb{K}^{I \times I}, \tag{2.9.11b}$$

$$r(A) = \|A\|_2 = \varrho(A) \qquad \textit{for all normal } A, \tag{2.9.11c}$$

$$\|A\|_2 \leqslant 2r(A) \qquad \textit{for all } A \in \mathbb{K}^{I \times I}. \tag{2.9.11d}$$

Proof. (i) (11a) follows from (7.6) and (7.1c): $|\langle Ax, x\rangle| = |\langle A^H x, x\rangle|$.

(ii) One concludes from $|\langle Ax, x\rangle| \leqslant \|Ax\|_2 \|x\|_2 \leqslant \|A\|_2 \|x\|_2^2$ that $r(A) \leqslant \|A\|_2$. Let x be the eigenvector of A associated with the eigenvalue λ with $|\lambda| = \varrho(A)$. $|\langle Ax, x\rangle| = |\lambda| \langle x, x\rangle$ leads to $r(A) \geqslant |\lambda| = \varrho(A)$.

(iii) (11b) and (4b) prove (11c).

(iv) An arbitrary matrix A has the unique decomposition

$$A = A_0 + iA_1, \qquad A_0 := \frac{1}{2}(A + A^H), \qquad A_1 := \frac{1}{2i}(A - A^H) \tag{2.9.12}$$

into the symmetric part A_0 and the skew-symmetric part iA_1 of A. Obviously, A_0 and A_1 are Hermitian:

$$A_0 = A_0^H, \qquad A_1 = A_1^H.$$

Hermitian matrices are normal, hence, $\|A_k\|_2 = r(A_k)$ $(k = 0, 1)$ and

$$\|A\|_2 \leqslant \|A_0\|_2 + \|A_1\|_2 = r(A_0) + r(A_1). \tag{2.9.11d'}$$

$\langle By, y\rangle$ has a *real* value for all Hermitian matrices B and all y. Hence, $\langle Ax, x\rangle$ consists of the real part $\langle A_0 x, x\rangle = \lambda\langle x, x\rangle$ and of the imaginary part $\langle A_1 x, x\rangle$. Setting $\zeta := \langle Ax, x\rangle / \|x\|_2^2$, we obtain $\langle A_k x, x\rangle / \|x\|_2^2 \leqslant |\zeta| \leqslant r(A)$ for $k = 0, 1$ and all x. Maximisation over all x yields $r(A_k) \leqslant r(A)$. From (11d') one concludes (11d). □

$r(\cdot)$ is a matrix norm, but the submultiplicativity is restricted to powers of A (cf. Pearcy [1]):

$$r(A) = 0 \text{ only for } A = 0, \tag{2.9.11e}$$

$$r(A + B) \leqslant r(A) + r(B) \qquad \text{for } A, B \in \mathbb{K}^{I \times I}, \tag{2.9.11f}$$

$$r(\lambda A) = |\lambda| r(A) \qquad \text{for } \lambda \in \mathbb{K}, A \in \mathbb{K}^{I \times I}, \tag{2.9.11g}$$

$$r(A^n) \leqslant r(A)^n \qquad \text{for } n \in \mathbb{N}_0, A \in \mathbb{K}^{I \times I}. \tag{2.9.11h}$$

Exercise 2.9.12. (a) Let A be decomposed as in (12). Prove

$$r(A) \leqslant \sqrt{r(A_0)^2 + r(A_1)^2} = \sqrt{\|A_0\|^2 + \|A_1\|^2}. \tag{2.9.13a}$$

(b) Given $\vartheta \in \mathbb{C}$, we decompose ϑA according to (12) into $\vartheta A = A_{\vartheta,0} + iA_{\vartheta,1}$. Prove

$$r(A) = \inf\{\sqrt{\|A_{\vartheta,0}\|^2 + \|A_{\vartheta,1}\|^2} : \vartheta \in \mathbb{C}, |\vartheta| = 1\}. \tag{2.9.13b}$$

2.10 Positive Definite Matrices

2.10.1 Definition and Notations

Definition 2.10.1. Let $\langle \cdot, \cdot \rangle$ denote the Euclidean scalar product on $\mathbb{K}^I$ and assume $A \in \mathbb{K}^{I \times I}$. Then A is called

positive definite if A is Hermitian and $\langle Ax, x \rangle > 0$ for all $0 \neq x \in \mathbb{K}^I$, (2.10.1a)

positive semi-definite if A is Hermitian and $\langle Ax, x \rangle \geqslant 0$ for all $x \in \mathbb{K}^I$, (2.10.1b)

negative definite if $-A$ is positive definite, (2.10.1c)

negative semi-definite if $-A$ is positive semi-definite. (2.10.1d)

The terms «positive [negative] (semi-)definite» define a partial ordering in the set of Hermitian matrices. In the case of (1a) (or (1b, c, d), respectively), we write $A > 0$ (or $A \geqslant 0$, $A < 0$, $A \leqslant 0$, resp.). For arbitrary Hermitian matrices A, B we define:

$$A > B \text{ if } A - B > 0, \quad \text{i.e., } A - B \text{ is positive definite,} \tag{2.10.2}$$

($A \geqslant B$, $A < B$, $A \leqslant B$, analogously). **Implicitly, any inequality like $A > B$ implies that the matrices involved are Hermitian.**

The definitions (1a–d) depend on the choice of the scalar product (cf. later Exercise 10c). Furthermore, we emphasize that other authors use the term «positive definite» for *non*-Hermitian matrices with $\langle Ax, x \rangle > 0$ for all $0 \neq x \in \mathbb{K}^I$.

2.10.2 Rules and Criteria for Positive Definite Matrices

Lemma 2.10.2. *The following rules are valid:*

$$A > 0 \Leftrightarrow CAC^H > 0 \qquad \textit{for all regular } C \in \mathbb{K}^{I \times I}, \tag{2.10.3a}$$

$$A > B \Leftrightarrow CAC^H > CBC^H \qquad \textit{for all regular } C \in \mathbb{K}^{I \times I}, \tag{2.10.3a'}$$

$$A \geqslant 0 \Rightarrow CAC^H \geqslant 0 \qquad \textit{for all } C \in \mathbb{K}^{I \times I}, \tag{2.10.3b}$$

$$A \geqslant B \Rightarrow CAC^H \geqslant CBC^H \qquad \textit{for all } C \in \mathbb{K}^{I \times I}, \tag{2.10.3b'}$$

$$A, B \geqslant 0 \Rightarrow A + B \geqslant 0 \textit{ and } A + B > 0 \textit{ if moreover } A > 0 \textit{ or } B > 0, \tag{2.10.3c}$$

$$A > 0 \Leftrightarrow \xi A > 0 \qquad \textit{for all } \xi > 0, \tag{2.10.3d}$$

$$\zeta I \leqslant A \leqslant \xi I \Leftrightarrow \sigma(A) \subset [\zeta, \xi] \qquad \textit{for Hermitian } A \in \mathbb{K}^{I \times I}. \tag{2.10.3e}$$

$$-\xi I \leqslant A \leqslant \xi I \Leftrightarrow \|A\|_2 \leqslant \xi \qquad \textit{for Hermitian } A \in \mathbb{K}^{I \times I}. \tag{2.10.3f}$$

$$A \geqslant B > 0 \Leftrightarrow 0 \leqslant A^{-1} \leqslant B^{-1}. \tag{2.10.3g}$$

Proof. (i) $x \neq 0$ implies $y := C^H x \neq 0$. Hence, the inequality $0 < \langle Ay, y \rangle = \langle AC^H x, C^H x \rangle = \langle CAC^H x, x \rangle$ shows $CAC^H > 0$. A second application with C^{-1} instead of C yields the reverse implication.
(ii) The proof of (3b) is analogous to (3a). (3c) and (3d) follow immediately from the definition (1a, b).
(iii) A can be diagonalised by a unitary $Q := A = QDQ^H$. (3b′) with $C = Q^H$ brings (3e) into the form $\zeta I \leqslant D \leqslant \xi I$, where the diagonal matrix D contains the eigenvalues $\lambda \in \sigma(A)$ as diagonal entries. The equivalence of $\zeta I \leqslant D \leqslant \xi I$ and $\sigma(A) \subset [\zeta, \xi]$ is easy to see.
(iv) Choosing $\zeta = -\xi$ in (3e) and exploiting the equivalence of $\sigma(A) \subset [-\xi, \xi]$ with $\varrho(A) = \|A\|_2 \leqslant \xi$, one obtains (3f).
(v) The proof of (3g) is postponed (after Remark 6). □

Lemma 2.10.3. *A matrix A is positive definite (semi-definite) if and only if A is Hermitian and all eigenvalues are positive (nonnegative).*

Proof. The demonstration of this assertion is elementary for a diagonal matrix D. Let $A = QDQ^H$ be the diagonalisation of A (cf. (8.8)). By Lemma 2, the positive definiteness of A is equivalent to the positive definiteness of D. Since both matrices have the same eigenvalues, the assertion is proved. □

2.10.3 Remarks Concerning Positive Definite Matrices

Lemma 2.10.4. (a) *Any positive definite matrix is regular.*
(b) *A is positive definite if and only if A^{-1} is positive definite:*

$$A > 0 \Leftrightarrow A^{-1} > 0. \tag{2.10.4a}$$

(c) *Each principal submatrix $(a_{\alpha\beta})_{\alpha,\beta \in J}$ $(J \subset I)$ of a positive (semi-)definite matrix is again positive (semi-)definite:*

$$A > 0 \Rightarrow (a_{\alpha\beta})_{\alpha,\beta \in J} > 0 \text{ and } A \geqslant 0 \Rightarrow (a_{\alpha\beta})_{\alpha,\beta \in J} \geqslant 0 \text{ for } J \subset I. \tag{2.10.4b}$$

(d) *All diagonal elements of a positive (semi-)definite matrix are positive (nonnegative):*

$$A > 0 \Rightarrow a_{\alpha\alpha} > 0 \text{ and } A \geqslant 0 \Rightarrow a_{\alpha\alpha} \geqslant 0 \quad \text{for all } \alpha \in I. \tag{2.10.4c}$$

(e) *Let A be positive (semi-)definite. The diagonal part $D = \operatorname{diag}\{A\}$ as well as each block-diagonal part $D = \operatorname{blockdiag}\{A\}$ of A are again positive (semi-)definite.*

Proof. (a, b) Consequence of Lemma 3, because A^{-1} has the inverse eigenvalues of A (cf. Remark 4.11b).
(c) Consequence of definition (1a) if one restricts x to the subspace with $x_\alpha = 0$ for $\alpha \notin J$.
(d) Special case of (c) for $J := \{\alpha\}$. (e) Consequence of (d) and (c). □

Lemma 2.10.5. (a) $0 \leqslant A \leqslant B$ *implies* $\|A\|_2 \leqslant \|B\|_2$ *and* $\varrho(A) \leqslant \varrho(B)$.
(b) $0 \leqslant A < B$ *implies* $\|A\|_2 < \|B\|_2$ *and* $\varrho(A) < \varrho(B)$.

Proof. Because of $\varrho(A) = \|A\|_2$ and $\varrho(B) = \|B\|_2$, it is sufficient to show $\varrho(A) \leqslant \varrho(B)$. $A \geqslant 0$ has an eigenvalue $\lambda = \varrho(A)$ and a corresponding eigenvector x with $\|x\|_2 = 1$. $\varrho(A) = \langle Ax, x\rangle \leqslant \langle Bx, x\rangle \leqslant r(B) = \varrho(B)$ (cf. (9.11c)) proves the assertion (a). Part (b) is analogous. □

Assume $A > 0$. The application of Remark 8.8 and Lemma 3 to the nonnegative square root $f(\xi) = \sqrt{\xi}$ (well-defined in $[0, \infty)$) yields the matrix $A^{1/2} := f(A)$. More generally, A^α is well-defined for $\alpha > 0$.

Remark 2.10.6. (a) If A is positive definite, then $A^{1/2}$ represents again a positive definite matrix. For its inverse we use the notation $A^{-1/2}$. $A^{-1/2} = (A^{1/2})^{-1}$ holds. For a positive semi-definite A, the matrix $A^{1/2}$ is well-defined as positive semi-definite matrix.
(b) $A^{1/2}$ commutes with A and any polynomial (function) of A.
(c) $A^{1/2}$ is the unique positive semi-definite solution of the matrix equation $X^2 = A \geqslant 0$.

Proof of Lemma 2, (3g). (3b′) for $C = B^{-1/2}$ yields $X := B^{-1/2}AB^{-1/2} \geqslant I$. By (3e), all eigenvalues of X are $\geqslant 1$. Therefore, the eigenvalues of X^{-1} are $\leqslant 1$. According to (3e), we conclude $X^{-1} \leqslant I$, hence $B^{1/2}A^{-1}B^{1/2} \leqslant I$. A further application of (3b′) shows $A^{-1} \leqslant B^{-1/2}IB^{-1/2} = B^{-1}$. □

(3c) implies that the positive (semi-)definite matrices form a semi-group with respect to the matrix addition. This does not hold for the multiplication: In general, AB is no longer positive (semi-)definite. However, we have

Remark 2.10.7. If A and B are positive (semi-)definite, the product AB is real diagonalisable and has only positive (nonnegative) eigenvalues.

Proof. The proof is simple if one of the factors is regular. Then we use the similarity transformation $AB \mapsto A^{-1/2}ABA^{1/2} = A^{1/2}BA^{1/2}$. The general proof breaks down as follows.
(i) If X and Y are similar matrices, then the real diagonalisability of X and of Y is equivalent.
(ii) A unitary transformation maps the positive semi-definite matrix B into diagonal form: $B = QDQ^H$ with $D = \operatorname{diag}\{d_\alpha : \alpha \in I\}$, $d_\alpha \geqslant 0$. Therefore, AB is similar to $A'D$ with the positive semi-definite matrix $A' := Q^H AQ$.
(iii) Define $J := \{\alpha \in I : d_\alpha = 0\}$, $d'_\alpha = 1$ for $\alpha \in J$, and $d'_\alpha = d_\alpha$ for $\alpha \notin J$. The diagonal matrix $D' := \{d'_\alpha : \alpha \in I\}$ is regular and satisfies $D = D'P = PD'$ and $D^{1/2} = D'^{1/2}P$, where $P := \operatorname{diag}\{p_\alpha : \alpha \in I\}$ with $p_\alpha = 0$ for $\alpha \in J$ and $p_\alpha = 1$ for $\alpha \notin J$.
(iv) $A'D = A'PD'$ is similar to $D'^{1/2}A'DP'^{1/2} = D'^{1/2}A'D'^{1/2}P = A''P$ with the positive semi-definite matrix $A'' = D'^{1/2}A'D'^{1/2}$.

(v) Without loss of generality, we can enumerate I so that $I\backslash J = \{1,\ldots,k\}$ and $J = \{k+1,\ldots,n\}$. Then, A'' has the block representation $A'' = \begin{bmatrix} C & E^H \\ E & F \end{bmatrix}$ with positive semi-definite block C (cf. (4b)). $T = \begin{bmatrix} I & 0 \\ S & I \end{bmatrix}$ with $S := -EC^{-1}$ transforms $A''P = \begin{bmatrix} C & 0 \\ E & 0 \end{bmatrix}$ into $A''' := TA''PT^{-1} = \begin{bmatrix} C & 0 \\ 0 & 0 \end{bmatrix}$. Since $C \geqslant 0$, A''' is also positive semi-definite and thereby real diagonalisable. (ii) to (v) prove the similarity of A''' and AB. By (i), AB is also real diagonalisable. As $A''' \geqslant 0$, AB has only nonnegative eigenvalues.

(vi) The positive definiteness of A and B implies that AB is regular and, thanks to (v), it has only positive eigenvalues. □

Let A be a positive definite matrix. As explained in Exercise 6.2c,

$$\|x\|_A := \|A^{1/2}x\|_2 \qquad (x \in \mathbb{K}^I) \tag{2.10.5a}$$

describes again a norm, the so-called *energy norm* (with respect to A). The notations in (5a) and (6.10) are related via $\|\cdot\|_A = |||\cdot|||_{A^{1/2}}$. Using Definition (5a) and Exercise 6.13c, one proves the following

Remark 2.10.8. Let A be positive definite. The norm $\|\cdot\|_A$ from (5) is generated by the («energy») scalar product

$$\langle x, y\rangle_A := \langle Ax, y\rangle. \tag{2.10.5b}$$

An equivalent representation of $\|\cdot\|_A$ is

$$\|x\|_A := \langle Ax, x\rangle^{1/2} \qquad (x \in \mathbb{K}^I). \tag{2.10.5c}$$

The corresponding matrix norm $\|\cdot\|_A$ is related to the spectral norm by

$$\|B\|_A = \|A^{1/2}BA^{-1/2}\|_2 \qquad (B \in \mathbb{K}^{I\times I}). \tag{2.10.5d}$$

There is a one-to-one correspondence between positive definite matrices and scalar products. In (5b), each matrix $A > 0$ is associated with a scalar product. The reverse direction is described in

Remark 2.10.9. Let $\langle\langle\cdot,\cdot\rangle\rangle$ be an arbitrary scalar product in $\mathbb{K}^I$ and denote the Euclidean scalar product by $\langle\cdot,\cdot\rangle$. Then there is a positive definite matrix A with

$$\langle\langle x, y\rangle\rangle = \langle Ax, y\rangle = \langle A^{1/2}x, A^{1/2}y\rangle \quad \text{for all } x, y \in \mathbb{K}^I. \tag{2.10.6}$$

Proof. Using the unit vectors e_α, define $a_{\alpha\beta} := \langle\langle e_\beta, e_\alpha\rangle\rangle$. □

Exercise 2.10.10. Prove: (a) $CC^H \geqslant 0$ and $C^HC \geqslant 0$ are always true.
(b) If C is regular, $CC^H \geqslant 0$ and $C^HC \geqslant 0$ are even positive definite.
(c) The *adjoint* C^* of a matrix C with respect to the scalar product (6) is defined by $\langle\langle Cx, y\rangle\rangle = \langle\langle x, C^*y\rangle\rangle$ for all x, $y \in \mathbb{K}^I$. The identity $C^* = A^{-1}C^HA$ holds. C is *self-adjoint* with respect to $\langle\langle\cdot,\cdot\rangle\rangle$ if $C^HA = AC$.

The *condition* of a regular matrix (with respect to the norm $\|\cdot\|$) is defined by

$$\operatorname{cond}(A) := \|A\| \, \|A^{-1}\|. \tag{2.10.7}$$

In particular, $\operatorname{cond}_2(\cdot)$ denotes the *spectral condition* (belonging to the Euclidean norm). In order to be independent of a reference norm, we define the *spectral number*

$$\kappa(A) := \varrho(A)\varrho(A^{-1}). \tag{2.10.8}$$

Exercise 2.10.11. Prove: (a) $\operatorname{cond}_2(A) = \kappa(A)$ for normal matrices A.
(b) $\kappa(A) = \max\{|\lambda|: \lambda \in \sigma(A)\}/\min\{|\lambda|: \lambda \in \sigma(A)\}$ holds for regular A.
(c) In case A has a positive spectrum with the minimal eigenvalue $\lambda_{\min}$ and the maximal one $\lambda_{\max}$, definition (8) reduces to

$$\kappa(A) = \lambda_{\max}/\lambda_{\min}. \tag{2.10.9}$$

(d) Positive definite matrices fulfil $\operatorname{cond}_2(A) = \kappa(A)$ with the representation (9), where the extreme eigenvalues are given by

$$\lambda_{\max} := \|A\|_2, \qquad \lambda_{\min} := 1/\|A^{-1}\|_2. \tag{2.10.10}$$

3
Iterative Methods

In this chapter we consider the general properties of iterative methods. Such properties are *consistency*, ensuring the connection between the iterative method and the given system of equations, as well as *convergence*, guaranteeing the success of the iteration. The most important result of this chapter is the characterisation of the convergence of linear iterations by the spectral radius of the iteration matrix (cf. §3.1.3). Since we consider iterative methods for systems with *regular* matrices only, iterative methods for singular systems or those with rectangular matrices will not be studied. Concerning this topic, we refer to Maess [2], Marek [1], and Kosmol–Zhou [1].

3.1 General Statements Concerning Convergence

3.1.1 Notations

We want to solve the *linear equation*

$$Ax = b \quad (A \in \mathbb{K}^{I \times I}, b \in \mathbb{K}^I \text{ given}). \tag{3.1.1}$$

To guarantee solvability for all $b \in \mathbb{K}^I$, we generally assume:

$$A \text{ is regular.} \tag{3.1.2}$$

An iterative method producing the iterates $x^1, x^2, \ldots,$ from the starting value x^0 can be characterised by a prescription $x^{m+1} := \Phi(x^m)$. Φ depends on the data A and b from (1). In particular, dependence on b will be expressed explicitly in the notation

$$x^{m+1} := \Phi(x^m, b) \qquad (m \geqslant 0, b \text{ from (1)}). \tag{3.1.3}$$

Definition 3.1.1. An *iterative method* is a (linear or nonlinear) mapping

$$\Phi\colon \mathbb{K}^I \times \mathbb{K}^I \to \mathbb{K}^I. \tag{3.1.4}$$

By $x^m(x^0, b)$ we denote the members (the so-called *iterates*) of the sequence generated by the prescription (3) from a starting value $x^0 \in \mathbb{K}^I$:

$$x^0(y,b) := y; \qquad x^{m+1}(y,b) := \Phi(x^m(y,b),b) \quad \text{for } m \geqslant 0. \tag{3.1.5}$$

3.1.2 Fixed Points

Definition 3.1.2. $x^* = x^*(b)$ is called a *fixed point* of the iterative method Φ belonging to $b \in \mathbb{K}^I$, if

$$x^* = \Phi(x^*, b). \tag{3.1.6}$$

If the sequence $\{x^m\}$ of the iterates from (3) converges, we may form the limit in (3) and obtain

Lemma 3.1.3. *Let the iteration Φ be continuous with respect to the first argument. If $x^* := \lim_{m\to\infty} x^m(y,b)$ exists (cf. (5)), x^* is a fixed point of Φ belonging to the right-hand side $b \in \mathbb{K}^I$.*

3.1.3 Consistency

Lemma 3 states that possible results of the iteration method are to be sought in the set of fixed points. Therefore, a minimum condition is that the solution of system (1) with the right-hand side $b \in \mathbb{K}^I$ is a fixed point with respect to b. This property is the subject of

Definition 3.1.4. The iterative method Φ is called *consistent* with the system of equations (1), if for all right-hand sides $b \in \mathbb{K}^I$ any solution of (1): $Ax = b$ is a fixed point of Φ with respect to b.

According to Definition 4, consistency means: For all $b, x \in \mathbb{K}^I$, the implication $Ax = b \Rightarrow x = \Phi(x, b)$ holds. The reverse implication would yield an *alternative* definition:

$$Ax = b \quad \text{for all fixed points } x \text{ of } \Phi \text{ with } b \text{ and all } b \in \mathbb{K}^I. \tag{3.1.7}$$

3.1.4 Convergence

A natural definition of the convergence of an iterative method Φ would be

$$\lim_{m\to\infty} x^m(y,b) \quad \text{exists for all } y, b \in \mathbb{K}^I, \tag{3.1.8}$$

where $x^m(y,b)$ are the iterates defined in (5) corresponding to the starting value $x^0 := y$. Since the starting value is not part of the iterative method Φ, it may happen that an iteration satisfying (8) converges, but to a limit *depending* on the starting value. Therefore, the independence of the limit is additionally required in the

Definition 3.1.5. An iterative method Φ is called *convergent*, if for all $b \in \mathbb{K}^I$, there is a limit $x^*(b)$ of the iterates (5) independent of the starting value $x^0 = y \in \mathbb{K}^I$.

3.1.5 Convergence and Consistency

In the following, we assume that the iterative method Φ is convergent *and* consistent. It will turn out that the chosen definitions of the terms «convergence» and «consistency» of Φ are almost equivalent to the combinations of the alternatives (7) and (8).

Theorem 3.1.6. *Let Φ be continuous in the first argument. Then Φ is consistent and convergent if and only if A is regular and Φ fulfils conditions* (7) *and* (8).

Proof. (i) Assume Φ to be consistent and convergent. (8) follows from the convergence definition 5. If A is singular, the equation $Ax = 0$ would have a non-trivial solution $x^{**} \neq 0$ besides $x^* = 0$. By consistency both are fixed points of Φ with respect to $b = 0$. Therefore, choosing the starting values $x^0 = x^*$ and $x^0 = x^{**}$, we obtain the constant sequences $x^m(x^*, 0) = x^*$ and $x^m(x^{**}, 0) = x^{**}$. The convergence definition states that the limits x^* and x^{**} coincide contrary to the assumption. Hence, A is regular. It remains proving (7). The preceding argument shows that a convergent iterative method can have only one fixed point with respect to b. Because of the regularity of A, there is a solution of $Ax = b$ that thanks to consistency is the unique fixed point of Φ with b. Hence, (7) is proved.
(ii) Assume $\Phi(x,b)$ to be continuous in x and that (7) and (8) are fulfilled. Furthermore, let A be regular. Due to Lemma 3, $x^* := \lim x^m(y,b)$ is a fixed point of Φ with respect to b and therefore, by (7), a solution of $Ax = b$. Because of the regularity of A, the solution of the system is unique and hence also the limit of $x^m(y,b)$, which thereby cannot depend on y. Hence, Φ is convergent in the sense of Definition 5. Convergence leads to the uniqueness of the fixed point with respect to b (cf. (i)). Since, by (7), this fixed point is the uniquely determined solution of $Ax = b$, Φ is consistent. □

3.2 Linear Iterative Methods

3.2.1 Notations, First Normal Form

One would expect iterative methods solving linear equations to be linear again. In fact, most of the methods described in this book are linear, but there are also important nonlinear iterations, e.g., the gradient method described in §9.

Definition 3.2.1. An iterative method Φ is called *linear* if $\Phi(x, b)$ is linear in x and b, i.e., if there are matrices M and N such that

$$\Phi(x, b) = Mx + Nb. \tag{3.2.1}$$

Here, the matrix M is called the *iteration matrix* of the iteration Φ.

The iteration (1.3) takes the form (2), which represents the *first normal form* of the method:

$$x^{m+1} := Mx^m + Nb \qquad (m \geqslant 0,\ b \text{ from (1.1)}). \tag{3.2.2}$$

3.2.2 Consistency, Second and Third Normal Form

For a linear and consistent iteration Φ, each solution of $Ax = b$ must be a fixed point with respect to b: $x = Mx + Nb$. Each $x \in \mathbb{K}^I$ can be the solution of $Ax = b$ (namely, for $b := Ax$). Hence, $x = Mx + Nb = Mx + NAx$ holds for all x and leads to the matrix equation

$$M + NA = I, \tag{3.2.3}$$

establishing a relation between M and N from (2). This proves

Theorem 3.2.2. *A linear iteration Φ is consistent if and only if the iteration matrix M results from N by*

$$M = I - NA. \tag{3.2.3'}$$

If, in addition, A is regular, one can solve (3) *with respect to N:*

$$N = (I - M)A^{-1}. \tag{3.2.3''}$$

Together, formulae (2) and (3′) allow us to represent linear and consistent iterations in their *second normal form* (4):

$$x^{m+1} := x^m - N(Ax^m - b) \qquad (m \geqslant 0,\ b \text{ from (1.1)}). \tag{3.2.4}$$

In the sequel, the matrix N will be called the «matrix of the second normal form of Φ». Equation (4) shows the x^{m+1} is obtained from x^m by a correction that is the *defect* $Ax^m - b$ of x^m multiplied by N. Since the defect vanishes for x^m if it is a solution of $Ax = b$, we have proved

Remark 3.2.3. The second normal form (4) with arbitrary $N \in \mathbb{K}^{I \times I}$ represents all linear and consistent iterations.

The *third normal form* of a linear iteration reads:

$$W(x^m - x^{m+1}) = Ax^m - b \qquad (m \geqslant 0, b \text{ from } (1.1)). \tag{3.2.5}$$

W will be called the «matrix of the third normal form of Φ». Equation (5) should be understood in the following algorithmic form:

$$\text{solve } W\delta = Ax^m - b \quad \text{and define} \quad x^{m+1} := x^m - \delta. \tag{3.2.5$'$}$$

This represents a definition of x^{m+1} as long as W is regular. Under this assumption one can solve for x^{m+1}. A comparison with (4) proves

Remark 3.2.4. If W from (5) is regular, the iteration (5) coincides with the second normal form (4), where N is defined as

$$N = W^{-1}. \tag{3.2.6}$$

Vice versa, the representation (4) with regular N can be rewritten as (5) with $W = N^{-1}$. We will see that for the interesting cases, N will be regular (cf. Remark 9).

3.2.3 Representation of the Iterates x^m

By the notation $x^m(x^0, b)$ in (1.5) we express dependency from the starting value x^0 and from the right-hand side b of the system of equations. The explicit representation is given in

Theorem 3.2.5. *The linear iteration* (1) *produces the iterates*

$$x^m(x^0, b) = M^m x^0 + \sum_{k=0}^{m-1} M^k N b \quad \text{for } m \geqslant 0. \tag{3.2.7}$$

Proof by induction. For $m = 0$ Eq. (7) takes the form $x^0(x^0, b) = x^0$ in accordance with (1.5). Assuming (7) for $m - 1$, we obtain from (1)

$$x^m(x^0, b) = Mx^{m-1} + Nb = M\left(M^{m-1}x^0 + \sum_{k=0}^{m-2} M^k N b\right) + Nb$$

$$= M^m x^0 + \sum_{k=1}^{m-1} M^k N b + Nb. \qquad \square$$

In the following, e^m denotes the (*iteration*) *error* of x^m:

$$e^m := x^m - x, \quad \text{where } x \text{ is the solution of } Ax = b. \tag{3.2.8}$$

Assuming consistency, we have $x = Mx + Nb$ for the solution x from (8). Forming the difference with (2): $x^{m+1} = Mx^m + Nb$, we attain the simple relation

$$e^{m+1} = Me^m \quad (m \geqslant 0), \qquad e^0 = x^0 - x, \tag{3.2.9a}$$

between two successive errors. A trivial conclusion is

$$e^m = M^m e^0 \qquad (m \geqslant 0). \tag{3.2.9b}$$

Following (4) the term of the *defect* $A\bar{x} - b$ of avector $\bar{x}$ has already been used. In particular,

$$d^m := Ax^m - b \tag{3.2.10}$$

denotes the defect of the mth iterate x^m.

Exercise 3.2.6. Prove: (a) The defect $\bar{d} = A\bar{x} - b$ and the error $\bar{e} = \bar{x} - x$ fulfil the equation

$$A\bar{e} = \bar{d}. \tag{3.2.11}$$

(b) Let the iteration be linear and consistent and assume A to be regular. Then the defects satisfy the equations

$$d^{m+1} = AMA^{-1}d^m, \quad d^0 = Ax^0 - b, \quad d^m = (AMA^{-1})^m d^0. \tag{3.2.12}$$

3.2.4 Convergence

A convergence criterion being necessary and sufficient can be formulated by means of the spectral radius of the iteration matrix:

Theorem 3.2.7. *A linear iterative method* (1) *with the iteration matrix M is convergent if and only if*

$$\varrho(M) < 1. \tag{3.2.13}$$

$\varrho(M)$ is called the convergence rate *of the iteration* (1).

In the sequel, the terms *convergence rate*, *convergence speed*, and *iteration speed* are used synonymously for $\varrho(M)$. This naming is not uniform: Many authors define the negative logarithm $-\log(\varrho(M))$ as the convergence rate (cf. (3.3a) and Varga [2], Young [2]).

Proof. (i) Let iteration (1) be convergent. In Definition 1.5 we may choose $b := 0$ and exploit the representation (7): $x^m = M^m x^0$. The starting value

$x^0 := 0$ yields the limit $x^* = 0$, which by the convergence definition must hold for any starting value. If $\varrho(M) \geqslant 1$, one could choose $x^0 \neq 0$ as the eigenvector corresponding to an eigenvalue λ with $|\lambda| = \varrho(M) \geqslant 1$. The resulting sequence $x^m = \lambda^m x^0$ would not converge to $x^* = 0$. Hence, inequality (13) is necessary for convergence.

(ii) Now let (13): $\varrho(M) < 1$ be valid. By Lemma 2.9.9, $M^m x^0$ converges to zero, while Theorem 2.9.10 proves $\sum_{k=0}^{m-1} M^k \to (I - M)^{-1}$. Thanks to the representation (7), x^m tends to $(I - M)^{-1} Nb$. Since this limit does not depend on the starting value x^0, iteration (1) is convergent. □

The proof already contains the first statement of

Corollary 3.2.8. (a) *If the iterative method* (1) *is convergent, the iterates converge to* $(I - M)^{-1} Nb$.
(b) *If the iteration is also consistent, then A and N are regular and the iterates x^m converge to the unique solution $x = A^{-1}b$.*

Proof for part (b). The regularity of A follows from Theorem 1.6. However, a more direct demonstration can be given by the consistency condition (3) rewritten in the form $NA = I - M$. Because of $\varrho(M) < 1$, 1 is not an eigenvalue of M and hence $I - M$ is regular. Therefore, both factors N and A in $NA = I - M$ must also be regular: $(I - M)^{-1} N = A^{-1}$ proves part (b). □

Remark 3.2.9. Since only convergent *and* consistent iterations are of interest and since by Corollary 8b in this case A and N are regular, the representation (3″) of N and the third normal form (5) hold with the matrix $W = N^{-1}$.

The convergence $x^m \to x$ is an asymptotical statement for $m \to \infty$ that allows no conclusion concerning the error $e^m = x^m - x$ for some fixed m. The values of $u_{16,16}$ given in Tables 1.4.1/2 even deteriorate during the first steps before they converge monotonically to the limit $\frac{1}{2}$. Often, one would like to have a statement for a fixed iteration number m. In this case, the convergence criterion (13) is to be strengthened by a norm estimate.

Theorem 3.2.10. Let $\|\cdot\|$ be a corresponding matrix norm. A sufficient condition for the convergence of an iteration is the estimate

$$\|M\| < 1 \tag{3.2.14}$$

of the iteration matrix M. If the iteration is consistent, the error estimates (15) hold:

$$\|e^{m+1}\| \leqslant \|M\| \, \|e^m\|, \qquad \|e^m\| \leqslant \|M\|^m \|e^0\|. \tag{3.2.15}$$

Proof. (14) implies (13) (cf. (2.9.1b)). (15) is a consequence of (9a, b). □

$\|M\|$ is called the *contraction number* of the iteration (with respect to the norm $\|\cdot\|$). In the case of (14), the iteration is called *monotonically convergent* with respect to the norm $\|\cdot\|$, since $\|e^{m+1}\| < \|e^m\|$. If the norm $\|\cdot\|$ fulfils the equality $\varrho(M) = \|M\|$, the terms «convergence» and «monotone convergence» coincide.

3.2.5 Convergence Speed

Inequality (15): $\|e^{m+1}\| \leqslant \zeta\|e^m\|$ with $\zeta := \|M\| < 1$ describes *linear convergence.* Faster convergence than a linear one is attainable by nonlinear methods only (cf. §9.4.3). The contraction number ζ depends on the choice of the norm. Thanks to (2.9.1b), the contraction number ζ is always larger or equal to the convergence rate $\varrho(M)$. On the other hand, Lemma 2.9.7 ensures that for a suitable choice of the norm, the contraction number ζ approximates the convergence rate $\varrho(M)$ arbitrarily close.

The contraction number as well as the convergence rate determine the quality of an iterative method. Both quantities can be determined from the error e^m as follows.

Remark 3.2.11. The contraction number is the maximum of the ratios $\|e^1\|/\|e^0\|$ taken over all starting values x.

Proof. Use (9b) and Exercise 2.6.10d. □

Exercise 3.2.12. Prove: (a) In general, Remark 11 would become wrong if $\|e^1\|/\|e^0\|$ were replaced by $\|e^{m+1}\|/\|e^m\|$.
(b) The later quotient takes the maximum

$$\zeta_{m+1} := \begin{cases} \max\{\|Mx\|/\|x\|: 0 \neq x \in \operatorname{range}(M^m)\} & \text{if } M^m \neq 0, \\ 0 & \text{otherwise,} \end{cases} \tag{3.2.16}$$

which can be interpreted as the matrix norm of the mapping $x \mapsto Mx$ restricted to the subspace $V_m := \operatorname{range}(M^m) := \{y\colon y = M^m x, x \in \mathbb{K}^I\}$.
(c) The inclusion $V_{m+1} \subset V_m$ holds with the equality sign at least for $m \geqslant \#I$.
(d) $\varrho(M) \leqslant \zeta_{m+1} \leqslant \zeta_m \leqslant \zeta_1 = \zeta := \|M\|$ holds for $m \geqslant 1$.
(e) For regular M one has $\zeta_m = \zeta$ for all m.

Exercise 12 demonstrates that the contraction number is a somewhat too coarse term: It may happen that the contraction number gives too pessimistic a prediction of the convergence speed. A somewhat more favourable estimate can be obtained by means of the numerical radius $r(\cdot)$ of the matrix M^m (cf. §2.9.5). The inequalities

$$\|M^m\|_2 \leqslant 2r(M^m) \leqslant 2r(M)^m \qquad \text{(cf. (2.9.11d, e))} \tag{3.2.17a}$$

and (9b) yield the error estimate

$$\|e^m\|_2 \leqslant 2r(M)^m \|e^0\|_2 \qquad (m \geqslant 0) \tag{3.2.17b}$$

with respect to the Euclidean norm. If $\|\cdot\|_C$ is the norm defined by (2.10.5a) with $C > 0$, one analogously proves the inequality

$$\|e^m\|_C \leqslant 2r(C^{1/2}MC^{-1/2})^m \|e^0\|_C. \tag{3.2.17c}$$

For the practical judgment of the convergence speed from «experimental data», i.e., from a sequence of errors e^m belonging to a special starting value x, one may use the *reduction factor*

$$\varrho_{m+1,m} := \|e^{m+1}\|/\|e^m\|. \tag{3.2.18a}$$

These numbers can, e.g., be found in the last column of Tables 1.4.1/2. More interesting than a single value $\varrho_{m+1,m}$ is the geometric mean $\varrho_{m+k,m} := [\varrho_{m+k,m+k-1} \cdot \ldots \cdot \varrho_{m+1,m}]^{1/k}$, which due to definition (18a) can more easily be represented by

$$\varrho_{m+k,m} := [\|e^{m+k}\|/\|e^m\|]^{1/k}. \tag{3.2.18b}$$

The properties of $\varrho_{m+k,m}$ are summarised in

Remark 3.2.13. (a) Denote the dependency of the magnitude $\varrho_{m+k,m}$ upon the starting value x^0 by $\varrho_{m+k,m}(x^0)$. Then

$$\lim_{k\to\infty} \max\{\varrho_{m+k,m}(x^0) \colon x^0 \in \mathbb{K}^I\} = \varrho(M) \quad \text{for all } m. \tag{3.2.18c}$$

(b) Even without maximisation over all $x^0 \in \mathbb{K}^I$

$$\lim_{k\to\infty} \varrho_{m+k,m}(x^0) = \varrho(M) \quad \text{for all } m \tag{3.2.18d}$$

holds, if x^0 does not lie in a certain subspace $U \subset \mathbb{K}^I$ of dimension $< \#I$ spanned by all eigenvectors and possibly existing main vectors of the matrix M corresponding to eigenvalues λ with $|\lambda| < \varrho(M)$. (18d) holds almost always, because only with probability zero does a stochastically chosen starting value x^0 lie in a fixed lower dimensional subspace.
(c) The reduction factors $\varrho_{m+1,m}$ tend to the spectral radius of M:

$$\lim_{m\to\infty} \varrho_{m+1,m}(x^0) = \varrho(M) \tag{3.2.18e}$$

for all $x^0 \notin U$ with $\dim U < \#I$ if and only if there is exactly one eigenvalue $\lambda \in \sigma(M)$ with $|\lambda| = \varrho(M)$, and if for this eigenvalue, the geometrical and algebraic multiplicities coincide. Sufficient conditions are: (i) $\lambda \in \sigma(M)$ with $|\lambda| = \varrho(M)$ is a single eigenvalue, or (ii) $M \geqslant 0$.
(d) Choose the norm in (18a) as $\|\cdot\| = \|\cdot\|_C$ ($C > 0$, cf. (17c)). If $C^{1/2}MC^{-1/2}$ is Hermitian, $\varrho_{m+1,m}(x^0)$ ($x^0 \notin U$) converges *monotonically increasing* to $\varrho(M)$.

Proof of (a). Note

$$\varrho(M) \leqslant \max\{\varrho_{m+k,m}(x^0): x^0 \in \mathbb{K}^I\}$$
$$\leqslant \max\{\varrho_{k,0}(x^0): x^0 \in \mathbb{K}^I\} = \|M^k\|^{1/k}$$

and Theorem 2.9.8: $\|M^k\|^{1/k} \to \varrho(M)$.

(b). Let $I_0 \subset I$ be the non-empty index subset $I_0 := \{i \in I: |J_{ii}| = \varrho(M)\}$, where J_{ii} are the diagonal elements of the Jordan normal form according to (2.8.3a, b): $M = TJT^{-1}$. The subspace $U := \{x: (T^{-1}x)_i = 0 \text{ for all } i \in I_0\}$ is the maximal subspace with the property $\lim[\|M^m x\|/\|x\|]^{1/m} < \varrho(M)$. Furthermore, $\dim(U) = \#I - \#I_0 < \#I$.

(d). Define $\hat{M} := C^{1/2}MC^{-1/2}$ and $\hat{e}^m := C^{1/2}e^m$. Since the norms are related by $\|e^m\|_C = \|\hat{e}^m\|_2$, we obtain for $m \geqslant 1$:

$$\|\hat{e}^m\|_2^2 = \|\hat{M}^m\hat{e}^0\|_2^2 = \langle \hat{M}^m\hat{e}^0, \hat{M}^m\hat{e}^0\rangle = \langle \hat{M}^{m+1}\hat{e}^0, \hat{M}^{m-1}\hat{e}^0\rangle$$
$$= \langle \hat{e}^{m+1}, \hat{e}^{m-1}\rangle \leqslant \|\hat{e}^{m+1}\|_2\|\hat{e}^{m-1}\|_2.$$

Hence,

$$\varrho_{m+1,m} = \|e^{m+1}\|/\|e^m\| = \|\hat{e}^{m+1}\|_2/\|\hat{e}^m\|_2$$
$$\geqslant \|\hat{e}^m\|_2/\|\hat{e}^{m-1}\|_2 = \varrho_{m,m-1}. \qquad \square$$

Remark 13 allows us to view the value $\varrho_{m+k,m}$ and eventually also $\varrho_{m+1,m}$ for sufficiently large m as a good approximation of the spectral radius. This viewpoint can be reversed:

Remark 3.2.14. The convergence rate $\varrho(M)$ is a suitable measure for judging (asymptotically) the convergence speed. This holds even if convergence is required with respect to a specific norm.

Proof. By Theorem 2.9.8, for each $\varepsilon > 0$ there is m_0 such that $\varrho(M) \leqslant \|M^m\|^{1/m} \leqslant \varrho(M) + \varepsilon$ for $m \geqslant m_0$, from which $\|e^m\| \leqslant (\varrho(M) + \varepsilon)^m \|e^0\|$. $\square$

3.2.6 Remarks Concerning the Matrices M, N, and W

The considerations in §§3.2.4–5 have shown the immediate connection between the iteration matrix M and the convergence speed. M directly describes the error reduction or amplification (cf. (9a)). Roughly speaking, the convergence is better the smaller M is. $M = 0$ would be optimal: Then Φ is a direct method, since x^1 would be the exact solution.

The matrix N transforms the defect $Ax^m - b$ into the correction $x^m - x^{m+1}$. The optimal case $M = 0$ mentioned above corresponds to $N = A^{-1}$. Therefore, one may regard N as an *approximate inverse of A*.

Concerning the implementation, often the matrix W of the third normal form (5) is the important one. By the relation (6): $W = N^{-1}$, $W = A$ would be optimal. However, then the computation of the correction $x^m - x^{m+1}$ is equivalent to the direct solution of the original equation. Therefore, one has to find approximations W of A, so that the solution of the system $W\delta = d$ for δ is *sufficiently easy*.

3.2.7 Product Iterations

Definition 3.2.15. Given two iterations Φ and Ψ, the *product iteration* $\Phi \circ \Psi$ is defined by

$$x^{m+1} = (\Phi \circ \Psi)(x^m, b) := \Phi(\Psi(x^m, b), b). \tag{3.2.19}$$

Special product iterations will be studied, e.g., in §4.8.

Exercise 3.2.16. (a) If Φ and Ψ are consistent, then $\Phi \circ \Psi$ is also.
(b) The iteration matrices of Φ, Ψ, and $\Phi \circ \Psi$ are related by

$$M_{\Phi \circ \Psi} = M_\Phi M_\Psi. \tag{3.2.20a}$$

The convergence properties of $\Phi \circ \Psi$ and $\Psi \circ \Phi$ (with regard to their rates) are identical.
(c) Let N_Φ, N_Ψ, and $N_{\Phi \circ \Psi}$ be the respective matrices of the second normal form of Φ, Ψ, $\Phi \circ \Psi$ and W_Φ, W_Ψ, $W_{\Phi \circ \Psi}$ be those of the third one. Then $N_{\Phi \circ \Psi}$ is given by

$$N_{\Phi \circ \Psi} = M_\Phi N_\Psi + N_\Phi = N_\Phi + N_\Psi - N_\Phi A N_\Psi. \tag{3.2.20b}$$

(d) Let W_Φ, W_Ψ, and $W_{\Phi \circ \Psi}$ be the respective matrices of the third normal form of Φ, Ψ, $\Phi \circ \Psi$. Assume that W_Φ and W_Ψ are regular. If $W_\Phi + W_\Psi - A$ is singular, $\Phi \circ \Psi$ diverges. Otherwise,

$$W_{\Phi \circ \Psi} = W_\Psi (W_\Phi + W_\Psi - A)^{-1} W_\Phi. \tag{3.2.20c}$$

Remark 3.2.17. In contrast to consistency, the convergence of factors Φ and Ψ is not inherited by product methods. However, a sufficient condition for the convergence of $\Phi \circ \Psi$ is that Φ and Ψ satisfy the criterion (14) of Theorem 10 with respect to the same norm: $\|M_\Phi\| < 1$ and $\|M_\Psi\| < 1$, where in one of the inequalities "<1" may be replaced by "$\leqslant 1$".

On the other hand, products of divergent methods may converge.

Definition 3.2.18. (a) A mapping $P\colon \mathbb{K}^I \to \mathbb{K}^I$ is a *projection* onto $U \subset \mathbb{K}^I$, if $P^2 = P$ and U is the range $\{Px\colon x \in \mathbb{K}^I\}$ of P.
(b) P is an *orthogonal projection* (onto U) if P is a Hermitian projection: $P = P^H$.
(c) A linear iteration Φ is called an [orthogonal] projection if Φ coincides with the product method $\Phi \circ \Phi$ [and if the iteration matrix M is Hermitian].

Exercise 3.2.19. Prove: (a) $P = P^2$ is an orthogonal projection if and only if $P = 0$ or $\|P\|_2 = 1$ with respect to the spectral norm.
(b) Let P be an orthogonal projection onto U. $w := Px$ and $w^\perp := x - w$ describe the unique decomposition of $x = w + w^\perp$ into $w \in U$ and $w^\perp \in U^\perp$.
(c) A linear iteration Φ is an orthogonal projection if and only if M is an orthogonal projection and $MN = 0$.
(d) An orthogonal projection P satisfies $0 \leqslant P \leqslant I$.

3.2.8 Three-Term Recursions (Two-Step Iterations)

Sometimes linear iterations occur, in which the computation of x^{m+1} involves x^m *and* x^{m-1}:

$$x^{m+1} = M_0 x^m + M_1 x^{m-1} + N_0 b \qquad (m \geqslant 1) \tag{3.2.21}$$

For the starting phase, one needs two initial values x^0 and x^1. Such three-term recursions can formally be reduced to the already studied iterations, which now should be denoted as «two-term recursions» or «one-step iterations». This one-step iteration acts in the space $(\mathbb{K}^I)^2$:

$$\begin{pmatrix} x^{m+1} \\ x^m \end{pmatrix} = \mathbf{M} \begin{pmatrix} x^m \\ x^{m-1} \end{pmatrix} + \begin{pmatrix} Nb \\ 0 \end{pmatrix} \quad \text{with } \mathbf{M} := \begin{pmatrix} M_0 & M_1 \\ I & 0 \end{pmatrix}. \tag{3.2.22}$$

Hence, the convergence condition becomes: $\varrho(\mathbf{M}) < 1$.

Exercise 3.2.20. The matrices M_0, M_1, N_0 from (21) corresponding to the iteration $x^{m+1} = Mx^m + Nb$ can be defined by

$$M_0 := \Theta M + \vartheta I, \qquad M_1 := (1 - \Theta - \vartheta) I, \qquad N_0 := \Theta N$$

with $\vartheta, \Theta \in \mathbb{R}$. The three-term recursion (21) then reads

$$x^{m+1} = \Theta\{(Mx^m + Nb) - x^{m-1}\} + \vartheta(x^m - x^{m-1}) + x^{m-1}. \tag{3.2.23}$$

Prove: (a) $\mathbf{M}$ has the spectrum

$$\sigma(\mathbf{M}) = \{\tfrac{1}{2}(\Theta\lambda + \vartheta)^{\pm}\sqrt{1 - \Theta - \vartheta + \tfrac{1}{4}(\Theta\lambda + \vartheta)^2} \colon \lambda \in \sigma(M)\}.$$

(b) From $\varrho(M) < 1$ and $\Theta > 0$, $\vartheta \geqslant 0$, $\Theta + \vartheta \leqslant 1$. One concludes that $\varrho(\mathbf{M}) < 1$.

3.3 Effectiveness of Iterative Methods

The convergence rate cannot be the only criterion for the quality of iterative methods, because one has also to take into account the possibly different amount of computational work.

3.3.1 Amount of Computational Work

The representation (2.5′) suggests that any iteration requires at least the computation of the defect $Ax^m - b$. For a general $n \times n$-matrix $A \in \mathbb{K}^{I \times I}$ ($n := \#I$) the multiplication $A * x^m$ would require $2n^2$ operations. However, it is more realistic to assume that A is *sparse*; i.e., the number $s(n)$ of the nonzero elements of A is distinctly smaller than n^2. For matrices arising from discretisations of partial differential equations, one has

$$s(n) \leqslant C_A n. \tag{3.3.1}$$

(cf. Hackbusch [15]). For the five-point formula (1.2.4a) of the model problem, inequality (1) holds with $C_A = 5$. Under the assumption (1), one can perform the matrix-vector multiplication in $2C_A n$ operations.

After the evaluation of $d := Ax^m - b$, one has still to solve the system $W\delta = d$ in (2.5′). For any practical iterative method, we should require that solving $W\delta = d$ consumes only $O(n)$ operations, so that the total amount of work is also of order $O(n)$. We relate the constant in $O(n)$ to C_A from (1) and obtain the following formulation:

The number of the arithmetical operations per iteration step of the method Φ is $Work(\Phi, A) \leqslant C_\Phi C_A n.$ (3.3.2)

Here, «$Work(\Phi, A)$» is the amount of work of the Φ-iteration under the application to $Ax = b$. Note that C_Φ is an *iteration-specific* constant, whereas $C_A n$ indicates the degree of sparsity of A. Therefore, the constant C_Φ may be called the *cost factor of the iteration* Φ.

3.3.2 Effectiveness

An iteration Φ can be called «more effective» than Ψ if for the same amount of work Φ is faster, or if Φ has the same convergence rate, but consumes less work than Ψ. To obtain a measure, we ask for the amount of work that is necessary to reduce the error by a fixed factor. The factor is chosen as $1/e$, since the (natural) logarithm is involved. According to Remark 2.14, we use the convergence rate $\varrho(M)$ for the (asymptotical) description of the error reduction per iteration step. After m iteration steps, the asymptotical error reduction is $\varrho(M)^m$. In order to ensure $\varrho(M)^m \leqslant 1/e$, we have to choose $m \geqslant -1/\log(\varrho(M))$, provided that convergence holds: $\log(\varrho(M)) < 0$. Therefore, we define

$$It(\Phi) := -1/\log(\varrho(M)). \tag{3.3.3a}$$

$It(\Phi)$ represents the (asymptotical) *number of the iteration steps for an error reduction by the factor* $1/e$.

Remark 3.3.1. (a) The convergence of Φ is equivalent to $0 \leqslant It(\Phi) < \infty$.
(b) Let Φ be convergent and consistent. To reduce the iteration error

(asymptotically) by a factor $\varepsilon < 1$, one needs

$$It(\Phi, \varepsilon) := -It(\Phi)\log(\varepsilon) \tag{3.3.3b}$$

iteration steps.
(c) If $\varrho(M) = \|M\|$ or $\varrho(M)$ in (3a) is replaced by $\|M\| < 1$, one can guarantee (not only «asymptotically»)

$$\|e^{m+k}\| \leqslant \varepsilon \|e^m\| \quad \text{for } k \geqslant It(\Phi, \varepsilon). \tag{3.3.3c}$$

(d) If $r(M) < 1$ holds for the numerical radius of M (cf. §2.9.5), definition (3b) can be replaced by $It(\Phi, \varepsilon) := \log(\varepsilon/2)/\log(r(M))$. Then, inequality (3c) holds with respect to the Euclidean norm.

The amount of work corresponding to the error reduction by $1/e$ is the product $It(\Phi) * Work(\Phi, A) \leqslant It(\Phi) C_\Phi C_A n$ (cf. (2)). As characteristic quantity we choose the «*effective amount of work*»

$$Eff(\Phi) := It(\Phi) C_\Phi = -C_\Phi/\log(\varrho(M)) \tag{3.3.4a}$$

$Eff(\Phi)$ measures the amount of work for a $(1/e)$-error reduction in the unit «$C_A n$ arithmetical operations». Correspondingly, the effective amount of work for the error reduction by the factor ε is given by

$$Eff(\Phi, \varepsilon) := -It(\Phi) C_\Phi \log(\varepsilon) = C_\Phi \log(\varepsilon)/\log(\varrho(M)). \tag{3.3.4b}$$

Example 3.3.2. In the case of the model problem, the cost factor of the Gauß-Seidel iteration is $C_\Phi = 1$ (because of $C_A = 5$, cf. Remark 1.4.3). The numerical values from Table 1.4.1 suggest $\varrho(M) = 0.99039$ for the grid size $h = 1/32$. Thus, the effective amount of work equals $Eff(\Phi) = 103.6$. Estimating $\varrho(M) = 0.82$ for the SOR method by means of Table 1.4.2 and using $C_\Phi = 7/5$, we deduce an effective amount of work of $Eff(\Phi) = 7.05$ for the SOR method with $h = 1/32$.

3.3.3 Order of the Linear Convergence

The convergence rates $\varrho(M)$ in Example 2 are typically close to one. Therefore, we may use the ansatz

$$\varrho(M) = 1 - \eta \qquad (\eta \text{ small}). \tag{3.3.5a}$$

Taylor expansion around $\eta = 0$ yields $\log(1 - \eta) = -\eta + O(\eta^2)$ and $-1/\log(1 - \eta) = 1/[\eta(1 + O(\eta)] = 1/\eta + O(1)$, since $1/(1 - \zeta) = 1 + \zeta + O(\zeta^2)$. Assuming (5a), we obtain the following effective amount of work:

$$Eff(\Phi) = C_\Phi/\eta + O(1). \tag{3.3.5b}$$

The respective numbers from Example 2 yield $C_\Phi/\eta = 104$ and 7.8.

For most of the methods we are going to discuss assumption (5a) holds in the case of the model problem. Then η is related to the grid size $h = 1/N =$

$1/(1+\sqrt{n})$ by

$$\eta = C_\eta h^\varkappa + O(h^{2\varkappa}), \text{ i.e., } \varrho(M) = 1 - C_\eta h^\varkappa + O(h^{2\varkappa}) \quad \text{with } \varkappa > 0, \tag{3.3.5c}$$

(5b) yields

$$\mathit{Eff}(\Phi) = C_{\text{eff}} h^{-\varkappa} + O(1) \quad \text{with } C_{\text{eff}} := C_\Phi / C_\eta. \tag{3.3.5d}$$

Remark 3.3.3. (a) $\varkappa$ from (5c) may be called the *order of the convergence rate.* If Φ has a higher order than Ψ, Φ is more expensive than Ψ for sufficiently small h. *The smaller the order, the better the method.* (b) If Φ_1 and Φ_2 have the same order but different constants $C_{\text{eff},1} < C_{\text{eff},2}$, then Φ_2 is more expensive by the factor $C_{\text{eff},2}/C_{\text{eff},1}$.

A comment is to be given concerning the desirable size of the iteration error $\|e^m\|$. For an unlimited iterative process, the rounding errors prevent the iteration error from converging to zero. Instead, the error will oscillate around «const $\cdot \|x\| \cdot$ eps» (eps = relative machine precision). For testing iterations, one may approach this lower limit. In practice, however, there is almost never a reason for such high accuracy.

Remark 3.3.4. The (exact) solution x of the Poisson-model problem from §1.2 is only an approximation of the true solution of the boundary value problem (1.2.1a, b). The latter differs from the discrete solution by a discretisation error, which in this case has the order $O(h^2)$ (cf. Hackbusch [15, §4.5]). Therefore, an additional iteration error of the same order $O(h^2)$ is acceptable. In practice, this means that an iteration can be stopped as soon as $\|e^m\|/\|x\|$ is smaller than 1/100 to 1_{10}−4.

3.4 Test of Iterative Methods

In later chapters numerous iterative methods will be defined. For the presentation of characteristic numerical results, one may ask how iterations should be tested. The quality of an iteration is (at least asymptotically) determined by the effective amount of work $\mathit{Eff}(\Phi)$. The amount of computational work per iteration is obtained by counting the operations. It remains to determine the convergence speed experimentally. The following trivial remark clarifies the fact that on no account does one need to test the method with *different* right-hand sides b and thereby with different solutions x.

Remark 3.4.1. A linear iteration applied to the two equations $Ax = b$ and $Ax' = b'$ results in the same errors $x^m - x$ and $x'^m - x'$, if the starting values x^0 and x'^0 are related by $x^0 - x = x'^0 - x'$.

Remark 3.4.2. Without loss of generality, one may always choose $x = b = 0$ together with an *arbitrary* starting value $x^0 \neq 0$.

Deviating from the proposal $x = b = 0$ but according to the choice in §1.4, we define the solution x of the *Poisson-model problem* as a grid function with the components

$$u_{ij} = (ih)^2 + (jh)^2 \qquad (1 \leqslant i, j \leqslant N-1) \tag{3.4.1a}$$

corresponding to the right-hand side (1b) (cf. Remark 1.4.4):

$$b \text{ defined by (1.2.6) with } f = -4. \tag{3.4.1b}$$

According to Remark 1, a test is sufficient if the errors $e^m = x^m - x$ and the ratio of their norms

$$\varrho_{m+1,m} := \|e^{m+1}\|/\|e^m\| \qquad \text{(cf. (2.18a))}$$

are computed for one or more starting vectors x^0.

Different starting values result in different errors. However, since the geometrical mean (2.18b): $\varrho_{m+k,m} := [\|e^{m+k}\|/\|e^m\|]^{1/k}$ converges to $\varrho(M)$ for $k \to \infty$, the ratios show remarkable deviations only during the first iterations. However, note the following

Remark 3.4.3. If the starting error $e^0 = x^0 - x$ lies in the subspace spanned by all eigen- and main vectors of the matrix M corresponding to eigenvalues λ with $|\lambda| < \varrho(M)$, the numbers $\varrho_{m+k,m}$ approximate a value $<\varrho(M)$.

In practice, meeting this exceptional case is unlikely, in particular, when the solution x is unknown.

The computation of $\varrho_{m+1,m} = \|e^{m+1}\|/\|e^m\|$ requires the knowledge of the exact solution. If we choose $b = 0$ and $x = 0$ according to Remark 2, $\varrho_{m+1,m} = \|x^{m+1}\|/\|x^m\|$ holds. If, contrarily, one wishes to estimate the convergence rate during the iterative computation of an unknown solution x, one can use the quantities

$$\hat{\varrho}_{m+1,m} := \|x^{m+1} - x^m\|/\|x^m - x^{m-1}\|$$

and $\hat{\varrho}_{m+k,m} := (\hat{\varrho}_{m+k,m+k-1} \times \cdots \times \hat{\varrho}_{m+1,m})^{1/k}$ instead of $\varrho_{m+k,m}$.

Exercise 3.4.4. Prove: In spite of $1 \in \sigma(M)$, it can happen that $\hat{\varrho}_{m+k,m} \to \varrho < 1 \leqslant \varrho(M)$ holds for $k \to \infty$. If $1 \notin \sigma(M)$, $\hat{\varrho}_{m+k,k} \to \varrho(M)$ is valid for all starting errors e^0 not lying the special subspace from Remark 3.

3.5 Comments Concerning the Pascal Procedures

A program indicates how simple or complicated a method really is. Therefore, we present Pascal procedures for all iterations mentioned in the sequel.

3.5.1 PASCAL

The following explanations in this section are addressed to readers without knowledge of the programming language PASCAL (cf. Jensen–Wirth [1]). Most of the program texts are self-explanatory. In PASCAL it is possible to define further types besides the standard ones («real», «integer», «Boolean»). First, one may form *arrays* of already defined variables (e.g.. `array [3:10] of ...`). In (7) *scalar type* is used, an example being `type Boolean = (false, true)`. The third important construction method is a record, which compounds variables of already defined types into a new one. In the example

```
type R = record A: real; B: Boolean; C: type1 end;     (3.5.1)
```

three variables are gathered. «`type1`» stands for an already defined type. A variable `x` of the above record, type `R`, is defined by

```
var x: R;
```

To address the single components «`A`», «`B`», «`C`» of `x`, they are indexed in the form

```
x.A:=3.14; x.B:=true;
```

If `type1` in (1) is again a record with the components «`P`», «`Q`», «`R`», ..., these are written as «`x.C.P`», «`x.C.Q`», «`x.C.R`», ... The `with` statement allows a simplification of the notation (namely, the omission of the prefix «`x.`»):

```
with x do begin A:=3.14; B:=true; C.P:=... end;
```

3.5.2 Concerning the Test Examples

To design programs for iterative methods, one has to know the structure of the system of equations. The simpler the structure of the matrix A the simpler the programs. The model problem from §1.2 has the simplest structure under all nontrivial examples. However, we do not want to restrict the programs to this case. The next level would be a system of equations in which the coefficients 4, -1 of the five-point formula (1.2.4a) are replaced by five other constants. By the «five-point star»

$$\begin{bmatrix} & \alpha_{01} & \\ \alpha_{-1,0} & \alpha_{00} & \alpha_{10} \\ & \alpha_{0,-1} & \end{bmatrix} \tag{3.5.2}$$

we characterise the matrix A that belongs to the equations

$$\alpha_{00}u_{ij} + \alpha_{-1,0}u_{i-1,j} + \alpha_{01}u_{i,j+1} + \alpha_{0,-1}u_{i,j-1} + \alpha_{10}u_{i+1,j} = f_{ij} \qquad (1 \leqslant i,j \leqslant N-1) \tag{3.5.3}$$

(cf. §2.1.3). For $\alpha_{00} = 4h^{-2}, \alpha_{-1,0} = \alpha_{10} = \alpha_{01} = \alpha_{0,-1} = -h^{-2}$, we regain the five-point formula (2.1.9) of the Poisson problem (1.2.4a).

As soon as the differential equation contains mixed derivatives $\partial^2 u/\partial x \partial y$, the five-point star is not sufficient (cf. Hackbusch [15, §5.1.4]). Therefore, as the most general case, we consider the «nine-point star» (2.1.8a) describing the equations (2.1.8b). Concerning nine-point formulae for the Poisson equation, we refer to Hackbusch [15, §4.6 and Exercise 8.3.16 therein].

Since the nine-point star covers the preceding cases, one might restrict programs only to these problems. On purpose we do not follow this approach. *To solve sparse systems means always to exploit the given structure.* Avoiding vanishing terms $\alpha_{11} = \cdots = 0$ in the five-point formula and multiplications by $h^2\alpha_{01} = 1$ in the case of the Poisson-model problem, we can essentially reduce the amount of work. Therefore, in the following procedures will will always make the distinction between the cases:

Poisson-model problem: (1.2.4a) and (2.1.9) (3.5.4a)

five-point formula: (2) (3.5.4b)

nine-point formula: (2.1.8a, b). (3.5.4c)

Using a corresponding case statement is shorter than presenting three different procedures specialised for each case. The user may shorten the source code by omitting noninteresting cases.

We may replace the square $\Omega = (0,1) \times (0,1)$ (cf. (1.2.1c)) by a rectangle. Then the grid Ω_h contains inner knots (ih, jh) with

$$1 \leqslant i \leqslant Nx - 1, \qquad 1 \leqslant j \leqslant Ny - 1. \tag{3.5.5}$$

This generalisation helps to indicate whether the loops of the programs run over row or column indices.

3.5.3 Constants and Types

The constants and types needed in the sequel are the following:

constants (3.5.6)

```
const Nmax=64; Lmax=5; ADImax=16; ITmax=300;
      pi=3.14159265358979323846264338;
```

types (3.5.7)

```
type column = array [0:Nmax] of real;
gridfunction = array [0:Nmax] of column;
kind_of_matrix = (Poisson_model_problem, fivepoint_formula,
  ninepoint_formula);
```

```
kind_of_restriction = (trivial_restriction, fivepoint_
  restriction, sevenpoint_restriction, ninepoint_restriction,
  general_restriction, matrixdependent_restriction);
kind_of_prolongation = (fivepoint_prolongation, sevenpoint_
  prolongation, ninepoint_prolongation, general_prolongation,
  matrixdependent_prolongation);
data_for_comparison = record z: ^gridfunction; start,actual:
  integer end;
column_star = array[-1:1] of real;
star = array[-1:1] of column_star:
data_of_restriction = record kind: kind_of_restriction; S:
  star end;
data_of_prolongation = record kind: kind_of_prolongation; S:
  star end;
ADIparameters = array [1:ADImax] of real;
CR_parameter = record a,a_old,p_old,w: gridfunction end;
cg_parameter = record r,p: gridfunction; rho: real; CR:
  ^CR_parameter end;
parameter_array = record length, actual: integer; om:
  ADIparameters end;
tridiag = record decomposition_computed: Boolean; lower,diag,
  upper: column end;
ABFG = record A,B,F,G: gridfunction end;
iterationparameter = record Nr: integer; omega,theta,gmg,gpg,
  sigma,diag: real; previous,Res: ^gridfunction; cycle:
  ^parameter_array; cg: ^cg_parameter end;
MG_parameter = record gamma,ny1,ny2,lm,lmin,actual_level:
  integer; direct, Galerkin: Boolean; R: data_of_restriction; P:
  data_of_prolongation end;
data_of_discretisation = record kind: kind_of_matrix; S: star;
  T: ^tridiag; ILUD: ^gridfunction; ILU7: ^ABFG; nx,ny: integer;
  h,h2: real end;
hierarchy_of_discretisation = array[0:Lmax] of data_of_
  discretisation;
H_star = array [0:2] of star;
MG_data = record AL: hierarchy_of_discretisation; SH: H_star;
  MI: MG_parameter; IP: array[0:Lmax] of iterationparameter end;
data_of_iteration = record x,b: gridfunction; A:
  data_of_discretisation; IP: iterationparameter end;
history_of_iteration = record value: array[0:Itmax] of real;
  last: integer end;
```

Nmax from (6) denotes the upper bound of Nx and Ny from (5). Also, ITmax, Lmax, and ADImax are used as bounds of arrays. Their values can be changes according to the available storage.

At present, the only interesting type is `gridfunction`, which is already introduced in (1.4.6). `kind_of_matrix` contains the cases mentioned in (4a–c).

3.5.4 Format of the Iteration Procedures

All iteration methods have the following procedure head:

```
procedure name_of_iteration(var new: gridfunction;
          var A: data_of_discretisation;
           var x,b: gridfunction;
          var IP: iterationparameter);                    (3.5.8)
```

A call of `name_of_iteration(...)` produces the new iterate x^m = `new` from x^{m-1} = `x`. Here, identical variables may be inserted for `new` and `x`, i.e., `name_of_iteration(x,A,x,b,IP)` with `x` $= x^{m-1}$ as input results in `x` $= x^m$. To be on the safe side, care is taken that the variable `x` remains unchanged if different variables for `new` and `x` are inserted. The parameter `IP` contains the parameters that are needed by some of the methods (e.g., the relaxation parameter ω of the SOR method). Since the same procedure head is used always, the parameter formally appears also in those iteration procedures that need no parameters.

3.5.5 Test Environment

To test the iteration (8), the following steps are to be performed: (i) definition of the discretisation data, (ii) definition of the right-hand side and the boundary values, (iii) definition of the starting iterate, (iv) definition of the exact solution for comparison purposes, and (v) computation of the norms $\|x^m - x\|$ and their ratios (cf. §3.4). The program call using the procedures available in [Prog] can be as follows:

Example for a frame program (3.5.9)

```
program frame_program;
{Insert the necessary constants, types, and procedures,
 e.g., "lex_SOR" and "maximum_norm"}
var it: data_of_iteration; i,itnr: integer;
    v: data_for_comparison; max: history_of_iteration;

function righthandside(x,y: real): real;
                           begin righthandside:=-4 end;
function zerofunction(x,y: real): real;
                           begin zerofunction:=0 end;
function exact_solution(x,y: real): real;
                           begin exact_solution:=x*x+y*y end;
```

```
function boundaryvalue(x,y: real): real;
 begin boundaryvalue:=x*x+y*y end;

begin initialise_IT(it); initialise_comparison(v);
 repeat release_IT(it); release_comparison(v);
  define_problem(it,boundaryvalue,righthandside); writeln
    ('Problem defined.');
  define_SOR_parameter(it); writeln('SOR parameter is ω=',
    it.IP.omega);
  define_starting_iterate(it,zerofunction); writeln
    ('Starting iterate defined.');
  define_comparison_solution(v,it.A,exact_solution);
    writeln('Comparison.');
  comparison_with_exact_solution(max,v,it,maximum_norm);
  write('--> Number of the iterations ='); readln(itnr);
  for i:=1 to itnr do
  begin iteration_1(it,lex_SOR);
        comparison_with_exact_solution
          (max,v,it,maximum_norm);
        writeln('Nr.', it.IP.Nr, 'maximum norm:',
          max.value[max.last])
  end;
  writeln('*** Run completed ***');
    {The results are listed in the following}
  writeln('lexicographical SOR method/maximum norm');
  writeln('SOR parameter equals omega=',it.IP.omega);
  write_discretisation_screen(IT.A);
  output_history_of_iteration_screen(max,it.A);
  writeln(F, 'Nr maximum norm ratio averaged');
  for i:=0 to max.last do
  begin writeln; write(i:3,' '); evaluation_history_of_
    iteration_screen(max,i)
  end; writeln; {Here, it is possible to store the results
    on a file}
  write('Should the run be repeated?')
  until not yes_no
  end.
```

The procedures and functions to be specified (`initialise_IT`, `initialise_comparison`, etc.) can be found in the PASCAL index on page 426.

The quantities Nx, Ny (cf. (5), and $h :=$ interval length$/Nx$ can be determined interactively by calling `define_problem`. According to §1.4, the right-hand side b of the system can be defined from the boundary values and the right-hand side of the differential equation. For this purpose the functions `boundary_values` and `right_hand_side` can be used.

By means of define_starting_iterate, the starting iterate x^0 is determined (cf. (11c)). In the given example, $x^0 := 0$ holds (cf. Remark 1.4.4).

```
procedure define_data_of_system(var it: data_of_iteration;
                    function boundary_values(x,y: real): real;
                    function right_hand_side(x,y: real): real);
begin with it do with A do
  begin set_right_hand_side(it,right_hand_side);
    if kind=Poisson_model_problem then
     factor_x_vector(nx,ny,b,h2,b);
    define_boundary_values(nx,ny,x,boundary_values)
end end;

procedure define_problem(var it: data_of_iteration;
                         function boundary_values(x,y: real):
                          real;
                         function right_hand_side(x,y: real):
                          real);
begin with it do with A do
  begin determine_nxnyh(nx,ny,h); h2:=h*h;
    determine_discretisation(A)
    end;
  define_data_of_system (it, boundary_values,
     right_hand_side)
end;

procedure define_starting_iterate(var it: data_of_iteration;
                                function startvalue(x,y:
                                 real): real);
begin define_inner_points(it.A.nx,it.A.ny,it.x,startvalue);
 it.IP.Nr:=0
end;
```

When necessary, parameters are to be defined. In the given example, ω can be selected interactively by means of define_SOR_parameter.

If desired, define_comparison_solution provides the definition of the exact solution x stored in the variable named v. This enables the computation of the error norms and their ratios for each new iterate. The corresponding variable is max, since maximum_norm is used. In [Prog] further functions Euclidean_norm, A_norm (=energy norm), and value_in_the_middle are available. The variable max can store at most ITmax iterates. write_discretisation_screen lists the data contained in it.A. By output_history_of_iteration_screen, the computed error norms and their ratios can be shown.

Iteration_1 performs on iteration step of the method that is specified as a parameter. Function yes_no asks interactively for true or false.

4
Methods of Jacobi and Gauß-Seidel and SOR Iteration in the Positive Definite Case

The methods of Jacobi and Gauß-Seidel and the SOR method are closely connected and therefore they will be analysed simultaneously. The analysis, however, is essentially different for the case of positive definite matrices A discussed below and other cases studied in §§5–6. The introductory Section 4.1 underlines the fact that the positive definite case is of practical interest: The Poisson-model matrix is positive definite. Chapter 4.2 is a recapitulation and generalisation of the algorithms already known from §1.4. Simple modifications of these iterations are discussed in §4.3 and §4.6. The convergence results of the qualitative and quantitative kind are given in §4.4. Section 4.8 introduces the symmetric iterations and, in particular, the SSOR method.

4.1 Eigenvalue Analysis of the Model Problem

The eigenvalues of the matrix A from §1.2 (e.g., in the representation (1.2.8)) can be described explicitly.

Theorem 4.1.1. *Then $n \times n$ matrix A of the Poisson-model problem from* §1.2 *with $n = (N-1)^2$ has the following n eigenvalues*:

$$\lambda_{ij} = 4h^{-2}[\sin^2(i\pi h/2) + \sin^2(j\pi h/2)] \qquad (1 \leqslant i, j \leqslant N-1), \quad (4.1.1a)$$

not all being different. The multiplicity of $\lambda = \lambda_{ij}$ is given by the number of indices $i', j' \in [1, N-1]$ with coinciding values $\lambda_{ij} = \lambda_{i'j'}$. The minimal eigenvalue is attained for $i = j = 1$, the maximal one for $i = j = N-1$:

$$\lambda_{\min} = \|A^{-1}\|_2^{-1} = 8h^{-2}\sin^2(\pi h/2), \tag{4.1.1b}$$

$$\lambda_{\max} = \|A\|_2 = 8h^{-2}\cos^2(\pi h/2). \tag{4.1.1c}$$

In particular, A is a positive definite matrix.

Proof. (1.2.8) shows $A = A^H$. The positive definiteness is a consequence of (1a) and Lemma 2.10.3 (1b, c) also follows from (1a). λ_{ij} from (1a) depends monotonically on $1 \leqslant i, j \leqslant N-1$; hence, $\lambda_{\max} = \lambda_{N-1,N-1} = \varrho(A) = \|A\|_2$ and $\lambda_{\min} = \lambda_{1,1} = 1/\varrho(A^{-1}) = 1/\|A^{-1}\|_2$. The assertion (1a) follows from

Lemma 4.1.2. *The n linear independent and even orthonormal eigenvectors of the matrix A from* §1.2 *corresponding to the eigenvalues* λ_{ij} *from* (1a) *are the vectors* e^{ij} *with*

$$(e^{ij})_{\nu\mu} = 2h\sin(ih\nu\pi)\sin(jh\mu\pi) \qquad (1 \leqslant i, j, \nu, \mu \leqslant N-1). \tag{4.1.2}$$

Proof. e^{ij} can be viewed as tensor product $e^i \otimes e^j$, where

$$e^k \in \mathbb{R}^{N-1}, \quad (e^k)_\nu = \sqrt{2h}\sin(kh\nu\pi) \qquad (1 \leqslant k, \nu \leqslant N-1)$$

since

$$(e^{ij})_{\nu\mu} = (e^i)_\nu (e^j)_\mu.$$

The scalar product in $\mathbb{R}^n$ is

$$\langle e^{ij}, e^{k\ell}\rangle = \sum_{\nu,\mu} e^{ij}_{\nu\mu} e^{k\ell}_{\nu\mu} = \sum_{\nu,\mu} e^i_\nu e^j_\mu e^k_\nu e^\ell_\mu = \sum_\nu e^i_\nu e^k_\nu \sum_\mu e^j_\mu e^\ell_\mu$$
$$= \langle e^i, e^k\rangle\langle e^j, e^\ell\rangle,$$

where the latter scalar products are those of $\mathbb{R}^{N-1}$. This identity shows that $\{e^{ij}\colon 1 \leqslant i, j \leqslant N-1\}$ is an orthonormal basis, provided that $\{e^i\colon 1 \leqslant i \leqslant N-1\}$ is an orthonormal basis of $\mathbb{R}^{N-1}$. The latter statement is the subject of

Exercise 4.1.3. Assume $hN = 1$ and $1 \leqslant k, l \leqslant N-1$. Prove

$$\sum_{\nu=1}^{N-1} \sin kh\nu\pi \sin \ell h\nu\pi = \begin{cases} 1/(2h) & \text{for } k = \ell, \\ 0 & \text{otherwise.} \end{cases} \tag{4.1.3}$$

For the model matrix A, the values of the grid function Ae^{ij} at the inner grid points $(x, y) = (kh, \ell h) \in \Omega_h$ are

$$(Ae^{ij})(x, y) = h^{-2}\frac{h}{2}[4\sin ix\pi \sin jy\pi$$
$$- \sin i(x+h)\pi \sin jy\pi - \sin i(x-h)\pi \sin jy\pi$$
$$- \sin ix\pi \sin j(y+h)\pi - \sin ix\pi \sin j(y-h)\pi]. \tag{4.1.4}$$

By the sine addition theorem, one obtains

$$\begin{aligned} \sin i(x+h)\pi + \sin i(x-h)\pi &= 2\sin ix\pi \cos ih\pi, \\ \sin j(y+h)\pi + \sin j(y-h)\pi &= 2\sin jy\pi \cos jh\pi \end{aligned} \tag{4.1.5a}$$

and therefore,

$$(Ae^{ij})(x,y) = h^{-2}\frac{h}{2}\sin ix\pi \sin jy\pi[4 - 2\cos ih\pi - 2\cos jh\pi].$$

From the identity

$$1 - \cos\xi = 2\sin^2\frac{\xi}{2}, \tag{4.1.5b}$$

one concludes the assertion (1a): $Ae^{ij} = \lambda_{ij}e^{ij}$. Equation (4) requires an additional consideration. If all neighbours $(x \pm h, y)$ and $(x, y \pm h)$ of the grid point (x, y) belong again to Ω_h, Eq. (4) represents directly the component of the vector Ae^{ij} corresponding to the point (x, y). If, however, one neighbour Q [e.g., $Q = (x - h, y)$ because of $x = h$] is not an inner grid point, the term $-h^{-2}e^{ij}(Q)$ should not appear. However in this case, $e^{ij}(Q) = 0$ holds (why?), and therefore, Eq. (4) is still valid. □

In the Pascal procedures the matrix A of the Poisson-model problem has no factor h^{-2} (instead, the right-hand side is multiplied by h^2). Therefore, h^{-2} is also omitted in (1a–c). $\lambda_{\min}$ and $\lambda_{\max}$ correspond to the following functions.

```
function minimal_ev(var A: data_of_discretisation): real;
begin minimal_ev:=4*(sqr(sin(pi/(2*A.nx)))+
  sqr(sin(pi/(2*A.ny))))
end;

function maximal_ev(var A: data_of_discretisation): real;
begin maximal_ev:=8-minimal_ev(A)
end;
```

4.2 Construction of Iterative Methods

4.2.1 Jacobi Iteration

4.2.1.1 Additive Splitting of the Matrix *A*. Given the system of equations

$$Ax = b \qquad (A \in \mathbb{K}^{I \times I},\ b \in \mathbb{K}^I), \tag{4.2.1}$$

one splits A into

$$A = W - R \qquad (W \text{ regular}). \tag{4.2.2}$$

The system of equations (1) is equivalent to

$$Wx = Rx + b. \tag{4.2.1'}$$

This suggests the iterative method

$$Wx^{m+1} = Rx^m + b. \tag{4.2.3}$$

Lemma 4.2.1. (a) *Assume* (2): $A = W - R$. *Then the iterative method* (3) *is consistent. The matrices of the first normal form* (3.2.2) *are*

$$M = W^{-1}R, \qquad N = W^{-1}.$$

The notation «W» for the matrix in (2) *is chosen, because the third normal form* (3.2.5):

$$W(x^m - x^{m+1}) = Ax^m - b \tag{4.2.3'}$$

is valid with the same matrix W.
(b) *Vice versa, any consistent iterative method* Φ *with regular N can be obtained from an additive splitting* (2).

Proof. (a) A comparison of the representation $x^{m+1} = W^{-1}Rx^m + W^{-1}b$ following from (3) with (3.2.2) shows that $M = W^{-1}R$ and $N = W^{-1}$.
(b) Choose $W := N^{-1}$ and $R := W - A$ in (2). □

4.2.1.2 Definition of the Jacobi Method

The *Jacobi method* results from (3) by the choice

$$W := D := \operatorname{diag}\{A\} \quad \text{and} \quad R := D - A, \tag{4.2.4}$$

i.e., D is the diagonal part $\operatorname{diag}\{a_{\alpha\alpha}: \alpha \in I\}$ of the matrix A, whereas R contains the off-diagonal part with reverse sign.

Remark 4.2.2. (a) The Jacobi method is well-defined if and only if $a_{\alpha\alpha} \neq 0$ for all $\alpha \in I$.
(b) The normal forms of the Jacobi method are

$$\begin{aligned} x^{m+1} &= M^{\mathrm{Jac}}x^m + N^{\mathrm{Jac}}b \quad \text{with} \\ M^{\mathrm{Jac}} &= D^{-1}(D - A) = I - D^{-1}A, \qquad N^{\mathrm{Jac}} = D^{-1}, \end{aligned} \tag{4.2.5a}$$

$$x^{m+1} = x^m - D^{-1}(Ax^m - b), \tag{4.2.5b}$$

$$D(x^m - x^{m+1}) = Ax^m - b. \tag{4.2.5c}$$

(c) The Jacobi method does not depend on the ordering of the indices.

The statement (c) causes renewed interest in the Jacobi method, because it allows a parallel computation. All operations in $x^m \mapsto D^{-1}[(A - D)x^m + b]$ can be performed by arithmetical *vector operations*.

In §1.1 we referred to the original paper of Jacobi from 1845. In 1918 H. Liebmann wrote a paper («Die angenäherte Ermittelung harmonischer Funktionen und konformer Abbildungen» [The approximation of harmonic functions and conformal mappings], Sitzungsber. der Math.-Phys. Klasse der K. B. Akad. der Wiss. zu München, vol. 47), in which the Jacobi iteration is applied to the Poisson problem. In this paper, reference is given to the lectures of L. Boltzmann on this topic in the 1892/93 winter semester.

4.2.1.3 Pascal Procedure. Since `new` and `x` might be identical variables, `new[i,j]` must not be immediately overwritten by the new result. In the following procedure, the (i − 1)st column `v` is transferred to `new[i-1]` as soon as the computation of the *i*th column is completed.

`procedure Jacobi_iteration` (4.2.6)

```
procedure Jacobi_iteration(var new: gridfunction;
                           var A: data_of_discretisation;
                           var x,b: gridfunction;
                            var IP: iterationparameter);
var i,j: integer; v,z: column;
begin with A do begin v:=x[0];
   for i:=1 to nx-1 do
   begin z[0]:=x[i,0]; z[ny]:=x[i,ny]; case kind of
Poisson_model_problem:
     for j:=1 to ny-1 do z[j]:=(b[i,j]+x[i-1,j]+x[i+1,j]
      +x[i,j-1]+x[i,j+1])/4;
fivepoint_formula:
     for j:=1 to ny-1 do
     z[j]:=(b[i,j]-S[0,-1]*x[i,j-1]-S[-1,0]*x[i-1,j]
           -S[1,0]*x[i+1,j]
          -S[0,1]*x[i,j+1])/S[0,0];
ninepoint_formula:
     for j:=1 to ny-1 do
     z[j]:=(b[i,j]-S[-1,-1]*x[i-1,j-1]-S[0,-1]*x[i,j-1]
           -S[1,-1]*x[i+1,j-1]
          -S[-1,0]*x[i-1,j]-S[1,0]*x[i+1,j]
           -S[-1,1]*x[i-1,j+1]
          -S[0,1]*x[i,j+1]-S[1,1]*x[i+1,j+1])/S[0,0]
     end {case};
     new[i-1]:=v; v:=z
    end;
    new[nx-1]:=v; new[nx]:=x[nx]
end end;
```

4.2.2 Gauß-Seidel Method

4.2.2.1 Definition. Let the index set I be ordered and identified with $\{1,\ldots,n\}$. Then the following splitting of A:

$$A = D - E - F \qquad \text{with} \tag{4.2.7a}$$

$$D\text{: diagonal matrix (diagonal part of } A), \tag{4.2.7b}$$

$$E\text{: strictly lower triangular matrix,} \tag{4.2.7c}$$

$$F\text{: strictly upper triangular matrix} \tag{4.2.7d}$$

is uniquely determined (cf. (1.4.2)).

> The *Gauß-Seidel method* results from (3) by the choice
>
> $$W = D - E \qquad (D, E \text{ from (7a)}). \tag{4.2.8}$$

The following obvious properties have already been mentioned, in part, in Remark 1.4.1.

Remark 4.2.3. (a) The Gauß-Seidel method is well-defined if and only if I is ordered and $a_{ii} \neq 0$ holds for all $i \in I$.
(b) The normal forms of the Gauß-Seidel method are

$$\begin{aligned} x^{m+1} &= M^{GS}x^m + N^{GS}b \quad \text{with} \\ M^{GS} &= (D-E)^{-1}F, \qquad N^{GS} = (D-E)^{-1}, \end{aligned} \tag{4.2.9a}$$

$$x^{m+1} = x^m - (D-E)^{-1}(Ax^m - b), \tag{4.2.9b}$$

$$(D-E)(x^m - x^{m+1}) = Ax^m - b, \qquad \text{i.e., } W^{GS} = D - E. \tag{4.2.9c}$$

(c) Differently from the Jacobi method, the Gauß-Seidel iteration depends on the ordering of the indices.

Because of part (c), a precise description of the Gauß-Seidel method must characterise the ordering, if this is not already fixed by the problem. In the model case (1.2.4a), one has, e.g., the *lexicographical Gauß-Seidel method* and the *chequer-board Gauß-Seidel method*, referring to the corresponding ordering of the grid points.

The componentwise representation

$$\text{for } i := 1 \text{ to } n \text{ do } x_i^{m+1} := \left(b_i - \sum_{j=1}^{i-1} a_{ij}x_j^{m+1} - \sum_{j=i+1}^{n} a_{ij}x_j^m \right) \Big/ a_{ii} \tag{4.2.10}$$

shows that the algorithm works sequentially. In contrast to the Jacobi iteration (5), one needs no intermediate storage. The vector components may be overwritten immediately (cf. Remark 1.4.1b). On the other hand, the parallel computation by vector operations is more difficult (for the chequer-board variant of the model problem the parallel treatment of both colours is possible; cf. also Niethammer [31]).

4.2.2.2 Pascal Procedure. The Poisson-model case is already given in (1.4.6). The lexicographical variant now reads as follows:

procedure lex_Gauss_Seidel (4.2.11)

```
procedure lex_Gauss_Seidel(var new: gridfunction;
                           var A: data_of_discretisation;
                           var x,b: gridfunction;
                           var IP: iterationparameter);
var i,j: integer;
begin with A do begin case kind of
Poisson_model_problem:
   for j:=1 to ny-1 do for i:=1 to nx-1 do
   new[i,j]:=(b[i,j]+new[i-1,j]+x[i+1,j]
    +new[i,j-1]+x[i,j+1])/4;
fivepoint_formula:
   for j:=1 to ny-1 do for i:=1 to nx-1 do
   new[i,j]:=(b[i,j]-S[0,-1]*new[i,j-1]
    -S[-1,0]*new[i-1,j]-S[1,0]*x[i+1,j]
    -S[0,1]*x[i,j+1])/S[0,0];
ninepoint_formula:
   for j:=1 to ny-1 do for i:=1 to nx-1 do
   new[i,j]:=(b[i,j]-S[-1,-1]*new[i-1,j-1]-S[0,-1]*new[i,j-1]
                    -S[1,-1]*new[i+1,j-1]
                     -S[-1,0]*new[i-1,j]-S[1,0]*x[i+1,j]
                    -S[-1,1]*x[i-1,j+1]-S[0,1]*x[i,j+1]
                     -S[1,1]*x[i+1,j+1])/S[0,0]
   end {case};
   transfer_boundary_values(nx,ny,x,new)
end end;
```

Here, the procedure `transfer_boundary_values` is needed to transfer the boundary values of the old iterate x onto the new iterate. Note that the boundary values appear on the right-hand side b of the system.

```
procedure transfer_boundary_values
                (nx,ny: integer; var from,onto: gridfunction);
var i: integer;
begin for i:=0 to nx do
    begin onto[i,0]:=from[i,0]; onto[i,ny]:=from[i,ny] end;
    for i:=1 to ny-1 do begin onto[0,i]:=from[0,i];
     onto[nx,i]:=from[nx,i] end
end;
```

The chequer-board numbering can be generalised to the case of the nine-point formula by introducing *four-colour numbering*:

$$\begin{aligned}
\Omega_h^1 &:= \{(x,y) = (ih,jh) \in \Omega_h \colon i,j \text{ even}\},\\
\Omega_h^2 &:= \{(x,y) = (ih,jh) \in \Omega_h \colon i,j \text{ odd}\},\\
\Omega_h^3 &:= \{(x,y) = (ih,jh) \in \Omega_h \colon i \text{ even}, j \text{ odd}\},\\
\Omega_h^4 &:= \{(x,y) = (ih,jh) \in \Omega_h \colon i \text{ odd}, j \text{ even}\}.
\end{aligned} \tag{4.2.12}$$

First, the points of Ω_h^1 are numbered, then those of Ω_h^2, Ω_h^3, and finally Ω_h^4. Since $\Omega_h^1 \cup \Omega_h^2$ represents the black squares and $\Omega_h^3 \cup \Omega_h^4$ the red ones and since for the five-point formula the ordering inside of one colour is irrelevant, the four-colour numbering defined above coincides with the chequer-board ordering for the five-point formulae. One single step in Ω_h^j corresponding to a quarter of the complete iteration is performed by

`procedure chequerboard_Gauss_Seidel_quarterstep` (4.2.13)

```
procedure chequerboard_Gauss_Seidel_quarterstep
(var new: gridfunction; var A: data_of_discretisation;
var x,b: gridfunction; instart,jstart: integer;
UB: Boolean);
var i,j: integer;
begin with A do
 begin i:=istart; if UB then
   transfer_boundary_values(nx,ny,x,new);
  while i<nx do begin j:=jstart; case kind of
Poisson_model_problem:
    while j<ny do
    begin new[i,j]:=(b[i,j]+x[i-1,j]+x[i+1,j]+x[i,j-1]
      +x[i,j+1])/4; j:=j+2 end;
fivepoint_formula:
    while j<ny do
    begin new[i,j]:=(b[i,j]-S[0,-1]*x[i,j-1]-S[-1,0]*x[i-1,j]
           -S[1,0]*x[i+1,j]-S[0,1]*x[i,j+1])/S[0,0];
        j:=j+2
    end;
ninepoint_formula:
    while j<ny do
```

```
    begin new[i,j]:=(b[i,j]-S[-1,-1]*x[i-1,j-1]
           -S[0,-1]*x[i,j-1]
          -S[1,-1]*x[i+1,j-1]-S[-1,0]*x[i-1,j]
           -S[1,0]*x[i+1,j]
          -S[-1,1]*x[i-1,j+1]-S[0,1]*x[i,j+1]
           -S[1,1]*x[i+1,j+1])/S[0,0];
        j:=j+2
    end end {case};
    i:=i+2
end end end;
```

Therefore, the chequer-board Gauß-Seidel iteration reads as follows:

`procedure chequerboard_Gauss_Seidel` (4.2.14)

```
procedure chequerboard_Gauss_Seidel(var new: gridfunction;
          var A: data_of_discretisation;
          var x,b: gridfunction; var IP: iterationparameter);
begin {black squares:}
    chequerboard_Gauss_Seidel_quarterstep (new,A,x,b,1,1,true);
    chequerboard_Gauss_Seidel_quarterstep
     (new,A,new,b,2,2,false);
    {red squares:}
    chequerboard_Gauss_Seidel_quarterstep
     (new,A,new,b,1,2,false);
    chequerboard_Gauss_Seidel_quarterstep
     (new,A,new,b,2,1,false);
end;
```

Reversing the lexicographical numbering (i.e., the previous index i becomes $n+1-i$) and basing the Gauß-Seidel method on this ordering, one obtains the *backward lexicographical Gauß-Seidel iteration.* In [Prog] the corresponding Pascal procedure is called

`procedure lex_Gauss_Seidel_backward.` (4.2.15)

4.3 Damped Iterative Methods

4.3.1 Damped Jacobi Method

4.3.1.1 Damping of a General Iterative Method. Let a linear iterative method Φ be given by its second normal form

$$x^{m+1} = x^m - N(Ax^m - b). \tag{4.3.1a}$$

Adding a scaling factor ϑ, we obtain the corresponding *damped method*

$$x^{m+1} = x^m - \vartheta N(Ax^m - b) \qquad (\vartheta \in \mathbb{R}), \tag{4.3.1b}$$

which we have symbolised by Φ_ϑ. To be precise, the term «damped» holds in a proper sense only for $0 < \vartheta < 1$. For $\vartheta = 1$ one regains the (nondamped) original method, whereas for $\vartheta > 1$ the iteration Φ_ϑ may be called an «*extrapolated method*».

Remark 4.3.1. The iteration (1b) is consistent for all ϑ.

In the case of the Jacobi method with $N^{\text{Jac}} = D^{-1}$ (cf. (2.5b)), the *damped Jacobi method* takes the form

$$x^{m+1} = x^m - \vartheta D^{-1}(Ax^m - b) \qquad (\vartheta \in \mathbb{R}), \tag{4.3.2}$$

where D is the diagonal part of A.

4.3.1.2 Pascal Procedures. The damping parameter ϑ can be determined by the respective procedures `set_theta` or `determine_theta` explicitly or interactively:

```
procedure set_theta(var IP: iterationparameter; theta: real);
begin IP.theta:=theta end;

procedure determine_theta(var IP: iterationparameter);
begin write ('--> theta ='); readln(IP.theta) end;
```

The damping of an arbitrary iteration $\Phi(x, b)$ yields

$$\Phi_\vartheta(x, b) := x + \vartheta(\Phi(x, b) - x) = (1 - \vartheta)x + \vartheta\Phi(x, b).$$

Φ_ϑ can be performed by means of

```
procedure damp_iteration(var new: gridfunction;
      var A: data_of_discretisation; var x,b:gridfunction;
      var IP: iterationparameter;
      procedure iteration(var new: gridfunction;
                           var A: data_of_discretisation;
                           var x,b: gridfunction;
                            var IP: iterationparameter));
var y: gridfunction;
begin iteration(y,A,x,b,IP);
   factor_x_vector_plus_factor_x_vector
                       (A.nx,A.ny,new,IP.theta,y,1-IP.theta,x);
   transfer_boundary_values(A.nx,A.ny,x,new)
end;
```

where the procedures `transfer_boundary_values` (cf. §4.2.1.3) and `factor_x_vector_plus_factor_x_vector` are to be specified:

```
procedure factor_x_vector_plus_factor_x_vector
          (nx,ny: integer; var result: gridfunction;
          factor1: real; var x: gridfunction; factor2: real;
           var y: gridfunction);
var i,j: integer;
begin
  for i:=1 to nx-1 do for j:=1 to ny-1 do result[i,j]
    :=factor1*x[i,j]+factor2*y[i,j]
end;
```

The damped Jacobi method then takes the following form:

```
procedure damped_Jacobi_iteration(var new: gridfunction;
                                  var A: data_of_discretisation;
                                  var x,b: gridfunction;
                                   var IP: iterationparameter);
begin damped_iteration(new,A,x,b,IP,Jacobi_iteration) end;
```

A procedure determining the optimal damping parameters will be outlined in Remark 9.2.9.

4.3.2 Richardson Iteration

4.3.2.1 Definition. If, as in the model case, the diagonal D is a multiple of the identity matrix: $D = dI$, one can combine the factors d and ϑ in $\Theta := \vartheta/d$. One obtains a method identical to (2):

$$x^{m+1} = x^m - \Theta(Ax^m - b) \qquad (\Theta \in \mathbb{R}). \tag{4.3.3}$$

The value $\Theta := 1/d$ reproduces the (undamped) Jacobi method (5a). For matrices with nonconstant diagonal elements, (3) represents a new iteration, the (stationary) *Richardson iteration*. Just like the Jacobi method, the Richardson method is independent of the ordering of indices.

Naming (3) the «Richardson method» is not completely correct, since this name is often used for a modified method with parameters $\Theta = \Theta_m$ being dependent on the iteration index m (cf. §7). The latter method will be called the «instationary Richardson iteration» or the «semi-iterative Richardson method». In contrast to this method, (3) is the «stationary Richardson iteration» or simply the «Richardson method».

In a certain sense, the Richardson iteration with $\Theta = 1$ is the *prototype of any linear iteration.* Let Φ be an arbitrary consistent iteration with regular N (concerning the regularity of N, compare Remark 3.2.9). The iteration Φ has the second normal form (1a): $x^{m+1} = x^m - N(Ax^m - b)$. Transforming the system of equations $Ax = b$ into the equivalent system

$$\hat{A}x = \hat{b} \quad \text{with } \hat{A} := NA, \quad \hat{b} := Nb, \tag{4.3.4}$$

one can rewrite (1a) as

$$x^{m+1} = x^m - (\hat{A}x^m - \hat{b}). \tag{4.3.5}$$

Remark 4.3.2. Hence, any linear and consistent iterative method with regular N can be regarded as a Richardson iteration with $\Theta = 1$ applied to the transformed system (4).

4.3.2.2 Pascal Procedures. We need the following procedures:

```
procedure vector_plus_factor_x_vector(nx,ny: integer;
                    var result,x: gridfunction; factor:
                     real; var y: gridfunction);
var i,j: integer;
begin for i:=1 to nx-1 do for j:=1 to ny-1 do
  result[i,j]:=x[i,j]+factor*y[i,j]
end;

procedure zero_boundary_values(nx,ny: integer;
  var x: gridfunction);
var i: integer;
begin for i:=0 to nx do  begin x[i,0]:=0; x[i,ny]:=0
end;
        for i:=1 to ny-1 do begin x[0,i]:=0, x[nx,i]:=0
end
end;

procedure residual(var R: gridfunction;
                       var A: data_of_discretisation;
                        var x,b: gridfunction);
{Computation of r:=b-Ax, where the grid function x may have
  inhomogeneous boundary values.}
var i,j: integer; v,z: column;
begin with A do begin v:=x[0];
   for i:=1 to nx-1 do
   begin case kind of
Poisson_model_problem:
     for j:=1 to ny-1 do
     z[j]:=b[i,j]-4*x[i,j]+x[i,j-1]+x[i,j+1]+x[i-1,j]
       +x[i+1,j];
```

```
fivepoint_formula:
    for j:=1 to ny-1 do
    z[j]:=b[i,j]-S[-1,0]*x[i-1,j]-S[1,0]*x[i+1,j]
          -S[0,-1]*x[i,j-1]
        -S[0,1]*x[i,j+1]-S[0,0]*x[i,j];
ninepoint_formula:
    for j:=1 to ny-1 do
    z[j]:=b[i,j]-S[-1,-1]*x[i-1,j-1]-S[0,-1]*x[i,j-1]
          -S[1,-1]*x[i+1,j-1]
        -S[-1,0]*x[i-1,j]-S[0,0]*x[i,j]-S[1,0]*x[i+1,j]
        -S[-1,1]*x[i-1,j+1]-S[0,1]*x[i,j+1]
          -S[1,1]*x[i+1,j+1]
    end {case};
    R[i-1]:=v; v:=z
    end;
    R[nx-1]:=v; zero_boundary_values(nx,ny,R)
end end;
```

The procedures `set_theta` and `determine_theta` from §4.3.1.2 serve to define the factor Θ. By `residual` the residue (residual) $r := b - Ax$ is computed. The iteration procedure reads as follows:

```
procedure Richardson_iteration(var new: gridfunction;
                               var A: data_of_discretisation;
                               var x,b: gridfunction;
                               var IP: iterationparameter);
var r: gridfunction;
begin residual(r,A,x,b);
      vector_plus_factor_x_vector
        (A.nx,A.ny,new,x,IP.theta,r);
      transfer_boundary_values(A.nx,A.ny,x,new)
end;
```

4.3.3 SOR Method

4.3.3.1 Definition. The *SOR method* (*s*uccessive *o*ver*r*elaxation) has already been defined in (1.4.9) by

$$\text{for } i := 1 \text{ to } n \text{ do } x_i^{m+1} := x_i^m - \omega\left(\sum_{j=1}^{i-1} a_{ij}x_j^{m+1} + \sum_{j=i}^{n} a_{ij}x_j^m - b_i\right)\bigg/ a_{ii}. \tag{4.3.6}$$

Although each correction of x_i^m is damped (or extrapolated) by the (relaxation) factor ω, iteration (6) has not the structure (1b). The reason is that the right-hand side in (6), in part, contains the already damped x^{m+1} components.

Exercise 4.3.3. Prove by means of the representation (6) that the SOR method is consistent.

Remark 4.3.4. (a) For $\omega = 1$ the SOR method (6) coincides with the Gauß-Seidel method (2.10), which therefore is also called «*relaxation*». For $0 < \omega < 1$, the precise name of (6) is «*underrelaxation method*», whereas the term «*overrelaxation method*» suits $\omega > 1$.
(b) Like the Gauß-Seidel method, the SOR method depends on the ordering of indices.

The representation of the SOR iteration as matrix formulation is less simple than the componentwise representation (6).

Lemma 4.3.5. *In a matrix formulation the SOR iteration* (6) *reads*

$$x^{m+1} = M_\omega^{\mathrm{SOR}} x^m + N_\omega^{\mathrm{SOR}} b \tag{4.3.7a}$$

with

$$M_\omega^{\mathrm{SOR}} := (I - \omega L)^{-1}\{(1-\omega)I + \omega U\} \tag{4.3.7b}$$
$$= (D - \omega F)^{-1}\{(1-\omega)D + \omega F\},$$

$$N_\omega^{\mathrm{SOR}} := \omega(I - \omega L)^{-1} D^{-1} = \omega(D - \omega E)^{-1}, \quad \textit{where} \tag{4.3.7c}$$

$$L := D^{-1}E, \quad U := D^{-1}F \quad \text{with} \quad A = D - E - F \textit{ according } (2.7\text{a–d}). \tag{4.3.7d}$$

The matrix of the third normal form is

$$W_\omega^{\mathrm{SOR}} := \frac{1}{\omega}(D - \omega E) = \frac{1}{\omega}D - E. \tag{4.3.7e}$$

Proof. Multiplication by $I - \omega L$ brings (7a) into the form

$$(I - \omega L)x^{m+1} = \{(1-\omega)I + \omega U\}x^m + \omega D^{-1} b$$

or

$$x^{m+1} = x^m - \omega[-Lx^{m+1} + \{I - U\}x^m - D^{-1}b]. \tag{4.3.7f}$$

Moving D^{-1} in front of the bracket, we obtain the expression $[-Ex^{m+1} + \{D - F\}x^m - b]$ according to definition (7d) of L and U. The componentwise interpretation of this equation coincides with (6). □

4.3.3.2 Pascal Procedures. The relaxation parameter ω can be defined by means of `set_omega` or `determine_omega` (cf. §4.3.1.3). An alternative is the procedure `define_SOR_parameter` based on the results from §5.6.2 (cf. also the example (3.5.9)).

```
procedure set_omega(var IP: iterationparameter; omega: real);
begin IP.omega:=omega end;

procedure determine_omega(var IP: iterationparameter);
begin write('--> omega ='); readln(IP.omega) end;

function Jacobi_convergence_factor(var it: data_of_
 iteration): real; {Th. 7.2}
begin with it.A do Jacobi_convergence_factor:=
 (cos(pi/nx)+cos(pi/ny))/2
end;

function optimal_omega_for_SOR(beta: real): real;
begin optimal_omega_for_SOR:=2/(1+sqrt((1+beta)*
 (1-beta)))
end;

procedure define_SOR_parameter(var it: data_of_
 iteration);
var beta: real;
begin with it do with A do with IP do
  begin writeln('Choice of the SOR_parameter:');
   writeln(' Omega may be defined via the Jacobi convergence
    rate.');
   write(' Is the direct input of omega desired?');
   if yes_no then determine_omega(IP) else
   begin if kind=Poisson_model_problem then
                                        beta:=Jacobi_
                                         convergence_factor(it)
      else
      begin write('--> Jacobi convergence factor =');
       readln(beta)
      end;
    set_omega(IP,optimal_omega_for_SOR(beta))
   end
  end
end;
```

Analogously to the Gauß-Seidel method, the SOR procedures for lexicographical and chequer-board orderings are organised as follows:

```
procedure lex_SOR(var new: gridfunction; var A: data_
 of_discretisation;
                  var x,b: gridfunction; var IP:
                   iterationparameter);
```

```
var i,j: integer; om1,om4: real;
begin with A do with IP do
 begin om1:=1-omega; om4:=omega-S[0,0];
   transfer_boundary_values(nx,ny,x,new);
   case kind of
Poisson_model_problem:
   begin om4:=omega/4;
      for j:=1 to ny-1 do for i:=1 to nx-1 do
      new[i,j]:=om1*x[i,j]+om4*(b[i,j]+new[i-1,j]+x
        [i+1,j]+new[i,j-1]
             +x[i,j+1]) end;
fivepoint_formula:
   for j:=1 to ny-1 do for i:=1 to nx-1 do
   new[i,j]:=om1*x[i,j]+om4*(b[i,j]-S[0,-1]*
    new[i,j-1]-S[-1,0]*new[i-1,j]
          -S[1,0]*x[i+1,j]-S[0,1]*x[i,j+1]);
ninepoint_formula:
   for j:=1 to ny-1 do for i:=1 to nx-1 do
   new[i,j]:=om1*x[i,j]+om4*(b[i,j]-S[-1,-1]*
    new[i-1,j-1]-S[0,-1]*new[i,j-1]
          -S[1,-1]*new[i+1,j-1]-S[-1,0]*new[i-1,j]-
           S[1,0]*x[i+1,j]
          -S[-1,1]*x[i-1,j+1]-S[0,1]*x[i,j+1]-S[1,1]*
           x[i+1,j+1])
   end {case}
 end
end;

procedure chequerboard_SOR_quarterstep(var new:
 gridfunction;
                      var A: data_of_discretisation;
                       var x,b: gridfunction;
                      var omega: real; istart,jstart:
                       integer; UB: Boolean);
var i,j: integer; om1,om4: real;
begin with A do
 begin i:=istart; if UB then transfer_boundary_
   values(nx,ny,x,new);
   om1:=1-omega; om4:=omega/S[0,0];
   while i<nx do begin j:=jstart; case kind of
Poisson_model_problem:
    while j<ny do
    begin new[i,j]:=om1*x[i,j]+om4*(b[i,j]+x[i-1,j]+
     x[i+1,j]+x[i,j-1]+x[i,j+1]);
        j:=j+2
    end;
```

```
fivepoint_formula:
    while j<ny do
    begin new[i,j]:=om1*x[i,j]+om4*(b[i,j]-S[0,-1]*x[i,j-1]
              -S[-1,0]*x[i-1,j]
              -S[1,0]*x[i+1,j]-S[0,1]*x[i,j+1]);
       j:=j+2
    end;
ninepoint_formula:
    while j<ny do
    begin new[i,j]:=om1*x[i,j]+om4*(b[i,j]-S[-1,-1]*
     x[i-1,j-1]
             -S[0,-1]*x[i,j-1]-S[1,-1]*x[i+1,j-1]-
              S[-1,0]*x[i-1,j]
             -S[1,0]*x[i+1,j]-S[-1,1]*x[i-1,j+1]
             -S[0,1]*x[i,j+1]
             -S[1,1]*x[i+1,j+1]);
        j:=j+2
    end end {case};
    i:=i+2
end end end;

procedure chequerboard_SOR(var new: gridfunction; var A:
    data_of_discretisation; var x,b: gridfunction; var IP:
     iterationparameter);
begin {black squares:}
    chequerboard_SOR_quarterstep(new,A,x,b,IP.omega,
     1,1,true);
    chequerboard_SOR_quarterstep(new,A,new,b,IP.omega,
     2,2,false);
    {red squares:}
    chequerboard_SOR_quarterstep(new,A,new,b,IP.omega,
     1,2,false);
    chequerboard_SOR_quarterstep(new,A,new,b,IP.omega,
     2,1,false)
end;
```

Reversing the lexicographical numbering, one obtains in analogy to (2.15) the *backward lexicographical SOR iteration.* In [Prog] it is included under the name

```
procedure lex_SOR_backward
```

(cf. (2.15)). The *backward chequer-board SOR variant* is obtained by reversing the four «colours» of Ω_h^1, Ω_h^2, Ω_h^3, and Ω_h^4:

```
procedure chequerboard_SOR_backward
```

4.4 Convergence Analysis

In the following, the convergence considerations are based on the assumption that A is positive definite or has weaker but related properties. In §§5–6 other assumptions will be posed.

4.4.1 Richardson Iteration

The iteration matrix of the Richardson method (1) is M_Θ^{RICH} from (2):

$$x^{m+1} = x^m - \Theta(Ax^m - b) \qquad (\Theta \in \mathbb{C}), \tag{4.4.1}$$

$$M_\Theta^{\mathrm{Rich}} = I - \Theta A. \tag{4.4.2}$$

We may write $M_\Theta^{\mathrm{Rich}} = P(A)$ with the polynomial $P(\xi) = 1 - \Theta\xi$. If $\lambda_\nu \in \sigma(A)$ are the eigenvalues of A, $\mu_\nu = P(\lambda_\nu) = 1 - \Theta\lambda_\nu$ are those of M_Θ^{Rich} (cf. Lemma 2.4.7a). Since the function $|1 - \Theta\xi|$ has no local maxima, one obtains

Lemma 4.4.1. *Assume that A possesses only* real *eigenvalues. Let* $\lambda_{\min} := \min\{\lambda\colon \lambda \in \sigma(A)\}$ *and* $\lambda_{\max} := \max\{\lambda\colon \lambda \in \sigma(A)\}$ *denote the extreme eigenvalues of A. Then the spectrum of M_Θ^{Rich} for $\Theta \in \mathbb{R}$ is real: $\sigma(M_\Theta^{\mathrm{Rich}}) \subset \mathbb{R}$. For all $\Theta \in \mathbb{C}$ the convergence rate is given by*

$$\varrho(M_\Theta^{\mathrm{Rich}}) = \max\{|1 - \Theta\lambda_{\min}|, |1 - \Theta\lambda_{\max}|\}. \tag{4.4.3}$$

Theorem 4.4.2. *Assume that A has only* positive *eigenvalues. $\lambda_{\max}(A)$ is the maximal eigenvalue of A. Let Θ be real. Then the Richardson method converges if and only if*

$$0 < \Theta < 2/\lambda_{\max}(A). \tag{4.4.4}$$

The convergence rate is given by (3).

Proof. (i) For $0 < \Theta < 2/\lambda_{\max}$, we have $-1 < 1 - \Theta\lambda_{\max} \leqslant 1 - \Theta\lambda_{\min} < 1$. (3) yields $\varrho(M_\Theta^{\mathrm{Rich}}) < 1$, i.e., convergence.
(ii) Now assume convergence: $\varrho(M_\Theta^{\mathrm{Rich}}) < 1$. One concludes from $1 > \varrho(M_\Theta^{\mathrm{Rich}}) \geqslant |1 - \Theta\lambda_{\max}| \geqslant 1 - \Theta\lambda_{\max}$ that $\Theta\lambda_{\max} > 0$ and therefore $\Theta > 0$ because of $\lambda_{\max} > 0$. Similarly, $-1 < -\varrho(M_\Theta^{\mathrm{Rich}}) \leqslant -|1 - \Theta\lambda_{\max}| \leqslant 1 - \Theta\lambda_{\max}$ shows $\Theta\lambda_{\max} < 2$, also proving that the second inequality in (4) is necessary. □

Representation (3) allows us to determine the factor Θ so that $\varrho(M_\Theta^{\mathrm{Rich}})$ becomes *minimal*. The *optimal damping factor* Θ results as the intersection (cf. Fig. 1) of the lines $y = \Theta\lambda_{\max} - 1$ and $y = 1 - \Theta\lambda_{\min}$.

Theorem 4.4.3. *Assume that A has only* positive *eigenvalues.* $\lambda_{\max}$ *and* $\lambda_{\min}$ *are the respective maximal and minimal eigenvalues of A. The optimal convergence rate of Richardson's method is attained for*

$$\Theta_{\text{opt}} = \frac{2}{\lambda_{\max} + \lambda_{\min}}, \qquad \varrho(M^{\text{Rich}}_{\Theta_{\text{opt}}}) = \frac{\lambda_{\max} - \lambda_{\min}}{\lambda_{\max} + \lambda_{\min}}. \tag{4.4.5}$$

In particular, the assumption that A has only positive eigenvalues is satisfied for positive definite matrices.

Corollary 4.4.4. *Assume that A is positive definite and* Θ *is real. Then the Richardson method converges if and only if*

$$0 < \Theta < 2/\|A\|_2. \tag{4.4.6a}$$

The convergence is monotone with respect to the Euclidean norm $\|\cdot\|_2$ *and to the energy norm* $\|\cdot\|_A$ *defined in* (2.10.5a, c) *by* $\|x\|_A := \|A^{1/2}x\|_2$. *Furthermore, the convergence rate and contraction number coincide*:

$$\varrho(M^{\text{Rich}}_{\Theta}) = \|M^{\text{Rich}}_{\Theta}\|_2 = \|M^{\text{Rich}}_{\Theta}\|_A. \tag{4.4.6b}$$

The optimal convergence rate (5) *can be expressed as a function of the condition number* $\varkappa(A) = \text{cond}_2(A) := \|A\|_2\|A^{-1}\|_2$:

$$\varrho(M^{\text{Rich}}_{\Theta_{\text{opt}}}) = \frac{\varkappa(A) - 1}{\varkappa(A) + 1} \quad \text{for } \Theta_{\text{opt}} = \frac{2\|A^{-1}\|_2}{\varkappa(A) + 1}. \tag{4.4.6c}$$

Proof. (6a) follows from $\lambda_{\max} = \|A\|_2$. Using $\lambda_{\min} = 1/\|A^{-1}\|_2$ (cf. (2.10.10)) and $\varkappa(A) = \lambda_{\max}/\lambda_{\min}$, we can transform (5) into (6c). As A is normal, $M := M^{\text{Rich}}_{\Theta}$ is also normal, proving $\varrho(M) = \|M\|_2$ (cf. (2.9.4b)). The second equality in (6b) follows from $\|M\|_A = \|A^{1/2}MA^{-1/2}\|_2$ and the commutativity $A^{1/2}M = MA^{1/2}$ (cf. (2.10.5d), Remark 2.10.6b). □

The importance of the assumption that the eigenvalues are positive is confirmed by counter-examples in

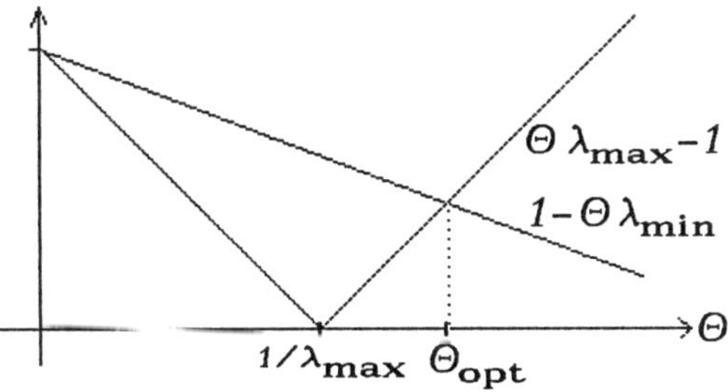

Fig. 4.4.1 Optimal Θ

Exercise 4.4.5. Prove: (a) If A has at least one positive and one negative eigenvalue, the Richardson method diverges for any choice of $\Theta \in \mathbb{C}$. (b) If besides other eigenvalues, A has two complex eigenvalues λ_1, λ_2 with opposite signs: $\lambda_1/|\lambda_1| = -\lambda_2/|\lambda_2|$, the Richardson method diverges.

Nevertheless, the assumption of positivity can be weakened.

Exercise 4.4.6. (a) Let the spectrum $\sigma(A)$ of A lie in a closed circle around $\mu \in \mathbb{C}\backslash\{0\}$ with radius $r < |\mu|$. Prove that the choice $\Theta = 1/\mu$ leads to convergence of the Richardson method:

$$\varrho(M_\Theta^{\text{Rich}}) \leqslant r/|\mu| < 1.$$

(b) Let g be an arbitrary line in the complex plane passing through the origin $z = 0$ and cutting $\mathbb{C}\backslash g$ into two half-planes. If $\sigma(A)$ lies in one of the half-planes, then Richardson's method converges for suitable Θ.

In the non-Hermitian case, A can be split into symmetric and skew-symmetric parts (cf. (2.9.12)):

$$A = A_0 + iA_1 \quad \text{with } A_0 := \frac{1}{2}(A + A^H), \quad A_1 := \frac{1}{2i}(A - A^H). \tag{4.4.7}$$

Together with a suitable estimate of the Hermitian matrices A_0 and A_1 convergence statements for the Richardson method can be formulated (cf. Theorems 7, 8 and (10c, d), Samarskii–Nikolaev [1, §6.4]).

Theorem 4.4.7. *For A and A_0 from* (7) *constants $0 < \lambda \leqslant \Lambda$ are assumed to exist such that*

$$0 \leqslant \lambda I \leqslant A_0, \tag{4.4.8a}$$

$$A^H A \leqslant \Lambda A_0. \tag{4.4.8b}$$

Then for the parameters

$$0 < \Theta < \frac{2}{\Lambda} \tag{4.4.8c}$$

the Richardson method converges monotonically with respect to the Euclidean norm:

$$\varrho(M_\Theta^{\text{Rich}}) \leqslant \|M_\Theta^{\text{Rich}}\|_2 \leqslant \sqrt{1 - \Theta\lambda(2 - \Theta\Lambda)} < 1. \tag{4.4.9a}$$

The bound of the right-hand side is minimal for $\Theta' := 1/\Lambda$:

$$\varrho(M_{\Theta'}^{\text{Rich}}) \leqslant \|M_{\Theta'}^{\text{Rich}}\|_2 \leqslant \sqrt{1 - \lambda/\Lambda}. \tag{4.4.9b}$$

Proof. (8a, b) leads to the estimate

$$(M_\Theta^{\text{Rich}})^H(M_\Theta^{\text{Rich}}) = (I - \Theta A)^H(I - \Theta A) = I - \Theta(A + A^H) + \Theta^2 A^H A \underset{(8b)}{\leqslant}$$
$$\leqslant I - 2\Theta A_0 + \Theta^2 \Lambda A_0 = I - \Theta(2 - \Theta\Lambda)A_0 \underset{(8a)}{\leqslant} I - \Theta\lambda(2 - \Theta\Lambda)I,$$

entailing $\|M_\Theta^{\mathrm{Rich}}\|_2^2 = \|(M_\Theta^{\mathrm{Rich}})^H(M_\Theta^{\mathrm{Rich}})\|_2 \leqslant 1 - \Theta\lambda(2 - \Theta\Lambda)$ and therefore (9a) (cf. (2.10.3f)). The convergence $1 - \Theta\lambda(2 - \Theta\Lambda) < 1$ is ensured by (8c). (9b) is easy to verify. □

Condition (8b) can also be written as (8b′):

$$\langle Ax, Ax\rangle \leqslant \Lambda\langle A_0 x, x\rangle \quad \text{for all } x \in \mathbb{K}^I. \tag{4.4.8b′}$$

In the positive definite case, the inequalities (8a, b) are satisfied with $\lambda = \lambda_{\min}$, $\Lambda = \lambda_{\max}$. However, the optimal parameters Θ_{opt} and Θ' from Theorems 3 and 7 are different and lead to different bounds in (6c) and (9b). A stronger estimate than (9b) is possible if, in addition, the skew-symmetric part A_1 is estimated.

Theorem 4.4.8. *Assume that for A_0 and A_1 from* (7) *there are constants $0 < \lambda \leqslant \Lambda$, $\tau \geqslant 0$ with*

$$\lambda I \leqslant A_0 \leqslant \Lambda I, \tag{4.4.10a}$$

$$\|A_1\|_2 \leqslant \tau. \tag{4.4.10b}$$

Then the Richardson method converges for

$$0 < \Theta < \frac{2\lambda}{\lambda\Lambda + \tau^2} \tag{4.4.11a}$$

monotonically with respect to the Euclidean norm:

$$\|M_\Theta^{\mathrm{Rich}}\|_2 \leqslant \frac{1}{2}\Theta(\Lambda - \lambda) + \sqrt{\left[1 - \frac{1}{2}\Theta(\Lambda + \lambda)\right]^2 + \Theta^2\tau^2} < 1. \tag{4.4.11b}$$

The best upper bound is

$$\|M_{\Theta'}^{\mathrm{Rich}}\|_2 \leqslant \frac{1 - \xi}{1 + \xi} \quad \text{for } \Theta' = \frac{2}{\lambda + \Lambda}\left(1 - s\frac{1 - \xi}{1 + \xi}\right), \tag{4.4.11c}$$

where $s := \tau/\sqrt{\lambda\Lambda + \tau^2}$, $\xi := \dfrac{1 - s\,\lambda}{1 + s\,\Lambda}$.

Proof. Let $\vartheta \in (0, 1)$ be arbitrary. In analogy to (3), the first term in

$$\|M_\Theta^{\mathrm{Rich}}\|_2 = \|I - \Theta A\|_2 = \|[\vartheta I - \Theta A_0] + [(1 - \vartheta)I - i\Theta A_1]\|_2$$
$$\leqslant \|\vartheta I - \Theta A_0\|_2 + \|(1 - \vartheta)I - i\Theta A_1\|_2$$

is bounded by $\|\vartheta I - \Theta A_0\|_2 \leqslant \max\{|\vartheta - \Theta\lambda|, |\vartheta - \Theta\Lambda|\}$. Since the matrix $C := (1 - \vartheta)I - i\Theta A_1$ is normal, $\|C\|_2 = \varrho(C)$ holds. From $\sigma(C) = \{1 - \vartheta - i\Theta\mu \colon \mu \in \sigma(A_1)\}$ and $\sigma(A_1) \subset [-\tau, \tau]$ (cf. (2.10.3e)), one concludes that $\varrho(C) \leqslant [(1 - \vartheta)^2 + \Theta^2\tau^2]^{1/2}$. Together, one obtains

$$\|M_\Theta^{\mathrm{Rich}}\|_2 \leqslant \max\{|\vartheta - \Theta\lambda|, |\vartheta - \Theta\Lambda|\} + [(1 - \vartheta)^2 + \Theta^2\tau^2]^{1/2}.$$

The optimal choice $\vartheta = \frac{1}{2}\Theta(\Lambda + \lambda)$ yields (11b). One verifies that under condition (11a) this bound remains below one. □

For the Hermitian case ($\tau = 0$), the estimate (11c) corresponds exactly to the convergence rate (6c). For $\tau \neq 0$ the convergence rate can be estimated even better than by means of the $\|M_\Theta^{\text{Rich}}\|_2$ bound from (11c).

Theorem 4.4.9. *Under the assumption* (10a, b) *the estimate*

$$\varrho(M_\Theta^{\text{Rich}}) \leqslant r_\Theta := \sqrt{[\max\{|1-\Theta\lambda|, |1-\Theta\Lambda|\}]^2 + \Theta^2\tau^2}. \tag{4.4.12a}$$

holds. The convergence is ensured in the form $r_\Theta < 1$ *if*

$$0 < \Theta < \overline{\Theta} \quad \text{with } \overline{\Theta} := \begin{cases} 2\Lambda/(\Lambda^2+\tau^2) & \text{if } \tau^2 < \lambda\Lambda, \\ 2\lambda/(\lambda^2+\tau^2) & \text{if } \tau^2 > \lambda\Lambda. \end{cases} \tag{4.4.12b}$$

r_Θ *is minimal for* $\Theta' := \min\left\{\dfrac{\lambda}{\lambda^2+\tau^2}, \dfrac{2}{\lambda+\Lambda}\right\}$. Further, the norm estimate (12c) holds:

$$\|(M_\Theta^{\text{Rich}})^m\|_2 \leqslant 2r_\Theta^m \qquad (m \geqslant 0). \tag{4.4.12c}$$

Proof. (2.9.13a) shows that r_Θ is an upper bound of the numerical radius $r(M_\Theta^{\text{Rich}})$ of the iteration matrix. The analysis of r_Θ as a function of Θ yields (12b) and the value Θ'. (12c) follows from (2.9.11d). □

While (10b) represents the inequality $-\tau I \leqslant A_1 \leqslant \tau I$, it is also possible to estimate A_1 in relation to A_0 by

$$A_1^2 \leqslant \tau A_0 \tag{4.4.10c}$$

or even only by

$$-\sqrt{\tau}A_0^{1/2} \leqslant A_1 \leqslant \sqrt{\tau}A_0^{1/2}. \tag{4.4.10d}$$

(10c) implies (10d). From (10d), using (10a), one arrives at the estimate (10b) with $\sqrt{\tau\Lambda}$ instead of τ. An estimate based *directly* on (10a) and (10c) can be found in Samarskii–Nikolaev [1, page 101].

Corollary 4.4.10. *Let* $K > 0$ *be an arbitrary positive definite matrix and* $\|\cdot\|_K$ *the related norm* (2.10.5a, c). *In order to obtain the monotone convergence of the Richardson iteration with respect to* $\|\cdot\|_K$, *one has to replace assumption* (8a, b) *in Theorem* 7 *by*

$$A^H K + KA \geqslant 2\lambda K, \qquad A^H KA \leqslant \tfrac{1}{2}\Lambda(A^H K + KA). \tag{4.4.13}$$

These inequalities are equivalent to

$$\operatorname{Re}\langle Ax, Kx\rangle \geqslant \lambda\langle Kx, x\rangle, \qquad \text{for all } x \in \mathbb{K}^I, \tag{4.4.13'a}$$

$$\langle Ax, KAx\rangle \leqslant \Lambda\operatorname{Re}\langle Ax, Kx\rangle \qquad \text{for all } x \in \mathbb{K}^I. \tag{4.4.13'b}$$

Under assumption (13) *the corresponding estimates* (9a, b) *hold with* $\|M_\Theta^{\mathrm{Rich}}\|_K$ *instead of* $\|M_\Theta^{\mathrm{Rich}}\|_2$.

Proof. Let $M := M_\Theta^{\mathrm{Rich}} = I - \Theta A$. Using (13), one concludes from

$$\begin{aligned} M^H K M &= K - \Theta(A^H K + KA) + \Theta^2 A^H K A \\ &\leqslant K - \Theta(1 - \tfrac{1}{2}\Theta\Lambda)(A^{HK} + KA) \\ &\leqslant K - \Theta(2 - \Theta\Lambda)K \end{aligned}$$

that $I - \Theta(2 - \Theta\Lambda)I \geqslant K^{-1/2}M^H K M K^{-1/2} = (K^{1/2}MK^{-1/2})^H(K^{1/2}MK^{-1/2})$. This is equivalent to $1 - \Theta(2 - \Theta\Lambda) \geqslant \|K^{1/2}MK^{-1/2}\|_2^2 = \|M\|_K^2 \geqslant \varrho(M)^2$ (cf. (2.10.3f)) and shows (9a) with respect to the $\|\cdot\|_K$ norm. □

4.4.2 Jacobi Iteration

Theorem 4.4.11. *Sufficient for the convergence of the Jacobi method* (2.5b) *are the conditions* (14), *which also imply* $\sigma(M^{\mathrm{Jac}}) \subset (-1, 1)$.

$$A \text{ and } 2D - A \text{ are positive definite:} \quad 2D > A > 0. \tag{4.4.14}$$

The contraction numbers with respect to the norms $\|\cdot\|_A$ *and* $\|\cdot\|_D$ *coincide with the convergence rate*:

$$\varrho(M^{\mathrm{Jac}}) = \|M^{\mathrm{Jac}}\|_A = \|M^{\mathrm{Jac}}\|_D < 1. \tag{4.4.15}$$

Proof. Choose $W := D$ in Criterion 12a that follows. □

Criterion 4.4.12. *Let* W *be the matrix of the third normal form* (3.2.5), *i.e.*, $M = I - W^{-1}A$ *is the iteration matrix.*

(a) *Under the assumption*

$$2W > A > 0, \tag{4.4.16a}$$

the iteration $x^{m+1} = x^m - W^{-1}(Ax^m - b)$ *converges. Furthermore, the convergence is monotone with respect to the energy norm* $\|\cdot\|_A$ *and norm* $\|\cdot\|_W$:

$$\varrho(M) = \|M\|_A = \|M\|_W < 1. \tag{4.4.16b}$$

(b) *For real* λ, Λ *with* $0 < \lambda \leqslant \Lambda$, *assume*

$$0 < \lambda W \leqslant A \leqslant \Lambda W. \tag{4.4.16c}$$

Then the spectrum of M *is real and is contained in*

$$\sigma(M) \subset [1 - \Lambda, 1 - \lambda] \tag{4.4.16d}$$

and the convergence rate is

$$\varrho(M) = \|M\|_A = \|M\|_W \leqslant \max\{1 - \lambda, \Lambda - 1\}, \tag{4.4.16e}$$

where equality instead of «$\leqslant$» holds if λ and Λ are the optimal bounds in (16c). *These optimal bounds can be expressed as*

$$\lambda = 1/\|W^{1/2}A^{-1}W^{1/2}\|_2, \qquad \Lambda = \|W^{-1/2}AW^{-1/2}\|_2. \tag{4.4.16f}$$

(c) *Assume $W > 0$, $A = A^H$, and $0 < \lambda \leqslant \Lambda$. The conditions* (16c) *and* (16d) *are equivalent. In particular,* (16a) *is equivalent to $\sigma(M) \subset (-1, 1)$ and the equivalence* (16g) *holds:*

$$W \geqslant A > 0 \Leftrightarrow \sigma(M) \subset [0, 1). \tag{4.4.16g}$$

Proof. (i) The following matrices are similar to the iteration matrix M:

$$M' := A^{1/2}MA^{-1/2} = I - A^{1/2}W^{-1}A^{1/2},$$
$$M'' := W^{1/2}MW^{-1/2} = I - W^{-1/2}AW^{-1/2},$$

implying $\varrho(M) = \varrho(M') = \varrho(M'')$. M' and M'' are Hermitian; hence, $\varrho(M') = \|M'\|_2 = \|M\|_A$ and $\varrho(M'') = \|M''\|_2 = \|M\|_W$ (cf. (2.9.4b), (2.10.5d)).
(ii) According to (2.10.3a′, b′), multiplication of (16a) and (16c) by $W^{-1/2}$ from both sides yields the inequalities

$$2I > A' := W^{-1/2}AW^{-1/2} > 0 \quad \text{and} \quad \lambda I \leqslant A' \leqslant \Lambda I, \text{ respectively.}$$

Following (2.10.3e), A' has a spectrum in $(0, 2)$ or $[\lambda, \Lambda]$, respectively. From $M'' = I - A'$, one derives the respective inconclusion $\sigma(M'') \subset (-1, 1)$ and $\sigma(M'') \subset [1 - \Lambda, 1 - \lambda]$. Together with part (i), the first case proves the assertion (16b). The second case leads to (16d).
(iii) (16e) is a consequence of $\varrho(M) \leqslant \max\{|\xi|: \xi \in [1 - \Lambda, 1 - \lambda]\} = \max\{|1 - \Lambda|, |1 - \lambda|]\}$. $0 < \lambda \leqslant \Lambda$ leads us to $\ldots = \max\{1 - \lambda, \Lambda - 1\}$.
(iv) Since the spectra of M and $M'' = W^{1/2}MW^{-1/2}$ coincide and the latter is real under the assumptions $W > 0$ and $A = A^H$, there are a minimal eigenvalue $\varrho_{\min}$ and a maximal one $\varrho_{\max}$. $\lambda := 1 - \varrho_{\max}$ and $\Lambda := 1 - \varrho_{\min}$ are the minimal and maximal eigenvalues of $A' := W^{-1/2}AW^{-1/2}$. From $\sigma(A') \subset [\lambda, \Lambda]$, using (2.10.3e), one concludes $\lambda I \leqslant A' \leqslant \Lambda I$ and therefore (16c) with the extreme eigenvalues of A' specified above. (16f) follows from $\|A'\|_2 = \varrho(A') = \Lambda$ and $0 < \lambda = 1/\|A'^{-1}\|_2$. Hence, the equivalence of (16d) and (16c) claimed in part (c) of the criterion is also shown. □

The optimal bounds λ and Λ from (16c) are the minimal and maximal eigenvalues of the *generalised eigenvalue problem*

$$Ae = \lambda We \qquad (W > 0, e \neq 0). \tag{4.4.17}$$

Remark 4.4.13. (a) The matrices A and $2D - A$ have identical diagonal entries, whereas their off-diagonal entries have opposite signs.
(b) The statements $A > 0$ and $2D - A > 0$ are identical for 2×2 matrices, but they may differ in higher dimensions.

The assumption $2D - A > 0$ in (14) can be omitted for suitable damping:

Theorem 4.4.14. *Let A be positive definite. The* Jacobi iteration (3.2) damped *by the factor ϑ converges for*

$$0 < \vartheta < 2/\Lambda \quad \text{with } \Lambda := \|D^{-1/2}AD^{-1/2}\|_2 = \varrho(D^{-1}A). \tag{4.4.18}$$

An equivalent formulation of condition (18) *is*

$$0 < \vartheta A < 2D. \tag{4.4.18'}$$

Proof. The matrix of the third normal form is $W := \frac{1}{\vartheta}D$. Criterion 12a leads to $2W > A > 0$; hence, we have (18′). □

The statement of Theorem 14 corresponds to Theorem 2, if the Jacobi method is regarded as the Richardson method for $\hat{A}x = \hat{b}$ with $\hat{A} := D^{-1}A$ (cf. Remark 3.2). In analogy to Theorem 3, one proves

Exercise 4.4.15. Assume that A is positive definite and that $0 < \lambda \leqslant \Lambda$ are the best bounds in

$$\lambda D \leqslant A \leqslant \Lambda D. \tag{4.4.19}$$

(a) For all $\vartheta \in \mathbb{C}$, the *damped Jacobi method* has the convergence rate (20a), where $0 < \vartheta < 2/\Lambda$ (cf. (18)) guarantees $\varrho(M_\vartheta^{\mathrm{Jac}}) < 1$:

$$\varrho(M_\vartheta^{\mathrm{Jac}}) = \|M_\vartheta^{\mathrm{Jac}}\|_A = \|M_\vartheta^{\mathrm{Jac}}\|_D = \max\{|1 - \vartheta\lambda|, |1 - \vartheta\Lambda|\}. \tag{4.4.20a}$$

(b) The optimal convergence rate is

$$\varrho(M_\vartheta^{\mathrm{Jac}}) = \|M_\vartheta^{\mathrm{Jac}}\|_A = \|M_\vartheta^{\mathrm{Jac}}\|_D = \frac{\Lambda - \lambda}{\Lambda + \lambda} \quad \text{for } \vartheta = \frac{2}{\Lambda + \lambda}. \tag{4.4.20b}$$

(c) A lower estimate of λ from (19) is $\lambda \geqslant 1/\varkappa(A) = 1/\mathrm{cond}_2(A)$. *Hint*: $\|D\|_2 \leqslant \|A\|_2$, $D \leqslant \|A\|_2 I$, $A \geqslant \|A^{-1}\|_2^{-1} I$.

If A is not Hermitian but real, $D = \mathrm{diag}\{A\} = \mathrm{diag}\{A_0\}$ holds for A_0 from the splitting (7). The statement of Theorem 8 can be carried over as follows.

Theorem 4.4.16. *Let the symmetric part A_0 from* (7), $A = A_0 + iA_1$, *be positive definite*: $A_0 > 0$ *and set* $D = \mathrm{diag}\{A_0\}$. *Assume that there are constants* $0 < \lambda \leqslant \Lambda$ *and* $\tau \geqslant 0$ *with*

$$\lambda D \leqslant A_0 \leqslant \Lambda D, \qquad -\tau D \leqslant A_1 \leqslant \tau D. \tag{4.4.21}$$

Then the damped Jacobi method *converges for* $0 < \vartheta < \dfrac{2\lambda}{\lambda\Lambda + \tau^2}$ *monotonically with respect to the norm* $\|\cdot\|_D$:

$$\|M_\vartheta^{\mathrm{Jac}}\|_D \leqslant \frac{1}{2}\vartheta(\Lambda - \lambda) + \sqrt{\left[1 - \frac{1}{2}\vartheta(\Lambda + \lambda)\right]^2 + \vartheta^2\tau^2} < 1. \tag{4.4.22}$$

The optimal ϑ can be determined as in (11c).

Proof. $M_\vartheta^{\mathrm{Jac}}$ is similar to $M := D^{1/2}M_\vartheta^{\mathrm{Jac}}D^{-1/2} = I - \vartheta D^{-1/2}AD^{-1/2}$. M can be regarded as the iteration matrix of the Richardson method for $\Theta = \vartheta$ and $A' := D^{-1/2}AD^{-1/2}$ instead of A. The splitting (7) for A induces the splitting $A' = A'_0 + iA'_1$ with the Hermitian matrices

$$A'_0 = D^{-1/2}A_0D^{-1/2}, \qquad A'_1 = D^{-1/2}A_1D^{-1/2}. \tag{4.4.23}$$

The inequalities (10a, b) applied to A' are equivalent to (21). The estimate (11b) following from Theorem 8 refers to the iteration matrix M: $\|M\|_2 = \|D^{1/2}M_\vartheta^{\mathrm{Jac}}D^{-1/2}\|_2 = \|M_\vartheta^{\mathrm{Jac}}\|_D$. □

Remark 4.4.17. In none of the proofs here has use been made of the diagonal form of the matrix D. We have only used the fact that $D > 0$ follows from $A > 0$. If, therefore, $D = \operatorname{diag}\{A\}$ is replaced by some other positive definite matrix W (cf. Criterion 12), all statements of Sec. 4.4.2 remain valid. *Exception*: Exercise 15c remains valid only for block-diagonals of A.

4.4.3 Gauß-Seidel and SOR Methods

Theorem 4.4.18. *The Gauß-Seidel method converges for positive definite matrices A. The convergence is monotone with respect to the energy norm*:

$$\varrho(M^{\mathrm{GS}}) \leqslant \|M^{\mathrm{GS}}\|_A < 1. \tag{4.4.24}$$

Proof. $A > 0$ implies $D > 0$. The matrix $W = W^{\mathrm{GS}}$ from (2.8) satisfies

$$W + W^H = D - E + (D - E)^H = 2D - E - F = D + A > A.$$

Hence, the following convergence criterion is satisfied. □

Criterion 4.4.19. *Let the matrix W of the third normal form satisfy*

$$W + W^H > A > 0. \tag{4.4.25}$$

Then W is regular and the iteration converges monotonically with respect to the energy norm $\|\cdot\|_A$:

$$\varrho(M) \leqslant \|M\|_A < 1 \quad \text{for } M = I - W^{-1}A.$$

Proof. (a) To prove the regularity of the matrix W, assume $Wx = 0$. $0 = \langle Wx, x\rangle + \langle x, Wx\rangle = \langle (W + W^H)x, x\rangle$ entails $x = 0$, since $W + W^H > 0$.
(b) As $\varrho(M) \leqslant \|M\|_A$ by (2.9.1b), only $\|M\|_A < 1$ is to be shown. By (2.10.5d), we have $\|M\|_A = \|A^{1/2}MA^{-1/2}\|_2 = \|\hat{M}\|_2$ for $\hat{M} := I - A^{1/2}W^{-1}A^{1/2}$. One

verifies that

$$\begin{aligned}\hat{M}^H\hat{M} &= (I - A^{1/2}W^{-H}A^{1/2})(I - A^{1/2}W^{-1}A^{1/2})\\ &= I - A^{1/2}(W^{-H} + W^{-1})A^{1/2} + A^{1/2}W^{-H}AW^{-1}A^{1/2}\\ &= I - A^{1/2}W^{-H}(W + W^H)W^{-1}A^{1/2} + A^{1/2}W^{-H}AW^{-1}A^{1/2}\\ &< I - A^{1/2}W^{-H}AW^{-1}A^{1/2} + A^{1/2}W^{-H}AW^{-1}A^{1/2} = I.\end{aligned} \tag{4.4.26}$$

Hence, by Lemma 2.10.5b,

$$\|M\|_A = \|\hat{M}\|_2 = \varrho(\hat{M}^H\hat{M})^{1/2} < \varrho(I)^{1/2} = 1. \qquad \square$$

Note that the condition (25) coincides with (14) in the case of $W = D$.

Since the Gauß-Seidel method is the special case $\omega = 1$ of the SOR iteration, we formulate further convergence statements, e.g., quantitative estimates, only for the SOR method. When asking for real values of ω leading to convergence of the SOR method, we obtain the necessary condition $0 < \omega < 2$ from the following

Lemma 4.4.20. *Without any assumption on A, the estimate* (27) *holds*:

$$\varrho(M_\omega^{\mathrm{SOR}}) \geqslant |\omega - 1| \quad \text{for all } \omega \in \mathbb{C}. \tag{4.4.27}$$

Proof. Let $n := \#I$ be the matrix size. Since $I - \omega L$ and $(1 - \omega)I + \omega U$ are triangular matrices, $\det(I - \omega L) = 1$ and $\det((1 - \omega)I + \omega U) = (1 - \omega)^n$ hold; hence,

$$\det(M_\omega^{\mathrm{SOR}}) = \det((1 - \omega)I + \omega U)/\det(I - \omega L) = (1 - \omega)^n.$$

On the other hand, each determinant is the product of all eigenvalues of the matrix (cf. Jordan normal form (2.8.3a, b)): $\det(M_\omega^{\mathrm{SOR}}) = \prod_{\nu=1}^{n} \lambda_\nu$ (λ_ν: eigenvalues of M_ω^{SOR}). Together, we obtain $\prod_{\nu=1}^{n} \lambda_\nu = (1 - \omega)^n$ or $\prod_{\nu=1}^{n} |\lambda_\nu| = |1 - \omega|^n$. Therefore, at least one factor (eigenvalue) λ_ν must exist with $|\lambda_\nu| \geqslant |1 - \omega|$. This eigenvalue proves the assertion (27). $\square$

Inequality (27) allows $\varrho(M_\omega^{\mathrm{SOR}}) < 1$ only for $0 < \omega < 2$. The next theorem shows that $0 < \omega < 2$ is not only necessary but also sufficient for convergence.

Theorem 4.4.21 (Ostrowski [2]). *Assume that A is positive definite and A can be split into*

$$A = D - E - E^H \tag{4.4.28a}$$

with the properties (28b, c):

$$E \text{ is a strictly lower triangular matrix,} \tag{4.4.28b}$$

$$D \text{ is the diagonal of } A. \tag{4.4.28c}$$

Furthermore, assume

$$0 < \omega < 2. \tag{4.4.28d}$$

Then the SOR iteration (3.7a–c) *converges*:

$$\varrho(M_\omega^{\text{SOR}}) < 1. \tag{4.4.28e}$$

The convergence is monotone with respect to the energy norm:

$$\varrho(M_\omega^{\text{SOR}}) \leqslant \|M_\omega^{\text{SOR}}\|_A < 1. \tag{4.4.28f}$$

The splitting (28a) *does not differ from* $A = D - E - F$ *in* (2.7a–d), *since* $F = E^H$ *holds for any Hermitian matrix* A. *The assumptions of Theorem* 21 *can be weakened. Then they refer to a method more general than the standard SOR method. The proof of Theorem* 21 *is superfluous, when we prove the following*

Corollary 4.4.22. *Let* $A > 0$. *The statements* (28e, f) *of Theorem* 21 *remain valid if, instead of* (28b) *and* (28c), *we only assume*:

$$E \text{ is arbitrary,} \tag{4.4.28b'}$$

$$D \text{ is an arbitrary positive definite matrix.} \tag{4.4.28c'}$$

Under the conditions (28a, c′), *the matrix* $D - \omega E$ *is always regular.*

By Lemma 2.10.4e, *the diagonal* D *from* (28c) *is a positive definite matrix and hence also satisfies* (28c′). *We remark that the assumption «A is positive definite» from Theorem* 21 *is not only sufficient but also necessary* (*cf. Varga* [2, *page* 77]).

Proof. The matrix of the third normal form is $W = W_\omega^{\text{SOR}} = \frac{1}{\omega}D - E$ (cf. (3.7e)). Inequality (25) of the convergence Criterion 19 is satisfied:

$$W + W^H = \frac{2}{\omega}D - E - F = A + \left(\frac{2}{\omega} - 1\right)D > A > 0 \tag{4.4.29}$$

because of (28c′) and $\frac{2}{\omega} - 1 > 0$ ($\Leftrightarrow 0 < \omega < 2$). Concerning the regularity, compare Criterion 19. □

Theorem 21 does not state which ω is the most favourable one. This question will be answered in Theorem 5.6.5 for the spectral radius $\varrho(M_\omega^{\text{SOR}})$. Instead, one can also analyse the contraction number $\|M_\omega^{\text{SOR}}\|_A$ or its upper bound as a function of ω and look for an optimal ω in this sense.

Lemma 4.4.23. *Under the assumptions* $A > 0$ *and* (28a, b′, c′, d), *we have*

$$\|M_\omega^{\text{SOR}}\|_A = \sqrt{1 - \left(\frac{2}{\omega} - 1\right)\Big/ \|A^{-1/2} W_\omega^{\text{SOR}} D^{-1/2}\|_2^2}. \tag{4.4.30}$$

The norm $\|A^{-1/2}W_\omega^{\mathrm{SOR}}D^{-1/2}\|_2^2$ *in* (30) *can be estimated by*

$$\|A^{-1/2}W_\omega^{\mathrm{SOR}}D^{-1/2}\|_2^2 \leqslant 1/c \tag{4.4.31a}$$

if and only if (31b) *or* (31c) *are valid*:

$$WD^{-1}W^H \leqslant \frac{1}{c}A \qquad (W = W_\omega^{\mathrm{SOR}}), \tag{4.4.31b}$$

$$W^{-H}DW^{-1} \geqslant cA^{-1}. \tag{4.4.31c}$$

Proof. (i) The equivalence of (31b) and (31c) follows from (2.10.3g). The equivalence of (31b) and (31a) can be concluded from

$$\frac{1}{c}I \geqslant A^{-1/2}WD^{-1}W^HA^{-1/2}$$

$$= [A^{-1/2}WD^{-1/2}][A^{-1/2}WD^{-1/2}]^H \quad (\geqslant 0), \quad \text{(cf. (2.10.3b'))}$$

$$\frac{1}{c} \geqslant \|[A^{-1/2}WD^{-1/2}][D^{-1/2}W^HA^{-1/2}]\|_2$$

$$= \|A^{-1/2}WD^{-1/2}\|_2^2 \qquad \text{(cf. (2.10.3f))}.$$

(ii) Define $\hat{M} := A^{1/2}M_\omega^{\mathrm{SOR}}A^{-1/2}$. From inequality (26), the representation (29): $A - W - W^H = \left(1 - \frac{2}{\omega}\right)D$, and (31c) one obtains

$$\hat{M}^H\hat{M} = I + A^{1/2}W^{-H}[A - W - W^H]W^{-1}A^{1/2}$$

$$= I - \left(\frac{2}{\omega} - 1\right)A^{1/2}W^{-H}DW^{-1}A^{1/2}.$$

The largest eigenvalue of $\hat{M}^H\hat{M}$ is 1 minus the $\left(\frac{2}{\omega} - 1\right)$-fold of the smallest eigenvalue of $A^{1/2}W^{-H}DW^{-1}A^{1/2} = X^{-H}X^{-1} = (XX^H)^{-1}$, where $X := A^{-1/2}WD^{-1/2}$. The latter eigenvalue equals $1/\varrho(XX^H) = 1/\|X\|_2^2$. (30) follows from

$$\|M_\omega^{\mathrm{SOR}}\|_A^2 = \|\hat{M}\|_2^2 = \varrho(\hat{M}^H\hat{M}) = 1 - \left(\frac{2}{\omega} - 1\right)\Big/\|A^{-1/2}W_\omega^{\mathrm{SOR}}D^{-1/2}\|_2^2. \quad \square$$

The right-hand side in (30) depends via $\frac{2}{\omega} - 1$ explicitly on ω. However, $W_\omega^{\mathrm{SOR}} = \frac{1}{\omega}D - E$ also contains the parameter ω. The minimisation of the norm $\|M_\omega^{\mathrm{SOR}}\|_A$ is the subject of the following theorem (cf. Samarskii–Nikolaev [1], Young [2, page 464]).

Theorem 4.4.24. *Let the constants* $\gamma, \Gamma > 0$ *fulfil*

$$0 < \gamma D \leqslant A, \tag{4.4.32a}$$

$$\left(\frac{1}{2}D - E\right) D^{-1} \left(\frac{1}{2}D - E^H\right) \leqslant \frac{1}{4}\Gamma A. \tag{4.4.32b}$$

Further, assume (28a, d). *Then,* (31a–c) *holds with the value*

$$c = 1 \Big/ \left[\frac{\Omega^2}{\gamma} + \Omega + \frac{\Gamma}{4}\right] \quad \text{with } \Omega := \frac{2-\omega}{2\omega} \in (0, \infty). \tag{4.4.32c}$$

The SOR contraction number can be estimated by

$$\|M_\omega^{\mathrm{SOR}}\|_A \leqslant \sqrt{1 - 2\Omega \Big/ \left[\frac{\Omega^2}{\gamma} + \Omega + \frac{\Gamma}{4}\right]}. \tag{4.4.33a}$$

The right-hand side takes the following minimum:

$$\|M_{\omega'}^{\mathrm{SOR}}\|_A \leqslant \sqrt{\frac{\sqrt{\Gamma} - \sqrt{\gamma}}{\sqrt{\Gamma} + \sqrt{\gamma}}} \quad \text{for } \omega' := 2/(1 + \sqrt{\gamma\Gamma}). \tag{4.4.33b}$$

Proof. We rewrite $W := W_\omega^{\mathrm{SOR}} = \frac{1}{\omega}D - E$ as

$$W = \Omega D + \left(\frac{1}{2}D - E\right) \quad \text{with } \Omega := \frac{2-\omega}{2\omega} = \frac{1}{\omega} - \frac{1}{2} \tag{4.4.34}$$

and estimate as follows:

$$\begin{aligned} WD^{-1}W^H &= \left[\Omega D + \left(\frac{1}{2}D - E\right)\right] D^{-1} \left[\Omega D + \left(\frac{1}{2}D - E^H\right)\right] \\ &= \Omega^2 D + \Omega\left(\frac{1}{2}D - E + \frac{1}{2}D - E^H\right) + \left(\frac{1}{2}D - E\right)D^{-1}\left(\frac{1}{2}D - E^H\right) \\ &= \Omega^2 D + \Omega A + \left(\frac{1}{2}D - E\right)D^{-1}\left(\frac{1}{2}D - E^H\right) \\ &\underset{(28a),(32a,b)}{\leqslant} \left(\frac{\Omega^2}{\gamma} + \Omega + \frac{\Gamma}{4}\right)A. \end{aligned}$$

Hence, (31b) holds with $\frac{1}{c} = \frac{\Omega^2}{\gamma} + \Omega + \frac{\Gamma}{4}$. Inserting inequality (31a) into (30), we get (33a). The function $\Omega \Big/ \left[\frac{\Omega}{\gamma} + \Omega + \frac{\Gamma}{4}\right]$ attains its global maximum in $(0, \infty)$ at $\Omega = \frac{1}{2}\sqrt{\gamma\Gamma}$ corresponding to ω' from (33b). The evaluation of this expression yields the bound in (33b). □

Concerning the constants γ and Γ, we can make the following comments.

Corollary 4.4.25. *Assume* (28a, b). (a) *Let the Jacobi method be defined by means of D from* (28a): $M^{\mathrm{Jac}} := D^{-1}(E + E^H)$. *The optimal bound in* (32a) *is*

$$\gamma = 1 - \varrho(M^{\mathrm{Jac}}). \tag{4.4.35a}$$

(b) *Set* $d := \varrho(D^{-1}ED^{-1}E^H) = \|D^{-1/2}ED^{-1/2}\|_2^2$. *Then,* (32b) *holds with*

$$\Gamma = 2 + \frac{4d-1}{\gamma} \quad \left\{ \leqslant 2 \text{ if } d \leqslant \frac{1}{4} \right\}. \tag{4.4.35b}$$

Proof. (a) The best bound in (32a) is the smallest eigenvalue of $D^{-1}A = I - D^{-1}(E + E^H) = I - M^{\mathrm{Jac}}$.

(b) Forming the products in (32b) and using $E + E^H = D - A$ and $D \leqslant \frac{1}{\gamma}A$ yields

$$\begin{aligned} \frac{1}{4}D - \frac{1}{2}(E + E^H) + ED^{-1}E^H &\leqslant \frac{1}{4}D - \frac{1}{2}(E + E^H) + dD \\ &= \frac{1}{4}\{(4d+1)D - 2(E + E^H)\} \\ &= \frac{1}{4}\{(4d-1)D - 2A\} \\ &\leqslant \frac{1}{4}\left\{2 + \frac{4d-1}{\gamma}\right\}A. \end{aligned}$$

□

The notations M_ω^{SOR} in (33a, b) and M^{Jac} in (35a) are justified only if D is the diagonal or block-diagonal of A. If D from Theorem 24 is another matrix, a new method is defined and the iteration matrix in (33a, b) should be denoted differently.

Remark 4.4.26 (order improvement). Assume (28a, b) and define $d := \varrho(D^{-1}ED^{-1}E^H) \leqslant 1/4$. Let $\varkappa$ be the order of the Jacobi method: $\varrho(M^{\mathrm{Jac}}) = 1 - \gamma = 1 - Ch^\varkappa + O(h^{2\varkappa})$. In the case of the Gauß-Seidel iteration ($\omega = 1$), the bound (33a) has the same order:

$$\|M_1^{\mathrm{SOR}}\|_A = \|M^{\mathrm{GS}}\|_A \leqslant \sqrt{1 - 4\Big/\left(\Gamma + 2 + \frac{1}{\gamma}\right)} = (1 + 4\gamma)^{-1/2}. \tag{4.4.36a}$$

On the contrary, the order is improved (halved) for $\omega = \omega'$ from (33b):

$$\|M_{\omega'}^{\mathrm{SOR}}\|_A \leqslant 1 - \sqrt{\gamma/\Gamma} + O(\gamma/\Gamma) = 1 - \sqrt{\frac{C}{2}}h^{\varkappa/2} + O(h^\varkappa). \tag{4.4.36b}$$

(36b) holds (with another constant) even if the condition $d \leqslant 1/4$ is weakened into $d \leqslant 1/4 + O(h^\varkappa)$. $d = O(1)$ suffices for (36a).

Proof. Insert the values (35a, b) into (33a, b). □

The improvement of order will become even clearer and more transparent in Sec. 5.6.3.

A discussion of the optimal choice of ω for *generalised SOR methods*, in which L and U may not necessarily possess strictly triangular form, can be found in Hanke–Neumann–Niethammer [1].

According to the title of this chapter, Theorems 21 and 24 describe the convergence only for positive definite matrices A. In chapters §5 and §6, the matrices of other structure (e.g., nonsymmetric ones) are discussed. However, these chapters will not cover all matrices. Therefore, we now mention the results due to Niethammer [1] for the nonsymmetric and, in particular, the skew-symmetric case. In the following statements, the normalisation

$$D = \operatorname{diag}\{A\} = I \tag{4.4.37}$$

is required. This condition can always be obtained by the transformations $A \mapsto D^{-1}A$ or $A \mapsto D^{-1/2}AD^{-1/2}$, if D is regular.

Theorem 4.4.27. *For the real matrix A assume* (37) *and $A + A^T > 0$. Then Λ and $\tilde{\Lambda}$ from*

$$\Lambda := \lambda_{\max}(\tfrac{1}{2}(L + L^T + U + U^T)), \qquad \tilde{\Lambda} := \lambda_{\max}(\tfrac{1}{2}(L + L^T - U - U^T)),$$

$$\sigma := \varrho(\tfrac{1}{2}(L - L^T + U - U^T)), \qquad \tilde{\sigma} := \varrho(\tfrac{1}{2}(L - L^T - U + U^T))$$

satisfy $0 \leqslant \Lambda < 1$ and $\tilde{\Lambda} \geqslant 0$. The SOR method converges for ω with

$$0 < \omega < 2/[1 + \tilde{\Lambda} + \sigma\tilde{\sigma}/(1 - \Lambda)]. \tag{4.4.38}$$

For $A > 0$ and $L = U^T$, one obtains $\sigma = \tilde{\Lambda} = 0$; hence, (38) *becomes $0 < \omega < 2$ (cf. Theorem 21). If $A - I$ is skew-symmetric, i.e., $L = -U^T$, one proves the following corollary because of $\Lambda = \tilde{\sigma} = 0$, $\tilde{\Lambda} = \varrho(U - L)$.*

Corollary 4.4.28. *Assume that $A = I - L + L^T$ (L lower triangular matrix). Then the SOR method converges for ω with $0 < \omega < 2/(1 + \varrho(L + L^T))$. If, furthermore, L is componentwise $\geqslant 0$ and $\varrho(L + L^T) < 1$, the SOR iteration* diverges *for all other real ω.*

A similar divergence statement can also be shown for $L \neq -U^T$, if $L - U$ is componentwise $\geqslant 0$. $\omega_{\text{opt}} < 1$ can be proved for the optimal relaxation parameter.

Convergence results for *complex matrices* can be found in Niethammer [2].

4.5 Block Versions

4.5.1 Block-Jacobi Method

4.5.1.1 Definition. Let a block structure $\{I_\varkappa : \varkappa \in B\}$ be given as described in §2.5. In the following, D does not denote the diagonal but the *block-diagonal* of A:

$$D := \text{blockdiag}\{A\} = \text{blockdiag}\{A^{\varkappa\varkappa}: \varkappa \in B\} = \begin{bmatrix} * & & & 0 \\ & * & & \\ & & * & \\ 0 & & & \ddots \; * \end{bmatrix} \tag{4.5.1}$$

Here, $A^{\varkappa\varkappa}$ are the diagonal-blocks of A.

The *block-Jacobi method* is the iteration (2.3) with

$$W := D \quad \text{from (1)}, \qquad R := D - A. \tag{4.5.2}$$

Remark 4.5.1. (a) The block-Jacobi method is well-defined if and only if all diagonal-blocks $A^{\varkappa\varkappa}$ ($\varkappa \in B$) are regular.
(b) If A is positive definite, D and all diagonal-blocks $A^{\varkappa\varkappa}$ are positive definite and thereby regular (cf. Lemma 2.10.4e).
(c) The representations (2.5a–c) remain valid if D is defined by (1).
(d) The block-Jacobi method depends neither on the ordering of the blocks nor on the ordering or the indices inside of the blocks.
(e) If $\{1, 2, \ldots, \beta\}$ is the numbering of the blocks and $(x^m)^i$ and A^{ij} denote the blocks of x^m and A, the blockwise representation reads

$$\text{for } i := 1 \text{ to } \beta \text{ do } (x^{m+1})^i := (A^{ii})^{-1}\left\{b^i - \sum_{\substack{j=1 \\ j \neq i}}^{\beta} A^{ij}(x^m)^j\right\}. \tag{4.5.3}$$

The inverse $(A^{ii})^{-1}$ appearing in (3) indicates that for computing the ith block $(x^{m+1})^i$, one system $A^{ii}\delta = r$ is to be solved.

For the model problem, one can choose the columns ($x = ih$ constant)

$$u^i := (u_{i,1}, u_{i,2}, \ldots, u_{i,N-1})^T \qquad (1 \leqslant i \leqslant N-1)$$

of the unknowns as blocks. In the Poisson-model case, according to (1.2.8), the corresponding matrix blocks are

$$A^{ii} = h^{-2}\begin{bmatrix} 4 & -1 & & \\ -1 & 4 & -1 & \\ & \ddots & \ddots & \ddots \\ & & -1 & 4 \end{bmatrix}, \quad A^{i,i\pm 1} = -h^{-2}I, \quad A^{i,j} = 0 \text{ otherwise.}$$

If the «columns» form the blocks (as in this case), the iteration is named the *column-Jacobi method*. Analogously, one can define the *row-Jacobi iteration*.

4.5.1.2 Pascal Procedures. First, we provide a solver for the arising tridiagonal block-system:

```
procedure define_tridiag(var A: data_of_discretisation;
                          l,d,u: real; neu: Boolean);
var i: integer: q: real;
begin
 with A do if T=nil then begin new(T); T^.decomposition_
  computed:=false
end;
 with A do with T^ do
 if neu or not decomposition_computed then
 begin for i:=1 to ny-1 do begin lower[i]:=l; diag[i]:=d;
  upper[i]:=u
end;
   for i:=1 to ny-2 do
   begin q:=lower[i+1]/diag[i]; lower[i+1]:=q; diag[i+1]
    :=diag[i+1]-q*upper[i]
   end {tridiagonal LU-decomposition generated};
   decomposition_computed:=true
 end
end;

procedure solve_tridiag(var A: data_of_discretisation;
 var R,z: column);
var i: integer; label 1;
begin if A.T=nil then
1:  define_tridiag(A,A.S[0,-1],A.S[0,0],A.S[0,1],true);
     {standard choice}
    if not A.T^.decomposition_computed then goto 1;
    with A do with T^ do
    begin for i:=0 to ny-2 do R[i+1]:=z[i+1]-lower[i+1]*R[i];
          for i:=ny-1 downto 1 do R[i]:=(R[i]-upper[i]*
           R[i+1])/diag[i]
end end;
```

The component `A.T^.decomposition_computed` indicates whether the block-matrix is already decomposed into the factors `L` (lower triangular matrix) and `U` (upper triangular matrix). If a new problem with another matrix is treated, the flag `decomposition_computed:=false` has to signify that the splitting must be renewed. The required storage is installed by means of `new(T)` only if necessary and can be released again by `dispose(T)`. To avoid an undefined `T` in the beginning and to define an empty pointer `T=nil`, an initialisation of the variable `A` of type `data_of_discretisation` is necessary:

```
procedure initialise_discretisation(var A: data_of_
 discretisation);
begin A.T:=nil; A.ILUD:=nil; A.ILU7:=nil end;
```

If as in example (3.5.9), the component `A` is contained in the variable `it`, the call of `initialise_IT(it)` involves the initialisation of `it.A`.

The block-Jacobi method (more precisely, the column-Jacobi method) takes the following form:

```
procedure column_Jacobi(var new: gridfunction;
              var A: data_of_discretisation;
              var x,b: gridfunction; var IP:
               iterationparameter);
var i,j: integer; v,z: column;
begin with A do
 begin v:=x[0];
  for i:=1 to nx-1 do
  begin case kind of
Poisson_model_problem: for j:=1 to ny-1 do z[j]:=b[i,j]+
 x[i-1,j]+x[i+1,j];
fivepoint_formula:
   for j:=1 to ny-1 do z[j]:=b[i,j]-S[-1,0]*x[i-1,j]-S[1,0]*
    x[i+1,j];
ninepoint_formula:
   for j:=1 to ny-1 do
   z[j]:=b[i,j]-S[-1,-1]*x[i-1,j-1]-S[1,-1]*x[i+1,j-1]-
    S[-1,0]*x[i-1,j]
         -S[1,0]*x[i+1,j]-S[-1,1]*x[i-1,j+1]-S[1,1]*
          x[i+1,j+1]
   end {case};
   new[i-1]:=v; v[0]:=x[i,0]; v[ny]:=x[i,ny]; solve_
    tridiag(A,v,z)
  end;
  new[nx-1]:=v; transfer_boundary_values(nx,ny,x,new)
 end
end;
```

Note that the boundary values at $j = 0$ and $j = N$ are contained on the right-hand side of the block-system. In the Poisson-model case, the tridiagonal system of equations corresponding to the ith column takes the form

$$\begin{bmatrix} 4 & -1 & & \\ -1 & 4 & -1 & \\ & \ddots & \ddots & \ddots \\ & & -1 & 4 \end{bmatrix} v = z' + \begin{bmatrix} u[i,0] \\ 0 \\ \vdots \\ u[i,N] \end{bmatrix} =: z \qquad (N-1 \text{ equations}) \tag{4.5.4a}$$

with $z'[j] = h^2 f[i,j] + u[i-1,j] + u[i+1,j]$. Since the factors L and U of the LU decomposition have only two nonzero diagonals and since, moreover, $L_{ii} = 1$ holds, the solution of $LU\, v = z$ requires only

$$5N \text{ arithmetical operations per problem (4a).} \tag{4.5.4b}$$

4.5.2 Block-Gauß-Seidel and Block-SOR Method

4.5.2.1 Definition. To obtain the *block-Gauß-Seidel method*, only the conditions (2.7b–d) are to be changed:

$$A = D - E - F, \tag{4.5.5a}$$

$$D\text{: block-diagonal matrix blockdiag}\{A\}, \tag{4.5.5b}$$

$$E\text{: strictly lower block-triangular matrix,} \tag{4.5.5c}$$

$$F\text{: strictly upper block-triangular matrix.} \tag{4.5.5d}$$

With this meaning for the matrices D, E, F, formulae (2.9a–c) yield the normal forms of the block-Gauß-Seidel method.

Remark 4.5.2. (a) The block-Gauß-Seidel method is well-defined under the same assumptions as the block-Jacobi iteration (cf. Remark 1a).
(b) The block-Gauß-Seidel method depends on the ordering of the blocks, but not that of the indices inside the blocks.
(c) The blockwise description of the method reads as follows (cf. (3)):

for $i := 1$ to β do

$$(x^{m+1})^i := (A^{ii})^{-1}\left\{b^i - \sum_{j=1}^{i-1} A^{ij}(x^{m+1})^j - \sum_{j=i+1}^{\beta} A^{ij}(x^m)^j\right\}. \tag{4.5.6}$$

(d) To emphasize the fact that not the block-version but the standard Gauß-Seidel method from §4.2.2 is meant, we will use the term *pointwise* Gauß-*Seidel method.*

In the model case, one may introduce the rows or columns as blocks and define the corresponding *row-Gauß-Seidel* or *column-Gauß-Seidel method.* In analogy to the lexicographical and chequer-board ordering, we present the *lexicographical column-Gauß-Seidel method* (8) and the «*zebra*»-*column-Gauß-*

Seidel iteration (9) in the next section. The latter means that first the columns with odd numbers («black») and then those with even numbers («white») are enumerated.

Having defined the matrices in $A = D - E - F$ according to (5b–d) and determined L and U by (3.7d): $L = D^{-1}E$, $U = D^{-1}F$, we find that Eqs. (3.7a–c) define the *block-SOR method*. The blockwise description reads

for $i := 1$ to β do

$$(x^{m+1})^i := (x^m)^i + \omega(A^{ii})^{-1}\left\{b^i - \sum_{j=1}^{i-1} A^{ij}(x^{m+1})^j - \sum_{j=i}^{\beta} A^{ij}(x^m)^j\right\}. \tag{4.5.7}$$

Remark 4.5.3. Since in all the block-versions mentioned above, the ordering of the indices inside of the blocks is arbitrary, one may use arithmetical *vector* operations at the block level as far as the computation of $z := h2 * f[i] - u[i-1] - u[i+1]$ in (4a) is concerned. The call of `solve_tridiag` in the proposed form is sequential. In the case of the zebra versions (9), (11), and (12b, d), a parallelisation inside the same «colour» is possible.

4.5.2.2 Pascal Procedures. The procedures

`procedure lex_column_Gauss_Seidel` (4.5.8)

`procedure zebra_column_Gauss_Seidel` (4.5.9)

are obtained from the SOR variants given below by choosing an `omega` equal to 1 (and therefore, `om1` := 0). Therefore, the statements

```
solve_tridiag(A,z,z); for j:=1 to ny-1 do new[i,j]
                                  :=om1*x[i,j]+omega*z[j]
```

in (10/11) can be simplified to `solve_tridiag(A,new[i],z)`.

The SOR versions read as follows:

`procedure lex_column_SOR` (4.5.10)

```
procedure lex_column_SOR(var new: gridfunction;
                         var A: data_of_discretisation;
                         var x,b:gridfunction; var IP:
                          iterationparameter);
var i,j: integer; z: column; om1; real;
begin with A do with IP do
 begin transfer_boundary_values(nx,ny,x,new); om1:=1-omega;
  for i:=1 to nx-1 do
  begin case kind of
Poisson_model_problem: for j:=1 to ny-1 do z[j]:=b[i,j]+
 new[i-1,j]+x[i+1,j];
```

```
fivepoint_formula:
   for j:=1 to ny-1 do
   z[j]:=b[i,j]-S[-1,0]*new[i-1,j]
        -S[1,0]*x[i+1,j];
ninepoint_formula:
   for j:=1 to ny-1 do
   z[j]:=b[i,j]-S[-1,-1]*new[i-1,j-1]-S[1,-1]*x[i+1,j-1]
        -S[-1,0]*new[i-1,j]-S[1,0]*x[i+1,j]-S[-1,1]*
         new[i-1,j+1]
        -S[1,1]*[i+1,j+1]
   end {case};
   z[0]:=x[i,0]; z[ny]:=x[i,ny]; solve_tridiag(A,z,z);
   for j:=1 to ny-1 do new[i,j]:=om1*x[i,j]+omega*z[j]
end end end;
```

procedure zebra_column_SOR (4.5.11)

```
procedure zebra_column_SOR(var new: gridfunction;
                         var A: data_of_discretisation;
                         var x,b: gridfunction; var IP:
                          iterationparameter);
var i,j,colour: integer; z: column; om1: real;
begin with A do with IP do
 begin transfer_boundary_values(nx,ny,x,new); om1:=1-omega;
 for colour:=1 to 2 do
 begin i:=colour; while i<nx do
  begin case kind of
Poisson_model_problem: for j:=1 to ny-1 do z[j]:=b[i,j]+
 new[i-1,j]+x[i+1,j];
fivepoint_formula:
   for j:=1 to ny-1 do z[j]:=b[i,j]-S[-1,0]*new[i-1,j]-S[1,0]*
    x[i+1,j];
ninepoint_formula:
   for j:=1 to ny-1 do
   z[j]:=b[i,j]-S[-1,-1]*new[i-1,j-1]-S[1,-1]*x[i+1,j-1]
        -S[-1,0]*new[i-1,j]-S[1,0]*x[i+1,j]
        -S[-1,1]*new[i-1,j+1]-S[1,1]*x[i+1,j+1]
   end {case};
   z[0]:=x[i,0]; z[ny]:=x[i,ny]; solve_tridiag(A,z,z);
   for j:=1 to ny-1 do new[i,j]:=om1*x[i,j]+omega*z[j];
   i:=i+2
end end end end;
```

The following procedures determine the relaxation parameters.

```
function column_Jacobi_convergence_factor(var it: data_of_
 iteration): real;
begin with it.A do
column_Jacobi_convergence_factor:=cos(pi/nx)/
 (1+2*sqr(sin(pi/ny)))
end;

procedure define_column_SOR_parameter(var it: data_of_
 iteration);
var beta: real;
begin with it do with A do with IP do
 begin writeln('Choice of the column-SOR parameter.');
  writeln(' Omega may be defined via the Jacobi convergence
   rate.');
  write(' Is the direct input of omega desired?');
  if yes_no then determine_omega(IP) else
  begin
   if kind=Poisson_model_problem then
     beta:=column_Jacobi_convergence_factor(it)
   else
   begin write('--> column-Jacobi convergence factor=');
    readln(beta)
   end;
   set_omega(IP,optimal_omega_for_SOR(beta))
end end end;
```

In [Prog] one finds the «row»-versions

`function row_Jacobi_convergence_factor`

`procedure define_row_SOR_parameter`

`procedure lex_row_Gauss_Seidel` (4.5.12a)

`procedure zebra_row_Gauss_Seidel` (4.5.12b)

`procedure lex_row_SOR` (4.5.12c)

`procedure zebra_row_SOR` (4.5.12d)

and the backward variants

`procedure lex_column_Gauss_Seidel_backward` (4.5.13a)

`procedure lex_row_Gauss_Seidel_backward` (4.5.13b)

`procedure lex_column_SOR_backward` (4.5.13c)

`procedure lex_row_SOR_backward.` (4.5.13d)

4.5.3 Convergence of the Block Variants

Theorem 4.5.4. *Theorems* 4.11, 4.14, 4.16 *and Corollary* 4.12 *are also valid for the block-Jacobi method, if D represents a block-diagonal in all the formulae* (4.14–23).

Proof. Not only the diagonal but also the block-diagonal is positive definite (cf. Remark 1b); hence, the assertion follows from Remark 4.17. □

Remark 4.13b corresponds to the following interesting statement:

Exercise 4.5.5. Let A be a 2×2 block matrix, i.e., assume $\# B = 2$. Prove that A and $2D - A$ (D: block-diagonal of A) have the same eigenvalues: $\sigma(A) = \sigma(2D - A)$.

From Exercise 5 one concludes that if A is positive definite, $2D - A$ is also. This proves

Corollary 4.5.6. *The block-Jacobi method converges for a positive definite 2×2 block-matrix A.*

In the case of the blockwise Gauß-Seidel and SOR methods, the block-diagonal D of a positive definite matrix A satisfies the conditions (4.28b′, c′). *Therefore, the statement of the Ostrowski theorem* (*Theorem* 4.21) *also holds for the block-SOR method and covers the case of the block-Gauß-Seidel iteration for $\omega = 1$.*

Theorem 4.5.7. *Theorems* 4.18, 4.21 *and* 4.24 *together with Corollary* 4.22 *and Lemmata* 4.20 *and* 4.23 *remain valid for the block-versions of the Gauß-Seidel and SOR iteration.*

The matrix

$$I_j := \operatorname{blockdiag}\{\underbrace{I, I, \ldots, I}_{j \text{ blocks}}, \underbrace{0, \ldots, 0}_{\beta - j \text{ blocks}}\} \quad \textit{for } 1 \leqslant j \leqslant \beta$$

is the unity matrix with respect to the first j blocks. The following convergence statement is proved by Bank–Dupont–Yserentant [1; *therein Theorem* 3.4, (3.42), (3.67)].

Theorem 4.5.8. *Let $A > 0$. The block-Gauß-Seidel iteration converges with the rate*

$$\varrho(M^{\text{blockGS}}) \leqslant \left(1 - 1 \Big/ \sum_{j=1}^{\beta} \|I_j\|_A^2\right)^{1/2}.$$

4.6 Computational Work of the Methods

4.6.1 Case of General Sparse Matrices

In the following let $s(n) \leqslant C_A n$ be the number of the nonzero entries of A (cf. (3.3.1)). Since the diagonal elements do not vanish, the iteration matrix $M^{\mathrm{Jac}} = D^{-1}(D - A)$ of the Jacobi method contains $s(n) - n \leqslant (C_A - 1)n$ nonzero elements. First, $\hat{b} := N^{\mathrm{Jac}} b = D^{-1} b$ is computed and stored instead of b. The multiplication $M^{\mathrm{Jac}} x$ requires $(C_A - 1)n$ multiplications and $(C_A - 2)n$ additions. The work of $x^{m+1} = M^{\mathrm{Jac}} x^m + \hat{b}$ amounts to

$$Work(\Phi^{\mathrm{Jac}}, A) \leqslant 2(C_A - 1)n. \tag{4.6.1a}$$

Since the Gauß-Seidel method differs from the Jacobi iteration only by the fact that, in part, x^{m+1} instead of x^m components are used, one obtains the same amount of work:

$$Work(\Phi^{\mathrm{GS}}, A) \leqslant 2(C_A - 1)n. \tag{4.6.1b}$$

To save as many operations in the SOR method as possible, we take the term $a_{ij} x_j^m / a_{ii}$ for $j = i$ out of the bracket and obtain

$$x_i^{m+1} := (1 - \omega) x_i^m - \omega \left(\sum_{j=1}^{i-1} a_{ij} x_j^{m+1} + \sum_{j=i+1}^{n} a_{ij} x_j^m - b_i \right) \bigg/ a_{ii},$$

where $\omega' := 1 - \omega$ is computed beforehand. Similarly, a_{ij}/a_{ii} and b_i/a_{ii} may be assumed to be already computed. This yields

$$Work(\Phi^{\mathrm{SOR}}, A) \leqslant (2C_A + 1)n \qquad (\text{or} = 2C_A n, \text{ respectively}\}. \tag{4.6.1c}$$

The case $\{\ldots\}$ refers to a further possibility. Beforehand, one may also multiply a_{ij}/a_{ii} and b_i/a_{ii} by ω. However note that, occasionally, ω can vary during the iteration (cf. §5.6.4).

For the Richardson method (3.3) one finds

$$Work(\Phi_\Theta^{\mathrm{Rich}}, A) \leqslant 2C_A n \qquad \{= (2C_A + 2)n\}, \tag{4.6.1d}$$

if $I - \Theta A$ is available in this form. The value in brackets is valid if $x^{m+1} = x^m - \Theta(Ax^m - b)$ is evaluated via the defect $d := Ax^m - b$ and $x^{m+1} = x^m - \Theta d$. Analogous results hold for the damped Jacobi method (3.1b):

$$Work(\Phi_\omega^{\mathrm{damped\ Jac}}, A) \leqslant 2C_A n \qquad \{= (2C_A + 2)n\}. \tag{4.6.1e}$$

The *cost factors* defined in §3.3.1 are

$$C_\Phi^{\mathrm{Jac}} = C_\Phi^{\mathrm{GS}} = 2 - \frac{2}{C_A}, \tag{4.6.2a, b}$$

$$C_\Phi^{\mathrm{SOR}} = 2 + \frac{1}{C_A} \qquad \{= 2\}, \tag{4.6.2c}$$

$$C_\Phi^{\mathrm{Rich}} = C_\Phi^{\mathrm{damped\ Jac}} = 2 \qquad \{= 2 + 2/C_A\}. \tag{4.6.2d}$$

The amount of work of the block-variants depends on the structure of the diagonal-blocks. For the additional considerations, we assume:

$$\begin{aligned}&\text{There are } \beta \text{ blocks of the size } n/\beta,\\ &\text{the amount of work for solving } A^{ii}u = z \text{ is } \leqslant C_B n/\beta,\\ &A - D \text{ has } s_1(n) \leqslant C_{AD} n \text{ nonzero elements,}\end{aligned} \tag{4.6.3}$$

where $D = \text{blockdiag}\{A\}$. Then, the operation count yields

$$Work(\Phi^{\text{blockJac}}, A) = Work(\Phi^{\text{blockGS}}, A) \leqslant (C_B + 2C_{AD})n, \tag{4.6.4a}$$

$$\begin{aligned}Work(\Phi^{\text{blockSOR}}, A) &= Work(\Phi^{\text{damped blockJac}}, A)\\ &= (C_B + 2C_{AD} + 3)n \qquad \{(C_B + 2C_{AD} + 2)n, \text{resp.}\},\end{aligned} \tag{4.6.4b}$$

where the bracket refers to the case when the damping factor is already contained in the matrix, so that no multiplication by ω or Θ is necessary.

4.6.2 Amount of Work in the Model Case

For the model problem for §1.2 one needs a smaller amount of work than given by (1) through (4) with

$$C_A = 5, \qquad C_B = 5, \qquad C_{AD} = 2.$$

The reason is that multiplication by the coefficients -1 can be omitted. Computing $D^{-1}b$ beforehand corresponds to the replacement of f by $h^2 f$. Counting the operations in (2.6), etc., one obtains

$$Work(\Phi^{\text{Jac}}, A) = Work(\Phi^{\text{GS}}, A) \leqslant 5n, \tag{4.6.5a}$$

$$\begin{aligned}Work(\Phi^{\text{SOR}}, A) &= Work(\Phi^{\text{Rich}}_{\Theta}, A)\\ &= Work(\Phi^{\text{damped Jac}}, A) \leqslant 7n.\end{aligned} \tag{4.6.5b}$$

By (5.4b), we obtain $C_B = 5$ for the block-variants and

$$Work(\Phi^{\text{blockJac}}, A) = Work(\Phi^{\text{blockGS}}, A) \leqslant 7n, \tag{4.6.6a}$$

$$Work(\Phi^{\text{blockSOR}}, A) = Work(\Phi^{\text{damped blockJac}}, A) \leqslant 9n \quad \{\leqslant 10n, \text{resp.}\}, \tag{4.6.6b}$$

where $9n$ holds for the case that instead of $h^2 A^{ii}$, the matrices $h^2 A^{ii}/\omega$ or $h^2 A^{ii}/\Theta$ are decomposed into triangular LU factors.

In the model case, the *cost factors* equal

$$C_\Phi^{\text{Jac}} = C_\Phi^{\text{GS}} = 1, \tag{4.6.7a}$$

$$\begin{aligned}C_\Phi^{\text{SOR}} &= C_\Phi^{\text{Rich}} = C_\Phi^{\text{damped Jac}} = C_\Phi^{\text{blockJac}} = C_\Phi^{\text{blockGS}}\\ &= 7/5 = 1.4,\end{aligned} \tag{4.6.7b}$$

$$C_\Phi^{\text{blockSOR}} = C_\Phi^{\text{damped blockJac}} = 9/5 = 1.8 \quad \{= 2\}. \tag{4.6.7c}$$

These numbers do not merit attention before we also know the respective convergence speeds. Then we are able to weigh whether, e.g., the block-Gauß-Seidel method is preferable to the pointwise Gauß-Seidel iteration in spite of its 1.4-fold amount of work.

4.7 Convergence Rates in the Case of the Model Problem

Up to now, we know the model case convergence rates only for the Richardson, Jacobi and block-Jacobi method. A discussion of the Gauß-Seidel iteration and SOR method is postponed until §5.6. There, we will also discuss the contraction numbers estimated in Theorem 4.24.

4.7.1 Richardson and Jacobi Iteration

The convergence rate $\varrho(M_\Theta^{\mathrm{Rich}}) = \max\{|1-\Theta\lambda_{\min}|, |1-\Theta\lambda_{\max}|\}$ of the Richardson method depends only on the extreme eigenvalues $\lambda_{\min}$, $\lambda_{\max}$ of the matrix A (cf. (4.3)). Inserting the values for $\lambda_{\min}$ and $\lambda_{\max}$ given in (1.1b, c) into (4.3–5), one obtains

Theorem 4.7.1. *In the model case, the Richardson method has the rate*

$$\varrho(M_\Theta^{\mathrm{Rich}}) = \max\{|1-8\Theta h^{-2}\sin^2(\pi h/2)|, |1-8\Theta h^{-2}\cos^2(\pi h/2)|\}. \tag{4.7.1}$$

Convergence holds for $0 < \Theta < h^2/[4\cos^2(\pi h/2)]$. *The optimal convergence rate is attained for* $\Theta = h^2/4$ *and equals*

$$\varrho(M_\Theta^{\mathrm{Rich}}) = 1 - 2\sin^2\left(\frac{\pi h}{2}\right) \quad \text{for } \Theta = \Theta_{\mathrm{opt}} = h^2/4. \tag{4.7.2}$$

In the model case $D = 4h^{-2}I$ *holds. Hence, the Jacobi method* $x^{m+1} = x^m - D^{-1}(Ax^m - b)$ *coincides for* $\Theta = h^2/4$ *with the Richardson iteration* $x^{m+1} = x^m - \Theta(Ax^m - b)$. *Equation* (2) *immediately implies*

Theorem 4.7.2. *In the model case, the Jacobi method leads to the convergence rate*

$$\varrho(M^{\mathrm{Jac}}) = 1 - 2\sin^2\left(\frac{\pi h}{2}\right) = \cos \pi h. \tag{4.7.3}$$

Replacing the square of the model problem by a rectangle, one obtains the rate $\varrho(M^{\mathrm{Jac}})$ *coded in the function* `Jacobi_convergence_factor` *from* §4.3.3.2.

Remark 4.7.3. The convergence rate (3) of the Jacobi methods has the form (3.3.5a): $\varrho(M^{\mathrm{Jac}}) = 1 - \eta^{\mathrm{Jac}}$ with

$$\eta^{\mathrm{Jac}} = 2\sin^2\left(\frac{\pi h}{2}\right) = \frac{\pi^2}{2}h^2 + O(h^4), \tag{4.7.4a}$$

i.e., the convergence is of order $\varkappa = 2$ and the constant in (3.3.5c) equals

$$C_\eta^{\mathrm{Jac}} = \pi^2/2. \tag{4.7.4b}$$

4.7.2 Block-Jacobi Iteration

The eigenvector e^{ij} of A defined in (1.2) is also an eigenvector of the block-diagonal matrix D of A, where the rows form the blocks. For symmetry reasons one obtains the same results, if the columns instead of the rows are taken as blocks.

Lemma 4.7.4. *Let A have a row-block structure and D be the corresponding block-diagonal matrix. Then e^{ij} is an eigenvector of D related to the eigenvalue d_{ij}:*

$$De^{ij} = d_{ij}e^{ij} \quad \textit{with} \quad d_{ij} = h^{-2}\left[2 + 4\sin^2\frac{ih\pi}{2}\right] \quad \textit{for } 1 \leqslant i, j \leqslant N-1. \tag{4.7.5}$$

Proof. For each grid point $(x, y) = (\nu h, \mu h) \in \Omega_h$, we have (cf. (1.4–5))

$$\begin{aligned}(De^{ij})(x, y) &= h^{-2}[4\sin ix\pi \sin jy\pi - \sin i(x+h)\pi \sin jy\pi - \sin i(x-h)\pi \sin jy\pi] \\ &= h^{-2}[4 - 2\cos ih\pi]e^{ij}(x, y) = h^{-2}\left[2 + 4\sin^2\frac{ih\pi}{2}\right]e^{ij}(x, y).\end{aligned}$$ □

The eigenvalues of $2D - A$ are $2d_{ij} - \lambda_{ij} = 4h^{-2}\left[\cos^2\frac{jh\pi}{2} + \sin^2\frac{ih\pi}{2}\right]$ (cf. λ_{ij} from (1.1a)). Their positivity proves $2D - A > 0$. Therefore, the block-Jacobi method converges (cf. Theorem 5.4). For determining the convergence speed, one has to study the eigenvalues of the iteration matrix $M = I - D^{-1}A$:

$$\sigma(M^{\mathrm{blockJac}}) = \left\{\frac{d_{ij} - \lambda_{ij}}{d_{ij}} : 1 \leqslant i, j \leqslant N-1\right\}.$$

Since

$$\left|\frac{d_{ij} - \lambda_{ij}}{d_{ij}}\right| = \left|\frac{1 - 2\sin^2(jh\pi/2)}{1 + 2\sin^2(ih\pi/2)}\right| \qquad (1 \leqslant i, j \leqslant N-1),$$

one may optimise the numerator and denominator separately. The numerator is maximal for $j = 1$, the denominator takes its minimum for $i = 1$. This yields

$$\varrho(M^{\text{blockJac}}) = \frac{1 - 2\sin^2(h\pi/2)}{1 + 2\sin^2(h\pi/2)}. \tag{4.7.6}$$

For the model problem with an underlying rectangle, we refer to the function `column_Jacobi_convergence_factor` from §4.5.2.2. The asymptotical expansion of (6) with respect to h yields

$$\begin{aligned}\varrho(M^{\text{blockJac}}) &= 1 - 4\sin^2(h\pi/2) + O(h^4) = 1 - \pi^2 h^2 + O(h^4)\\ &= 1 - \eta^{\text{blockJac}}\end{aligned}$$

with

$$\eta^{\text{blockJac}} = \pi^2 h^2 + O(h^4), \tag{4.7.7a}$$

i.e., its order equals $\varkappa = 2$, and the constant in (3.3.5c) equals

$$C_\eta^{\text{blockJac}} = \pi^2. \tag{4.7.7b}$$

By means of the cost factor C_Φ from (6.7a, b) and the quantities C_η from (4b/7b), one determines the coefficients C_{eff} of the effective amount of work (cf. (3.3.5d)):

$$Eff(\Phi^{\text{Jac}}) = \frac{2}{\pi^2} h^{-2} + O(1), \tag{4.7.8a}$$

$$Eff(\Phi^{\text{blockJac}}) = \frac{7}{5\pi^2} h^{-2} + O(1). \tag{4.7.8b}$$

This proves

Remark 4.7.5. For the Poisson-model problem from §1.2, the *block*-Jacobi method is more effective by a factor 0.7 than the *pointwise* Jacobi method.

4.7.3 Numerical Examples for the Jacobi Variants

Table 1 reports the results of the pointwise and blockwise Jacobi methods. As in Table 1.4.1, the numbers refer to the Poisson-model problem for the step size $h = 1/32$. For certain iteration numbers m, the table presents the value $u^m_{16,16}$ at midpoint, that should converge to $u(\frac{1}{2}, \frac{1}{2}) = 0.5$. Furthermore, it contains the maximum norm $\varepsilon_m := \|u^m - u_h\|_\infty$ of the errors $e^m = u^m - u_h$ and the reduction factor $\varrho_{m,m-1} = \varepsilon_m/\varepsilon_{m-1}$.

We observe that the reduction factors converge to different limits for odd and even m. The explanation is that with $r := \varrho(M^{[\text{block}]\text{Jac}})$, $-r$ is also an eigenvalue of the iteration matrix (cf. Remark 5.2.2). Hence, the dominating

Table 4.7.1 Results of the Jacobi iteration for $N = 32$ in the model case

pointwise Jacobi method				blockwise Jacobi method			
m	$u_{16,16}$	ε_m	$\varrho_{m,m-1}$	m	$u_{16,16}$	ε_m	$\varrho_{m,m-1}$
1	−0.0010	1.759		1	−0.0019	1.666	
2	−0.0019	1.644	0.93504	2	−0.0039	1.560	0.93621
3	−0.0029	1.588	0.96598	3	−0.0059	1.475	0.94605
62	−0.0480	0.795	0.99321	37	−0.0449	0.734	0.98597
63	−0.0480	0.789	0.99311	38	−0.0426	0.727	0.98953
64	−0.0480	0.784	0.99313	39	−0.0429	0.715	0.98478
100	−0.0230	0.629	0.99468	100	0.14077	0.374	0.98565
101	−0.0217	0.626	0.99462	101	0.14176	0.372	0.99433
102	−0.0205	0.623	0.99464	102	0.14713	0.367	0.98619
103	−0.0192	0.619	0.99458	103	0.14812	0.364	0.99376
200	0.14011	0.374	0.99497	200	0.36033	0.141	0.99008
201	0.14173	0.372	0.99493	201	0.36077	0.139	0.99077
202	0.14333	0.370	0.99497	202	0.36299	0.138	0.98996
203	0.14493	0.368	0.99493	203	0.36342	0.137	0.99090
297	0.27122	0.231	0.99508	297	0.44474	0.055	0.99411
298	0.27231	0.230	0.99512	298	0.44563	0.055	0.98671
299	0.27340	0.229	0.99508	299	0.44580	0.054	0.99414
300	0.27477	0.228	0.99512	300	0.44666	0.053	0.98668

error part has the form

$$r^m e_1 + (-r)^m e_2 = r^m[e_1 + (-1)^m e_2]$$

and oscillates with period 2. The geometric mean of two successive factors approximates the spectral radius $\varrho(M^{[\text{block}]\text{Jac}})$. These mean values are

$$\sqrt{\varepsilon_{300}/\varepsilon_{298}} = \begin{cases} 0.995099 & \text{for the pointwise method,} \\ 0.990401 & \text{for the blockwise method,} \end{cases}$$

and are in good agreement with the values $\varrho(M^{\text{Jac}}) = \cos \pi/32 = 0.99518$ and $\varrho(M^{\text{blockJac}}) = 0.990416$ that result from (3) and (6) for $h = 1/32$.

4.7.4 SOR and Block-SOR Iteration with Numerical Examples

For the evaluation of the SOR bounds from Theorem 4.24, the constants γ and Γ must be determined.

Lemma 4.7.6. *For the model problem, the pointwise SOR method with lexicographical ordering satisfies* (4.32a, b) *with* $\gamma = 2\sin^2(\pi h/2)$ *and* $\Gamma = 2$. *The* *optimal* ω' *from* (4.33b) *is*

$$\omega' = 2 \Big/ \left[1 + 2 \sin \frac{\pi h}{2} \right] = 2 - 2\pi h + O(h^2). \tag{4.7.9a}$$

The bounds for $\omega = 1$ *and* $\omega = \omega'$ *are*

$$\|M^{\mathrm{GS}}\|_A \leqslant \sqrt{1 \Big/ \left[1 + 8 \sin^2 \frac{\pi h}{2} \right]} = 1 - \pi^2 h^2 + O(h^4), \tag{4.7.9b}$$

$$\|M_{\omega'}^{\mathrm{SOR}}\|_A \leqslant \cos \frac{\pi h}{2} \Big/ \left[1 + \sin \frac{\pi h}{2} \right] = 1 - \frac{\pi h}{2} + O(h^2). \tag{4.7.9c}$$

Proof. (i) γ from (4.32a) is the smallest eigenvalue of $D^{-1}A = \frac{1}{4}h^2 A$; hence, $\gamma = \frac{1}{4}h^2 \lambda_{\min} = 2\sin^2(\pi h/2)$ (cf. 1.1b)).
(ii) For lexicographical ordering, the matrix E contains at most two entries $-h^{-2}$ per row and column; hence, $\|E\|_\infty \leqslant 2h^{-2}$ and $\|E^H\|_\infty \leqslant 2h^{-2}$ hold and imply $\varrho(EE^H) \leqslant \|EE^H\|_\infty \leqslant 4h^{-4}$. The inequality

$$\begin{aligned}(\tfrac{1}{2}D - E)D^{-1}(\tfrac{1}{2}D - E^H) &= \tfrac{1}{4}D - \tfrac{1}{2}(E + E^H) + ED^{-1}E^H \\ &= -\tfrac{1}{4}D - \tfrac{1}{2}A + ED^{-1}E^H \\ &= -h^{-2}I + \tfrac{1}{2}A + \tfrac{1}{4}h^2 EE^H \\ &\leqslant -h^{-2}I + \tfrac{1}{2}A + \tfrac{1}{4}h^2 4h^{-4} I = \tfrac{1}{2}A\end{aligned}$$

shows (4.32b) with $\Gamma = 2$.
(iii) The further statements (9a–c) follow by inserting this result into (4.33a, b).
□

The last estimate shows that the order of convergence improves from $1 - O(h^2)$ to $1 - O(h)$. However, the bound in (9c) is distinctly less favourable than the convergence rates $\varrho(M_{\omega'}^{\mathrm{SOR}})$. On the other side, the bound in (9b) and the convergence rate $\varrho(M^{\mathrm{GS}})$ coincide up to $O(h^4)$. Table 2 contrasts the bounds (9b, c) with the spectral radii that are determined in Theorem 5.6.5. Since the respective optimal parameters ω' from (9a) and ω_{opt} from (5.6.5b) differ slightly, the results for both of the values are reported.

In the case of the block-SOR method, γ is the smallest eigenvalue of $D^{-1}A$ with $D = \mathrm{diag}(A)$. Similar considerations as in §4.7.2 lead to $\gamma = 1 - [1 - 2\sin^2(\pi h/2)]/[1 + 2\sin^2(\pi h/2)]$. Lemma 4 shows $d_{ij} \geqslant 2h^{-2}$. This implies $D \geqslant 2h^{-2}I$ and $\|D^{-1}\|_2 = \varrho(D^{-1}) \leqslant \frac{1}{2}h^2$. The matrix E from $A = D - E - E^H$ contains only one entry $-h^{-2}$ per row and column; hence, $\|E\|_\infty = \|E^H\|_\infty = h^{-2}$ and $\varrho(EE^H) \leqslant \|EE^H\|_\infty \leqslant h^{-4}$ hold. As above, one concludes $\Gamma = 2$ from

$$\begin{aligned}(\tfrac{1}{2}D - E)D^{-1}(\tfrac{1}{2}D - E^H) &= -\tfrac{1}{4}D - \tfrac{1}{2}A + ED^{-1}E^H \\ &= \quad h^{-2}I + \tfrac{1}{2}A + \tfrac{1}{4}h^2 EE^H \leqslant \tfrac{1}{2}A\end{aligned}$$

because of $ED^{-1}E^H \leqslant \frac{1}{2}h^2 EE^H \leqslant \frac{1}{2}h^{-2}I \leqslant \frac{1}{4}D$. This proves

Table 4.7.2 Contraction bounds and convergence rates in the model case

h	1/8	1/16	1/32	1/64	1/128
bound (9b) for $\|M^{GS}\|_A$	0.8756	0.9637	0.9905	0.9975996	0.9993982
$\varrho(M^{GS})$	0.8536	0.9619	0.9904	0.9975924	0.9993977
ω'	1.4387	1.6722	1.8213	1.9064278	1.9520897
ω_{opt}	1.4465	1.6735	1.8215	1.9064547	1.9520932
bound (9c) for $\|M^{SOR}_{\omega'}\|_A$	0.8207	0.9063	0.9521	0.9757526	0.9878028
$\varrho(M^{SOR}_{\omega'})$	0.5174	0.6991	0.8293	0.9086167	0.9526634
bound for $\|M^{SOR}_{\omega_{opt}}\|_A$	0.8207	0.9063	0.9521	0.9757527	0.9878028
$\varrho(M^{SOR}_{\omega_{opt}})$	0.4465	0.6735	0.8215	0.9064547	0.9520932

Lemma 4.7.7. *For the model problem, the block-SOR method with lexicographical block ordering satisfies* (4.32a, b) *with* $\Gamma = 2$ *and* $\gamma = 1 - \left[1 - 2\sin^2\frac{\pi h}{2}\right] \Big/ \left[1 + 2\sin^2\frac{\pi h}{2}\right]$. *The optimal* ω' *from* (4.33b) *is*

$$\omega' = 2 \Big/ \left[1 + \sqrt{8}\sin\frac{\pi h}{2} \Big/ \sqrt{1 + 2\sin^2\frac{\pi h}{2}}\right] = 2 - 2\sqrt{2}\pi h + O(h^2). \qquad (4.7.10a)$$

The bounds at $\omega = 1$ *and* $\omega = \omega'$ *are*

$$\|M^{\text{blockGS}}\|_A \leqslant 1 - 2\pi^2 h^2 + O(h^4), \qquad (4.7.10b)$$

$$\|M_\omega^{\text{blockSOR}}\|_A \leqslant 1 - \frac{\pi h}{\sqrt{2}} + O(h^2). \qquad (4.7.10c)$$

4.8 Symmetric Iterations

4.8.1 General Form of the Symmetric Iteration

Even if A is Hermitian, the iteration matrix M is, in general, not necessarily Hermitian. While in the case of the Jacobi method, M^{Jac} still has positive eigenvalues, the spectrum of the SOR iteration matrix, in general, also contains complex eigenvalues.

In the following, we start from the third normal form

$$W(x^m - x^{m+1}) = Ax^m - b \qquad (4.8.1a)$$

and assume (1b):

$$W \text{ positive definite, } A \text{ Hermitian:} \quad W > 0,\ A = A^H. \tag{4.8.1b}$$

An iteration of the form (1a, b) is called a *symmetric iteration*. A more general definition reads as follows: The matrix N of the second normal form must be Hermitian. For convergent Φ and positive definite $A > 0$, one then attains (1b).

Example 4.8.1. Examples for (1a, b) are the pointwise and blockwise Jacobi method with $W = D$ if A is positive definite.

4.8.2 Convergence

Convergence has already been studied in Criterion 4.12. The essential statements are again repeated here.

Remark 4.8.2. (a) The iteration matrix of the symmetric iteration (1a) is

$$M = I - W^{-1}A \tag{4.8.2a}$$

(b) Assume (1b). The iteration matrix M is similar to

$$\hat{M} := W^{1/2}MW^{-1/2} = I - W^{-1/2}AW^{-1/2}. \tag{4.8.2b}$$

(c) If A is positive definite, the iteration matrix M is similar to

$$\check{M} := A^{1/2}MA^{-1/2} = I - A^{1/2}W^{-1}A^{1/2}. \tag{4.8.2c}$$

(d) Assume A, $W > 0$. The contraction numbers with respect to the norms $\|\cdot\|_A$ and $\|\cdot\|_W$ (cf (2.10.5a)) coincide with the convergence rate:

$$\varrho(M) = \|M\|_A = \|M\|_W. \tag{4.8.2d}$$

(e) Remark 3.2.13d is applicable to symmetric iterations with respect to the energy norm $\|\cdot\|_A$: $\varrho(M) \geqslant \varrho_{m+1,m} \geqslant \varrho_{m,m-1}$.

Under assumption (1b), the transformed matrix $\hat{M}$ is again Hermitian. The positive definiteness of W is required to be able to define $W^{1/2}$ (cf. Lemma 2.10.6). $\check{M}$ is Hermitian if W is Hermitian and A is positive definite.

Remark 4.8.3. Assume (1a, b). (a) The convergence of the symmetric iteration is equivalent to (3a) as well as (3b):

$$2W > A > 0, \tag{4.8.3a}$$

$$\sigma(M) = \sigma(\hat{M}) \subset (-1, 1). \tag{4.8.3b}$$

(b) The strenghtened inequality

$$W \geqslant A > 0 \tag{4.8.4a}$$

is equivalent to

$$\sigma(M) = \sigma(\hat{M}) \subset [0, 1). \tag{4.8.4b}$$

(c) Let $a < b$ and $W > 0$. The inclusion $\sigma(M) \subset [a, b]$ is equivalent to

$$\gamma W \leqslant A \leqslant \Gamma W \quad \text{with } \gamma := 1 - b, \Gamma := 1 - a. \tag{4.8.4c}$$

Concerning the *optimal damping* of a symmetric iteration, we refer to Exercise 8.3.1.

4.8.3 Symmetric Gauß-Seidel Method

The Gauß-Seidel iteration is not of the form (1a, b), since $W = D - E$ (apart from the uninteresting case $A = D$, $E = F = 0$) is not symmetric. There is no special reason that the splitting of $A = W - R$ is chosen as such that the matrix E appears in $W = D - E$ and F in R. Just as well, one could split the matrix $A = D - E - F$ into

$$W = D - F, \qquad R = E \qquad (A = W - R) \tag{4.8.5a}$$

and define the iteration

$$(D - F)x^{m+1} = Ex^m + b. \tag{4.8.5b}$$

If $D = \operatorname{diag}\{A\}$, the componentwise description of the iteration (5b) is

$$\text{for } i := n \text{ downto } 1 \text{ do } x_i^{m+1} := \left(b_i - \sum_{j=1}^{i-1} a_{ij}x_j^m - \sum_{j=i+1}^{n} a_{ij}x_j^{m+1}\right)\Big/ a_{ii}, \tag{4.8.5c}$$

i.e., (5b) describes the Gauß-Seidel method with the *backward ordering of the indices*. This is the so-called *backward Gauß-Seidel iteration*, whose Pascal implementation is already given in §4.3.3.2.

Remark 4.8.4. The backward Gauß-Seidel iteration is characterised by the matrices

$$M^{\text{bGS}} = (D - F)^{-1}E, \qquad N^{\text{bGS}} = (D - F)^{-1}, \qquad W^{\text{bGS}} = D - F. \tag{4.8.6}$$

Definition 4.8.5. Let Φ^{GS} and Φ^{bGS} be the standard and backward Gauß-Seidel iterations. The product iteration

$$\Phi^{\text{symGS}} := \Phi^{\text{bGS}} \circ \Phi^{\text{GS}} \tag{4.8.7}$$

defines the *symmetric Gauß-Seidel method.*

Lemma 4.8.6. *The iteration matrix of the symmetric Gauß-Seidel method is*

$$M^{\text{symGS}} = (D - F)^{-1}E(D - E)^{-1}F. \tag{4.8.8a}$$

The matrix of the second normal form reads

$$N^{\mathrm{symGS}} = (D - F)^{-1} D (D - E)^{-1}. \tag{4.8.8b}$$

The matrix of the third normal form is

$$W^{\mathrm{symGS}} = (D - E) D^{-1} (D - F) = A + E D^{-1} F. \tag{4.8.8c}$$

Proof. (8a) follows from (3.2.20a), (8c) from the following characterisation (11), and finally, (8b) from (8c) and (3.2.6). □

Theorem 4.8.7. *Let A be positive definite.* (a) *The matrix W^{symGS} of the third normal form is also positive definite, so that the symmetric Gauß-Seidel method takes the form* (1a, b).
(b) *The symmetric Gauß-Seidel iteration converges.*
(c) *The spectrum of the iteration matrix is nonnegative*:

$$\sigma(M^{\mathrm{symGS}}) \subset [0, 1).$$

Proof. (i) If A is positive definite, D and D^{-1} are also positive definite matrices. $ED^{-1}F = ED^{-1}E^H \geqslant 0$ follows from $D^{-1} > 0$, from which $W^{\mathrm{symGS}} = A + ED^{-1}F \geqslant A > 0$. (4a) proves the assertions (b), (c) of Theorem 7. □

Quantitative estimates of the convergence rate will follow in §4.8.5, where we discuss the SSOR method.

4.8.4 Adjoint and Corresponding Symmetric Iterations

The construction of a symmetric iteration from a given nonsymmetric iteration is possible not only for the Gauß-Seidel iteration but also in general.

We $W(A)$ be the matrix of the third normal form (1a) of Φ applied to $Ax = b$ and, analogously, $W(A^H)$ is the corresponding matrix for the application to $A^H x' = b'$. The *adjoint iteration* Φ^* is defined by (9):

$$W(A^H)^H (x^m - x^{m+1}) = Ax^m - b. \tag{4.8.9}$$

Exercise 4.8.8. Prove: (a) Let M_A^{Φ} be the iteration matrix of Φ applied to $Ax = b$ and $M_{A^H}^{\Phi^*}$ that of the adjoint iteration Φ^* applied to $A^H x' = b'$. Then, the similarity (10) holds:

$$M_A^{\Phi} = A^{-1} (M_{A^H}^{\Phi^*})^H A. \tag{4.8.10}$$

(b) Assume that $A = A^H$. Then, $\varrho(M_A^{\Phi}) = \varrho(M_{A^H}^{\Phi^*})$ holds; hence, Φ converges only if Φ^* does also, and vice versa.
(c) Assume that $A = A^H$. An iteration with $\Phi = \Phi^*$ has a Hermitian matrix N (of the second normal form). If N is regular, then W also is Hermitian.
(d) $\Phi^{**} = \Phi$ is always true.

Let $A > 0$. For each consistent linear iteration Φ, one can define the *corresponding symmetric iteration*

$$\Phi^{\text{sym}} := \Phi^* \circ \Phi. \tag{4.8.11}$$

Remark 4.8.9. Let $W = W(A)$ be the matrix of the third normal form of an iteration Φ applied to $Ax = b$. By W^H we abbreviate $W(A^H)^H$. The corresponding symmetric iteration (11) has the matrices

$$M^{\text{sym}} = (I - W^{-H}A)(I - W^{-1}A) = I - (W^{\text{sym}})^{-1}A, \tag{4.8.12a}$$

$$W^{\text{sym}} = W(W + W^H - A)^{-1}W^H \qquad \text{(if the inverse exists).} \tag{4.8.12b}$$

Proof. Apply Exercise 3.2.16b, d. □

Theorem 4.8.10. *Assume that $A > 0$. Let M and W be the matrices belonging to Φ. The corresponding symmetric iteration* (11) *converges if and only if*

$$W + W^H > A. \tag{4.8.13}$$

Then, Φ^{sym} from (11) *satisfies* (1a, b). *The convergence rate coincides with the contraction number with respect to the energy norm*:

$$\varrho(M^{\text{sym}}) = \|M^{\text{sym}}\|_A = \|M\|_A^2. \tag{4.8.14}$$

The spectrum of Φ^{sym} is nonnegative: $\sigma(M^{\text{sym}}) \subset [0, \varrho(M^{\text{sym}})]$. (13) *is also sufficient for the convergence of Φ: $\|M\|_A^2 < 1$.*

We remark that Ortega [1] calls the splitting $A = W - R$ *P-regular*, if (13) holds with regular W.

Proof. (i) The matrix M^{sym} from (12a) is similar to $A^{1/2}M^{\text{sym}}A^{-1/2} = (A^{1/2}M^{\Phi*}A^{-1/2})\ (A^{1/2}M^{\Phi}A^{-1/2})$. We transform the first factor into $A^{1/2}M^{\Phi*}A^{-1/2} = I - A^{1/2}W^{-H}A^{1/2} = (I - A^{1/2}W^{-1}A^{1/2})^H = (A^{1/2}M^{\Phi}A^{-1/2})^H$ and conclude that $A^{1/2}M^{\text{sym}}A^{-1/2}$ is Hermitian and positive semidefinite (i.e., $\sigma(M^{\text{sym}}) \subset [0, \varrho(M^{\text{sym}})]$. (14) follows from

$$\varrho(M^{\text{sym}}) = \varrho(A^{1/2}M^{\text{sym}}A^{-1/2}) = \|A^{1/2}M^{\text{sym}}A^{-1/2}\|_2 = \|M^{\text{sym}}\|_A$$

(cf. Theorem 2.9.5, (2.10.5d)) and

$$\begin{aligned}\|A^{1/2}M^{\text{sym}}A^{-1/2}\|_2 &= \|(A^{1/2}M^{\Phi}A^{-1/2})^H(A^{1/2}M^{\Phi}A^{-1/2})\|_2\\ &= \|A^{1/2}M^{\Phi}A^{-1/2}\|_2^2 = \|M^{\Phi}\|_A^2 \qquad (M^{\Phi} = M).\end{aligned}$$

(ii) Under the assumption (13), the estimate $\|M\|_A < 1$ follows from Criterion 4.19. The representation (14) guarantees the convergence of the symmetric iteration. Now we assume that (13) is not valid. Then $W + W^H - A$ and consequently also $X := A^{1/2}W^{-H}(W + W^H - A)W^{-1}A^{1/2}$ have a nonpositive eigenvalue (cf. Lemma 2.10.3). In the singular case, divergence is already mentioned in Remark 9. Otherwise, X has a negative eigenvalue $\mu < 0$, and because of its similarity to $I - X$, M^{sym} has an eigenvalue $1 - \mu > 1$, i.e., $\varrho(M^{\mathrm{sym}}) > 1$. □

Theorem 10 clarifies the fact that the convergence condition $\|M\|_A < 1$ (implying monotone convergence with respect to the energy norm), which before has been only a sufficient condition, is now a necessary requirement. Therefore, the energy norm estimate of $\|M\|_A$ in Corollary 4.4, Theorem 4.11, Corollary 4.12, (4.21a, b), Lemma 4.23, and Theorem 4.24 gains renewed importance.

4.8.5 SSOR: Symmetric SOR

One obtains the method adjoint to the SOR iteration by exchanging U and L (or E and F, respectively). It is the *backward SOR iteration* $\Phi_\omega^{\mathrm{bSOR}}$:

$$\text{for } i := n \text{ downto } 1 \text{ do} \qquad x_i^{m+1} := x_i^m - \omega\left(\sum_{j=1}^{i} a_{ij}x_j^m + \sum_{j=i+1}^{n} a_{ij}x_j^{m+1} - b_i\right)\Big/ a_{ii}. \tag{4.8.15}$$

The *symmetric SOR method* (abbreviated by SSOR) is the product

$$\Phi_\omega^{\mathrm{SSOR}} := \Phi_\omega^{\mathrm{bSOR}} \circ \Phi_\omega^{\mathrm{SOR}}. \tag{4.8.16}$$

Theorem 4.8.11. *Let A be positive definite. The symmetric SOR method* (15) *converges for* $0 < \omega < 2$. *The spectrum $\sigma(M_\omega^{\mathrm{SSOR}})$ of the iteration matrix is contained in* $[0, 1)$. *The same statement holds for the block-SSOR version.*

Proof. Since, according to Theorem 4.21 (Ostrowski), the SOR method converges monotonically with respect to the norm $\|\cdot\|_A$ (cf. (4.28f)), Theorem 10 can be applied. □

The SSOR method was first described in 1955 by Sheldon [1].

The amount of work required by the symmetric SOR iteration seems to be twice as large as that for the original SOR method, since one SSOR step consists of two SOR steps. However, this disadvantage can be overcome.

Remark 4.8.12 (Niethammer [2], [3]). The SSOR iteration requires essentially the *same amount of work* as the SOR method, if one tolerates the additional storage needed for an auxiliary vector. The cost factor (cf. §3.3) amounts to

$$C_\Phi^{\mathrm{SSOR}} = 2 + 6/C_A = C_\Phi^{\mathrm{SOR}} + 5/C_A \tag{4.8.17a}$$

for an optimal implementation instead of

$$C_\Phi^{\mathrm{SSOR}} = 2C_\Phi^{\mathrm{SSOR}} = 4 + 2/C_A \quad \text{for the naive implementation.} \tag{4.8.17b}$$

Proof. The first SSOR half-step $x^m \mapsto x^{m+1/2}$ can be rewritten as

$$x^{m+1/2} = x^m + \omega\{Lx^{m+1/2} - x^m + Ux^m + D^{-1}b\} \tag{4.8.17c}$$

(cf. (3.7f)). The second backward SOR step

$$x^{m+1} = x^{m+1/2} + \omega\{Ux^{m+1} - x^{m+1/2} + Lx^{m+1/2} + D^{-1}b\} \tag{4.8.17d}$$

contains the term $Lx^{m+1/2}$ already evaluated in (17c). Analogously, the term Ux^{m+1} computed in (17d) can be used in the following half-step:

$$x^{m+3/2} = x^{m+1} + \omega\{Lx^{m+3/2} - x^{m+1} + Ux^{m+1} + D^{-1}b\}.$$

Therefore, on the average, one SSOR step requires one evaluation of Lx and Ux. □

This and the following results carry over to the symmetric Gauß-Seidel method because of

Remark 4.8.13. For $\omega = 1$ the SSOR method coincides with the symmetric Gauß-Seidel method: $\Phi_1^{\text{SSOR}} = \Phi^{\text{symGS}}$.

The statements of Theorem 4.24 can be transcribed in the following form for the SSOR method.

Theorem 4.8.14. *Let* $A = D - E - E^H > 0$ *and* $0 < \omega < 2$. *Furthermore, assume that there are constants* $\gamma > 0$ *and* Γ *with* (18a, b) (cf. (4.32a, b)):

$$0 < \gamma D \leqslant A, \tag{4.8.18a}$$

$$(\tfrac{1}{2}D - E)D^{-1}(\tfrac{1}{2}D - E^H) \leqslant \tfrac{1}{4}\Gamma A. \tag{4.8.18b}$$

Then the following estimate holds:

$$\varrho(M_\omega^{\text{SSOR}}) = \|M_\omega^{\text{SSOR}}\|_A \leqslant 1 - 2\Omega\Big/\left[\frac{\Omega^2}{\gamma} + \Omega + \frac{\Gamma}{4}\right] \quad \textit{with } \Omega := \frac{2-\omega}{2\omega}. \tag{4.8.18c}$$

For $\omega' = 2/(1 + \sqrt{\gamma\Gamma})$, *the bound in* (18c) *becomes a minimum*:

$$\varrho(M_{\omega'}^{\text{SSOR}}) \leqslant \frac{\sqrt{\Gamma} - \sqrt{\gamma}}{\sqrt{\Gamma} + \sqrt{\gamma}} = \frac{1 - \sqrt{\gamma/\Gamma}}{1 + \sqrt{\gamma/\Gamma}}. \tag{4.8.18d}$$

Proof. Combine (14) with Theorem 4.24. □

The following remark is analogous to Remark 4.26.

Remark 4.8.15 (order improvement). In the case of $\varrho(D^{-1}ED^{-1}E^H) \leqslant 1/4$ $\{$or $\leqslant 1/4 + O(1 - \varrho(M^{\text{Jac}}))\}$, the choice $\omega = \omega'$ enables an improvement of order. If $\varkappa$ is the order of the Jacobi (and of the symmetric Gauß-Seidel) method, then $\varkappa/2$ is the order of the SSOR method with $\omega = \omega'$.

The condition $\varrho(D^{-1}ED^{-1}E^H) \leqslant 1/4$ is essential. It does not hold for the model problem with chequer-board ordering. Then, as we will see in §5.8.4, no order improvement is possible.

4.8.6 Pascal Procedures and Numerical Results for the SSOR Method

For the sake of simplicity, the following procedures make no use of Niethammer's technique mentioned in Remark 12. The symmetric lexicographical Gauß-Seidel method reads as follows:

$$\texttt{procedure symmetric\_lex\_Gauss\_Seidel} \quad (4.8.20a)$$

```
procedure symmetric_lex_Gauss_Seidel(var new: gridfunction;
                                     var A: data_of_
                                      discretisation;
                                     var x,b: gridfunction;
                                     var IP:
                                      iterationparameter);
begin lex_Gauss_Seidel(new,A,x,b,IP);
      lex_Gauss_Seidel_backward(new,A,new,b,IP)
end;
```

Table 4.8.1 Symmetric Gauß-Seidel method and SSOR for $h = 1/32$

symmetric Gauß-Seidel method					SSOR with $\omega = 1.8213$	
m	$\Vert e^m\Vert_\infty$	$\Vert e^m\Vert_A$	$\frac{\Vert e^m\Vert_\infty}{\Vert e^{m-1}\Vert_\infty}$	$\frac{\Vert e^m\Vert_A}{\Vert e^{m-1}\Vert_A}$	$\Vert e^m\Vert_A$	$\frac{\Vert e^m\Vert_A}{\Vert e^{m-1}\Vert_A}$
1	1.48	202			$2.3_{10}+02$	
2	1.35	159	0.91627	0.790646	$1.6_{10}+02$	0.71534
3	1.27	137	0.94025	0.858495	$1.2_{10}+02$	0.72622
4	1.20	122	0.94528	0.891046	$9.0_{10}+01$	0.73679
5	1.14	111	0.94734	0.910237	$6.7_{10}+01$	0.74876
94	0.158	11.2	0.98074	0.980884	$3.2_{10}-04$	0.87961
95	0.155	11.0	0.98075	0.980891	$2.8_{10}-04$	0.87961
96	0.152	10.8	0.98075	0.980897	$2.5_{10}-04$	0.87961
97	0.149	10.6	0.98076	0.980903	$2.2_{10}-04$	0.87961
98	0.146	10.4	0.98076	0.980909	$1.9_{10}-04$	0.87961
99	0.144	10.2	0.98077	0.980914	$1.7_{10}-04$	0.87961
100	0.141	10.0	0.98077	0.980919	$1.5_{10}-04$	0.87961

Table 4.8.2 SSOR method, convergence rates for $h = 1/32$

ω	$\varrho(M_\omega^{\mathrm{SSOR}})$
1	0.98092
1.8	0.88376
1.81	0.88163
1.8213	0.87962
1.83	0.87845
1.84	0.87765
1.8450	0.877529
1.8455	0.877528
1.8460	0.877528
1.847	0.877538
1.85	0.87762
1.86	0.87855
1.87	0.88066

Just as simple are the additional iterations, contained in [Prog]:

`procedure symmetric_column_Gauss_Seidel` (4.8.20b)

`procedure symmetric_row_Gauss_Seidel` (4.8.20c)

`procedure lex_SSOR` (4.8.20d)

`procedure column_SSOR` (4.8.20e)

`procedure row_SSOR` (4.8.20f)

When we choose ω as the optimal ω' from (4.33b), the following procedure is required:

```
function optimal_omega_for_SSOR(small_gamma,capital_gamma:
 real):real;
begin
   optimal_omega_for_SSOR:=2/(1+sqrt(small_gamma*capital_
    gamma))
end;

procedure define_optimal_SSOR_parameter(var it: data_of_
 iteration);
var smg,capg: real;
begin with it do with A do with IP do
  begin if kind=Poisson_model_problem then
   begin capg:=2; smg:=lower_gamma_Jacobi(it) {cf. §7.4.3}
   end else
```

```
    begin writeln('Optimal SSOR parameter to be determined.');
      write('--> upper bound capital gamma from (4.8.18b) =');
        readln(capg);
      write('--> lower bound small gamma from (4.8.18a) =');
        readln(smg)
    end;
    set_omega(IP,optimal_omega_for_SSOR(smg,capg))
  end
end;
```

For iterations with an iteration matrix satisfying $0 \leqslant M \leqslant \varrho(M)I$, Remark 3.2.13d is applicable: The quotients $\|e^{m+1}\|_A/\|e^m\|_A$ converge monotonically to $\varrho(M)$. Since $M = M_\omega^{\text{SSOR}}$ satisfies this assumption, one also observes this monotone behaviour for the SSOR method and with $\omega = 1$ for the symmetric Gauß-Seidel method. Table 1 contains the results of the symmetric Gauß-Seidel method with lexicographical ordering. For the step size $h = 1/32$, one obtains the convergence rate 0.98092. According to Table 7.2, $\omega = \omega' = 1.8213$ is the optimal value for the bound (7.9c), which becomes $\|M_\omega^{\text{SSOR}}\|_A \leqslant 0.9065$. Table 2 shows the convergence rates for different ω. Obviously, $\varrho(M_\omega^{\text{SSOR}})$ attains its minimum not at $\omega = \omega'$ but at $\omega = \omega_{\text{opt}}$ from the interval $[1.845, 1.846]$. The values of Table 2 demonstrate that, differently from the SOR method, the convergence rate has a flat minimum. Small errors in the choice of $\omega = \omega_{\text{opt}}$ deteriorate the convergence rate only insignificantly. So far, the choice $\omega = \omega'$ is quite sufficient.

5
Analysis in the 2-Cyclic Case

For the class of problems studied in this chapter, we succeed in obtaining *quantitative* convergence statements for the classical methods (Jacobi, Gauß-Seidel, and SOR iteration).

5.1 2-Cyclic Matrices

First, we define the term «weakly 2-cyclic» for matrices and for the pair $\{A, D\}$, where D is the diagonal part or block-diagonal part of A.

Definition 5.1.1. A matrix $A \in \mathbb{K}^{I \times I}$ is called *weakly 2-cyclic* (or *weakly cyclic of index 2*), if a block structure $\{I_1, I_2\}$ with nonempty index subsets $I_1, I_2 \subset I$ exists such that

$$a_{\alpha\beta} = 0 \quad \text{for } \alpha, \beta \in I_1 \text{ as well as for } \alpha, \beta \in I_2. \tag{5.1.1}$$

Condition (1) means that the diagonal-blocks are vanishing:

$$A^{11} = 0, \qquad A^{22} = 0. \tag{5.1.1'}$$

Often, not A but $A - D$ has the form required in (1). In this case, we introduce the same name for the pair $\{A, D\}$.

Definition 5.1.2. The pair $\{A, D\}$, A, $D \in \mathbb{K}^{I \times I}$, is called *weakly 2-cyclic*, if $A - D$ is weakly 2-cyclic. An equivalent statement is that a block structure $\{I_1, I_2\}$ with nonempty index subsets $I_1, I_2 \subset I$ exists such that

$$D = \text{blockdiag}\{A^{11}, A^{22}\}. \tag{5.1.2}$$

Let B be the block structure $\{I_1, I_2\}$ from Definition 1. Denote the block-diagonal part of a matrix (with respect to B) by $\text{blockdiag}_B\{\cdot\}$. Then A is weakly 2-cyclic if and only if

$$\text{blockdiag}_B\{A\} = 0. \tag{5.1.1''}$$

The pair $\{A, D\}$ is weakly 2-cyclic if and only if

$$\text{blockdiag}_B\{A\} = D. \tag{5.1.2'}$$

The additional term «weakly» in front of «2-cyclic» indicates that the ordering of the indices is irrelevant. This is different in

Definition 5.1.3. A [or $\{A, D\}$] is called *2-cyclic* if the index set I is ordered and the matrix A [or the pair $\{A, D\}$] is weakly 2-cyclic with respect to the blocks $I_1 = \{1, \ldots, n_1\}$ and $I_2 = \{n_1 + 1, \ldots, n\}$ for a suitable n_1 with $1 \leqslant n_1 \leqslant n - 1$.

$1 \leqslant n_1 \leqslant n - 1$ ensures that both I_1 and I_2 are nonempty. The property «2-cyclic» is *different* from the property «*cyclic of index 2*» as, e.g., introduced by Varga [2, page 35]. A 2-cyclic matrix A has the form

$$A = \underbrace{\begin{bmatrix} 0 & A_1 \\ A_2 & 0 \end{bmatrix}}_{I_1 \quad I_2} \begin{matrix} \}I_1 \\ \}I_2. \end{matrix} \tag{5.1.3a}$$

Note that, in general, $A_1 = A^{12} \in \mathbb{K}^{I_1 \times I_2}$ and $A_2 = A^{21} \in \mathbb{K}^{I_2 \times I_1}$ are not square block-matrices. The pair $\{A, D\}$ is 2-cyclic if

$$A = \begin{bmatrix} D_1 & A_1 \\ A_2 & D_2 \end{bmatrix} \quad \text{and} \quad D = \begin{bmatrix} D_1 & 0 \\ 0 & D_2 \end{bmatrix}, \quad A - D = \begin{bmatrix} 0 & A_1 \\ A_2 & 0 \end{bmatrix}. \tag{5.1.3b}$$

The definitions immediately imply the following remark.

Remark 5.1.4. (a) The property «2-cyclic» for a special ordering of the indices implies «weakly 2-cyclic» for any ordering of the indices.
(b) The property «weakly 2-cyclic» is independent of the ordering of the indices, whereas in the case of the term «2-cyclic» the indices may be permuted only *inside* of the respective blocks I_1, I_2.
(c) Let A [or $\{A, D\}$] be weakly 2-cyclic. If I is not ordered, then there is an ordering of the indices, so that A [or $\{A, D\}$] is 2-cyclic with respect to this ordering. If I is already ordered, there is a permutation of the indices with a corresponding permutation matrix P, so that $\hat{A} := PAP^T$ [or $\{\hat{A}, \hat{D} := PDP^T\}$] is 2-cyclic.

Examples of (weakly) 2-cyclic matrices are given below for the case of the model problem.

Example 5.1.5. Let A be the matrix of the model problem from §1.2.
(a) If $D = \operatorname{diag}\{a^{\alpha\alpha}: \alpha \in I\}$ is the (pointwise) diagonal of A, then $\{A, D\}$ is weakly 2-cyclic. If the *chequer-board ordering* from Fig. 1.2.1c is used, $\{A, D\}$ is even 2-cyclic. The exact definition of the chequer-board ordering (red-black ordering) reads as follows:

$$I_1 = I_{\text{black}} = \{(x, y) = (ih, jh) \in \Omega_h \colon i + j \text{ even}\}, \tag{5.1.4a}$$

$$I_2 = I_{\text{red}} = \{(x, y) = (ih, jh) \in \Omega_h \colon i + j \text{ odd}\}. \tag{5.1.4b}$$

(b) Let the rows (or columns) of the grid Ω_h form the block structure B and choose D as $D = \operatorname{blockdiag}\{A^{\alpha\alpha}: \alpha \in B\}$. Then $\{A, D\}$ is weakly 2-cyclic. If the rows (or columns) are ordered according to the zebra pattern (cf. §4.5.2), $\{A, D\}$ is even 2-cyclic. The exact definition of the *zebra-(row-)block structure* reads

$$I_1 = I_{\text{black}} = \{(x, y) = (ih, jh) \in \Omega_h \colon j \text{ even}\}, \tag{5.1.5a}$$

$$I_2 = I_{\text{white}} = \{(x, y) = (ih, jh) \in \Omega_h \colon j \text{ odd}\}. \tag{5.1.5b}$$

In the case of the *zebra-column-block structure*, one has to replace «j odd/even» in (5a, b) by «i odd/even».

Proof. (i) In the case of the chequer-board ordering, A has the block structure (1.2.9) with diagonal matrices $4h^{-2}I$ in the diagonal-blocks. Hence, the diagonal and block-diagonal part of A coincide: $D = 4h^{-2}I$. Therefore, (2) holds: $\{A, D\}$ is also 2-cyclic. For the chequer-board ordering, we have $I_1 = \{1, \ldots, n_1\}$ and $I_2 = \{n_1 + 1, \ldots, n\}$ with $n_1 := \# I_1 =$ number of the «black» grid points. For all $n > 1$ (i.e., $h < \frac{1}{2}$), $n_1 \in [1, n - 1]$ holds. According to Remark 4a, A is weakly 2-cyclic for an arbitrary ordering or even no ordering of the indices.
(ii) In the case of the row-block structure, the diagonal-blocks of A are given by the matrices $h^{-2}T$ from (1.2.8): $T = \operatorname{tridiag}\{-1, 4, -1\}$. The block structure of A illustrated in (1.2.8) corresponds to the lexicographical ordering of the rows. The zebra-ordering (5a, b) leads to

$$A = h^{-2} \left[\begin{array}{cccc|cccc} T & & & & -I & & & \\ & T & & & -I & -I & & \\ & & \ddots & & & \ddots & \ddots & \\ & & & T & & & \ddots & -I \\ & & & & & & & -I \\ \hline -I & -I & & & T & & & \\ & \ddots & \ddots & & & \ddots & & \\ & & -I & -I & & & & T \end{array} \right] \quad \text{with } T \text{ from (1.2.8).} \tag{5.1.6}$$

Composing the coarser zebra-block structure of the finer row-block structure, we obtain the following diagonal-blocks of A: $A^{ii} = h^{-2}$ blockdiag $\{T, \ldots, T\}$ for $i = 1, 2$, where the number of the diagonal-blocks T is determined by the number of the respective «black» ($i = 1$) or «white» ($i = 2$) rows. The block-diagonal parts D of A with respect to the row-block structure as well as the zebra-block structure coincide. This proves (2): $\{A, D\}$ is 2-cyclic. As in (i), we see that, independently of the ordering of the indices, $\{A, D\}$ is weakly 2-cyclic. □

The model equation (1.2.4a) is called a *five-point formula*, since the equation in (ih, jh) contains only the five unknowns u_{ij}, $u_{i+1,j}$, $u_{i-1,j}$, $u_{i,j+1}$ and $u_{i,j-1}$. For more general problems than the Poisson equation (1.2.1a), one needs other formulae, e.g., *nine-point formulae*. In the latter case, the equation in (ih, jh) contains the nine unknowns

$$\{u_{k\ell}: k = i - 1, i, i + 1, \ell = j - 1, j, j + 1\}.$$

Since we have permitted the corresponding matrix coefficients to vanish, the five-point formulae are a subset of the nine-point formulae.

Exercise 5.1.6. Prove: (a) If A represents a nine-point formula, then, in general, $\{A, D\}$ with $D := \text{diag}\{A\}$ is *not* weakly 2-cyclic. In particular, there is no ordering of the indices for which $\{A, D\}$ is 2-cyclic.
(b) If A represents a nine-point formula and D is the row- or column-block diagonal of A, then $\{A, D\}$ is weakly 2-cyclic as in Example 5b. In the case of zebra-block ordering, it is even 2-cyclic.

The statement of Exercise 6b can be generalised as follows.

Lemma 5.1.7. *Let A be either a tridiagonal matrix with the diagonal D or a block-tridiagonal matrix with respect to a block structure $\{I_1, I_2, \ldots\}$ with the block-diagonal D. Then $\{A, D\}$ is weakly 2-cyclic.*

Proof. It is sufficient to prove the block case. $J_1 := I_1 \cup I_3 \cup \cdots$ and $J_2 := I_2 \cup I_4 \cup \cdots$ define a coarser block structure. It is easy to see that the block-diagonal of A with respect to the block structure $\{J_1, J_2\}$ coincides with D. Hence, (2′) implies the assertion. □

5.2 Preparatory Lemmata

In this Section we study the properties of a weakly 2-cyclic matrix B. For a suitable ordering of the indices, B takes the form

$$B = \begin{bmatrix} 0 & B_1 \\ B_2 & 0 \end{bmatrix}. \tag{5.2.1}$$

Lemma 5.2.1. *The spectrum of a weakly 2-cyclic matrix B with the off-diagonal blocks $B_1 = B^{12}$ and $B_2 = B^{21}$ is given by*

$$\sigma(B) = \pm\sqrt{\sigma(B_1 B_2)} \cup \pm\sqrt{\sigma(B_2 B_1)}. \tag{5.2.2a}$$

Here, the notation $\pm\sqrt{\sigma(C)} := \{\lambda \in \mathbb{C}: \lambda^2 \in \sigma(C)\}$ is used. The spectra $\sigma(B_1 B_2)$ and $\sigma(B_2 B_1)$ coincide up to a possible zero-eigenvalue:

$$\sigma(B_1 B_2)\backslash\{0\} = \sigma(B_2 B_1)\backslash\{0\}. \tag{5.2.2b}$$

Proof. (i) Let e be an eigenvector of B, and e^1, e^2 the corresponding block vectors. Then the equivalence (3a) holds:

$$Be = \lambda e \Leftrightarrow \begin{cases} B_1 e^2 = \lambda e^1, \\ B_2 e^1 = \lambda e^2. \end{cases} \tag{5.2.3a}$$

Inserting one of the right-hand side equations into the other, we get

$$\lambda^2 e^1 = B_1 B_2 e^1, \qquad \lambda^2 e^2 = B_2 B_1 e^2. \tag{5.2.3b}$$

By $e \neq 0$, either $e^1 \neq 0$ or $e^2 \neq 0$ must hold and therefore $\lambda^2 \in \sigma(B_1 B_2)$ and $\lambda^2 \in \sigma(B_2 B_1)$. In any case, $\lambda \in \pm\sqrt{\sigma(B_1 B_2)} \cup \pm\sqrt{\sigma(B_2 B_1)}$ is valid. Since $\lambda \in \sigma(B)$ is arbitrary, $\sigma(B) \subset \pm\sqrt{\sigma(B_1 B_2)} \cup \pm\sqrt{\sigma(B_2 B_1)}$ is proved.
(ii) Assume that $0 \neq \lambda \in \pm\sqrt{\sigma(B_1 B_2)}$, i.e., $0 \neq \lambda^2 \in \sigma(B_1 B_2)$. Let $e^1 \neq 0$ the corresponding eigenvector: $\lambda^2 e^1 = B_1 B_2 e^1$. For $e^2 := \frac{1}{\lambda} B_2 e^1$, one observes

$$B_1 e^2 = \frac{1}{\lambda} B_1 B_2 e^1 = \frac{1}{\lambda} \lambda^2 e^1 = \lambda e^1.$$

By definition of e^2, $B_2 e^1 = \lambda e^2$ holds. Hence, $e := \begin{pmatrix} e^1 \\ e^2 \end{pmatrix}$ satisfies the equations (3a), i.e., $\lambda \in \sigma(B)$.
(iii) If $0 = \lambda^2 \in \sigma(B_1 B_2) \cup \sigma(B_2 B_1)$, one of the matrices B_1, B_2 has a nontrivial kernel. Without loss of generality, this might be B_1: $B_1 e^2 = 0$ for $e^2 \neq 0$. Since $e^1 := 0$ leads to $B_2 e^1 = 0$, $e := \begin{pmatrix} e^1 \\ e^2 \end{pmatrix}$ is the eigenvector corresponding to the eigenvalue $0 = \lambda \in \sigma(B)$.
(iv) The parts (ii) and (iii) prove $\sigma(B) \supset \pm\sqrt{\sigma(B_1 B_2)} \cup \pm\sqrt{\sigma(B_2 B_1)}$. Together with (i), one obtains the assertion (2a). (2b) follows from Theorem 2.4.6. □

The definition of $\pm\sqrt{\sigma(C)}$ leads to

Remark 5.2.2. If λ is an eigenvalue of a weakly 2-cyclic matrix, then $-\lambda$ is an eigenvalue also.

Lemma 5.2.3. *Under the assumptions of Lemma 1, the following identity holds for the spectral radii:*

$$\varrho(B) = \sqrt{\varrho(B_1 B_2)} = \sqrt{\varrho(B_2 B_1)}. \tag{5.2.4}$$

Proof. By Lemma 2.4.16, $\varrho(B_1 B_2) = \varrho(B_2 B_1)$ holds. Together with (2a), we arrive at the assertion. □

Remark 5.2.4. In the symmetric case $B = B^H$, the blocks from Lemma 1 satisfy $B_1 = B_2^H$. By Theorem 2.9.5, $\varrho(B_1 B_2) = \varrho(B_2 B_1) = \varrho(B_1^H B_1) = \varrho(B_2^H B_2)$ coincides with $\|B_1\|_2^2 = \|B_2\|_2^2$, so that

$$\varrho(B) = \|B_1\|_2 = \|B_2\|_2. \tag{5.2.5}$$

Exercise 5.2.5. For a general matrix of the form (1), prove that

$$\|B\|_2 = \max\{\|B_1\|_2, \|B_2\|_2\}. \tag{5.2.6}$$

5.3 Analysis of the Richardson Iteration

First, we study the case of the parameter choice $\Theta = 1$. Furthermore, we assume that the block-diagonal of A is the identity matrix I. For a suitable ordering of the indices, A has the form

$$A = \begin{array}{|c|c|} \hline I & A_1 \\ \hline A_2 & I \\ \hline \end{array}. \tag{5.3.1}$$

Theorem 5.3.1. *Let $\{A, I\}$ be weakly 2-cyclic with the off-diagonal blocks $A_1 = A^{12}$, $A_2 = A^{21}$ (cf. (1)). Then the Richardson iteration $x^{m+1} = x^m - \Theta(Ax^m - b)$ with $\Theta = 1$ has the convergence rate*

$$\varrho(M_1^{\text{Rich}}) = \sqrt{\varrho(A_1 A_2)} = \sqrt{\varrho(A_2 A_1)}. \tag{5.3.2}$$

Proof. The iteration matrix $M_1^{\text{Rich}} = I - A$ coincides with B from §5.2, if $B_i := -A_i$. Hence, (2.4) from Lemma 2.3 implies the result (2). □

If we admit $\Theta \neq 1$, $M_\Theta^{\text{Rich}} = I - \Theta A$ leads to

$$\sigma(M_\Theta^{\text{Rich}}) = \{\lambda = 1 - \Theta(1 - \mu) : \mu \in \sigma(B)\} \quad \text{with } B = I - A \tag{5.3.3}$$

and $\sigma(B)$ from (2.2a), where $B_1 := -A_1$ and $B_2 := -A_2$. For an arbitrary complex spectrum $\sigma(B)$, a simple characterisation of the spectral radius $\varrho(M_\Theta^{\text{Rich}})$ is not so easy. Therefore, we assume

$$\beta := \varrho(B) \in \sigma(B) \quad \text{for } B = I - A = -\begin{bmatrix} 0 & A_1 \\ A_2 & 0 \end{bmatrix}. \tag{5.3.4}$$

Condition (4) means that $\varrho(B)$ is not only the absolute value $|\lambda|$ of some eigenvalue $\lambda \in \sigma(B)$, but even an eigenvalue of B. Sufficient conditions for (4) will follow after Theorem 2.

Theorem 5.3.2. *Let $\{A, I\}$ be weakly 2-cyclic satisfying* (4). *Then the Richardson iteration $x^{m+1} = x^m - \Theta(Ax^m - b)$ has the convergence rate*

$$\varrho(M_\Theta^{\mathrm{Rich}}) = \begin{cases} 1 - \Theta(1 - \varrho(B)) & \text{for } 0 \leqslant \Theta \leqslant 1, \\ \Theta(1 + \varrho(B)) - 1 & \text{for } \Theta \geqslant 1, \\ 1 + |\Theta|(1 + \varrho(B)) & \text{for } \Theta \leqslant 0, \end{cases} \tag{5.3.5a}$$

with

$$\varrho(B) = \sqrt{\varrho(A_1 A_2)} = \sqrt{\varrho(A_2 A_1)} \quad \text{for } B := I - A. \tag{5.3.5b}$$

If $\varrho(B) \geqslant 1$, the method is divergent for all $\Theta \in \mathbb{R}$. If $\varrho(B) < 1$, the method converges for $0 < \Theta < 2/(1 + \varrho(B))$. $\Theta = 1$ yields the optimal convergence rate (2).

Proof. (i) Let $\Theta \in [0, 1]$, $\mu \in \sigma(B)$, and $\beta := \varrho(B)$. According to (3), we have to estimate $\lambda = 1 - \Theta(1 - \mu)$. Since $|\mu| \leqslant \beta$ by assumption (4),

$$\begin{aligned} |\lambda| &= |1 - \Theta(1 - \mu)| = |(1 - \Theta) + \Theta\mu| \leqslant 1 - \Theta + \Theta|\mu| \\ &\leqslant 1 - \Theta + \Theta\beta = 1 - \Theta(1 - \beta). \end{aligned} \tag{5.3.6}$$

holds. For $\mu = \beta \in \sigma(B)$, the equality holds in (6), so that $1 - \Theta(1 - \beta)$ is the smallest bound for $\varrho(M_\Theta^{\mathrm{Rich}})$. Hence, the first case in (5a) is proved.
(ii) The cases $\Theta \geqslant 1$ and $\Theta \leqslant 0$ in (5a) are treated analogously.
(iii) The further statements are a direct consequence of (5a). □

Condition (4) required in Theorem 2 is the subject of the following two criteria.

Criterion 5.3.3. If B has only *real* eigenvalues, condition (4) is satisfied. In particular, the symmetry of B is sufficient: $A_1 = A_2^H$.

Proof. Let $\beta_{\min}$ and $\beta_{\max}$ be the minimal and maximal eigenvalues of B. Since, by Remark 2.2, the spectrum $\sigma(B)$ is symmetric, $\beta_{\min} = -\beta_{\max} \leqslant 0 \leqslant \beta_{\max}$ proves $\varrho(B) = \max\{|\beta_{\min}|, |\beta_{\max}|\} = \beta_{\max} \in \sigma(B)$. □

Criterion 5.3.4. If all matrix entries of A_1 and A_2 are nonnegative (or all nonpositive), condition (4) is satisfied.

Proof. Theorem 6.3.10 will show $\beta := \varrho(-B) \in \sigma(-B)$. Since $\sigma(B) = \sigma(-B)$ (cf. Remark 2.2), $\beta = \varrho(B) \in \sigma(B)$ is also valid. □

5.4 Analysis of the Jacobi Method

In the following, D is assumed to be the (pointwise) diagonal or block-diagonal part of A. Let $\{A, D\}$ be weakly 2-cyclic with respect to a block structure $\{I_1, I_2\}$ with off-diagonal blocks $A_1 = A^{12}$ and $A_2 = A^{21}$. For a suitable numbering, A and D take the form

$$A = \begin{bmatrix} D_1 & A_1 \\ A_2 & D_2 \end{bmatrix} \quad \text{and} \quad D = \begin{bmatrix} D_1 & 0 \\ 0 & D_2 \end{bmatrix}. \tag{5.4.1}$$

The matrix $A' := D^{-1}A$ has the representation

$$D^{-1}A = \begin{bmatrix} I & A'_1 \\ A'_2 & I \end{bmatrix} = I - B \quad \text{with } B := -\begin{bmatrix} 0 & A'_1 \\ A'_2 & 0 \end{bmatrix}, \tag{5.4.2a}$$

where

$$A'_1 := D_1^{-1}A_1, \qquad A'_2 := D_2^{-1}A_2. \tag{5.4.2b}$$

Depending on whether D is the (pointwise) diagonal or block-diagonal of A, the following theorem describes the *pointwise* or *blockwise Jacobi method*, respectivey.

Theorem 5.4.1. *Let $\{A, D\}$ be weakly 2-cyclic. Then the Jacobi method $x^{m+1} = x^m - D^{-1}(Ax^m - b)$ has the convergence rate*

$$\varrho(M^{\mathrm{Jac}}) = \varrho(B) = \sqrt{\varrho(A'_1A'_2)} = \sqrt{\varrho(D_1^{-1}A_1D_2^{-1}A_2)}. \tag{5.4.3}$$

Proof. The Jacobi method is identical to the Richardson iteration applied to $A'x = b'$ with $A' := D^{-1}A$, $b' := D^{-1}b$, and $\Theta = 1$ (cf. Remark 4.3.2). Hence, (3) follows from Theorem 3.1. □

To obtain statements about the *damped Jacobi method*, one has to satisfy condition (3.4) for B from (2a). Besides Criteria 3.3 and 3.4, the following one can be applied.

Criterion 5.4.2. Let $\{A, D\}$ be weakly 2-cyclic. (a) If D is positive definite and A is Hermitian, then $B = M^{\mathrm{Jac}}$ has only real eigenvalues. Further,

$$\varrho(B) \in \sigma(B) \quad \text{for } B = I - D^{-1}A. \tag{5.4.4}$$

(b) If A is positive definite, besides (4), $\varrho(B) < 1$ is also valid, i.e., the Jacobi iteration converges.

Proof. (a) Since $\hat{B} := D^{-1/2}(D - A)D^{-1/2}$ is Hermitian, it has only real eigenvalues. Since $\hat{B}$ is similar to $B = D^{-1}(D - A)$, $\sigma(\hat{B}) = \sigma(B)$ holds. Therefore, all eigenvalues of B are real and Criterion 3.3 is applicable.

(b) By Corollary 4.5.6, the Jacobi iteration converges. Since $B = M^{\mathrm{Jac}}$, $\varrho(B) < 1$ follows. □

A further sufficient condition for (4) corresponding to Criterion 3.4 reads as follows: $D_1^{-1}, A_1, D_2^{-1}, A_2$ have only nonnegative matrix entries. According to Theorem 3.2, the following remark holds.

Remark 5.4.3. Assume that $\{A, D\}$ is weakly 2-cyclic and satisfies (4). If $\varrho(B) \geqslant 1$, the *damped Jacobi method* $x^{m+1} = x^m - \omega D^{-1}(Ax^m - b)$ diverges for all $\omega \in \mathbb{R}$. If $\varrho(B) < 1$, it converges for $0 < \omega < 2/(1 + \varrho(B))$, where the optimal convergence rate is attained for $\omega = 1$, i.e., the undamped Jacobi method is already optimal.

5.5 Analysis of the Gauß-Seidel Iteration

Let the index set I be ordered and assume that the matrix A has the form

$$A = \begin{bmatrix} D_1 & A_1 \\ A_2 & D_2 \end{bmatrix}. \tag{5.5.1a}$$

with respect to the block structure $\{I_1, I_2\}$. A allows the splitting $A = D - E - F$ with

$$D = \begin{bmatrix} D_1 & 0 \\ 0 & D_2 \end{bmatrix}, \quad E = \begin{bmatrix} 0 & 0 \\ -A_2 & 0 \end{bmatrix}, \quad F = \begin{bmatrix} 0 & -A_1 \\ 0 & 0 \end{bmatrix} \tag{5.5.1b}$$

(cf. (4.5.5a–d). Furthermore, $\{A, D\}$ is 2-cyclic. If D is the pointwise or blockwise diagonal of A, respectively, the respective pointwise or blockwise Gauß-Seidel methods are described in the following.

The Gauß-Seidel iteration matrix is $M^{\mathrm{GS}} = (D - E)^{-1}F$. By

$$(D - E)^{-1} = \begin{bmatrix} D_1 & 0 \\ A_2 & D_2 \end{bmatrix}^{-1} = \begin{bmatrix} D_1^{-1} & 0 \\ -D_2^{-1}A_2D_1^{-1} & D_2^{-1} \end{bmatrix},$$

one obtains

$$M^{\mathrm{GS}} = (D - E)^{-1}F = \begin{bmatrix} 0 & -D_1^{-1}A_1 \\ 0 & D_2^{-1}A_2D_1^{-1}A_1 \end{bmatrix}. \tag{5.5.1c}$$

Theorem 5.5.1. *The Gauß-Seidel method* $x^{m+1} = (D - E)^{-1}(Fx^m + b)$ *defined by means of the matrices* $A = D - E - F$ *according to* (1b) *has the convergence rate*

$$\varrho(M^{\mathrm{GS}}) = \varrho(D_1^{-1}A_1D_2^{-1}A_2). \tag{5.5.2}$$

Proof. Inspection of (1c) yields $\varrho(M^{\mathrm{GS}}) = \varrho(D_2^{-1}A_2D_1^{-1}A_1)$ (cf. (2.5.5b). By Lemma 2.4.16, $\varrho(D_2^{-1}A_2D_1^{-1}A_1) = \varrho(D_1^{-1}A_1D_2^{-1}A_2)$ holds. □

Remark 5.5.2. Let $\{A, D\}$ be 2-cyclic. (a) Then the Jacobi method converges if and only if the Gauß-Seidel method converges.
(b) *The convergence of the Gauß-Seidel iteration is exactly twice as fast as that of the Jacobi iteration*:

$$\varrho(M^{GS}) = \varrho(M^{Jac})^2. \tag{5.5.3}$$

(c) M^{GS} has at least an n_1-fold eigenvalue $\lambda = 0$ if the first block D_1 of A has the size $n_1 \times n_1$.

Proof. A comparison of (2) with (4.3) yields Eq. (3), proving the parts (a) and (b). Assertion (c) follows because of the zero blocks in (1c). □

Since the Jacobi and Gauß-Seidel iterations require the same amount of computational work, one concludes from (3) the following remark.

Remark 5.5.3. Let $\{A, D\}$ be 2-cyclic. For the Jacobi method, the effective amount of work defined in (3.3.4a) is twice as large as for the Gauß-Seidel method:

$$Eff(\Phi^{Jac}) = 2Eff(\Phi^{GS}). \tag{5.5.4}$$

The order of linear convergence defined in §3.3.3, however, coincides for both methods.

Exercise 5.5.4. Let $\{A, D\}$ be 2-cyclic. (a) What is the form of the iteration matrix M^{symGS} of the symmetric Gauß-Seidel method?
(b) Prove that $\varrho(M^{symGS}) = \varrho(M^{GS})$ (cf. also §5.8.3).

The results of Remarks 2 and 3 will be confirmed again in §5.6 under somewhat weaker assumptions.

5.6 Analysis of the SOR Method

5.6.1 Consistently Ordered Matrices

In §5.5, $\{A, D\}$ has been assumed to be 2-cyclic. This assumption almost corresponds to «property A» of Young [2]. This condition can be generalised by the term of the «consistent ordering» due to Varga [2].

Definition 5.6.1. Let the index set I be ordered. According to (4.2.7a–d) or (4.5.5a–d), A is split into $A = D - E - F$. Consequently, the matrices $L := D^{-1}E$ and $U := D^{-1}F$ are strictly triangular matrices. A is called *consistently ordered* if the eigenvalues of the matrix $zL + \frac{1}{z}U$ do not depend on $z \in \mathbb{C}\backslash\{0\}$.

In the following, we avoid the term «consistent ordering of A» by two reasons. First, it is somewhat inexact, since L and U depend on not only A, but also D, if matrices D of the diagonal or block-diagonal form are admitted. Second, contrary to its name, the addressed property is independent of the ordering of the indices if we admit other L and U than triangular matrices.

The pair $\{L, U\}$ is required to satisfy the following property:

$$\text{The eigenvalues of } zL + \frac{1}{z}U \text{ are not dependent of } z \in \mathbb{C}\backslash\{0\}. \tag{5.6.1}$$

Criterion 5.6.2. $\{L, U\}$ satisfies (1) and A is consistently ordered in the sense of Definition 1, if L and U are strictly lower or upper triangular matrices, respectively, satisfying one of the following conditions (2a–d):

$L + U$ is 2-cyclic, (5.6.2a)

$L + U$ is tridiagonal, (5.6.2b)

$L + U$ is block-tridiagonal with vanishing diagonal-blocks, (5.6.2c)

$L + U$ is block-tridiagonal, where the diagonal-blocks $(L + U)^{ii}$ are tridiagonal and the (possibly rectangular) off-diagonal blocks $(L + U)^{i,i\pm 1}$ are diagonal. (5.6.2d)

Here, a rectangular matrix A is called diagonal if at most the diagonal entries A_{ii} are different from zero.

Proof. (i) Set $B := L + U$. Since L, U are strictly triangular matrices, $b_{ii} = 0$ holds for the diagonal entries.
(ii) From (2a) one concludes the assertion by means of Lemma 5.1. Furthermore, (2a) is a special case of (2c).
(iii) Assume (2b). Let $z \in \mathbb{C}\backslash\{0\}$ and construct the diagonal matrix

$$\Delta := \operatorname{diag}\{1, z, z^2, z^3, \ldots, z^{n-1}\}.$$

$B' := \Delta B \Delta^{-1}$ has the entries $b'_{ij} = b_{ij} z^{i-j}$. Since $b'_{ij} \neq 0$ only for $|i - j| = 1$, B' is of the form $zL + \frac{1}{z}U$. Because B and B' are similar, $\{L, U\}$ has property (1).
(iv) Let $\{I_1, \ldots, I_b\}$ be the block structure in the case (2c). We define

$$\Delta_b := \operatorname{blockdiag}\{I, zI, z^2 I, z^3 I, \ldots, z^{b-1} I\},$$

where I are the identity matrices of the respective block-sizes: $I \in \mathbb{R}^{I_\kappa \times I_\kappa}$. $B' := \Delta_b B \Delta_b^{-1}$ contains the blocks $(B')^{ij} = z^{i-j} B^{ij}$, so that the assertion follows as in (iii).
(v) In the case of (2d), apply first the similarity transformation Δ_b from (iv): B' contains the factor z $\left(\text{or } \frac{1}{z}\right)$ in the blocks below (or above) the diagonal;

however, the tridiagonal diagonal-blocks $(B')^{ii} = B^{ii}$ are still unchanged. We define the diagonal matrix

$$\Delta_B \quad \text{with } (\Delta_B)^{ii} := \operatorname{diag}\{1, z, z^2, z^3, \ldots, z^{\#I_i - 1}\}.$$

The transformation $C' := \Delta_B C \Delta_B^{-1}$ does not change the blocks with diagonal structure. Hence,

$$B'' := \Delta_B B' \Delta_B^{-1} = \Delta_B \Delta_b B \Delta_b^{-1} \Delta_B^{-1}$$

has the off-diagonal blocks $(B'')^{i,i-1} = zB^{i,i-1}$, $(B'')^{i,i+1} = \frac{1}{z} B^{i,i+1}$, whereas, as in (iii), the diagonal-blocks have the entries zb_{ij} in the lower and $\frac{1}{z} b_{ij}$ in the upper triangular part, i.e., $B'' = zL + \frac{1}{z} U$. Since B'' is similar to B, the assertion follows. □

Remark 5.6.3. The properties (2a), (2b), (2c), or (2d) imply that $\{A, D\}$ is weakly 2-cyclic if D is the (pointwise) diagonal in the cases (2b, d) and, otherwise, the block-diagonal of A.

Proof. For (2a–c) apply Remark 1.4a or Lemma 1.7a, b, respectively. In the case of (2d), $L + U$ has a five-point structure admitting a chequer-board-like ordering, for which $A - D = -L - U$ is 2-cyclic. □

Lemma 5.6.4. *Let* $\{L, U\}$ *satisfy condition* (1). (a) *Then*

$$\sigma(\alpha L + \beta U) = \sigma(\pm\sqrt{\alpha\beta}(L + U)) \quad \text{for all } \alpha, \beta \in \mathbb{C}.$$

(b) *In particular,* $L + U$ *and* $-(L + U)$ *have identical spectra.*
(c) *If all eigenvalues of* $L + U$ *are real,* $\varrho(L + U) \in \sigma(L + U)$ *holds.*

Proof. (i) For $\alpha\beta \neq 0$, choose $z := \pm\sqrt{\alpha/\beta}$. Since $zL + \frac{1}{z} U$ and $L + U$ have the same eigenvalues, this is also true for the matrices

$$\alpha L + \beta U = \pm\sqrt{\alpha\beta}\left(zL + \frac{1}{z} U\right) \quad \text{and} \quad \pm\sqrt{\alpha\beta}(L + U).$$

(ii) Since $\alpha = \beta = 0$ represents a trivial case, assume that $\alpha = 0$ and $\beta \neq 0$. The assertion then reads $\sigma(\beta U) = \sigma(0(L + U)) = \sigma(0) = \{0\}$ and is satisfied, because U is a strictly triangular matrix (cf. Exercise 2.4.15b). The case $\beta = 0$ is analogous.
(iii) For $\alpha = \beta = 1$, the expression $\pm\sqrt{\alpha\beta}$ also takes the value -1, so that $\sigma(L + U) = \sigma(-(L + U))$ proves part (b). Part (c) is demonstrated as in Criterion 3.3. □

5.6.2 Theorem of Young

The following theorem, which in its basic form is due to Young [1], describes the pointwise as well as blockwise SOR iteration (4.3.7a):

$$x^{m+1} = M_\omega^{SOR} x^m + N_\omega^{SOR} b \quad \text{with} \tag{5.6.3a}$$

$$A = D - E - F, \qquad L := D^{-1}E, \qquad U := D^{-1}F, \tag{5.6.3b}$$

$$M_\omega^{SOR} := (I - \omega L)^{-1}\{(1-\omega)I + \omega U\}, \qquad N_\omega^{SOR} := \omega(I - \omega L)^{-1}\mathrm{D}^{-1}, \tag{5.6.3c}$$

where $A = D - E - F$ is split according to (4.2.7a–d) or (4.5.5a–d). The theorem is valid for any iteraton of the form (3a–c). The matrix

$$M^{Jac} := L + U = I - D^{-1}A \tag{5.6.3d}$$

represents the iteration matrix of the (pointwise or blockwise) Jacobi method only if D coincides with the diagonal or block-diagonal of A. Note that Theorem 5 admits more general matrices D.

Theorem 5.6.5. *For the iteration* (3a–c) *we assume*:

$$0 < \omega < 2, \tag{5.6.4a}$$

$$M^{Jac} \textit{ has only real eigenvalues,} \tag{5.6.4b}$$

$$\beta := \varrho(M^{Jac}) < 1, \tag{5.6.4c}$$

$$D \textit{ and } I - \omega L \textit{ are regular, } \{L, U\} \textit{ satisfies condition } (1). \tag{5.6.4d}$$

Then the following statements hold: (a) *Iteration* (3a–c) *converges.*
(b) *The convergence rate equals*

$$\varrho(M_\omega^{SOR}) = \begin{cases} 1 - \omega + \frac{1}{2}\omega^2\beta^2 + \omega\beta\sqrt{1 - \omega + \frac{\omega^2\beta^2}{4}} & \textit{for } 0 < \omega \leqslant \omega_{opt} \\ \omega - 1 & \textit{for } \omega_{opt} \leqslant \omega < 2, \end{cases} \tag{5.6.5a}$$

where

$$\omega_{opt} := \frac{2}{1 + \sqrt{1 - \beta^2}}. \tag{5.6.5b}$$

(c) *The convergence rate* $\varrho(M_\omega^{SOR})$ *is minimal for* $\omega = \omega_{opt}$.
(d) *For* $\omega \leqslant \omega_{opt}$, *the spectral radius* $\varrho(M_\omega^{SOR}) \in \sigma(M_\omega^{SOR})$ *is an eigenvalue.*
(e) *For* $\omega \geqslant \omega_{opt}$, *all eigenvalues of* M_ω^{SOR} *have coinciding absolute values* $\omega - 1$.

Before proving the theorem, we discuss its assumptions and results:

Concerning (4a). The assumption $0 < \omega < 2$ is necessary for convergence, as we know know from Lemma 4.4.20.

Concerning (4b). Because of Criterion 4.2a, M^{Jac} has the required real eigenvalues, if A is Hermitian and D is positive definite. Criterion 4.2b and Corollary 4.4.22 even provide a sufficient criterion for (4b–d):

Criterion 5.6.6. Let D be the (block-)diagonal of A. If A is positive definite, conditions (4b, c) and the first part of (4d) are satisfied.

Concerning (4c). $\beta < 1$ is equivalent to the convergence of the Jacobi method. The condition $\beta < 1$ is necessary because of

Exercise 5.6.7. If $\beta \geqslant 1$, the SOR iteration diverges for all $\omega \in \mathbb{R}$.

Concerning (4d). If (3a–c) represents a true SOR iteration, L must be a strictly triangular matrix. Then the regularity of $I - \omega L$ is trivial.

Concerning (5a, b). Except for the trivial case $\beta = 0$, it holds that

$$1 < \omega_{\text{opt}} < 2, \tag{5.6.5c}$$

so that the optimal convergence speed leads always to the *over*-relaxation method. Underrelaxation ($0 < \omega < 1$) is always slower than the Gauß-Seidel method, which is regained for $\omega = 1$. For $\omega = 1$ we again obtain the result (5.3). Figure 1 shows $\varrho(M_\omega^{\text{SOR}})$ as a function of ω for $\beta = \cos(\pi/8) = 0.92388$ with $\omega_{\text{opt}} = 2/(1 + \sin \pi/8) = 1.44646$. The latter value corresponds to the model problem with $h = 1/8$.

Proof of Theorem 5. (i) Let $\lambda \in \sigma(M_\omega^{\text{SOR}})$. The corresponding eigenvector e satisfies $\{(1 - \omega)I + \omega U\}e = \lambda(I - \omega L)e$, i.e.,

$$(\omega U + \lambda\omega L)e = (\lambda + \omega - 1)e,$$

so that $\lambda + \omega - 1 \in \sigma(\omega U + \lambda\omega L)$. Since $\sigma(\omega U + \lambda\omega L) = \sigma(\pm\sqrt{\lambda}\omega(L + U))$ holds by Lemma 4a, there is an eigenvalue $\mu \in \sigma(M^{\text{Jac}}) = \sigma(L + U)$ with

$$\lambda + \omega - 1 = \pm\sqrt{\lambda}\omega\mu, \quad \text{i.e.,} \tag{5.6.6a}$$

$$(\lambda + \omega - 1)^2 = \omega^2\lambda\mu^2. \tag{5.6.6b}$$

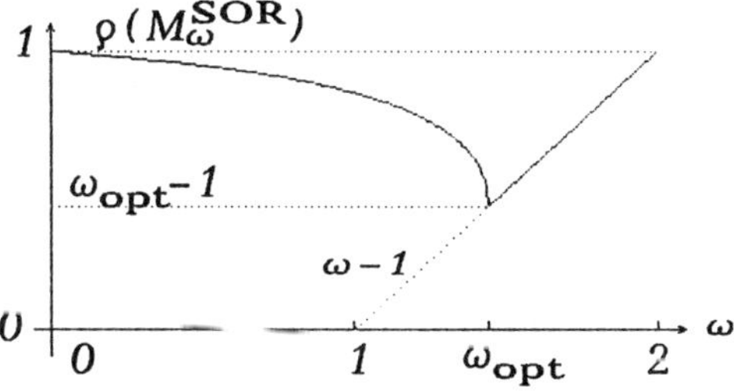

Fig. 5.6.1 The convergence rate $\varrho(M_\omega^{\text{SOR}})$ depending on ω

Since for any eigenvalue μ of $M^{\text{Jac}} = L + U$, $-\mu$ is also an eigenvalue (cf. Lemma 4b), any solution μ of (6b) belongs to $\sigma(M^{\text{Jac}})$. Vice versa, one concludes that for any $\mu \in \sigma(M^{\text{Jac}})$, both solutions

$$\lambda = 1 - \omega + \frac{1}{2}\omega^2\mu^2 \pm \omega\mu\sqrt{1 - \omega + \frac{\omega^2\mu^2}{4}} \tag{5.6.6c}$$

fulfil Eq. (6a) with a suitable sign in $\pm\sqrt{\lambda}$. Since $\pm\sqrt{\lambda}\omega\mu$ is an eigenvalue of $\pm\sqrt{\lambda}\omega(L + U)$, it is also an eigenvalue of $\omega U + \lambda\omega L$; hence, we arrive at $\lambda \in \sigma(M_\omega^{\text{SOR}})$ and obtain

$$\lambda \in \sigma(M_\omega^{\text{SOR}}) \Leftrightarrow \mu \in \sigma(M^{\text{Jac}}) \qquad (\lambda, \mu \text{ satisfy (6b)}). \tag{5.6.6d}$$

(ii) Let $\omega_{\text{opt}} \leqslant \omega < 2$. This inequality is equivalent to $1 - \omega + \dfrac{\omega^2\beta^2}{4} \leqslant 0$. By $-\beta \leqslant \mu \leqslant \beta$, we obtain for all $\mu \in \sigma(M^{\text{Jac}})$ the inequality

$$1 - \omega + \frac{\omega^2\mu^2}{4} \leqslant 0,$$

implying that Eq. (6b) has two complex conjugate roots:

$$\lambda = \lambda_{\text{Re}} \pm i\lambda_{\text{Im}}, \qquad \lambda_{\text{Re}} := 1 - \omega + \tfrac{1}{2}\omega^2\mu^2. \tag{5.6.6e}$$

Since the product of the roots of a quadratic equation coincides with the absolute term of the equation, we obtain

$$|\lambda|^2 = (\omega - 1)^2, \quad \text{i.e., } |\lambda| = |\omega - 1|. \tag{5.6.6f}$$

Hence, M_ω^{SOR} has only eigenvalues λ with an absolute value $\omega - 1$. This proves the second case in (5a), as well as the statements (a) and (e).
(iii) Assume the second case $0 < \omega < \omega_{\text{opt}}$. If $\omega \in (1, \omega_{\text{opt}})$, there may be eigenvalues $\mu \in \sigma(M^{\text{Jac}})$ with $\mu^2 < 4(\omega - 1)/\omega^2$, for which the radicand in (6c) is negative. As before, these μ generate eigenvalues $\lambda \in \sigma(M_\omega^{\text{SOR}})$ with $|\lambda| = |\omega - 1|$. This value, however, is smaller than the right-hand side in (5a). The latter proves to be an eigenvalue of M_ω^{SOR} by choosing $\mu := \beta \in \sigma(M^{\text{Jac}})$ in (6d) [concerning $\beta \in \sigma(M^{\text{Jac}})$, compare Lemma 4c]. Since the discussion can be reduced to the case of the real solutions of (6c), it is easy to see that $|\lambda|$ attains its maximum at $\mu = \beta$. □

Because of $\omega_{\text{opt}} \geqslant 1$ (cf. (5b)), $\omega = 1$ lies in the interval $(0, \omega_{\text{opt}}]$. Theorem 5 yields the following results for the Gauß-Seidel method, which is the special case of $\omega = 1$.

Remark 5.6.8 (Gauß-Seidel method). Under the assumptions (4b–d) the [block-]Gauß-Seidel method converges and has exactly the squared convergence speed of the [block-]Jacobi iteration:

$$\varrho(M^{[\text{block}]\text{GS}}) = \varrho(M_1^{[\text{block}]\text{SOR}}) = \beta^2 = \varrho(M^{[\text{block}]\text{Jac}})^2,$$

as already mentioned in (5.3) for the 2-cyclic case. Furthermore, $\varrho(M^{[\text{block}]\text{GS}})$ belongs to the spectrum $\sigma(M^{[\text{block}]\text{GS}})$. The statement of Remark 5.3 is still valid.

For the case of *complex relaxation parameters* ω with $|1-\omega|<1$, a convergence result is given by Niethammer–Varga [1, Theorem 12].

5.6.3 Order Improvement by SOR

We recall the term of the *order* of an iterative method from §3.3.3: Considering a family of systems corresponding to different step sizes h (and therefore to different dimensions) as in the model case, the Jacobi iteration has the order $\varkappa$ if

$$\varrho(M^{\text{Jac}}) = 1 - C_\eta^{\text{Jac}} h^\varkappa + O(h^{2\varkappa}) \quad \text{for } h \to 0. \tag{5.6.7a}$$

The model problem from §1.2 has the order $\varkappa = 2$. A comparison of Jacobi versus Gauß-Seidel by means of (5.3): $\varrho(M^{\text{GS}}) = \varrho(M^{\text{Jac}})^2$ shows that

$$\varrho(M^{\text{Jac}})^2 = (1 - C_\eta^{\text{Jac}} h^\varkappa + \cdots)^2 = 1 - 2C_\eta^{\text{Jac}} h^\varkappa + \cdots = 1 - C_\eta^{\text{GS}} h^\varkappa + \cdots.$$

Hence, only the coefficient $C_\eta^{\text{GS}} = 2C_\eta^{\text{Jac}}$ is improved, whereas the order remains unchanged.

In the weakly 2-cyclic case, the variation of the parameters ω of the damped (extrapolated) Jacobi method (4.3.2) is without success: By Remark 4.3, the choice $\omega = 1$ and therefore the standard Jacobi method are optimal. More notable is the possibility of improving the SOR convergence rate by the proper choice $\omega = \omega_{\text{opt}}$. The next theorem shows that in this way the order is also improved (halved).

Theorem 5.6.9. *Let $\varkappa > 0$ be the order of the Jacobi method (cf.* (7a)*). Under the assumptions of Theorem* 5*, the SOR method with $\omega = \omega_{\text{opt}}$ has the order $\varkappa/2$:*

$$\varrho(M_{\omega_{\text{opt}}}^{\text{SOR}}) = 1 - C_\eta^{\text{SOR}} h^{\varkappa/2} + O(h^\varkappa) \quad \textit{with} \tag{5.6.7b}$$

$$C_\eta^{\text{SOR}} := 2\sqrt{2C_\eta^{\text{Jac}}}. \tag{5.6.7c}$$

Proof. Following (5b), we have

$$\varrho(M_{\omega_{\text{opt}}}^{\text{SOR}}) = \omega_{\text{opt}} - 1 = \frac{2}{1+\sqrt{1-\beta^2}} - 1 = \frac{1-\sqrt{1-\beta^2}}{1+\sqrt{1-\beta^2}}. \tag{5.6.8}$$

By $1-\beta^2 = 1 - \varrho(M^{\text{Jac}})^2 = 1 - [1 - C_\eta^{\text{Jac}} h^\varkappa + O(h^{2\varkappa})]^2 = 2C_\eta^{\text{Jac}} h^\varkappa + O(h^{2\varkappa})$, the square root $\sqrt{1-\beta^2}$ has the expansion $\sqrt{2C_\eta^{\text{Jac}}} h^{\varkappa/2} + O(h^\varkappa)$. Inserting this expression into (8), we obtain $\varrho(M_{\omega_{\text{opt}}}^{\text{SOR}}) = 1 - 2\sqrt{2C_\eta^{\text{Jac}}} h^{\varkappa/2} + O(h^\varkappa)$, proving (7b, c). ☐

5.6.4 Practical Handling of the SOR Method

In general, the value $\beta = \varrho(M^{\mathrm{Jac}})$ is unknown. Thus, the optimal relaxation parameter ω_{opt} is also not available. Then, one may proceed as follows (cf. also Young [2, §6.6]).

Initially, choose some $\omega \leqslant \omega_{\mathrm{opt}}$, e.g., $\omega := 1$. Perform few SOR steps with this parameter ω and determine an approximation $\tilde{\lambda}$ to $\varrho(M_\omega^{\mathrm{SOR}})$ from the ratios of $\|x^{m+1} - x^m\|_2$ (see the end of §3.4). By means of $\tilde{\lambda}$, one can produce β via Eq. (5a) (case $\omega \leqslant \omega_{\mathrm{opt}}$):

$$\beta \approx \tilde{\beta} := |\tilde{\lambda} + \omega - 1|/(\omega\sqrt{\tilde{\lambda}})$$

(cf. (6a)). Using $\tilde{\beta}$, one determines an approximation ω to ω_{opt} by (5b). As long as $\omega \leqslant \omega_{\mathrm{opt}}$, it is possible to iterate the described approximation of ω_{opt}. Since the function $\varrho(M_\omega^{\mathrm{SOR}})$ has a vertical tangent at $\omega = \omega_{\mathrm{opt}}$ from the left, any deviation $\omega = \omega_{\mathrm{opt}} - \varepsilon$ ($\varepsilon > 0$) to the left leads to considerable worsening of the convergence. Therefore, one does better to choose $\omega \approx \omega_{\mathrm{opt}}$ too large. A program following this strategy can be found in Meis–Marcowitz [1, Appendix A.4].

5.7 Application to the Model Problem

5.7.1 Analysis in the Model Case

For the five-point formula of the model problem from §1.2, Criterion 6.2 is always applicable, since one of the following cases applies:

(i) *Pointwise variants* (i.e., D is diagonal):

For the *lexicographical* ordering of the indices, A has the form (1.2.8). The sum $L + U$ satisfies the condition (6.2d) of Criterion 6.2.

For *chequer-board* ordering, A takes the 2-cyclic form (1.2.9), so that $L + U$ fulfils the condition (6.2a).

(ii) *Blockwise variants* (i.e., D is block-diagonal):

Assume that the rows or columns of the grid constitute the block structure. If these blocks are ordered *lexicographically*, $L + U$ shows the block-tridiagonal shape required in (6.2c).

For the «zebra-ordering» of the blocks from Example 1.5, $L + U$ has a 2-cyclic form and satisfies (6.2a).

Besides the property (6.1), Criterion 6.2 also proves that $\{A, D\}$ is weakly 2-cyclic in all the cases mentioned above.

Theorem 5.7.1. *Assume the model problem from* §1.2 *with step size h.*
(a) *For lexicographical as well as chequer-board ordering, the pointwise* Gauß-Seidel method *shows the convergence speed*

$$\varrho(M^{\mathrm{GS}}) = \cos^2 \pi h = 1 - \sin^2 \pi h = 1 - \pi^2 h^2 + O(h^4). \qquad (5.7.1a)$$

(b) *The row- and column-block-Gauß-Seidel method with lexicographical or zebra ordering has the convergence rate*

$$\varrho(M^{\text{blockGS}}) = 1 - 8\sin^2\frac{\pi h}{2}\bigg/\left(1 + 2\sin^2\frac{\pi h}{2}\right)^2$$

$$= 1 - 2\pi^2 h^2 + O(h^4). \tag{5.7.1b}$$

(c) *For the pointwise* SOR methods *corresponding to case* (a) *from above, the optimal relaxation parameter is*

$$\omega_{\text{opt}} = 2/(1 + \sin\pi h) = 2 - 2\pi h + O(h^2), \tag{5.7.2a}$$

leading to the convergence rate

$$\varrho(M^{\text{SOR}}_{\omega_{\text{opt}}}) = \omega_{\text{opt}} - 1 = 1 - \frac{2\sin(\pi h)}{1 + \sin(\pi h)} = \frac{1 - \sin(\pi h)}{1 + \sin(\pi h)}. \tag{5.7.2b}$$

(d) *For the block-SOR versions corresponding to case* (b), *the following values apply*:

$$\omega_{\text{opt}} = 2\bigg/\left[1 + 2\sqrt{2}\sin\left(\frac{1}{2}\pi h\right)\bigg/\cos(\pi h)\right], \tag{5.7.3a}$$

$$\varrho(M^{\text{blockSOR}}_{\omega_{\text{opt}}}) = \omega_{\text{opt}} - 1 = 1 - \frac{4\sqrt{2}\sin(\pi h/2)}{\cos(\pi h) + 2\sqrt{2}\sin(\frac{1}{2}\pi h)}. \tag{5.7.3b}$$

Proof. According to Remark 6.8, $\varrho(M^{[\text{block}]\text{GS}})$ is the square of $\varrho(M^{[\text{block}]\text{Jac}})$, which is given in (4.7.3) and (4.7.6) for the model problem. Parts (c) and (d) result from (6.5b, a). □

Remark 5.7.2. The point- and blockwise Gauß-Seidel and SOR iterations described in Theorem 1 require the following *effective amount of work*:

$$\mathit{Eff}(\Phi^{\text{GS}}) = \pi^{-2}h^{-2} + O(1) = 0.101h^{-2} + O(1), \tag{5.7.4a}$$

$$\mathit{Eff}(\Phi^{\text{blockGS}}) = 0.7\pi^{-2}h^{-2} + O(1) = 0.0709h^{-2} + O(1), \tag{5.7.4b}$$

$$\mathit{Eff}(\Phi^{\text{SOR}}) = 0.7\pi^{-1}h^{-1} + O(1) = 0.2228h^{-1} + O(1), \tag{5.7.4c}$$

$$\mathit{Eff}(\Phi^{\text{blockSOR}}) = 0.9/(\sqrt{2}\pi)h^{-1} + O(1) = 0.2026h^{-1} + O(1). \tag{5.7.4d}$$

Proof. The cost factors C_Φ are already represented in (4.6.7a–c): $C_\Phi^{\text{GS}} = 1$, $C_\Phi^{\text{blockGS}} = C_\Phi^{\text{SOR}} = 7/5$, $C_\Phi^{\text{blockSOR}} = 9/5$. The convergence rates (1a, b), (2b), and (3b) have the form $1 - C_\eta h^{-\varkappa}$ with the constants

$$C_\eta^{\text{GS}} = \pi^2, \qquad C_\eta^{\text{blockGS}} = 2\pi^2, \qquad \varkappa^{[\text{block}]\text{GS}} = 2, \tag{5.7.5a}$$

$$C_\eta^{\text{SOR}} = 2\pi, \qquad C_\eta^{\text{blockSOR}} = 2\sqrt{2}\pi, \qquad \varkappa^{[\text{block}]\text{GS}} = 1 \tag{5.7.5b}$$

(cf. also (6.7c)). The assertion follows from representation (3.3.5d). □

The numbers in (4a–d) indicate, e.g., that the block-variants are more effective than the corresponding pointwise iterations. Although the SOR method is somewhat more expensive than the Gauß-Seidel iteration, the latter proves to be more effective than the SOR method already for $h \leqslant 0.7/\pi \approx 1/5$.

5.7.2 Gauß-Seidel Iteration: Numerical Examples

Table 1.4.1 contains the results of the lexicographical and chequer-board Gauß-Seidel method. After showing more favourable values in the beginning, the error reduction factors $\varepsilon_{m-1}/\varepsilon_m$ converge for both orderings to $\varrho(M^{GS}) = \cos^2 \pi/32 = 0.99039264$ (cf. (1a)).

In the following, let the block-variants be defined by means of the column-block structure. Table 1 contains the value of $x^m = u^m$ at the midpoint, the maximum norm $\varepsilon_m = \|u^m - u_h\|_\infty$, and the reduction factors $\varrho_{m,m-1} = \varepsilon_{m-1}/\varepsilon_m$, which in the examples approximate (almost monotonically increasing) the limit

$$\varrho(M^{\text{blockGS}} = 1 - 8\sin^2(\pi/64)/(1 + 2\sin^2(\pi/64))^2 \div 0.980923.$$

5.7.3 SOR Iteration: Numerical Examples

Table 1.4.2 contains the results of the SOR iteration for the relaxation parameter $\omega_{\text{opt}} = 2/(1 + \sin \pi h) = 1.821465$ that is optimal for $h = \frac{1}{32}$. The reduction factor that should converge to $\omega_{\text{opt}} - 1 = 0.821465$ behaves very irregularly. In particular, one observes the tendency that in the beginning, the reduction factors are distinctly worse than the asymptotical convergence rate. The same observation holds for the block-variants that are reported in Table 2 using the optimal relaxation parameter $\omega_{\text{opt}} = 1.7572848$. Because of the irregular behaviour of the factors $q_m := \varepsilon_{m-1}/\varepsilon_m$, an additional column

Table 5.7.1 Results of the block-Gauß-Seidel method for $N = 32$

m	lexicographical ordering			zebra ordering of the blocks		
	$u_{16,16}$	ε_m	factors	$u_{16,16}$	ε_m	factors
5	−0.01926	1.23834	0.939842	−0.01950	1.17160	0.958731
10	−0.03592	1.01501	0.965208	−0.03752	0.95064	0.968133
20	−0.04928	0.76180	0.974912	−0.04015	0.71340	0.976522
100	0.34781	0.15219	0.980968	0.36033	0.14097	0.980690
200	0.47781	0.02229	0.980934	0.47964	0.02046	0.980916
300	0.49677	0.00325	0.980924	0.49703	0.00298	0.980923

with the factors

$$\bar{q}_m := (\varepsilon_{m-10}/\varepsilon_m)^{1/10} = (q_{m-9}q_{m-8}\cdot\ldots\cdot q_m)^{1/10} \tag{5.7.6}$$

averaged over 10 values is presented in Table 2.

Since, initially, the convergence speed of the SOR method is slower, it is no contradiction when, as recommended in Meis–Marcowitz [1] and verified by examples, ω is chosen somewhat larger than ω_{opt} in order to reach a given error bound as soon as possible.

5.8 Supplementary Remarks

5.8.1 p-Cyclic Matrices

The property «weakly 2-cyclic» can be generalised. A is called «weakly p-cyclic» if a $p \times p$ block structure exists, so that only the blocks A^{1p}, A^{21}, $A^{32}, \ldots, A^{p,p-1}$ are nonvanishing. This case is discussed in detail by Varga [2, §4.2]. Under suitable further assumptions, the SOR method converges for

$$0 < \omega < \frac{p}{p-1}.$$

Optimal convergence holds for the unique positive root $\omega = \omega_{\text{opt}} < p/(p-1)$ of the polynomial

$$(p-1)^{p-1}\varrho(M^{\text{Jac}})^p\omega^p = p^p(\omega - 1).$$

The corresponding rate is $\varrho(M^{\text{SOR}}_{\omega_{\text{opt}}}) = (\omega_{\text{opt}} - 1)(p-1) < 1$. Also in this case, the order of linear convergence is halved (cf. Theorem 5.6.9). Compare also Eiermann–Niethammer–Ruttan [1].

Table 5.7.2 Results of the block-SOR method for $N = 32$ with $\omega = \omega_{\text{opt}}$

m	lexicographical ordering ε_m	q_m	$\bar{q}_m$	zebra ordering of the blocks ε_m	q_m	$\bar{q}_m$
10	0.6217327	0.9109	0.8954	0.2978516	0.7280	0.8318
20	0.2146420	0.8841	0.7142	0.0279097	0.7847	0.7892
30	0.0146717	0.4890	0.7647	0.0023936	0.7675	0.7822
40	0.0017416	0.8516	0.8081	0.0002034	0.7723	0.7815
50	0.0001095	0.7375	0.7583	0.0000144	0.8078	0.7672
60	0.0000119	0.7585	0.8007	$9.6527_{10}-7$	0.7777	0.7632
70	$6.4684_{10}-7$	0.7814	0.7477	$6.8937_{10}-8$	0.7684	0.7680
80	$5.6020_{10}-8$	0.8006	0.7830	$4.8121_{10}-9$	0.7545	0.7663
90	$3.5398_{10}-9$	0.7565	0.7587	$3.092_{10}-10$	0.7407	0.7600
100	$2.269_{10}-10$	0.7139	0.7598	$4.184_{10}-11$	—	—

5.8.2 Modified SOR

In the 2-cyclic case, one can regard the SOR method as a product iteration $\Phi_\omega^{\mathrm{SOR}} = \Phi_\omega^{(2)} \circ \Phi_\omega^{(1)}$, where $\Phi_\omega^{(1)}$ involves only the first block of the vector and $\Phi_\omega^{(2)}$ only the second one:

$$\Phi_\omega^{(1)}(x, b) = \begin{pmatrix} x^1 - \omega[x^1 - D_1^{-1}(A_1 x^2 - b^1)] \\ x^2 \end{pmatrix}, \tag{5.8.1a}$$

$$\Phi_\omega^{(2)}(x, b) = \begin{pmatrix} x^1 \\ x^2 - \omega[x^2 - D_2^{-1}(A_2 x^1 - b^2)] \end{pmatrix}, \tag{5.8.1b}$$

where $x = \begin{pmatrix} x^1 \\ x^2 \end{pmatrix}$, $b = \begin{pmatrix} b^1 \\ b^2 \end{pmatrix}$, and A is split as in (1.3b). The corresponding iteration matrices are

$$M_\omega^{(1)} = \begin{bmatrix} (1-\omega)I & \omega D_1^{-1} A_1 \\ 0 & I \end{bmatrix}, \qquad M_\omega^{(2)} = \begin{bmatrix} I & 0 \\ \omega D_2^{-1} A_2 & (1-\omega)I \end{bmatrix}. \tag{5.8.2}$$

Thus, we have $M_\omega^{\mathrm{SOR}} = M_\omega^{(2)} M_\omega^{(1)}$. The *modified SOR method* makes use of different relaxation parameters ω and ω' in both of the half-steps:

$$\Phi_{\omega,\omega'}^{\mathrm{modSOR}} := \Phi_{\omega'}^{(2)} \circ \Phi_\omega^{(1)}. \tag{5.8.3}$$

It is a trivial statement that the introduction of further parameters cannot result in the deterioration of the method: For the choice $\omega = \omega'$, the modified algorithm coincides with the original SOR method. However, one learns from the analysis of Young [2, §8] that hardly any improvement on the SOR iteration is achieved.

5.8.3 SSOR in the 2-Cyclic Case

In the 2-cyclic case, one can rewrite the backward SOR iteration as $\Phi_\omega^{\mathrm{rSOR}} = \Phi_\omega^{(1)} \circ \Phi_\omega^{(2)}$ with the partial steps defined in (1a, b). Therefore, the symmetric SOR iteration takes the form

$$\Phi_\omega^{\mathrm{SSOR}} = \Phi_\omega^{(1)} \circ \Phi_\omega^{(2)} \circ \Phi_\omega^{(2)} \circ \Phi_\omega^{(1)}. \tag{5.8.4}$$

Exercise 5.8.1. Prove that (a) the SSOR iteration matrix $M_\omega^{\mathrm{SSOR}} = M_\omega^{(1)} M_\omega^{(2)} M_\omega^{(2)} M_\omega^{(1)}$ leads to the rate

$$\varrho(M_\omega^{(1)} M_\omega^{(2)} M_\omega^{(2)} M_\omega^{(1)}) = \varrho(M_\omega^{(2)} M_\omega^{(2)} M_\omega^{(1)} M_\omega^{(1)}). \tag{5.8.5}$$

(b) $M_\omega^{(1)} M_\omega^{(1)} = M_{\omega'}^{(1)}$ and $M_\omega^{(2)} M_\omega^{(2)} = M_{\omega'}^{(2)}$ with $\omega' := \omega(2 - \omega)$.
(c) $0 < \omega < 2$ implies $0 < \omega' \leqslant 1$. $\omega' = 1$ holds only for $\omega = 1$.

Exercise 1 entails the following negative conclusion.

Remark 5.8.2. In the 2-cyclic case, $\varrho(M_\omega^{\mathrm{SSOR}}) = \varrho(M_{\omega'}^{\mathrm{SOR}})$ holds with $\omega' := \omega(2-\omega) \leqslant 1$ for all $0 < \omega < 2$. According to Theorem 6.5, underrelaxation ($\omega' < 1$) is always slower than the Gauß-Seidel method ($\omega' = 1$). Hence, $\omega = 1$ is the optimal parameter, for which the SSOR method simplifies to the symmetric Gauß-Seidel method (cf. Alefeld [2]).

The reason for the fact that, differently from the situation discussed in Remark 4.8.15, no order improvement can be achieved is that the condition $\varrho(D^{-1}ED^{-1}E^H) \leqslant 1/4$ is not satisfied. In the 2-cyclic case, we have $\varrho(D^{-1}ED^{-1}E^H) = \varrho(D_1^{-1}A_1D_2^{-1}A_2) = \varrho(M^{\mathrm{GS}}) \approx 1$ (cf. Theorem 5.1).

5.8.4 Unsymmetric SOR Method

In the following, A is not necessarily 2-cyclic. The only reason for mentioning the unsymmetric SOR method here is that it is constructed in analogy to the modified Gauß-Seidel method from §5.8.2. The SSOR method is the product $\Phi_\omega^{\mathrm{SSOR}} = \Phi_\omega^{\mathrm{bSOR}} \circ \Phi_\omega^{\mathrm{SOR}}$ (cf. (4.8.16)). One may choose different parameters in both factors. Accordingly, the unsymmetric SOR method reads

$$\Phi_{\omega,\omega'}^{\mathrm{unsymSOR}} := \Phi_{\omega'}^{\mathrm{bSOR}} \circ \Phi_\omega^{\mathrm{SOR}}. \tag{5.8.6}$$

Again, this method is not notably better than the SSOR method. In particular, the iteration matrix no longer has a real spectrum.

6
Analysis for M-Matrices

6.1 Positive Matrices

The positive definiteness $A > 0$ establishes a partial-order relation in the algebra of the Hermitian matrices. *Another* order relation in the algebra of all real matrices is defined by elementwise inequalities:

$$A > B :\Leftrightarrow a_{\alpha\beta} > b_{\alpha\beta} \quad \text{for all } \alpha, \beta \in I, \tag{6.1.1a}$$

$$A \geqslant B :\Leftrightarrow a_{\alpha\beta} \geqslant b_{\alpha\beta} \quad \text{for all } \alpha, \beta \in I. \tag{6.1.1b}$$

In this chapter, we use the signs «$>$» and «$\geqslant$» only in the componentwise sense (1a, b).

Remark 6.1.1. Since $A \geqslant B$ and $A \neq B$ do not imply $A > B$ $\left(\text{counterexample: } A = \begin{pmatrix} 0 & 1 \\ 0 & 0 \end{pmatrix}, B = \begin{pmatrix} 0 & 0 \\ 0 & 0 \end{pmatrix}\right)$, occasionally, the notation (1c) is used:

$$A \gneq B :\Leftrightarrow A \geqslant B \quad \text{and} \quad A \neq B. \tag{6.1.1c}$$

The *positive matrices* are characterised by

$$A > 0 \tag{6.1.2a}$$

(i.e., $a_{\alpha\beta} > 0$ for all $\alpha, \beta \in I$), the *nonnegative matrices* are defined by

$$A \geqslant 0. \tag{6.1.2b}$$

Remark 6.1.2. $A, B \geqslant 0$ yields $A + B \geqslant 0$ and $AB \geqslant 0$. $AB > 0$ follows from $A, B > 0$, whereas $A + B > 0$ holds if one of the matrices $A, B \geqslant 0$ is positive.

Analogous order relations can be defined for vectors:

$$x \geqslant y :\Leftrightarrow x_\alpha \geqslant y_\alpha \quad \text{for all } \alpha \in I,$$

and analogously $x > y$, $x \leqslant y$, $x < y$, $x \gneqq y$, etc.

Exercise 6.1.3. Prove: (a) $Ax \geqslant By$ for $A \geqslant B \geqslant 0$, $x \geqslant y \geqslant 0$. In particular, $Ax \geqslant 0$ holds for $A \geqslant 0$ and $x \geqslant 0$.
(b) A is positive if and only if $Ax > 0$ for all $x \gneqq 0$.

The «absolute value» of a matrix (or a vector) is again a matrix (vector) and is defined componentwise (do not confuse with a norm!):

$$|A| := (|a_{\alpha\beta}|)_{\alpha,\beta \in I} \in \mathbb{R}^{I \times I}, \tag{6.1.3a}$$

$$|x| := (|x_\alpha|)_{\alpha \in I} \in \mathbb{R}^I. \tag{6.1.3b}$$

The established order relations fit particularly well the maximum norm or row-sum norm, respectively:

Exercise 6.1.4. Prove that

$$\|x\|_\infty = \||x|\|_\infty, \qquad x \geqslant y \geqslant 0 \Rightarrow \|x\|_\infty \geqslant \|y\|_\infty \qquad (x, y \in \mathbb{K}^I), \tag{6.1.4a}$$

$$\|A\|_\infty = \||A|\|_\infty, \qquad A \geqslant B \geqslant 0 \Rightarrow \|A\|_\infty \geqslant \|B\|_\infty \qquad (A, B \in \mathbb{K}^{I \times I}). \tag{6.1.4b}$$

Many properties of positive matrices correspond to those of positive definite matrices. However, in contrast to (2.10.3f), we obtain

Exercise 6.1.5. Let $\#I > 1$. A and A^{-1} cannot be positive simultaneously.

6.2 Graph of a Matrix and Irreducible Matrices

Definition 6.2.1. Let $A \in \mathbb{K}^{I \times I}$ be a matrix corresponding to the index set I. The following subset of all pairs from $I \times I$ is denoted as *graph* $G(A)$ of the matrix A:

$$G(A) = \{(\alpha, \beta) \in I \times I \colon a_{\alpha\beta} \neq 0\}. \tag{6.2.1}$$

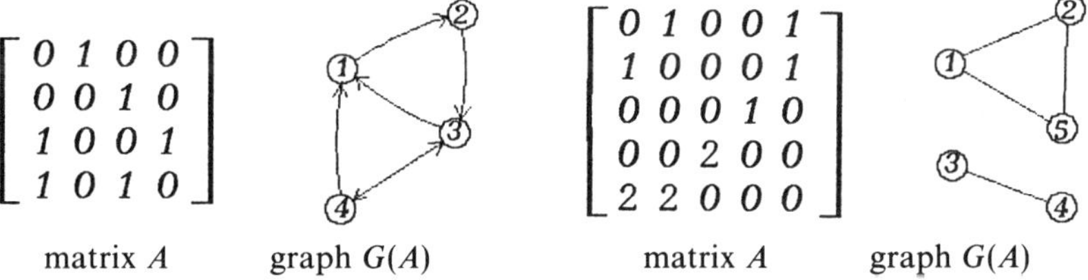

Fig. 6.2.1 directed graph **Fig. 6.2.2** symmetric structure

The set $G(A)$ can be visualised as follows. The indices $\alpha \in I$ are called *knots*, whereas $(\alpha, \beta) \in G(A)$ is called a (directed) *edge* from knot α to knot β and is graphically represented by an arrow pointing from α to β (cf. Figure 1). In the case of $a_{\alpha\beta} \neq 0$ and $a_{\beta\alpha} \neq 0$, the knots α and β are connected in both directions (cf. Figure 1: $\alpha = 3$, $\beta = 4$). If $a_{\alpha\beta} \neq 0$ holds always together with $a_{\beta\alpha} \neq 0$, the matrix A has a *symmetric structure*. In this case, all edges are bidirected and one can omit the specification of the directions (cf. Figure 2). The graph $G(A)$ is minimal for the zero matrix: $G(O) = \phi$ and maximal for full matrices: $G(A) = I \times I$.

If there is an edge $(\alpha, \beta) \in G(A)$, we say that «*α is directly connected to β*». The statement «*α is connected to β*» means that there is a chain of direct connections:

$$\alpha = \alpha_0, \alpha_1, \alpha_2, \ldots, \alpha_{k-1}, \alpha_k = \beta$$
$$\text{with } k \in \mathbb{N} \text{ and } (\alpha_{i-1}, \alpha_i) \in G(A) \text{ for all } i = 1, 2, \ldots, k. \tag{6.2.2}$$

Exercise 6.2.2. Prove: (a) The relation «α is connected to β» is transitive but not necessarily symmetric.
(b) Let $n := \#I$. If α is connected to β, there always exists a chain (2) of the length $k < n$.

Definition 6.2.3. A matrix $A \in \mathbb{K}^{I \times I}$ is called *irreducible* if any $\alpha \in I$ is connected to any $\beta \in I$. Otherwise, A is called *reducible*.

Remark 6.2.4. A is reducible if and only if there is an ordering of the indices such that A takes the block form

$$A = \underbrace{\begin{array}{|c|c|}\hline A^{11} & A^{12} \\ \hline 0 & A^{22} \\ \hline \end{array}}_{I_1 \quad I_2} \begin{array}{l} \}I_1 \\ \}I_2 \end{array} \qquad (A^{11}, A^{22}\text{: square blocks!}) \tag{6.2.3}$$

with nonempty index subsets $I_1, I_2 \subset I$.

Proof. (i) Let A have the form (3). Choose $\alpha \in I_2$, $\beta \in I_1$. Assume that a chain (2) exists connecting α to β. Then there must be an edge $(\alpha_{\ell-1}, \alpha_\ell) \in G(A)$ with $\alpha_{\ell-1} \in I_2$ and $\alpha_\ell \in I_1$. The contradiction results from $a_{\alpha_{\ell-1}, \alpha_\ell} = 0$, because this entry belongs to the block $A^{21} = 0$. Hence, α is not connected to β and therefore A is reducible.
(ii) Let A be reducible. Then there must be indices $\alpha, \beta \in I$, such that α is not connected to β. Choose

$$I_1 := \{\gamma \in I \colon \gamma \text{ connected to } \beta\}, \qquad I_2 := I \backslash I_1.$$

The sets are nonempty, since $\beta \in I_1$, $\alpha \in I_2$. Enumerate first I_1, then I_2. An entry $a_{\delta\gamma}$ from the block A^{21} has the indices $\delta \in I_2$, $\gamma \in I_1$. $a_{\delta\gamma} = 0$ must hold, since otherwise, δ is connected to γ, which by the definition of I_1 is connected to β; hence, $\delta \in I_1$ would follow in contradiction to $\delta \in I_2$. □

Exercise 6.2.5. Prove: (a) The matrix A from Figure 1 is irreducible, but the matrix from Figure 2 is not.
(b) If A has a symmetric structure, A is irreducible if and only if the graph $G(A)$ is connected.
(c) Triangular and diagonal matrices with $\# I > 1$ are reducible.
(d) The matrix of the Poisson-model problem is irreducible.

If the matrix of a linear system of equations is reducible, the system allows a *reduction*: If A is reducible and $\{I_1, I_2\}$ describes the block structure from (3), $Ax = b$ can be solved in two steps via smaller systems of equations:

$$A^{22}x^2 = b^2, \qquad A^{11}x^1 = b^1 - A^{12}x^2.$$

Exercise 6.2.6. Prove that $G(\alpha A + \beta B) = G(A) \cup G(B)$ for nonnegative matrices $A, B \geqslant 0$ and positive numbers α, β.

Remark 6.2.7. Define $G_k(A) := \{(\alpha, \beta) \in I \times I$: there is a chain (2) of the length $\leqslant k$ connecting α to $\beta\}$. Further let $n := \# I$. Then the following statements are valid:
(a) $G(A) = G_1(A) \subset G_2(A) \subset \cdots \subset G_{k-1}(A) \subset G_k(A) \subset \cdots$
(b) $G_k(A) = G_{n-1}(A)$ for all $k \geqslant n - 1$.
(c) A is irreducible if and only if $G_{n-1}(A) = I \times I$.
(d) $A \geqslant 0$ implies that $G_k(A) \subset G([I + A]^k)$.
(e) If $A \geqslant 0$ is irreducible, then $(I + A)^{n-1} > 0$ and $\sum_{\nu=0}^{n-1} A^\nu > 0$.

Proof. (a) A chain (2) of length $k = 1$ is a direct connection, i.e., an edge from $G(A)$. Vice versa, any edge from $G(A)$ belongs to $G_1(A)$; hence, $G(A) = G_1(A)$.
(b) It follows according to Exercise 2b.
(c) If α is connected to β, then according to (a) and (b), (α, β) must belong to $G_{n-1}(A)$. This proves assertion (c).
(d) Since only nonnegative matrices $A \geqslant 0$ appear in (d) and (e), the condition $a_{\alpha\beta} \neq 0$ from Definition 1 can be replaced by $a_{\alpha\beta} > 0$. Let $A' := I + A$ and $(\alpha_0, \alpha_k) \in G_k(A)$, i.e., there is a chain of direct connections $(\alpha_{\ell-1}, \alpha_\ell) \in G(A)$ for $1 \leqslant \ell \leqslant k$. Because of $A' \geqslant 0$, the coefficient

$$(A'^k)_{\alpha_0, \alpha_k} = \sum_{\beta_1, \ldots, \beta_{k-1} \in I} a'_{\alpha_0\beta_1} a'_{\beta_1\beta_2} \cdot \ldots \cdot a'_{\beta_{k-1}\alpha_k} \tag{6.2.4a}$$

of the matrix A'^k can be estimated by

$$(A'^k)_{\alpha_0, \alpha_k} \geqslant a'_{\alpha_0\alpha_1} a'_{\alpha_1\alpha_2} \cdot \ldots \cdot a'_{\alpha_{k-1}\alpha_k}. \tag{6.2.4b}$$

For $\alpha_{\ell-1} \neq \alpha_\ell$, we conclude $a'_{\alpha_{\ell-1}\alpha_\ell} > 0$ from $(\alpha_{\ell-1}, \alpha_\ell) \in G(A)$, whereas for $\alpha_{\ell-1} = \alpha_\ell$, the diagonal entry $a'_{\alpha_\ell\alpha_\ell} = 1 + a_{\alpha_\ell\alpha_\ell} \geqslant 1 > 0$ is positive also. Hence,

all factors $a'_{\alpha_{l-1}\alpha_l}$ appearing in (4b) are positive. This proves $(A'^k)_{\alpha_0,\alpha_k} > 0$, $(\alpha_0, \alpha_k) \in G(A'^k)$, and finally $G_k(A) \subset G(A'^k)$.
(e) Obviously, $B := (I + A)^{n-1} \geqslant 0$ holds (cf. Remark 1.2). An irreducible A satisfies

$$I \times I \underset{(c)}{=} G_{n-1}(A) \underset{(d)}{\subset} G((I + A)^{n-1}) = G(B).$$

Hence, $(\alpha, \beta) \in G(B)$ is always true, i.e., $B_{\alpha\beta} > 0$ holds. This proves $B > 0$. The case of $\sum A^\nu$ is analogous. □

6.3 Perron-Frobenius Theory of Positive Matrices

Theorem 6.3.1 (Perron [1], Frobenius [1]). *Let $n := \# I > 1$ and $A \geqslant 0$ be an irreducible matrix from $\mathbb{R}^{I \times I}$. Then the following statements hold:*

$$\varrho(A) > 0 \text{ is a simple eigenvalue of } A, \tag{6.3.1a}$$

$$\lambda = \varrho(A) \text{ is associated with a positive eigenvector } x > 0, \tag{6.3.1b}$$

$$\varrho(B) > \varrho(A) \quad \text{for all } B \underset{\neq}{\geqslant} A. \tag{6.3.1c}$$

The proof of this main theorem is prepared by Lemmata 2–6. We start with some auxiliary constructions. The set

$$E := \{x \in \mathbb{R}^I: \|x\|_\infty = 1, x \geqslant 0\}$$

consists of vectors with $0 \leqslant x_\beta \leqslant 1$ and at least one component $x_\alpha = 1$.

Lemma 6.3.2. *Assume that $A \geqslant 0$. The set*

$$K := \{(x, \varrho) \in E \times \mathbb{R}: \varrho \geqslant 0, Ax \geqslant \varrho x\} \tag{6.3.2a}$$

is compact (i.e., closed and bounded). The maximum

$$r := \max\{\varrho: (x, \varrho) \in K \text{ for some } x \in E\} \tag{6.3.2b}$$

is taken. For any pair $(y, r) \in K$, we have

$$Ay \geqslant ry \quad \text{and not } Ay > ry. \tag{6.3.2c}$$

Proof. (i) Let $(x_\nu, \varrho_\nu) \in K$ have the limit (x, ϱ). Then we conclude from $Ax_\nu \geqslant \varrho_\nu x_\nu$ that $Ax \geqslant \varrho x$. Therefore, $(x, \varrho) \in K$ proves that K is closed.
(ii) The boundedness of x is trivial because of $\|x\|_\infty = 1$. The component ϱ of $(x, \varrho) \in K$ is bounded by

$$0 \leqslant \varrho \leqslant \|A\|_\infty, \tag{6.3.2d}$$

because the index $\alpha \in I$ with $x_\alpha = 1$ satisfies $\varrho = \varrho x_\alpha \leqslant (Ax)_\alpha \leqslant \|Ax\|_\infty \leqslant \|A\|_\infty \|x\|_\infty \leqslant \|A\|_\infty$. This completes the proof of K being compact.

(iii) Let r be the supremum of $\{\varrho\colon (x,\varrho) \in K$ for some $x \in E\}$. There are $(x_\nu, \varrho_\nu) \in K$ with $\varrho_\nu \to r$. Since K is compact, a subsequence converges to $(y, r) \in K$. By definition of K, the inequality $Ay \geqslant ry$ must hold. If $Ay > ry$, r could be increased in contradiction to the maximality of r. □

Lemma 6.3.3. *Assume that $A \geqslant 0$ is irreducible with $n := \#I > 1$. Let r be defined according to* (2b) *and assume that $y \in E$ satisfies* (2c). *Then*

$$r > 0, \qquad y > 0, \qquad Ay = ry, \tag{6.3.3}$$

i.e., y is a positive eigenvector of A corresponding to the positive eigenvalue r.

Proof. (i) The residual vector $z := Ay - ry$ is nonnegative because of (2c). Under the assumption $z \neq 0$, Remark 2.7e yields $(I + A)^{n-1}z > 0$ and therefore,

$$\begin{aligned} 0 < (I + A)^{n-1}z &= (I + A)^{n-1}(Ay - ry) \\ &= (I + A)^{n-1}(A - rI)y \\ &= (A - rI)(I + A)^{n-1}y \\ &= Ay' - ry' \quad \text{for } y' := (I + A)^{n-1}y. \end{aligned} \tag{6.3.4}$$

From $y \geqslant 0$ one concludes again that $y' = (I + A)^{n-1}y > 0$. The normalised vector $y'' := y'/\|y'\|_\infty$ belongs to E. $Ay' > ry'$ implies $(y'', r) \in K$ and $Ay'' > ry''$, which contradicts (2c). Hence, the assumption $z \neq 0$ is not valid. Thus, $z = 0$ proves $Ay = ry$.
(ii) In (i) we have already used $(I + A)^{n-1}y > 0$. Therefore, the eigenvalue equation $Ay = ry$ yields $(1 + r)^{n-1}y > 0$. By $1 + r \geqslant 1 > 0$, $y > 0$ follows.
(iii) If $r = 0$, $Ay = ry = 0$ would follow. From $Ay = 0$ and $y > 0$, one concludes $A = 0$. By $n > 1$, A would be reducible. Hence, $r > 0$ must hold. □

Lemma 6.3.4. *Assume that A is irreducible and $|B| \leqslant A$. Then we have*

$$\varrho(B) \leqslant r \qquad (r \text{ according to (2b)}), \tag{6.3.5a}$$

$$\varrho(B) = r \Leftrightarrow |B| = A, \qquad B = \omega DAD^{-1}, \qquad |D| = I, \qquad |\omega| = 1. \tag{6.3.5b}$$

Proof. (i) Let y be the normalised eigenvector corresponding to $\beta \in \sigma(B)$: $By = \beta y$, $\|y\|_\infty = 1$. By

$$|\beta||y| = |\beta y| = |By| \leqslant |B||y| \leqslant A|y|,$$

$(|y|, |\beta|)$ belongs to K and proves that $|\beta| \leqslant r$. Since $\beta \in \sigma(B)$ is arbitrary, (5a): $\varrho(B) \leqslant r$ is shown.
(ii) Let $|\beta| = r$. The vector y from (i) satisfies $(|y|, r) \in K$. By Lemma 3, $|y| > 0$ is an eigenvector of A: $A|y| = r|y|$. The inequality

$$r|y| = |\beta||y| \leqslant |B||y| \leqslant A|y| = r|y|$$

implies that $|B||y| = A|y|$. Since $|y| > 0$ and $|B| \leqslant A$, $|B| = A$ follows. The definition $D := \operatorname{diag}\{y_\alpha/|y_\alpha|: \alpha \in I\}$ makes sense because of $|y| > 0$ and leads to $D|y| = y$. Define $\omega := \beta/r$ ($r > 0$ from Lemma 3). The conditions $|D| = I$ and $|\omega| = 1$ are satisfied. The eigenvalue equation $By = \beta y$ becomes

$$\frac{1}{\omega} D^{-1} B D|y| = r|y|.$$

The matrix $C := \frac{1}{\omega} D^{-1} B D$ fulfils $|C| = |B| = A$ and $C|y| = r|y| = A|y| = |C||y|$. $|y| > 0$ implies that $C = |C| = A$. This proves the direction «⇒» in (5b).

(iii) Now let the right-hand part of (5b) be valid. Then B has an eigenvalue $\beta = \omega r$ proving $|\beta| = r$ and, by part (i), also $\varrho(B) = r$. □

Lemma 6.3.5. *$r = \varrho(A)$ holds for any irreducible matrix $A \geqslant 0$.*

Proof. The right-hand part of (5b) is satisfied for $B := A$ with $D = I$ and $\omega = 1$. Hence, $r = \varrho(B) = \varrho(A)$ holds. □

Lemma 6.3.6. *Let $A \geqslant 0$ be irreducible and B a proper principal submatrix of A, i.e., $B = (a_{\alpha\beta})_{\alpha,\beta \in I'}$ for a nonempty index subset $I' \subsetneqq I$. Then $\varrho(B) < \varrho(A)$ holds.*

Proof. The matrix $B' := (b'_{\alpha\beta})_{\alpha,\beta \in I}$ with $b'_{\alpha\beta} = \begin{cases} a_{\alpha\beta} = b_{\alpha\beta} & \text{for } \alpha, \beta \in I' \\ 0 & \text{otherwise} \end{cases}$ is the block-diagonal matrix $\operatorname{blockdiag}(B, 0)$ with respect to the block structure $\{I', I \backslash I'\}$. The identity $\sigma(B') = \sigma(B) \cup \{0\}$ proves that $\varrho(B') = \varrho(B)$. Obviously, $|B'| = B' \leqslant A$ is valid. Since B' is reducible, the right-hand side in (5b) cannot be satisfied for B' and A; hence, $\varrho(B) = \varrho(B') < \varrho(A)$ follows. □

Proof of Theorem 1. (i) Lemma 5 shows that $r = \varrho(A)$, whereas Lemma 3 proves that $r = \varrho(A) > 0$ is an eigenvalue with a positive eigenvector.

(ii) If $B \gneqq A$, the irreducibility of A carries over to B because of $G(B) \supset G(A)$. Since $A = |A| \lneqq B$, one deduces from Lemma 4 with the interchanged rôles of A and B that $\varrho(A) < r_B$, where $r_B = \varrho(B)$ is the value r from (2b) belonging to B.

(iii) It remains to show (1a): $\lambda = \varrho(A)$ is a simple eigenvalue. Let A_γ for $\gamma \in I$ be the principal submatrices associated with the index set $I_\gamma := I \backslash \{\gamma\}$. The derivative of the determinant of $\lambda I - A$ equals

$$\frac{d}{d\lambda} \det(\lambda I - A) = \sum_{\gamma \in I} \det(\lambda I - A_\gamma). \tag{6.3.6}$$

Since $\varrho(A_\gamma) < \varrho(A)$ by Lemma 6, we have $\det(\lambda I - A_\gamma) \neq 0$ for all $\lambda \geqslant \varrho(A)$. The polynomial $\det(\lambda I - A_\gamma) = \lambda^{n-1} + \cdots$ tends to $+\infty$ for $\lambda \to \infty$. Hence, it must be positive in the interval $[\varrho(A), \infty)$. From $\det(\lambda I - A_\gamma) > 0$ and (6), one concludes that

$$\frac{d}{d\lambda}\det(\lambda I - A) > 0 \quad \text{for } \lambda \geqslant \varrho(A).$$

A double zero of $\det(\lambda I - A)$ at $\lambda = \varrho(A)$ would lead to a vanishing derivative; hence, $\lambda = \varrho(A)$ is only a simple root and thereby also a simple eigenvalue. □

Exercise 6.3.7. Prove that the eigenvalue $\lambda = \varrho(A)$ of an irreducible matrix $A \geqslant 0$ is the only one with the property $|\lambda| = \varrho(A)$. *Hint*: Prove that the eigenvector x belonging to λ with $|\lambda| = \varrho(A)$ yields a vector $y := |x|$ satisfying (2c). Apply Lemma 3.

Exercise 6.3.8. Prove that if $x > 0$ is the eigenvector of an irreducible matrix $A \geqslant 0$, then it belongs to the eigenvalue $\lambda = \varrho(A)$.

The irreducibility of A required in Theorem 1 is, in particular, ensured for positive $A > 0$. However, if one also acknowledges the reducible matrices $A \geqslant 0$, not all statements of the theorem remain valid.

Exercise 6.3.9. Prove that there are reducible matrices $A \geqslant 0$, such that $\varrho(A)$ is a multiple eigenvalue and the corresponding eigenvectors $x \geqslant 0$ have components $x_\alpha = 0$.

The properties remaining for possibly reducible matrices $A \geqslant 0$ are summarised in

Theorem 6.3.10. *Let* $A \geqslant 0$. *Then* (7a–c) *holds*:

$$0 \leqslant \varrho(A) \text{ is an eigenvalue of } A\text{: } \varrho(A) \in \sigma(A), \tag{6.3.7a}$$

$$\lambda = \varrho(A) \text{ corresponds to a nonnegative eigenvector } x \gneqq 0, \tag{6.3.7b}$$

$$\varrho(B) \geqslant \varrho(A) \quad \text{for all } B \geqslant A. \tag{6.3.7c}$$

Proof. (i) Since the case $n := \#I = 1$ is trivial, assume that $n > 1$. We define $A_\varepsilon := (a_{\alpha\beta} + \varepsilon)_{\alpha,\beta\in I}$ for $\varepsilon > 0$. A_ε is irreducible because of $G(A_\varepsilon) = I \times I$. By Theorem 1, $\lambda_\varepsilon = \varrho(A_\varepsilon)$ is an eigenvalue of A_ε with the eigenvector $x_\varepsilon > 0$, $\|x_\varepsilon\|_\infty = 1$. Since the eigenvalues (as zeros of a polynomial) vary continuously on A_ε, $\lambda := \lim_{\varepsilon\to 0} \lambda_\varepsilon = \lim_{\varepsilon\to 0} \varrho(A_\varepsilon) = \varrho(A)$ is an eigenvalue of A. The compactness of $\{x\colon \|x\|_\infty = 1\}$ implies the existence of a convergent subsequence $x_{\varepsilon_\nu} \to x$ with $\|x\|_\infty = 1$ and $x \geqslant 0$. $A_{\varepsilon_\nu} x_{\varepsilon_\nu} = \lambda_{\varepsilon_\nu} x_{\varepsilon_\nu}$ yields $Ax = \lambda x$, i.e., $x \geqslant 0$ is an eigenvector.
(ii) In analogy to A_ε, we define $B_{2\varepsilon}$. From $B_{2\varepsilon} \gneqq A_\varepsilon$ and $\varrho(B_{2\varepsilon}) \geqslant \varrho(A_\varepsilon)$, one concludes $\varrho(B) \geqslant \varrho(A)$ for $\varepsilon \to 0$. □

Exercise 6.3.11. Prove $\varrho(B) \leqslant \varrho(A)$ for all $|B| \leqslant A \in \mathbb{R}^{I\times I}$. *Hint*: Perform the limit $A_\varepsilon \to A$ in Lemmata 4 and 5.

6.4 M-Matrices

6.4.1 Definition

Definition 6.4.1. A matrix $A \in \mathbb{R}^{I \times I}$ is called an *M-matrix* if

$$a_{\alpha\alpha} > 0 \quad \text{for all } \alpha \in I, \tag{6.4.1a}$$

$$a_{\alpha\beta} \leqslant 0 \quad \text{for all } \alpha \neq \beta, \tag{6.4.1b}$$

$$A \text{ regular and } A^{-1} \geqslant 0. \tag{6.4.1c}$$

The properties (1a, b) are easy to check. Theorem 4 will show that condition (1a) can be omitted since it follows from (1b, c). The verification of $A^{-1} \geqslant 0$ is more difficult. For this purpose we will provide additional criteria. Matrices with the property (1c) are called *inverse positive*. Hence, M-matrices form a subclass of the inverse positive matrices. The name «M-matrix» was introduced by Ostrowski [1] in 1937 as an abbreviation for «Minkowskische Determinante».

Exercise 6.4.2. Assume (1c) and prove that if $b \leqslant b'$ holds for the right-hand sides of Ax = b and $Ax' = b'$, then $x \leqslant x'$ holds also.

Exercise 6.4.3. Prove that in general, the product $A = A_1 A_2$ of two M-matrices A_1 and A_2 is no more an M-matrix, although A is always inverse positive. *Hint*: Take tridiagonal matrices as an example.

6.4.2 Connection Between M-Matrices and the Jacobi Iteration

Theorem 6.4.4. *Let* $A \in \mathbb{R}^{I \times I}$ *satisfy* (1b): $a_{\alpha\beta} \leqslant 0$ *for all* $\alpha \neq \beta$. $D = \operatorname{diag}\{a_{\alpha\alpha}: \alpha \in I\}$ *denotes the diagonal of* A. (a) *Then*

$$A \text{ regular and } A^{-1} \geqslant 0 \tag{6.4.2a}$$

is equivalent to

$$a_{\alpha\alpha} > 0 \quad \text{for all } \alpha \in I, \tag{6.4.2b$_1$}$$

$$M := I - D^{-1}A \geqslant 0. \tag{6.4.2b$_2$}$$

$$\varrho(M) < 1. \tag{6.4.2b$_3$}$$

(b) *In the case of* (2a) *or* (2b$_{1-3}$), *A is an M-matrix. Vice versa,* (2b$_{2,3}$) *holds for any M-matrix.*

Since M from ($2b_2$) *coincides with the iteration matrix M^{Jac} of the pointwise Jacobi iteraton* (*cf.* (4.2.5a)), ($2b_3$) *describes the convergence of the Jacobi method.*

Proof. (i) First, we show that «(2a) $\Rightarrow$ ($2b_{1-3}$)». Let s^γ be the column of the matrix A corresponding to $\gamma \in I$. $A^{-1}A = I$ explains that $A^{-1}s^\gamma = e^\gamma :=$ unit vector of index $\gamma \in I$. If $a_{\gamma\gamma} \leqslant 0$, $s^\gamma \leqslant 0$ and (2a) would yield the contradiction $e^\gamma = A^{-1}s^\gamma \leqslant 0$. Hence, ($2b_1$) is shown. By ($2b_1$), $D \geqslant 0$ is regular. For $A' := D^{-1}A$, one finds the nonnegative inverse $A'^{-1} = A^{-1}D$. $M = I - A'$ has the diagonal entries $M_{\alpha\alpha} = 1 - 1 = 0$ and nonnegative off-diagonal entries $M_{\alpha\beta} = 0 - a_{\alpha\alpha}^{-1}a_{\alpha\beta} \geqslant 0$ for $\alpha \neq \beta$; hence, ($2b_2$): $M \geqslant 0$ is proved. By Theorem 3.10, $\lambda := \varrho(M) \in \sigma(M)$ belongs to an eigenvector $x \gneqq 0$. $Mx = \lambda x$ leads to $A'^{-1}(1 - \lambda)x = x$. Since $A'^{-1} \geqslant 0$ is regular and $x \gneqq 0$ holds, inequality $1 - \lambda > 0$ must hold, implying $0 \leqslant \varrho(M) = \lambda < 1$ and ($2b_3$).
(ii) For proving «($2b_{1-3}$) $\Rightarrow$ (2a)», we apply Theorem 2.9.10: Since $\varrho(M) < 1$, $(I - M)^{-1} = \sum M^\nu$ converges and is nonnegative because $M \geqslant 0$. $0 \leqslant (I - M)^{-1}D^{-1} = (D^{-1}A)^{-1}D^{-1} = A^{-1}DD^{-1} = A^{-1}$ proves (2a).
(iii) In the case of (2a) or ($2b_{1-3}$), both properties hold. The M-matrix properties are contained in ($2b_{1,2}$) and (2a).

Corollary 6.4.5. *Let $A \in \mathbb{R}^{I \times I}$ satisfy* (1b): *$a_{\alpha\beta} \leqslant 0$ for all $\alpha \neq \beta$. Define D and M as in Theorem* 4. *Then the following statements* (3a) *and* (3b) *are equivalent*:

$$A \text{ regular and } A^{-1} > 0, \tag{6.4.3a}$$

$$a_{\alpha\alpha} > 0\ (\alpha \in I),\ M \geqslant 0,\ \varrho(M) < 1,\ M \text{ irreducible}. \tag{6.4.3b}$$

Proof. «(3a) $\Rightarrow$ (3b)»: If A were reducible, there would be a block structure $\{I_1, I_2\}$ with $A^{21} = 0$. The inverse $C := A^{-1}$ would have the blocks $C^{ii} = (A^{ii})^{-1}$, $C^{12} = -(A^{11})^{-1}A^{12}(A^{22})^{-1}$ and, in particular, $C^{21} = 0$ in contradiction to $A^{-1} > 0$. Hence, A is irreducible. Since $G(A)$ and $G(M)$ coincide up to the diagonal pairs, M is also irreducible.
«(3b) $\Rightarrow$ (3a)»: Following part (ii), we have $A^{-1} = (\sum M^\nu)D^{-1}$. This proves $A^{-1} > 0$, since, by Remark 2.7e, $\sum M^\nu$ is positive. □

One learns from Theorem 4 that the first condition (1a): $a_{\alpha\alpha} > 0$ in the Definition 1 of the M-matrix can be omitted, since it follows necessarily from (1b, c). The following property is the discrete analogue of the maximum principle of second-order elliptic differential equations (cf. Hackbusch [15, Theorem 2.3.3]).

Exercise 6.4.6. Prove: (a) Irreducible M-matrices have a positive inverse.
(b) A regular matrix with (1b) and $\sum_\beta a_{\alpha\beta} \geqslant 0$ for all $\alpha \in I$ is an M-matrix.
(c) For any M-matrix A, there is $A' := \Delta^{-1}A\Delta$ with a diagonal matrix $\Delta \geqslant 0$ such that the inequality from (b): $\sum_\beta a'_{\alpha\beta} \geqslant 0$ holds for all $\alpha \in I$.

6.4.3 Diagonal Dominance

Definition 6.4.7. A matrix $A \in \mathbb{K}^{I \times I}$ is *strictly diagonally dominant* if

$$|a_{\alpha\alpha}| > \sum_{\substack{\beta \in I \\ \beta \neq \alpha}} |a_{\alpha\beta}| \quad \text{for all } \alpha \in I, \tag{6.4.4}$$

weakly diagonally dominant if

$$|a_{\alpha\alpha}| \geqslant \sum_{\substack{\beta \in I \\ \beta \neq \alpha}} |a_{\alpha\beta}| \quad \text{for all } \alpha \in I \tag{6.4.5}$$

and *irreducibly diagonally dominant* if A is an irreducible and weakly diagonally dominant matrix and if, furthermore, (6) holds:

$$|a_{\alpha\alpha}| > \sum_{\substack{\beta \in I \\ \beta \neq \alpha}} |a_{\alpha\beta}| \quad \text{for at least one } \alpha \in I. \tag{6.4.6}$$

If A is not irreducible, the following generalisation may help.

Definition 6.4.8. For $A \in \mathbb{K}^{I \times I}$ and $\gamma \in I$, define $G_\gamma := \{\beta \in I: \gamma \text{ connected to } \beta$ in the matrix graph $G(A)\}$. Then we call A *essentially diagonally dominant* if A is weakly diagonally dominant and if for all $\gamma \in I$ condition (7) applies:

$$|a_{\alpha\alpha}| > \sum_{\substack{\beta \in I \\ \beta \neq \alpha}} |a_{\alpha\beta}| \quad \text{for at least one } \alpha \in G_\gamma. \tag{6.4.7}$$

Exercise 6.4.9. Prove: (a) For irreducible matrices, the essential and irreducible diagonal dominance are equivalent.
(b) The implications «strictly diagonally dominant $\Rightarrow$ essentially diagonally dominant $\Rightarrow$ weakly diagonally dominant» hold.
(c) If A is strictly, irreducibly, or essentially diagonally dominant, the diagonal elements do not vanish: $a_{\alpha\alpha} \neq 0$.
(d) The matrix of the model problem from §1.2 is irreducibly diagonally dominant, but not strictly diagonally dominant.

The following theorem shows that the diagonal dominance of a matrix together with the sign conditions (1a, b) is sufficient for the M-matrix property. Usually, the theorem is proved by means of Gerschgorin circles (cf. Hackbusch [15, Criterion 4.3.4]). Here, however, we use the results from §6.3.

Theorem 6.4.10. (a) *Let the matrix $A \in \mathbb{K}^{I \times I}$ be strictly or essentially or irreducibly diagonally dominant. Then the Jacobi iteration matrix $M := I - D^{-1}A$ (D: diagonal of A) satisfies*

$$\varrho(M) < 1. \tag{6.4.8}$$

(b) *If furthermore, the sign conditions* (1a, b) *are satisfied, A is an M-matrix.*

Proof. (i) By Exercise 9c, D is regular. Hence, M is well-defined. $M' := |M|$ has the entries

$$M'_{\alpha\beta} = 0 \quad \text{for } \alpha = \beta, \qquad M'_{\alpha\beta} = |a_{\alpha\beta}/a_{\alpha\alpha}| \quad \text{for } \alpha \neq \beta.$$

In part (ii) we will show that $\varrho(M') < 1$. By Exercise 3.11, $\varrho(M) < 1$ also holds and proves (8). If furthermore, the conditions (1a, b) are satisfied, M fulfils condition $(2b_2)$: $M \geqslant 0$. Since $(2b_{1,3})$ are also vaid, Theorem 4b shows that A is an M-matrix.

(ii) By construction, $M' \geqslant 0$ holds. Hence, an eigenvector $x \geqslant 0$ with $\|x\|_\infty = 1$ belongs to $\lambda := \varrho(M') \in \sigma(M')$. Let $\alpha \in I$ be an index with $x_\alpha = 1$. We want to show that $\lambda < 1$ or $x_\gamma = 1$ for all $\gamma \in G_\alpha$, i.e., for all α connected to γ. Obviously, for an inductive proof it suffices to show this assertion for those γ that are *directly* connected to α, i.e., for γ with $(\alpha, \gamma) \in G(M')$. Because of weak diagonal dominance,

$$\lambda = \lambda x_\alpha = (M'x)_\alpha = \left(\sum_{\beta \neq \alpha} |a_{\alpha\beta}| x_\beta\right) \Big/ |a_{\alpha\alpha}| \leqslant \left(\sum_{\beta \neq \alpha} |a_{\alpha\beta}|\right) \Big/ |a_{\alpha\alpha}| \leqslant 1$$

holds. The equality $\lambda = 1$ can be valid only if $x_\beta = 1$ for all β with $a_{\alpha\beta} \neq 0$, i.e., for all β with $(\alpha, \beta) \in G(A) \supset G(M')$. This completes the induction proof. If $\lambda < 1$, $\varrho(M') < 1$ is shown. Otherwise, $x_\gamma = 1$ must hold for all $\gamma \in G_\alpha$. By Exercise 9b, A is essentially diagonally dominant. According to this definition, there is an index $\gamma \in G_\alpha$ with $|a_{\gamma\gamma}| > \sum_{\beta \neq \gamma} |a_{\gamma\beta}|$. Since $x_\gamma = x_\beta = 1$ for γ, $\beta \in G_\alpha$,

$$\lambda = \lambda x_\gamma = (M'x)_\gamma = \left(\sum_{\beta \neq \gamma} |a_{\gamma\beta}| x_\beta\right) \Big/ |a_{\gamma\gamma}| = \left(\sum_{\beta \neq \gamma} |a_{\gamma\beta}|\right) \Big/ |a_{\gamma\gamma}| < 1$$

proves the intermediate assertion $\lambda = \varrho(M') < 1$ from (i). □

The application of Corollary 5 yields

Corollary 6.4.11. *Assume that the irreducibly diagonally dominant matrix $A \in \mathbb{R}^{I \times I}$ satisfies the sign conditions* (1a, b). *Then A is an M-matrix with positive inverse $A^{-1} > 0$.*

Diagonal dominance is not only a criterion for the M-matrix property, but also positive definiteness (cf. Hackbusch [15, Criterion 4.3.24]).

Lemma 6.4.12. *Let $A \in \mathbb{K}^{I \times I}$ be Hermitian with a positive diagonal: $a_{\alpha\alpha} > 0$. If A is strictly, essentially, or irreducibly diagonally dominant, then A is also positive definite. A sufficient condition for positive* semidefiniteness *is that $A = A^H$ is weakly diagonally dominant with $a_{\alpha\alpha} \geqslant 0$.*

6.4.4 Further Criteria

In the following we describe situations in which M-matrices generate new ones.

Theorem 6.4.13. *Let $A \in \mathbb{R}^{I \times I}$ be an M-matrix and $B \geqslant A$ satisfy* (1b): $b_{\alpha\beta} \leqslant 0$ *for $\alpha \neq \beta$. Then B is also an M-matrix. Further, the inequalities*

$$0 \leqslant B^{-1} \leqslant A^{-1} \tag{6.4.9}$$

hold. If, in addition, A is irreducible and $B \neq A$, then even $0 \leqslant B^{-1} < A^{-1}$ is valid.

Proof. (i) Let $M = I - D^{-1}A$ and $M_B = I - D_B^{-1}B$ be the respective Jacobi iteration matrices. One verifies that $0 \leqslant D_B^{-1} \leqslant D^{-1}$ and $0 \leqslant M_B \leqslant M$. $A^{-1} = (\sum_{\nu=0}^{\infty} M^\nu)D^{-1}$ and $B^{-1} = (\sum M_B^\nu)D_B^{-1}$ prove (9).
(ii) According to Exercise 6, $A^{-1} > 0$ holds for irreducible A. Set $A(\lambda) := A + \lambda(B - A)$. For $0 \leqslant \lambda \leqslant 1$, we have $A = A(0) \leqslant A(\lambda) \leqslant A(1) = B$. The derivative

$$C(\lambda) := \frac{d}{d\lambda} A(\lambda)^{-1} = -A(\lambda)^{-1}(B - A)A(\lambda)^{-1}$$

is nonpositive: $C(\lambda) \leqslant 0$, because $A(\lambda)^{-1} \geqslant 0$ (cf. (9)) and $B - A \geqslant 0$. The particular choice $\lambda = 0$ yields $C(0) = -A^{-1}(B - A)A^{-1} < 0$, since each vector $x \gneqq 0$ leads to $A^{-1}x > 0$, $(B - A)A^{-1}x \geqslant 0$, and $A^{-1}(B - A)A^{-1}x > 0$; hence, $C(0) < 0$ (cf. Exercise 1.3b). $C(0) < 0$ and $C(\lambda) \leqslant 0$ prove $A^{-1} > A(\lambda)^{-1} \geqslant B^{-1}$ for all $0 < \lambda \leqslant 1$. ☐

Theorem 6.4.14. *Any principal submatrix of an M-matrix is again an M-matrix. More precisely*: *If* $B = (a_{\alpha\beta})_{\alpha,\beta \in I'}$ for $I' \subset I$, then B is an M-matrix with $0 \leqslant (B^{-1})_{\alpha\beta} \leqslant (A^{-1})_{\alpha\beta}$ for $\alpha, \beta \in I'$. If, furthermore, A is irreducible and $I' \subsetneqq I$ is a nonempty subset, then the strengthened inequality $(B^{-1})_{\alpha\beta} < (A^{-1})_{\alpha\beta}$ for $\alpha, \beta \in I'$ is valid.

Proof. Define $B' \in \mathbb{R}^{I \times I}$ by means of $b'_{\alpha\beta} := \begin{Bmatrix} a_{\alpha\beta} & \text{for } \alpha, \beta \in I' \text{ or } \alpha = \beta \in I \\ 0 & \text{otherwise} \end{Bmatrix}$.
B' has the form $\operatorname{blockdiag}\{B, D_2\}$, with D_2 being the diagonal of the block $A^{22} = (a_{\alpha\beta})_{\alpha,\beta \in I \setminus I'}$. Hence, $B'^{-1} = \operatorname{blockdiag}\{B^{-1}, D_2^{-1}\}$ holds. We can apply Theorem 13 to B' and obtain $0 \leqslant B'^{-1} \leqslant A^{-1}$ or $0 \leqslant B'^{-1} < A^{-1}$, respectively. A restriction to the first block yields the assertion. ☐

Exercise 6.4.15. Prove: (a) A 2×2 matrix A is an M-matrix if and only if (1a, b) and $\det A > 0$ hold.
(b) M-matrices A have a positive determinant: $\det A > 0$.
(c) All principal minors of an M-matrix are positive.

Hint for (b). Discuss the determinant of $A(\lambda) := D + \lambda(A - D)$ for $0 \leqslant \lambda \leqslant 1$ with $D := \operatorname{diag}\{A\}$. *For (c)* Apply Theorem 14.

Gauß elimination will play an important rôle in the following proof. Its basic operation is the elimination of an element $a_{\beta\alpha}$ ($\alpha \neq \beta$) by means of the αth row. The corresponding transformation $A \mapsto A'$ is described by the matrix $T^{\beta\alpha}$:

$$A' = T^{\beta\alpha}A \text{ with } T^{\beta\alpha}_{\nu\nu} = 1,\ T^{\beta\alpha}_{\beta\alpha} = -\frac{a_{\beta\alpha}}{a_{\alpha\alpha}},\ T^{\beta\alpha}_{\nu\mu} = 0 \text{ otherwise.} \tag{6.4.10}$$

The usual Gauß elimination (without pivoting) requires an ordered index set I and performs eliminations in the succession $(\beta, \alpha) = (2, 1), (3, 1), \ldots, (n, 1), (3, 2), (4, 2), \ldots, (n, 2), \ldots, (n, n-1)$ below the diagonal, resulting in an upper triangular matrix U. The diagonal elements $p_i = u_{ii}$ of U are the *pivot elements.* The following considerations are even simpler, if we eliminate all off-diagonal entries at $(\beta, \alpha) = (2, 1), (3, 1), \ldots, (n, 1), (1, 2), (3, 2), \ldots, (n, 2), (1, 3), \ldots$, leading to the diagonal matrix $D = \operatorname{diag}\{p_i: i \in I\}$ of the pivot elements. Let H_i be the *principal minor* of A:

$$H_i := \det((a_{k\ell})_{1 \leqslant k, \ell \leqslant i}) \quad \text{for } 1 \leqslant i \leqslant n,\ H_0 := 1. \tag{6.4.11a}$$

A simple consideration shows

$$p_i = H_i/H_{i-1} \qquad (1 \leqslant i \leqslant n), \tag{6.4.11b}$$

provided that $H_{i-1} \neq 0$. This implies that the elimination process described above can be performed *without* pivoting, if $H_i \neq 0$ for all i (cf. Gantmacher [1, page 36]).

The statement of Exercise 15c can be extended as follows:

Theorem 6.4.16. *Under assumption* (1b), *A is an M-matrix if and only if all principal minors are positive.*

Proof. (i) Since Exercise 15c proves one direction, showing that positive principal minors imply the M-matrix property remains. The diagonal elements $a_{\alpha\alpha}$ are the determinants of the principal 1×1 submatrices $(a_{\alpha\alpha})$. This ensures (1a): $a_{\alpha\alpha} > 0$. According to (11b), Gauß elimination can be performed without pivoting.

(ii) First we prove that the elimination step (10) preserves the sign conditions (1a, b). A and A' differ only in the row β. Since $\varkappa := T^{\beta\alpha}_{\beta\alpha} = -\dfrac{a_{\beta\alpha}}{a_{\alpha\alpha}} \geqslant 0$, the entries $a_{\beta\delta}$ for $\delta \neq \alpha$ become smaller: $a'_{\beta\delta} := a_{\beta\delta} + \varkappa a_{\alpha\delta} \leqslant 0$, while $a'_{\beta\alpha} = 0$ again satisfies (1b). The only problem is raised by condition (1a): Does $a'_{\beta\beta} > 0$ hold again? As seen above, the diagonal element decreases with each elimination step. Since at the end of the elimination, it represents the pivot element p_i, Eq. (11b) with $H_\beta, H_{\beta-1} > 0$ proves the inequality $a'_{\beta\beta} \geqslant p_\beta > 0$.

(iii) Performing the elimination (above and below the diagonal), one obtains the diagonal matrix D of the positive pivot elements. Denoting the elimination matrices $T^{\beta\alpha}$ for the respective index pairs by $T_1, T_2, \ldots, T_N$, one arrives at the representation

$$T_N T_{N-1} \cdot \ldots \cdot T_1 A = D, \quad \text{i.e., } A^{-1} = D^{-1} T_N T_{N-1} \cdot \ldots \cdot T_1. \tag{6.4.11c}$$

Since, according to (ii), all intermediate matrices fulfil the conditions (1a, b), $T_i \geqslant 0$ holds. Together with $D^{-1} \geqslant 0$, the missing M-matrix property (1c): $A^{-1} \geqslant 0$ is obtained. □

The close connection between M-matrices and positive definite matrices is underlined by

Remark 6.4.17. A Hermitian matrix is positive definite if and only if all principal minors are positive.

Proof. (i) By Lemma 2.10.4, all principal submatrices are positive definite. Hence, it is sufficient to show that $\det A > 0$ for positive definite A. This follows from $\det A = \Pi \lambda_i$ and the positivity $\lambda_i > 0$ of all eigenvalues $\lambda_i \in \sigma(A)$ (cf. Lemma 2.10.3).
(ii) The determinant of $A(\lambda) := A + \lambda I$ can be expanded into $\det A + \sum_i \lambda \det A_i(\lambda)$, where $A_i(\lambda)$ is the principal submatrix of $A(\lambda)$ for the index set $I_i := I \backslash \{i\}$. The analogous expansion of the determinants of $A_i(\lambda)$ yields a polynomial $p(\lambda) = \det A(\lambda) = \sum a_\nu \lambda^\nu$ with positive coefficients a_ν (e.g., $a_0 = \det A > 0$). Hence, $A + \lambda I$ is regular for all $\lambda \geqslant 0$. Since its eigenvalues are $\lambda_i + \lambda$, all eigenvalues $\lambda_i \in \sigma(A)$ must be positive. According to Lemma 2.10.3, A is positive definite. □

Remark 17 describes one of the numerous characterisations of the M-matrix property. The interested reader may find fifty different characterisations in the book of Berman–Plemmons [1].

The combination of Theorem 16 and Remark 17 yields

Theorem 6.4.18. (a) *A positive definite matrix satisfying the sign condition* (1b) *is an M-matrix.*
(b) *A Hermitian M-matrix is positive definite.*

The discussion of Gauß elimination is continued in

Lemma 6.4.19. *If A is an M-matrix and A' is obtained by one Gauß elimination step* (10), *then A' is again an M-matrix.*

Proof. Choose the ordering of the indices such that $\alpha = 1$ and $\beta = 2$. Then (10) describes the first step T_1 of the complete elimination process

$T_N T_{N-1} \cdot \ldots \cdot T_1 A = D$ (cf. (11c)). $A'^{-1} = (T_1 A)^{-1} \geqslant 0$ follows from $T_i \geqslant 0$, $D^{-1} \geqslant 0$, and $(T_1 A)^{-1} = D^{-1} T_N T_{N-1} \cdot \ldots \cdot T_2$. Since the conditions (1a, b) are already proved in part (ii) of the proof for Theorem 16, A' is an M-matrix. □

The *blockwise* elimination of the block-matrix $\begin{bmatrix} A & B \\ C & D \end{bmatrix}$ leads to $\begin{bmatrix} I & B' \\ 0 & S \end{bmatrix}$ with $B' := A^{-1}B$ and the so-called *Schur complement*

$$S := D - CA^{-1}B. \tag{6.4.12}$$

Blockwise elimination represents the product of all elementary eliminations (10) with indices α corresponding to the columns of the first block and $\beta \in I \backslash \{\alpha\}$. A multiple applicaton of Lemma 19 proves

Lemma 6.4.20. *The Schur complement S of an M-matrix $\begin{bmatrix} A & B \\ C & D \end{bmatrix}$ is again an M-matrix.*

6.5 Regular Splittings

The splitting

$$A = W - R \tag{6.5.1a}$$

induces the iterative method

$$Wx^{m+1} = Rx^m + b, \tag{6.5.1b}$$

if W is regular (cf. (4.2.1–3)). For characterising the splitting (1a), the specification of W is sufficient, since $R := W - A$. The following definition of «regular splitting» is due to Varga [2]. It allows not only qualitative convergence statements, but also a comparison of different iterative methods.

Definition 6.5.1. The matrix $W \in \mathbb{R}^{I \times I}$ describes a *regular splitting* of $A \in \mathbb{R}^{I \times I}$ if

$$W \text{ regular}, \qquad W^{-1} \geqslant 0, \qquad W \geqslant A. \tag{6.5.2}$$

Condition (2) compares with (4.8.3a) in the positive definite case. The iteration matrix of the iteration (1b) is

$$M = W^{-1}R \quad \text{with } R := W - A. \tag{6.5.1c}$$

Condition (2) implies that

$$M \geqslant 0 \quad \text{for regular splittings} \tag{6.5.3}$$

because of $R \geqslant 0$. Using (3), one can weaken Definition 1: (1a) is a *weakly regular splitting* (cf. Ortega [1]), if

$$W \text{ regular}, \qquad W^{-1} \geqslant 0, \qquad M = W^{-1}R \geqslant 0. \tag{6.5.4}$$

Theorem 6.5.2 (convergence). *Let A be inverse positive*: $A^{-1} \geqslant 0$ (*sufficient*: *A is an M-matrix*). *Assume that W describes a weakly regular splitting of A. Then the induced iteration* (1b) *converges*:

$$\varrho(M) = \varrho(W^{-1}R) = \frac{\varrho(A^{-1}R)}{1 + \varrho(A^{-1}R)} < 1. \tag{6.5.5}$$

Proof. (i) Obviously, it is sufficient to show $\varrho(W^{-1}R) = \varrho(C)/(1 + \varrho(C))$ for $C := A^{-1}R$. By (3) we have

$0 \leqslant M = W^{-1}R = [A^{-1}W]^{-1}A^{-1}R = [A^{-1}(A + R)]^{-1}A^{-1}R = [I + C]^{-1}C.$

By Theorem 3.10 and $M \geqslant 0$, an eigenvector $x \gneqq 0$ belongs to $\lambda = \varrho(M) \in \sigma(M)$. Rewriting $\lambda x = Mx = (I + C)^{-1}Cx$, we obtain

$$\lambda x + \lambda Cx = Cx. \tag{6.5.6a}$$

The value $\lambda = 1$ is excluded, since (6a) would yield $x = 0$. Hence,

$$Cx = \frac{\lambda}{1 - \lambda} x \tag{6.5.6b}$$

follows. In (iii) we will show $C \geqslant 0$. (6b) together with $x \gneqq 0$ and $Cx \geqslant 0$ ensures the inequality $\frac{\lambda}{1 - \lambda} \geqslant 0$, i.e., $0 \leqslant \lambda = \varrho(M) < 1$.

(ii) (6b) proves that λ is an eigenvalue of M, if and only if $\mu = \frac{\lambda}{1 - \lambda}$ is an eigenvalue of C. $0 \leqslant \lambda < 1$ shows $\mu \geqslant 0$. Since $\mu = \frac{\lambda}{1 - \lambda}$ increases monotonically in λ, $|\mu| = \mu$ is maximal for $\lambda = \varrho(M) \in \sigma(M)$. By Theorem 3.10, $\mu = \varrho(C) \in \sigma(C)$ is the maximal eigenvalue of C; thus, $\varrho(C) = \varrho(M)/[1 - \varrho(M)]$. Solving this equation for $\varrho(M)$, we arrive at assertion (5): $\varrho(M) = \varrho(C)/[1 + \varrho(C)]$.

(iii) $0 \leqslant \left(\sum_{\nu=0}^{m-1} M^\nu\right) W^{-1}$, $0 \leqslant W^{-1} = (I - M)A^{-1}$, and $\sum_{\nu=0}^{m-1} M^\nu(I - M) = I - M^m$ imply the inclusions $0 \leqslant (I - M^m)A^{-1} \leqslant A^{-1}$ and $0 \leqslant M^m A^{-1} \leqslant A^{-1}$. Therefore, M^m is bounded. This fact proves that $\varkappa = \varrho(M) \leqslant 1$. Since $\lambda = 1$ is already excluded, $\varrho(M) < 1$ holds and implies $C = A^{-1}R = [W(I - M)]^{-1}R = (I - M)^{-1}W^{-1}R = \left(\sum_{\nu=0}^{\infty} M^\nu\right) M \geqslant 0$. □

Obviously, the iteration should converge faster the closer W is to A, i.e., the smaller the remainder $R = W - A$ is. This presumption is precisely formulated in the following comparison theorem.

Theorem 6.5.3. *Let A be inverse positive: $A^{-1} \geqslant 0$. Let W_1, W_2 define two regular splittings. If W_1 and W_2 are comparable in the sense of*

$$A \leqslant W_1 \leqslant W_2, \tag{6.5.7a}$$

then the corresponding convergence rates can also be compared:

$$0 \leqslant \varrho(M_1) \leqslant \varrho(M_2) < 1, \quad \text{where } M_i := W_i^{-1}R_i,\ R_i := W_i - A. \tag{6.5.7b}$$

Proof. The matrices $B := A^{-1}R_1$ and $C := A^{-1}R_2$ satisfy $0 \leqslant B \leqslant C$ and therefore $0 \leqslant \varrho(B) \leqslant \varrho(C)$ (cf. (3.7c)). From representation (5), one obtains $0 \leqslant \varrho(M_1) = \varrho(B)/[1 + \varrho(B)] \leqslant \varrho(C)/[1 + \varrho(C)] = \varrho(M_2) < 1$. □

The comparisons (7a, b) can be strengthened into *strict* inequalities.

Theorem 6.5.4. *From $A^{-1} > 0$ and* (8a), *the strict inequality* (8b) *follows:*

$$A \lneqq W_1 \lneqq W_2, \qquad W_i\text{: regular splittings,} \tag{6.5.8a}$$

$$0 < \varrho(M_1) < \varrho(M_2) < 1, \quad \text{where } M_i := W_i^{-1}R_i,\ R_i := W_i - A. \tag{6.5.8b}$$

Proof. Define B and C as in the previous proof. Since $B = A^{-1}R_1$ may be reducible, Theorem 3.1 is not directly applicable. Define

$$I_+ := \{\beta \in I\colon R_{1,\alpha\beta} > 0 \text{ for some } \alpha \in I\} \quad \text{and} \quad I_0 := I \backslash I_+.$$

Any column s of R_1 corresponding to the index $\beta \in I_+$ satisfies $s \gneqq 0$ and therefore $A^{-1}s > 0$ by Exercise 1.3b. Hence, B has the form $B = \begin{bmatrix} B_1 & 0 \\ B_2 & 0 \end{bmatrix}$ with positive blocks $B_1 > 0$ and $B_2 > 0$ with respect to the block structure $\{I_+, I_0\}$. In particular,

$$\varrho(B) = \varrho(B_1) > 0 \tag{6.5.8c}$$

holds (cf. (3.1a)). Because of $R_2 - R_1 = W_2 - W_1 \gneqq 0$, there is a pair (α, β) with $(R_2 - R_1)_{\alpha\beta} > 0$. Hence, the column of $C - B = A^{-1}(R_2 - R_1)$ for the index β is positive. Assume $\beta \in I_+$. In this case, $C_1 > B_1$, $C_2 > B_2$ holds for the blocks in $C = \begin{bmatrix} C_1 & C_3 \\ C_2 & C_4 \end{bmatrix}$. Lemma 3.6 and (3.1c) yield the inequality

$$\varrho(C) \geqslant \varrho(C_1) > \varrho(B_1). \tag{6.5.8d}$$

In the remaining case $\beta \in I_0$, one concludes that $C_3 > B_3 = 0$, $C_4 > B_4 = 0$, and

$$\varrho(C) > \varrho(C_1) \geqslant \varrho(B_1) \tag{6.5.8e}$$

(cf. Lemma 3.6). In any case, using (8c), one arrives at the strict inequality $\varrho(C) > \varrho(B) > 0$, which via (5) leads to the assertion. □

6.6 Applications

Theorem 6.6.1. *Let A be an M-matrix. Then the pointwise as well as the blockwise* Jacobi method *converge, where the latter, however, is faster*:

$$\varrho(M^{\text{blockJac}}) \leqslant \varrho(M^{\text{Jac}}) < 1. \tag{6.6.1a}$$

Let D be the diagonal D^{ptw} or the block-diagonal D^{block} of A. Then

$$D \textit{ describes a regular splitting.} \tag{6.6.1b}$$

Assuming explicitly (1b), *we may replace the assumption «A is an M-matrix» by the inverse positivity: $A^{-1} \geqslant 0$. In* (1a) *the strict inequality $0 < \varrho(M^{\text{blockJac}}) < \varrho(M^{\text{Jac}}) < 1$ holds if $A^{-1} > 0$ and $D^{\text{ptw}} \neq D^{\text{block}} \neq A$.*

Proof. For an M-matrix A, the diagonals $D = D^{\text{ptw}}$ and $D = D^{\text{block}}$ satisfy the inequality $D \geqslant A$ and sign condition (4.1b). By Theorem 4.13, D is again an M-matrix, from which $D^{-1} \geqslant 0$ and (1b) follow. Because of $D^{\text{ptw}} \geqslant D^{\text{block}}$, Theorem 5.3 proves inequality (1a). For the strict inequality, compare Theorem 5.4. □

Theorem 6.6.2. *Split $A = D - E - F$ according to* (4.2.7a–d) or (4.5.5a–d). *The statements of Theorem* 1 *carry over to analogous ones for the pointwise and blockwise* Gauß-Seidel method, *where* (1a, b) *become*

$$\varrho(M^{\text{blockGS}}) \leqslant \varrho(M^{\text{GS}}) < 1, \tag{6.6.2a}$$

$$D - E \textit{ describes a regular splitting.} \tag{6.6.2b}$$

We omit the proof, since it is completely analogous to the previous one. The comparison between the Jacobi and Gauß-Seidel iteration is more interesting. The quantitative relation $\varrho(M^{\text{GS}}) = \varrho(M^{\text{Jac}})^2$, which according to Remark 5.6.8 holds for consistent orderings, can no longer be shown for the general case. However, a corresponding qualitative statement following from $D - E \leqslant D$ is valid.

Theorem 6.6.3. *For an M-matrix A, the following inequalities hold*:

$$\varrho(M^{\text{GS}}) \leqslant \varrho(M^{\text{Jac}}) < 1, \qquad \varrho(M^{\text{blockGS}}) \leqslant \varrho(M^{\text{blockJac}}) < 1. \tag{6.6.3}$$

This statement can be generalised for other than M-matrices.

Theorem 6.6.4 (Stein–Rosenberg [1]). *Exactly one of the following alternatives* (4a–d) *hold for the pointwise Jacobi and Gauß-Seidel iteration, if A fulfils the sign condition* (4.1b): *$a_{\alpha\beta} \leqslant 0$ for $\alpha \neq \beta$.*

$$0 = \varrho(M^{\text{GS}}) = \varrho(M^{\text{Jac}}), \tag{6.6.4a}$$

$$0 < \varrho(M^{\text{GS}}) < \varrho(M^{\text{Jac}}) < 1, \tag{6.6.4b}$$

$$\varrho(M^{\text{GS}}) = \varrho(M^{\text{Jac}}) = 1, \tag{6.6.4c}$$

$$\varrho(M^{\text{GS}}) > \varrho(M^{\text{Jac}}) > 1. \tag{6.6.4d}$$

In particular, both methods converge or diverge simultaneously. The statement of the theorem remains valid, if M^{Jac} *and* M^{GS} *are replaced by* $L + U$ *and* $(I - L)^{-1}U$, *where* $L \geqslant 0$ *is an arbitrary, strictly lower triangular matrix and* $U \geqslant 0$ *is a strictly upper one.*

Proof. Compare Varga [2, §3.3] or the original paper.

In the case of *over*relaxation (i.e., for $\omega > 1$), the SOR iteration does not lead to regular splitting. In order to ensure the regularity of the splitting, one has to restrict ω to $0 < \omega < 1$ (underrelaxation).

Exercise 6.6.5. Prove that the SOR method arises from a splitting (5.1a) with $W = \frac{1}{\omega}D - E$. Let A be an M-matrix and D its diagonal. For $0 < \omega \leqslant 1$, the matrix W describes regular splitting. What conclusion can be drawn from $\frac{1}{\omega}D - E \geqslant D - E$?

In the case of regular splitting, the property (5.4): $M \geqslant 0$ allows an inclusion of the solution $x = A^{-1}b$, if one succeeds in finding suitable starting iterates.

Theorem 6.6.6. *Let* $M \geqslant 0$ *be the iteration matrix of a convergent iteration. Starting with initial iterates* x^0 *and* y^0 *satisfying*

$$x^0 \leqslant x^1, \qquad x^0 \leqslant y^0, \qquad y^1 \leqslant y^0, \tag{6.6.5a}$$

one obtains iterates x^m *and* y^m *with the* inclusion *property*

$$x^0 \leqslant x^1 \leqslant \cdots \leqslant x^m \leqslant \cdots \leqslant x = A^{-1}b \leqslant \cdots \leqslant y^m \leqslant \cdots \leqslant y^1 \leqslant y^0. \tag{6.6.5b}$$

Proof. It follows from the estimates $x^{m+1} - x^m = M^m(x^1 - x^0) \geqslant 0$ and $y^m - y^{m+1} = M^m(y^0 - y^1) \geqslant 0$ and $y^m - x^m = M^m(y^0 - x^0) \geqslant 0$ (cf. (3.2.9b)). □

The term «M-matrix» can be generalised as follows:

Definition 6.6.7. $A \in \mathbb{K}^{I \times I}$ is called an *H-matrix* if $B := |D| - |A - D|$ with $D := \operatorname{diag}\{A\}$ is an M-matrix.

The construction of $B := |D| - |A - D|$ changes the signs of the entries $a_{\alpha\beta}$ in such a way that $b_{\alpha\alpha} \geqslant 0$ and $b_{\alpha\beta} \leqslant 0$ for $\alpha \neq \beta$ results (cf. (4.1a, b)). The letter H stands for Hadamard (cf. Ostrowski [1]).

Theorem 6.6.8. *Each of the following conditions* (6a, b) *is sufficient for the convergence of the pointwise Jacobi and Gauß-Seidel iterations*:

A is an H-matrix, (6.6.6a)

A is strictly diagonally dominant, irreducibly diagonally dominant, or essentially diagonally dominant. (6.6.6b)

Proof. (i) The case (6b) is reduced to (6a) by means of

Exercise 6.6.9. Prove that (6b) implies (6a) and $\|M^{\mathrm{Jac}}\|_\infty < 1$, $\|M^{\mathrm{GS}}\|_\infty < 1$.

(ii) Define $B := |D| - |A - D|$ as in Definition 7 and denote the iteration matrix of the Jacobi iteration for B by $M_B^{\mathrm{Jac}} := I - |D|^{-1}B = |D|^{-1}|A - D|$. Theorem 1 yields $\varrho(M_B^{\mathrm{Jac}}) < 1$. By $|M^{\mathrm{Jac}}| = M_B^{\mathrm{Jac}}$, the convergence $\varrho(M^{\mathrm{Jac}}) < 1$ follows from the next lemma, which remains to be proved.

Lemma 6.6.10. $\varrho(A) \leqslant \varrho(|A|)$ *for all* $A \in \mathbb{K}^{I \times I}$.

(iii) Split $A = D - E - F$ according to (4.2.7a–d) and define $L := D^{-1}E$, $U := D^{-1}F$. Since $B = |D| - |E| - |F| = |D|(I - |L| - |U|)$, the iteration matrices belonging to A and B are:

$$M^{\mathrm{GS}} = (I - L)^{-1}U = \sum_{\nu=0}^{\infty} L^\nu U, \qquad M_B^{\mathrm{GS}} = (I - |L|)^{-1}|U| = \sum_{\nu=0}^{\infty} |L|^\nu |U|$$

(cf. Lemma 2.4.8); hence, $|M^{\mathrm{GS}}| = |\sum L^\nu U| \leqslant \sum |L|^\nu |U| = M_B^{\mathrm{GS}}$. From Lemma 10 and Theorem 2, one concludes that $\varrho(M^{\mathrm{GS}}) \leqslant \varrho(M_B^{\mathrm{GS}}) < 1$. □

Proof of Lemma 10. By $\|A^\nu\|_\infty = \||A^\nu|\|_\infty \leqslant \||A|^\nu\|_\infty$, Theorem 2.9.8 yields

$$\varrho(A) = \lim_{\nu\to\infty} \|A^\nu\|_\infty^{1/\nu} \leqslant \lim_{\nu\to\infty} \||A|^\nu\|_\infty^{1/\nu} = \varrho(|A|). \qquad \square$$

Concerning the convergence of the SSOR method for H-matrices, we refer to Alefeld–Varga [1] and Neumaier–Varga [1].

7
Semi-Iterative Methods

7.1 First Formulation

7.1.1 The Semi-Iterative Sequence

Let Φ be a linear and consistent but not necessarily convergent iteration (in the following it will also be called the *basic iteration*) with an iteration matrix M. Assume that for a starting iterate x^0, the iterates

$$x^{m+1} = Mx^m + Nb = \Phi(x^m, b) \tag{7.1.1}$$

are computed. Up to now, the last computed iterate x^m has been regarded as the result of an iteration Φ. The previously calculated x^j $(0 \leqslant j \leqslant m-1)$ are forgotten.

The semi-iterative method is based on a different view. Now, the result of m steps by the basic iteration Φ is the complete sequence

$$X_m := (x^0, x^1, \ldots, x^m) \in (\mathbb{K}^I)^{m+1}. \tag{7.1.2}$$

It has to be investigated whether a better result than x^m can be constructed from X_m. A *semi-iterative method* is a mapping

$$\sum\colon \bigcup_{m=0}^{\infty} (\mathbb{K}^I)^{m+1} \to \mathbb{K}^I. \tag{7.1.3a}$$

The results

$$y^m := \sum(X_m) \qquad (m = 0, 1, 2, \ldots) \tag{7.1.3b}$$

yield a new sequence: the *semi-iterative sequence*. We will see that in some cases, $\{y^m\}$ converges much faster than $\{x^m\}$.

Remark 7.1.1. The simple example $y^m = \sum(x^0, x^1, \ldots, x^m) := x^m$ shows that an optimally chosen semi-iterative method cannot be worse than the basic iteration.

7.1.2 Consistency and Asymptotical Convergence Rate

Similarly as in Definition 3.1.4, a semi-iterative method $\sum$ is called *consistent* if Eq. (4) holds for all solutions of $Ax = b$:

$$x = \sum(\underbrace{x, x, \ldots, x}_{(m+1)\text{-fold}} \qquad (m = 0, 1, 2, \ldots) \tag{7.1.4}$$

Up to now, the convergence rate $\varrho = \varrho(M)$ could be characterised as minimal ϱ with $\lim_{m\to\infty} (\|x^m - x\|/\|x^0 - x\|)^{1/m} \leqslant \varrho$ for all $x^0 \neq x$ (cf. Remark 3.2.13b). This characterisation can be transfered.

Definition 7.1.2. The semi-iterative method $\sum$ has the *asymptotical convergence rate* ϱ, if ϱ is the smallest number with

$$\overline{\lim_{m\to\infty}} (\|y^m - x\|/\|y^0 - x\|)^{1/m} \leqslant \varrho \qquad (x = A^{-1}b) \tag{7.1.5}$$

for all semi-iterative sequences $\{y^m\}$ corresponding to arbitrary starting iterates $y^0 = x^0$.

We will restrict our consideration to linear semi-iterations. $\sum$ is called *linear* if $y^m = \sum(X_m)$ has the representation

$$y^m = \sum_{j=0}^{m} \alpha_{mj} x^j \qquad (m = 0, 1, \ldots) \tag{7.1.6}$$

with coefficients $\alpha_{mj} \in \mathbb{C}$ $(m \in \mathbb{N}_0,\ 1 \leqslant j \leqslant m)$. Obviously, *a linear semi-iterative method is consistent if and only if*

$$\sum_{j=0}^{m} \alpha_{mj} = 1 \quad \text{for all } m = 0, 1, \ldots. \tag{7.1.7}$$

Applying condition (7) to $m = 0$, one finds that a consistent semi-iterative method satisfies the start condition (8):

$$y^0 = x^0. \tag{7.1.8}$$

7.1.3 Error Representation

Theorem 7.1.3. *Let x be the solution of $Ax = b$ with regular A. The basic iteration Φ is linear and consistent with the iteration matrix M. The semi-iterative method $\sum$ is linear and consistent. Then the error*

$$\eta^m := y^m - x \qquad (x = A^{-1}b) \tag{7.1.9a}$$

admits the representation

$$\eta^m = p_m(M)e^0 \quad \text{with } e^0 := x^0 - x, \tag{7.1.9b}$$

where $y^0 = x^0$ *(cf. (8)) is the starting iterate and* p_m *is the polynomial*

$$p_m(\zeta) = \sum_{j=0}^{m} \alpha_{mj}\zeta^j \tag{7.1.9c}$$

with the coefficients α_{mj} *from* (7).

Proof. Subtracting the equation $A^{-1}b =: x = \sum_{j=0}^{m} \alpha_{mj}x$ from $y^m = \sum_{j=0}^{m} \alpha_{mj}x^j$ (cf. (4), (7)), one obtains

$$\eta^m = y^m - x = \sum_{j=0}^{m} \alpha_{mj}(x^j - x) = \sum_{j=0}^{m} \alpha_{mj}e^j$$

with the errors $e^j = x^j - x$ of the basic iteration. Inserting (3.2.9b): $e^j = M^j e^0$, we arrive at

$$\eta^m = \sum_{j=0}^{m} \alpha_{mj}(M^j e^0) = \left(\sum_{j=0}^{m} \alpha_{mj}M^j\right)e^0 = p_m(M)e^0. \qquad \square$$

In Theorem 3, we have associated the linear semi-iteration $\sum$ with polynomials $\{p_m: m = 0, 1, \dots\}$ of degree$(p_m) \leqslant m$. Vice versa, any sequence $\{p_m\}$ of polynomials of degree$(p_m) \leqslant m$ defines a semi-iterative method by means of its coefficients α_{mj}. This is stated in

Remark 7.1.4. A linear semi-iterative method $\sum$ is uniquely described by the associated polynomial sequence $\{p_m\}$, where degree$(p_m) \leqslant m$. $\sum$ is consistent if and only if

$$p_m(1) = 1 \quad \text{for } m = 0, 1, \dots. \tag{7.1.9d}$$

Let the basic iteration with the iteration matrix M be consistent. Then the semi-iterates y^m have the representation

$$y^m = M_m x^0 + N_m b \quad \text{with } M_m := p_m(M),\ N_m := (I - M_m)A^{-1}. \tag{7.1.10}$$

The *asymptotical convergence rate* equals

$$\overline{\lim_{m\to\infty}}\ \varrho(p_m(M))^{1/m} = \overline{\lim_{m\to\infty}}\ \|p_m(M)\|^{1/m}. \tag{7.1.11}$$

7.2 Second Formulation of a Semi-Iterative Method

7.2.1 General Representation

In its general form, the semi-iteration presented in §7.1 is not practical, since all iterates $\{x^0, x^1, \dots, x^m\}$ must be stored. Thus, for large m and high-dimensional systems, considerable difficulties arise. The definition of $y^m = \sum(X_m)$ is completely independent of the previous $y^j = \sum(X_j)$ $(0 \leqslant j \leqslant m-1)$. Therefore, in general, it is not possible to use the semi-iterative results $y^0, \ldots, y^{m-1}$ for computing y^m.

This situation will change in the *second formulation*. Let Φ be the consistent basic iteration. After starting with

$$y^0 = x^0 \qquad \text{(cf. (1.8))}, \tag{7.2.1a}$$

we compute the iterates recursively by

$$y^m = \Theta_m \Phi(y^{m-1}, b) + (1 - \Theta_m) y^{m-1} \qquad (m \geqslant 1) \tag{7.2.1b}$$

with *extrapolation factors* $\{\Theta_m : m \in \mathbb{N}\}$, which may be chosen arbitrarily.

Exploiting the normal forms $\Phi(x, b) = Mx + Nb = x - W^{-1}(Ax - b)$, (1b) can be written in the form (1b′) or (1b″):

$$y^m = \Theta_m(My^{m-1} + Nb) + (1 - \Theta_m) y^{m-1}, \tag{7.2.1b′}$$

$$y^m = y^{m-1} - \Theta_m W^{-1}(Ay^{m-1} - b) = \Phi_{\Theta_m}(y^{m-1}, b). \tag{7.2.1b″}$$

For $0 < \Theta_m < 1$, formulae (1b′, b″) represent one step of the damped version of the basic iteration (cf. §4.3.1.1), however, with a parameter Θ_m depending on m. Other Θ_m lead to an extrapolation of the basic iteration.

Definition (1a, b) represents a semi-iterative method as stated in

Theorem 7.2.1. *For arbitrary* $\{\Theta_m : m \in \mathbb{N}\}$, *the algorithm* (1a, b) *defines a linear and consistent semi-iteration* $\sum$. *The polynomials* $\{p_m\}$ *describing* $\sum$ *are recursively defined by*

$$p_0(\zeta) = 1, \quad p_m(\zeta) = [\Theta_m \zeta + 1 - \Theta_m] p_{m-1}(\zeta) \qquad (m \in \mathbb{N}). \tag{7.2.2}$$

Proof. (i) One shows by induction that the polynomials p_m from (2) satisfy the consistency condition (1.9d): $p_m(1) = 1$.
(ii) The basic iteration Φ is assumed to be consistent. By construction (1b′), the matrix M_m from the representation $y^m = M_m x^0 + N_m b$ has the form $M_m = \Theta_m M M_{m-1} + (1 - \Theta_m) M_{m-1}$, where $M_0 = I$. According to (1.10), the polynomials from (2) lead to the same $M_m = p_m(M)$. Since these matrices uniquely determine y^m as $N_m := (I - M_m)A^{-1}$ (consistency of Φ!), the method (1a, b) coincides with the semi-iteration defined by the polynomials (2). The case of a nonconsistent basic iteration is left to the reader (induction proof). □

The case $\Theta_m = 0$ is uninteresting because of $y^m = y^{m-1}$. Therefore, assume that $\Theta_m \neq 0$. The set of all methods representable by (1a, b) is characterised in

Lemma 7.2.2. *Assume that the basic iteration is consistent and that* $\Theta_m \neq 0$ *in* (1b). *Then the second formulation* (1a, b) *represents exactly those linear and*

consistent semi-iterations for which the associated polynomials p_m satisfy the conditions (3a, b) *besides* (1.9d):

$$degree(p_m) = m, \tag{7.2.3a}$$

$$p_{m-1} \text{ is divisor of } p_m \quad \text{for } m \geqslant 1. \tag{7.2.3b}$$

Given polynomials $\{p_m\}$ with (1.9d) *and* (3a, b), *the extrapolation factors Θ_m of the equivalent representation* (1a, b) *are determined by*

$$\frac{p_m(\zeta)}{p_{m-1}(\zeta)} = 1 + \Theta_m(\zeta - 1). \tag{7.2.3c}$$

Proof. In the case $\Theta_m \neq 0$, the method (1a, b) leads to polynomials (2) of degree$(p_m) = m$; hence, (3a, b) are satisfied. Vice versa, under the assumptions (3a,b), p_m/p_{m-1} must be a polynomial of the form (3c). □

7.2.2 Pascal Implementation of the Second Formulation

Given the extrapolation factors Θ_m, in principle, the semi-iteration with the basic iteration «**iteration**» can be implemented as follows:

```
for m:=1 to maximal_iteration_number do
begin set_theta(IP,Θm); damped_iteration(x,A,x,b,IP,iteration)
end;
```

Here, `set_theta` and `damped_iteration` are explained in §4.3.1.2. However, this algorithm may lead to numerical instabilities (cf. explanation following Remark 3.6).

7.2.3 Three-Term Recursion

Algorithm (1b) computes y^m from y^{m-1}. An alternative is a three-term recursion connecting y^m, y^{m-1}, and y^{m-2} (cf. §3.2.8):

$$y^0 = x^0, \tag{7.2.4a}$$

$$y^1 = (1 - \tfrac{1}{2}\vartheta_1)x^1 + \tfrac{1}{2}\vartheta_1 x^0 = (1 - \tfrac{1}{2}\vartheta_1)\Phi(x^0, b) + \tfrac{1}{2}\vartheta_1 x^0, \tag{7.2.4b}$$

$$y^m = \Theta_m\{\Phi(y^{m-1}, b) - y^{m-2}\} + \vartheta_m(y^{m-1} - y^{m-2}) + y^{m-2}. \tag{7.2.4c}$$

From the normal forms $\Phi(x, b) = Mx + Nb = x - W^{-1}(Ax - b)$ one obtains the representations

$$y^1 = (1 - \tfrac{1}{2}\vartheta_1)(Mx^0 + Nb) + \tfrac{1}{2}\vartheta_1 x^0 = x^0 - \tfrac{1}{2}\vartheta_1 W^{-1}(Ax^0 - b), \tag{7.2.4b'}$$

$$\begin{aligned} y^m &= \Theta_m\{(My^{m-1} + Nb) - y^{m-2}\} + \vartheta_m(y^{m-1} - y^{m-2}) + y^{m-2} \\ &= y^{m-2} + (\vartheta_m + \Theta_m)(y^{m-1} - y^{m-2}) - \Theta_m W^{-1}(Ay^{m-1} - b). \end{aligned} \tag{7.2.4c'}$$

Analogously to Theorem 1, one proves

Theorem 7.2.3. *For arbitrary Θ_m and ϑ_m the algorithm* (4a, b, c) *defines a linear and consistent semi-iteration $\sum$. The polynomials $\{p_m\}$ describing $\sum$ are recursively defined by*

$$p_0(\zeta) = 1, \qquad p_1(\zeta) = (1 - \tfrac{1}{2}\vartheta_1)\zeta + \tfrac{1}{2}\vartheta_1, \tag{7.2.5a}$$

$$p_m(\zeta) = (\Theta_m \zeta + \vartheta_m) p_{m-1}(\zeta) + (1 - \Theta_m - \vartheta_m) p_{m-2}(\zeta). \tag{7.2.5b}$$

The particular recursion for $\vartheta_m = 0$ reads

$$p_0(\zeta) = 1, \qquad p_1(\zeta) = \zeta, \tag{7.2.5c}$$

$$p_m(\zeta) = \Theta_m[\zeta p_{m-1}(\zeta) - p_{m-2}(\zeta)] + p_{m-2}(\zeta). \tag{7.2.5d}$$

We remark that all orthogonal polynomials can be generated by a recursion of the form (5a, b) (*cf. Stoer* [1, §3.5], *Stoer–Bulirsch* [1, §3.6]).

7.3 Optimal Polynomials

7.3.1 Minimisation Problem

Let $\sum$ be a linear and consistent semi-iteration. By Theorem 1.3, the error $\eta^m = y^m - x$ has the representation (1.9b):

$$\eta^m = p_m(M)e^0.$$

Therefore, it seems reasonable to pose the following problem.

1st Minimisation Problem

Given $m \in \mathbb{N}$, determine a polynomial p_m of degree$(p_m) \leqslant m$ such that

$$\|p_m(M)e^0\|_2 = \min, \tag{7.3.1}$$

i.e., $\|p_m(M)e^0\|_2 \leqslant \|q_m(M)e^0\|_2$ for all admissible polynomials q_m. Note that the zero polynomial $p_m = 0$ is no solution, because the polynomials p_m (and q_m) have to satisfy the *consistency condition*

$$p_m(1) = 1. \tag{7.3.2}$$

The solution of the problem (1–2) seems hopeless, since the unknown error $e^0 = x^0 - x$ is involved in the problem (if e^0 were known, $x = x^0 - e^0$ would already represent the solution). Nevertheless, we will solve this problem in some modified form in §9.3 (cf. Remark 9.4.9).

Even if e^0 is unknown, $\|p_m(M)e^0\|_2$ can be estimated by

$$\|p_m(M)e^0\|_2 \leqslant \|p_m(M)\|_2 \, \|e^0\|_2$$

and the factor $\|p_m(M)\|_2$ can be minimised separately:

2nd Minimisation Problem

Given $m \in \mathbb{N}$, determine a polynomial p_m with (2) and degree$(p_m) \leqslant m$ such that

$$\|p_m(M)\|_2 = \min. \tag{7.3.3}$$

7.3.2 Discussion of the Second Minimisation Problem

A partial answer to the minimisation problems is given by

Theorem 7.3.1. *Assume that M has no eigenvalue $\lambda = 1$ (sufficient: $\varrho(M) < 1$). For all $m \geqslant n := \#I$, the choice $p_m(\lambda) = \chi(\lambda) := \det(\lambda I - M)/\det(I - M)$ leads to a polynomial with the properties (2) and degree $p_m \leqslant m$ that solves the problems (1) and (3). In particular, (4) holds:*

$$p_m(M) = 0 \quad and \quad \|p_m(M)\|_2 = 0. \tag{7.3.4}$$

The minimum function $p_m(\lambda) = \mu(\lambda)$ of M (cf. (2.8.4d)) satisfies (4) already for $m \geqslant m_0 :=$ degree μ. If a polynomial of degree $p_m = m$ with $m > n$ ($m > m_0$, respectively) is desired for (1) and (3), one may choose $p_m(\zeta) = \zeta^{m-n}\chi(\zeta)$ or $p_m(\zeta) = \zeta^{m-m_0}\mu(\zeta)$, respectively.

Proof. Theorem 2.8.9 (Cayley-Hamilton) guarantees $\chi(M) = \mu(M) = 0$. □

The solution given in Theorem 1 is unsatisfactory for two reasons. First, the characteristic polynomial χ (more precisely, its coefficients) is not so easy to compute; second, the case $m \geqslant n$ is rather uninteresting.

Intermediately, we assume:

$$M \text{ is normal,} \tag{7.3.5}$$

i.e., $MM^H = M^H M$; M being Hermitian would suffice. Since then $p_m(M)$ is also normal, Theorem 2.9.5 implies that

$$\|p_m(M)\|_2 = \varrho(p_m(M)) = \max\{|p_m(\lambda)|: \lambda \in \sigma(M)\}. \tag{7.3.6}$$

Therefore, the minimisation in (3) is equivalent to the determination of a polynomial, whose absolute value is minimal in the set $\sigma(M)$. Even when the normality (5) does not hold, the minimisation of $\max\{|p_m(\lambda)|: \lambda \in \sigma(M)\}$ makes sense. First, we can understand the minimisation as

$$\varrho(p_m(M)) = \min,$$

i.e., the spectral radius is minimised instead of the spectral norm $\|p_m(M)\|_2$. For the next interpretation, we assume the diagonalisability of M, which leads to $p_m(M) = p_m(TDT^{-1}) = Tp_m(D)T^{-1}$ (D diagonal). Using the norm

$|||\cdot|||_T$ defined in Exercise 2.6.13c, one obtains

$$|||p_m(M)|||_T = \|Tp_m(M)T^{-1}\|_2 = \|p_m(D)\|_2 = \varrho(p_m(D)) = \varrho(p_m(M)) = \max\{|p_m(\lambda)|: \lambda \in \sigma(M)\}. \tag{7.3.6'}$$

In §4.8.1 we discussed symmetric iterations for which not M but $A^{1/2}MA^{-1/2}$ is Hermitian. Then the energy norm of $p_m(M)$ equals

$$\|p_m(M)\|_A = \max\{|p_m(\lambda)|: \lambda \in \sigma(M)\}. \tag{7.3.6''}$$

Assuming again diagonalisability: $T^{-1}MT = D$, one may also estimate by

$$\|p_m(M)\|_2 \leqslant \|p_m(D)\|_2 \operatorname{cond}_2(T) \tag{7.3.6''}$$

The minimisation of

$$\max\{|p_m(\lambda)|: \lambda \in \sigma(M)\} = \|p_m(D)\|_2$$

instead of the expression $\|p_m(M)\|_2$ then minimises the upper bound $\|p_m(D)\|_2 \operatorname{cond}_2(T)$ in (6‴).

The minimisation of $\max\{|p_m(\lambda)|: \lambda \in \sigma(M)\}$ can be solved only with the knowledge of the spectrum $\sigma(M)$. The computation of the complete spectrum, however, would be by far more expensive than the solution of the system.

As a remedy, we assume that there is an *a priori known* set

$$\sigma_M \supset \sigma(M) \tag{7.3.7}$$

containing the spectrum $\sigma(M)$. Then $\sigma(M)$ will be replaced by σ_M. An example for the larger set σ_M is the complex circle

$$\sigma_M = \{\lambda \in \mathbb{C}: |\lambda| \leqslant \bar{\varrho}\} \quad \text{with } \bar{\varrho} \geqslant \varrho(M). \tag{7.3.8a}$$

Unfortunately, this circle is inappropriate for our purposes as we will see in Theorem 9. If, however, M has only *real* eigenvalues, the interval

$$\sigma_M = [-\bar{\varrho}, \bar{\varrho}] \quad \text{with } \bar{\varrho} \geqslant \varrho(M) \tag{7.3.8b}$$

is a candidate. In some cases, it is known in addition that M has only non-negative eigenvalues (cf. Theorem 4.8.3b). Then one may choose

$$\sigma_M = [0, \bar{\varrho}] \quad \text{with } \bar{\varrho} \geqslant \varrho(M). \tag{7.3.8c}$$

In all cases, it is sufficient to know an upper bound $\bar{\varrho}$ of $\varrho(M)$, where $\bar{\varrho} = \varrho(M)$ would be optimal and $\bar{\varrho} < 1$ must hold as we will see. For instance, one may choose $\bar{\varrho}$ as $\varrho_{m+k,m}$ from (3.2.18b) for suitable m and k (cf. Remark 3.4.3).

According, the minimisation of the expression (6) is replaced by

3rd Minimisation Problem
Given $m \in \mathbb{N}$, determine a polynomial p_m with (2) and degree $p_m \leqslant m$ such that $$\max\{|p_m(\lambda)|: \lambda \in \sigma_M\} = \min. \tag{7.3.9}$$

Finally, we briefly discuss the choice of the norm in (1) and (3). A non-Hilbert norm (as, e.g., the maximum or row-sum norm $\|\cdot\|_\infty$) leads to a considerably more complicated minimisation problem. It would be possible to replace the Euclidean norm $\|x\|_2$ by $|||x|||_T = \|Tx\|_2$ or $\|x\|_K = \|K^{1/2}x\|_2$ (K positive definite) as already done in (6′) and (6″). Examples for K would be A and the matrix W of the third normal form (cf. Remark 4.8.2d and (6″)).

7.3.3 Chebyshev Polynomials

As preparation for the next section we discuss the Chebyshev polynomials (other spellings are Tschebyscheff, Čebyšev, etc.; Chebychev, however, is an incorrect transcription).

Definition 7.3.2. The Chebyshev polynomials T_m are defined by

$$T_m(x) := \cos(m \arccos x) \quad \text{for } m \in \mathbb{N}_0,\ |x| \leqslant 1. \tag{7.3.10}$$

Part (a) of the following theorem summarising all properties needed later shows that the functions T_m are, in fact, polynomials of degree m.

Lemma 7.3.3. (a) *The functions T_m from* (10) *fulfil the recursion*

$$T_0(x) = 1, \quad T_1(x) = x, \quad T_{m+1}(x) = 2xT_m(x) - T_{m-1}(x). \tag{7.3.11a}$$

(b) *For $|x| \geqslant 1$ the polynomials T_m have the representation*

$$T_m(x) = \cosh(m \operatorname{Arcosh} x) \quad \textit{for } m \in \mathbb{N}_0,\ |x| \geqslant 1, \tag{7.3.11b}$$

where $\cosh(x) = \frac{1}{2}(e^x + e^{-x})$ *is the* hyperbolic cosine *and Arcosh* (area-hyperbolic cosine) *its inverse function.*

(c) *For all $x \in \mathbb{C}$, the representation* (11c) *holds*:

$$T_m(x) = \tfrac{1}{2}[(x + \sqrt{x^2 - 1})^m + (x + \sqrt{x^2 - 1})^{-m}]. \tag{7.3.11c}$$

Proof. (11a) follows from the cosine addition theorem. For (11b) it suffices to prove that the functions defined there also satisfy the recursion (11a). The substitution $x = \cos\zeta$ shows that (11c) coincides with $\cos m\zeta = T_m(x)$. □

In addition, we mention that $\{T_m\}$ are the orthogonal polynomials with respect to the weight function $(1 - x^2)^{-1/2}$ (cf. Stoer [1, §3.5], Stoer–Bulirsch [1, §3.6]).

7.3.4 Chebyshev Method (Solution of the 3rd Minimisation Problem)

As in the examples (8b, c), we assume that σ_M is a real interval. The solution of the third minimisation problem (9) is given in

Lemma 7.3.4. *Let $[a, b]$ be an interval with $-\infty < a < b < 1$. The problem*

$$\begin{aligned}&\textit{minimise } \max\{|p_m(\lambda)|: a \leqslant \lambda \leqslant b\}\\ &\textit{over all polynomials } p_m \textit{ with degree } p_m \leqslant m \textit{ and } p_m(1) = 1\end{aligned} \tag{7.3.12a}$$

has the unique solution

$$p_m(\zeta) = T_m\left(\frac{2\zeta - a - b}{b - a}\right) \Big/ C_m \tag{7.3.12b}$$

with $C_m := T_m\left(\frac{2 - a - b}{b - a}\right)$ *and the Chebyshev polynomial T_m from* (10). *The minimising polynomial p_m has the degree m and leads to the minimum*

$$\max\{|p_m(\lambda)|: a \leqslant \lambda \leqslant b\} = 1/C_m \quad \text{for } p_m \textit{ from } (12b). \tag{7.3.12c}$$

Proof. (i) The constant C_m does not vanish, since the argument $(2 - a - b)/(b - a)$ lies outside of $[-1, 1]$ and the representation (11b) applies. By construction, $p_m(1) = 1$ and degree $p_m = m$ hold. For $a \leqslant \zeta \leqslant b$, the argument $(2\zeta - a - b)/(b - a)$ belongs to $[-1, 1]$. There, definition (10) shows that $|T_m| \leqslant 1$. Since T_m attains the bounds ± 1, (12c) follows.

(ii) It remains to show that for any other polynomial the maximum in (12c) is larger than $1/C_m$. Let q_m be a polynomial with $q_m(1) = 1$, degree $q_m \leqslant m$, and $\max\{|q_m(\zeta)|: \zeta \in [a, b]\} \leqslant 1/C_m$. One verifies that the Chebyshev polynomial $T_m(x) = \cos(m \arccos x)$ meets the values ± 1 in alternating ordering at $x = \cos(\nu\pi/m)$ for $\nu = -m, 1 - m, \ldots, 0$. The function p_m arising from T_m by the transformation $x \mapsto \zeta = \frac{1}{2}[a + b + x(b - a)]$ attains the values

$$p_m(\zeta_\nu) = (-1)^\nu/C_m \qquad (-m \leqslant \nu \leqslant 0)$$

as $\zeta_\nu = \frac{1}{2}[a + b + (b - a)\cos(\nu\pi/m)]$. From $|q_m(\zeta_\nu)| \leqslant 1/C_m = |p_m(\zeta_\nu)|$, one concludes that the difference $r := p_m - q_m$ satisfies

$$r(\zeta_\nu) \geqslant 0 \quad \text{for even } \nu, \qquad r(\zeta_\nu) \leqslant 0 \quad \text{for odd } \nu.$$

By the intermediate value theorem, there exists at least one zero of r in each subinterval $[\zeta_{\nu-1}, \zeta_\nu]$ $(1 - m \leqslant \nu \leqslant 0)$. Even when the zeros from $[\zeta_{\nu-1}, \zeta_\nu]$ and $[\zeta_\nu, \zeta_{\nu+1}]$ coincide at the common point ζ_ν, this is a double zero. Hence, counted with respect to multiplicity, r has at least m zeros in $[a, b]$. By $p_m(1) = q_m(1) = 1$, the value 1 represents the $(m + 1)$st zero of r. Hence, $r = 0$ follows from degree $r \leqslant m$, proving uniqueness $p_m = q_m$. □

Exercise 7.3.5. Prove by means of (11a) that the polynomials p_m from (12b) can be obtained by the recursion

$$p_0(\zeta) = 1, \qquad p_1(\zeta) = \frac{2\zeta - a - b}{2 - a - b}, \tag{7.3.13a}$$

$$C_{m+1}p_{m+1}(\zeta) = 2\frac{2\zeta - a - b}{b - a}C_m p_m(\zeta) - C_{m-1}p_{m-1}(\zeta). \tag{7.3.13b}$$

To estimate the quality of the minimum $1/C_m = 1/T_m\left(\frac{2-a-b}{b-a}\right)$ reached in (12c), we have to evaluate (11c) at $x_0 = (2-a-b)/(b-a)$. One verifies that $x_0^2 - 1 = 4(1-a)(1-b)/(b-a)^2 > 0$ and $x_0 + \sqrt{x_0^2-1} = (\sqrt{1-a} + \sqrt{1-b})^2/(b-a)$. The representation (11c) shows that

$$C_m = \frac{1}{2}\left\{\left(\frac{(\sqrt{1-a}+\sqrt{1-b})^2}{b-a}\right)^m + \left(\frac{(\sqrt{1-a}+\sqrt{1-b})^2}{b-a}\right)^{-m}\right\}.$$

The bracketed $\left(\frac{(\sqrt{1-a}+\sqrt{1-b})^2}{b-a}\right)$ can be rewritten as $\frac{1-a}{b-a}\left(1 + \sqrt{\frac{1-b}{1-a}}\right)^2$. To simplify the expression, we introduce

$$\varkappa := \frac{1-a}{1-b}, \qquad c := \left(1 - \frac{1}{\sqrt{\varkappa}}\right)\bigg/\left(1 + \frac{1}{\sqrt{\varkappa}}\right).$$

Since $\frac{b-a}{1-a} = 1 - \frac{1}{\varkappa} = \left(1 + \frac{1}{\sqrt{\varkappa}}\right)\left(1 - \frac{1}{\sqrt{\varkappa}}\right)$, we arrive at

$$\left(\frac{(\sqrt{1-a}+\sqrt{1-b})^2}{b-a}\right) = \left(1 - \frac{1}{\sqrt{\varkappa}}\right)^{-1}\left(1 + \frac{1}{\sqrt{\varkappa}}\right)^2$$
$$= \left(1 + \frac{1}{\sqrt{\varkappa}}\right)\bigg/\left(1 - \frac{1}{\sqrt{\varkappa}}\right) = \frac{1}{c}.$$

Hence, the expression for $1/C_m$ reduces to

$$\boxed{\frac{1}{C_m} = \frac{2c^m}{1+c^{2m}}, \text{ where } c := \frac{1-1/\sqrt{\varkappa}}{1+1/\sqrt{\varkappa}} = \frac{\sqrt{\varkappa}-1}{\sqrt{\varkappa}+1} \text{ and } \varkappa := \frac{1-a}{1-b}} \quad (7.3.13c)$$

For the interpretation of $\varkappa$ as a condition number, compare §7.3.8.

Remark 7.3.6. (a) For the case (8c): $\sigma_M = [-\bar{\varrho}, \bar{\varrho}]$ with $0 < \bar{\varrho} < 1$ (i.e., $a = -\bar{\varrho}$, $b = \bar{\varrho}$), the solution of the 3rd minimisation problem (9) is:

$$p_m(\zeta) = T_m(\zeta/\bar{\varrho})/C_m \quad \text{with } C_m := T_m(1/\bar{\varrho}) \tag{7.3.14a}$$

(b) For the case (8b): $\sigma_M = [0, \bar{\varrho}]$ with $0 < \bar{\varrho} < 1$ (i.e., $a = 0$, $b = \bar{\varrho}$), the respective solution reads

$$p_m(\zeta) = T_m\left(\frac{2\zeta - \bar{\varrho}}{\bar{\varrho}}\right)\bigg/C_m \quad \text{with } C_m := T_m\left(\frac{2-\bar{\varrho}}{\bar{\varrho}}\right). \tag{7.3.14b}$$

(c) The respective attained minima are $\frac{1}{C_m} = \frac{2c^m}{1+c^{2m}}$ with

$$c = \frac{2\bar{\varrho}}{(\sqrt{1+\bar{\varrho}} + \sqrt{1-\bar{\varrho}})^2} \quad \text{for (14a)}, \qquad c = \frac{\bar{\varrho}}{(1+\sqrt{1-\bar{\varrho}})^2} \quad \text{for (14b)}. \tag{7.3.14c}$$

(d) For the semi-iterates y^m, the following error estimates hold:

$$\|y^m - x\|_2 \leqslant \eta_m \operatorname{cond}_2(T)\|x^0 - x\|_2 \quad \text{with} \tag{7.3.14d$_1$}$$

$$\eta_m = 2\left(1 - \frac{1}{\varkappa}\right)^m \Big/ \left[\left(1 + \frac{1}{\sqrt{\varkappa}}\right)^{2m} + \left(1 - \frac{1}{\sqrt{\varkappa}}\right)^{2m}\right], \tag{7.3.14e}$$

where $\varkappa$ is defined by (13c) and T is the transformation from (6‴). In the case of a symmetric iteration (cf. §4.8.1), the estimate (14d$_2$) holds with respect to the energy norm:

$$\|y^m - x\|_A \leqslant \eta_m \|x^0 - x\|_A. \tag{7.3.14d$_2$}$$

For the implementation of the Chebyshev method, one could, in principle, compute the coefficients α_{mj} of $p_m(\zeta) = \sum \alpha_{mj}\zeta^j$ and use the first formulation (1.6) of the semi-iteration. In the case of a *fixed m*, one may use the second formulation. The Chebyshev polynomial T_m has the zeros $x_\nu = \cos([\nu + \frac{1}{2}]\pi/m)$ $(1 \leqslant \nu \leqslant m)$. Hence, the transformed polynomial p_m from (12b) admits the factorisation $\prod_{\nu=1}^{m} (\zeta - \zeta_\nu)/(1 - \zeta_\nu)$ with $\zeta_\nu = \frac{1}{2}[a + b + (b - a)\cos([\nu + \frac{1}{2}]\pi/m)]$. The auxiliary polynomials

$$\hat{p}_k(\zeta) := \prod_{\nu=1}^{k} (\zeta - \zeta_\nu)/(1 - \zeta_\nu) \qquad (0 \leqslant k \leqslant m)$$

satisfy (2.3a, b) and lead to $\Theta_k := 1/(1 - \zeta_k)$ in (2.3c). Therefore, the method (2.1a, b) (the «second formulation») can be performed with these Θ_k for $k = 0, 1, \ldots, m$. Since $\hat{p}_m = p_m$ for the fixed index m, one ends up with the optimal semi-iterative solution y^m. This approach, however, has severe disadvantages:

(i) For the following computation of y^{m+1}, one has to perform (2.1a, b) again from $k = 0$ to $k = m + 1$, since then other auxiliary polynomials $\hat{p}_k$ are needed.

(ii) In general, the second formulation (2.1a, b) is *unstable*. Already for relative small m, the rounding error influence of the iteration errors $y^m - x$ can predominate.

It is possible to avoid instability by a suitable renumbering of the Θ_ν. Concerning stability analysis and the choice of an appropriate ordering, we refer the interested reader to Lebedev–Finogenov [1] (cf. also Samarskij–Nikolaev [1, §6.2.4]).

The only elegant and practical implementation is the use of the *three-term recursion* (2.4a–c), since the recursion (13a, b) is a particular case of (2.5a, b). The coefficients Θ_m and ϑ_m required in (2.4a–c) are provided by

Exercise 7.3.7. Prove that (a) for the case $\sigma_M = [a, b]$ with $a < b < 1$, the recursion (2.5a, b) for p_m from (13a, b) has the factors

$$\Theta_m = 4C_{m-1}/[(b - a)C_m], \tag{7.3.15a}$$

$$\vartheta_m = -2(a + b)C_{m-1}/[(b - a)C_m]. \tag{7.3.15b}$$

(b) In the case $\sigma_M = [-\varrho, \varrho]$, (13b) leads to the recursion (2.5c, d) with

$$\Theta_m = 2C_{m-1}/[\varrho C_m] = 1 + C_{m-2}/C_m. \qquad (7.3.15a')$$

(c) Which coefficients result in the case of $\sigma_M = [0, \varrho]$?
(d) Use the equation (13b) at $\zeta = 1$: $C_{m+1} = AC_m - C_{m-1}$ with $A := 2(2-a-b)/(b-a)$ and prove for the general case $\sigma_M = [a, b]$ that

$$\Theta_m = 16/[8(2-a-b) - (b-a)^2\Theta_{m-1}], \quad \Theta_1 = 4/(2-a-b), \qquad (7.3.15c)$$

$$\vartheta_m = -\tfrac{1}{2}(a+b)\Theta_m. \qquad (7.3.15d)$$

(e) The coefficients converge monotonically to $\lim \Theta_m = 4c/(b-a)$ and $\lim \vartheta_m = -2c(a+b)/(b-a)$ with c from (13c).
Hint for (a): For $m \geqslant 2$, compare the coefficients in (2.5b) and (13b). For $m = 1$, compare (2.5a) with (13a), taking notice of $C_0 = 1$ and $C_1 = (2-a-b)/(b-a)$ according to (13c). (e) Insert (13c) into (15a, b).

Instead of the quantities Θ_m and ϑ_m, one can also compute the sum $\sigma_m := \Theta_m + \vartheta_m$ recursively from

$$\sigma_m = 4 \Big/ \left\{4 - \left(\frac{1 - 1/\varkappa}{1 + 1/\varkappa}\right)^2 \sigma_{m-1}\right\}, \qquad \sigma_1 = 2 \qquad (7.3.15e)$$

(derived from (15c, d)). Equation (15d) yields the values

$$\Theta_m = 2\sigma_m/(2-a-b), \qquad \vartheta_m = -(a+b)\sigma_m/(2-a-b). \quad (7.3.15f)$$

The coefficients σ_m can also be used directly for the three-term recursion. Given the matrix W of the third normal form of Φ, the formulae (2.4a–c) with the coefficients (15a, b) are equivalent to

$$y^0 = x^0, \qquad (7.3.16a)$$

$$y^1 = y^0 - \frac{2}{2-a-b} W^{-1}(Ay^0 - b), \qquad (7.3.16b)$$

$$y^m = \sigma_m \left\{ y^{m-1} - \frac{2}{2-a-b} W^{-1}(Ay^{m-1} - b) \right\} + (1 - \sigma_m) y^{m-2}. \quad (7.3.16c)$$

7.3.5 Order Improvement by the Chebyshev Method

Theorem 7.3.8. (a) *Assume that $\sigma(M) \subset \sigma_M = [a, b]$ with $a < b < 1$. The Chebyshev method has the asymptotical convergence rate c from* (13c):

$$\lim_{m\to\infty} (1/C_m)^{1/m} = c = \frac{b-a}{2 - b - a + 2\sqrt{(1-a)(1-b)}}. \qquad (7.3.17a)$$

Particular values are

$$\lim_{m\to\infty} (1/C_m)^{1/m} = \frac{\varrho}{1+\sqrt{1-\varrho^2}} \qquad \textit{for } \sigma_M = [-\varrho,\varrho],\ \varrho < 1, \tag{7.3.17b}$$

$$\lim_{m\to\infty} (1/C_m)^{1/m} = \frac{\varrho}{(1+\sqrt{1-\varrho})^2} \qquad \textit{for } \sigma_M = [0,\varrho],\ \varrho < 1. \tag{7.3.17c}$$

(b) *Let $\varkappa$ be the order of the basic iteration: $\varrho(M) = 1 - Ch^{\varkappa} + O(h^{2\varkappa})$. Then the Chebyshev method has the order $\varkappa/2$. The asymptotical convergence rate equals*

$$1 - 2\sqrt{C/(1-a)}h^{\varkappa/2} + O(h^{\varkappa}) \quad \textit{for (17a) with } b = \varrho(M), \tag{7.3.18a}$$

$$1 - \sqrt{2C}h^{\varkappa/2} + O(h^{\varkappa}) \quad \textit{for } \sigma_M = [-\varrho(M),\varrho(M)], \tag{7.3.18b}$$

$$1 - 2\sqrt{C}h^{\varkappa/2} + O(h^{\varkappa}) \quad \textit{for } \sigma_M = [0,\varrho(M)]. \tag{7.3.18c}$$

Proof. Since $0 \leqslant c \leqslant 1$, (13c) shows $(1/C_m)^{1/m} = c[2/(1+c^{2m}]^{1/m} \to c$. □

Therefore, the Chebyshev method achieves a halving of the order similar to the SOR iteration. Concerning the connection of both methods, compare §7.4.3 and Varga [2, §5.2].

7.3.6 Optimisation over Other Sets

Up to now, we have considered an interval $[a,b]$ with $a < b < 1$. If, for instance, no eigenvalue of M lies in $(c,d) \subset [a,b]$, one may replace σ_M by the smaller set

$$\sigma_M = [a,c] \cup [d,b] \qquad (a \leqslant c < d \leqslant b).$$

Obviously, the minimum $\min_{p_m} \max_{\sigma_M} |p(\zeta)|$ can only become smaller. In the case of $c - a = b - d$, it is easy to describe the optimal polynomial (cf. Axelsson–Barker [1, p. 26f]). Concerning the determination of optimal polynomials, we refer to deBoor–Rice [1]. The case $\sigma_M = [a,c] \cup [d,b]$ is interesting, in particular, if $a \leqslant c < 1 < d \leqslant b$. This situation occurs for indefinite A (see also §8.3.2).

If *one* extreme eigenvalue c of M is known and the others are enclosed by $[a,b]$, one arrives at

$$\sigma_M = [a,b] \cup \{c\} \quad \text{with } c \notin [a,b],\ 1 \notin [a,b],\ c \neq 1.$$

Let q_{m-1} be optimal for $[a,b]$. A simple but not optimal proposal for a polynomial p_m suited to σ_M is

$$p_m(\zeta) := q_{m-1}(\zeta)(\zeta - c)/(1-c).$$

Concerning the construction of asymptotical optimal polynomials for arbitrary compact sets σ_M with $1 \notin \sigma_M$, we refer to Niethammer–Varga [1] and Eiermann–Niethammer–Varga [1]. The simplest set σ_M that is more general than the interval $[a, b]$ is the ellipse (cf. Fischer–Freund [1, 2], Niethammer–Varga [1], Manteuffel [1]). Since, in general, a suitable ellipse enclosing the eigenvalues of M is not known a priori, one has to improve its parameters adaptively (cf. Manteuffel [1]). The fact that the ellipse lies in the complex plane does not imply that the optimal polynomials have complex parameters also. As long as σ_M is symmetric with respect to the real axis, one can find an optimal polynomial with real coefficients (cf. Opfer–Schober [1]).

In any case, the spectrum $\sigma(M)$ is enclosed by the complex circle

$$\sigma_M = \{z = x + iy \in \mathbb{C}: x^2 + y^2 \leqslant \varrho(M)^2\}.$$

Unfortunately, this choice does not lead to an interesting solution.

Theorem 7.3.9. *Let σ_M be the circle around $z_0 \in \mathbb{C}\backslash\{1\}$ with radius $r < |1 - z_0|$. The optimal polynomial for σ_M is $p_m(\zeta) = [(\zeta - z_0)/(1 - z_0)]^m$. In particular, for $z_0 = 0$ the corresponding semi-iteration coincides with the basic iteration Φ. In the general case, the semi-iteration corresponds to the damped method Φ_Θ with $\Theta := 1/|1 - z_0|$.*

Proof (cf. Opfer–Schober [1]). The maximum $\varrho := \max\{|p_m(\zeta)|: \zeta \in \sigma_M\} = r/|1 - z_0|$ is taken by p_m on the whole boundary of σ_M. If p_m is not optimal, there would be some q_m of degree $\leqslant m$ with $q_m(1) = 1$ and $\max\{|q_m(\zeta)|: \zeta \in \sigma_M\} < \varrho$. Hence, $q_m(\zeta) < \varrho = p_m(\zeta)$ would hold for all boundary values $\zeta \in \partial\sigma_M$, so that the theorem of Rouché is applicable: The holomorphic functions p_m and $p_m - q_m$ have the same number of zeros in σ_M. Since p_m has an m-fold zero at z_0, $p_m - q_m$ also has m zeros in σ_M. Since $(p_m - q_m)(1) = p_m(1) - q_m(1) = 1 - 1 = 0$, the polynomial $p_m - q_m$ of degree $\leqslant m$ has even $m + 1$ zeros, implying $p_m = q_m$. Hence, p_m is already optimal. □

7.3.7 Cyclic Iteration

Following Remark 6, it has been mentioned that, in principle, it would be possible to apply the second formulation (2.1b) with the factors $\Theta_\nu := -\zeta_\nu/(1 - \zeta_\nu)$, $\zeta_\nu = \cos([\nu + \frac{1}{2}]\pi/m)$ for $\nu = 1, \ldots, m$. The result y^m (for this *fixed* m) is the desired Chebyshev semi-iterate. However, by this approach the Chebyshev method cannot be continued. In order to obtain nonetheless an infinite iterative process, one repeats the extrapolation factors in a cyclic manner:

$$\Theta_1, \Theta_2, \ldots, \Theta_m \quad \text{given}, \tag{7.3.19a}$$

$$\Theta_i := \Theta_{i-m} \quad \text{for } i > m. \tag{7.3.19b}$$

The semi-iterative method (2.1a, b) with these parameters is called the *cyclic iteration*. The restriction to the iterates $y^0, y^m, y^{2m}, y^{3m}, \ldots$ produces a proper iteration. The related iteration matrix is $p_m(M)$. The convergence rate of the cyclic iteration is not described by $\varrho(p_m(M))$ but by $\varrho(p_m(M))^{1/m}$, since one cycle $y^0 \mapsto y^m$ is thought to consist of m and not of one step. The cyclic iteration also runs the risk of *numerical instabilities* which were already discussed after Remark 6.

Exercise 7.3.10. Prove that if one views the cyclic iteration as a semi-iteration $\{y^0, y^1, y^2, \ldots\}$ of all iterates, the asymptotical convergence rate from Definition 1.2 also coincides with $\varrho(p_m(M))^{1/m}$.

7.3.8 Reformulation

The matrices M and W of the first and third normal form are connected via $M = I - W^{-1}A$ (cf. (3.2.3′/6)). In the present formulation, we are looking for suitable polynomials $p(\zeta)$ with the side condition $p(1) = 1$. A polynomial in $M = I - W^{-1}A$ can be expanded as a polynomial in $W^{-1}A$:

$$p(M) = q(W^{-1}A) \tag{7.3.20a}$$

with q from

$$q(\mu) := p(1 - \mu). \tag{7.3.20b}$$

The side condition $p(1) = 1$ becomes

$$q(0) = 1. \tag{7.3.20c}$$

Since (20c) can also be expressed by the factorisation $q(\mu) = (1 - \mu)\hat{q}(\mu)$ with degree $(\hat{q})$ = degree$(q) - 1$, (21) is an alternative representation of (20a–c):

$$p(M) = I - W^{-1}A\hat{q}(W^{-1}A) \quad \text{with } \hat{q}(\mu) := [p(1 - \mu) - 1]/\mu. \tag{7.3.21}$$

It is easy to see that

(i) M has a real spectrum $\sigma(M)$ if and only if $\sigma(W^{-1}A)$ is also real;

(ii) a is a lower and b an upper bound for $\sigma(M)$ if

$$\Gamma := 1 - a, \qquad \gamma := 1 - b \tag{7.3.22}$$

are the upper and lower bounds of $\sigma(W^{-1}A)$:

$$\sigma(W^{-1}A) \subset [\gamma, \Gamma]. \tag{7.3.23a}$$

The expression $\varkappa$ from (13c) can be rewritten as

$$\varkappa = \Gamma/\gamma. \tag{7.3.23b}$$

The number $\varkappa = \Gamma/\gamma$ with the best constants γ and Γ satisfying (23a) coincides with the condition number (23c), provided that $\gamma > 0$:

$$\varkappa = \varkappa(W^{-1}A) \qquad (\varkappa(\cdot) \text{ defined in (2.10.8)}). \tag{7.3.23c}$$

We recall that a sufficient condition for a symmetric iteration is that W and A are positive definite (cf. §4.8.1)). These assumptions are made in

Lemma 7.3.11. *For a symmetric iteration,* (23a) *is equivalent to*

$$\gamma W \leqslant A \leqslant \Gamma W. \tag{7.3.23a'}$$

The optimal bounds $0 < \gamma \leqslant \Gamma$ *from* (23a') *yield the condition number*

$$\varkappa = \Gamma/\gamma = \varkappa(W^{-1}A) = \mathrm{cond}_2(W^{-1/2}AW^{-1/2}). \tag{7.3.23c'}$$

Proof. Since $\sigma(W^{-1}A) = \sigma(W^{-1/2}AW^{-1/2})$ (cf. Lemma 2.4.16), (23a) is equivalent to $\gamma I \leqslant W^{-1/2}AW^{-1/2} \leqslant \Gamma I$ (cf. (2.10.3e)). Multiplication by $W^{1/2}$ from both sides yields (23a') (cf. (2.10.3b')). □

The optimal polynomial is $p_m(\zeta) = T_m\left(\frac{2\zeta - a - b}{b - a}\right)\Big/ T_m\left(\frac{2 - a - b}{b - a}\right)$ (cf. (12b)). Inserting the expressions $1 - \Gamma$ and $1 - \gamma$ into $\frac{2\zeta - a - b}{b - a}$ for a and b, one obtains $[(\Gamma + \gamma)I - 2W^{-1}A]/(\Gamma - \gamma)$. Analogously, $\frac{2 - a - b}{b - a}$ becomes $(\Gamma + \gamma)/(\Gamma - \gamma)$. Expressed as polynomial in $W^{-1}A$, the optimal polynomial now reads

$$p_m(M) = q_m(W^{-1}A) = T_m\left(\frac{\Gamma + \gamma}{\Gamma - \gamma}I - \frac{2}{\Gamma - \gamma}W^{-1}A\right)\Big/ T_m\left(\frac{\Gamma + \gamma}{\Gamma - \gamma}\right). \tag{7.3.24}$$

The asymptotical convergence speed (17a) expressed by γ and Γ is

$$\lim_{m\to\infty} (1/C_m)^{1/m} = c = \frac{\sqrt{\Gamma} - \sqrt{\gamma}}{\sqrt{\Gamma} + \sqrt{\gamma}} = \frac{1 - \sqrt{\gamma/\Gamma}}{1 + \sqrt{\gamma/\Gamma}}. \tag{7.3.25}$$

7.3.9 Multi-Step Iterations

In Exercise 7e we determined the limits $\Theta = \lim \Theta_m$ and $\vartheta = \lim \vartheta_m$. Hence, the three-term recursion (2.4c) converges to the (stationary) two-step iteration (3.2.23):

$$y^m = \Theta\{\Phi(y^{m-1}, b) - y^{m-2}\} + \vartheta(y^{m-1} - y^{m-2}) + y^{m-2}.$$

As described in §3.2.8, the convergence of iteration (3.2.23) can be reduced to the convergence of a one-step iteration with the iteration matrix

$$\mathbf{M} = \begin{bmatrix} \mu_0 M + \mu_1 I & \mu_2 I \\ I & O \end{bmatrix}, \qquad \mu_0 = \frac{4c}{b - a}, \quad \mu_1 - 2c\frac{a + b}{b - a},$$

$$\mu_2 = 1 - \mu_0 - \mu_1$$

(c from (13c)). From these coefficients, assuming that $\sigma(M) \subset \sigma_M$ and using Exercise 3.2.20, one obtains the value $\varrho(\mathbf{M}) = c$, i.e., the (stationary) two-step iteration (3.2.23) achieves the same convergence rate as the semi-iterative method.

More generally, one can consider the k-step iteration

$$x^m = \mu_0 \Phi(x^m, b) + \sum_{i=1}^{k} \mu_i x^{m-i} \quad \text{with} \sum_{i=0}^{k} \mu_i = 1.$$

The connection between k-step iterations and semi-iterative methods is described by Niethammer–Varga [1].

7.3.10 Pascal Procedures

The Chebyshev method requires information about the spectrum bounds a and b, which have to satisfy $a \leqslant \lambda_{\min} \leqslant \lambda_{\max} \leqslant b < 1$, where $\lambda_{\min}$ and $\lambda_{\max}$ are the extreme eigenvalues of the iteration matrix M. The quantities a and b are not stored directly but via γ and Γ from (22) as components gmg:= $\Gamma - \gamma$ and gpg:= $\Gamma + \gamma$ of the record of type `iterationparameter`. The procedures `set_lambda_bounds` and `determine_lambda_bounds` are available for the direct or interactive input. The function `asymptotical_semi-iterative_rate` determines the rate (25) from gmg and gpg.

```
function check_lambda(a,b: real): Boolean; var ok: Boolean;
begin ok:=b<1;
  if not ok then writeln('b must be <1!');
  if a>b then begin ok:=false; writeln('a must be <b=',b,'!')
  end;
  check_lambda:=ok
end;

procedure set_lambda_bounds(var IP: iterationparameter; a,b:
 real);
begin if not check_lambda(a,b) then
  message('Inconsistent parameters for semi-iteration!');
  IP.gmg:=b-a; IP.gpg:=2-a-b; IP.Nr:=0
end;

procedure determine_lambda_bounds(var IP:
 iterationparameter); var a,b: real;
begin writeln('Parameters for semi-iteration:');
  writeln('*** Input of a upper bound for lambda-max.');
  repeat write('--> b='); readln(b); a:=b until check_
   lambda(a,b);
  writeln('*** Input of a lower bound for lambda-min.');
  repeat write('--> a='); readln(a) until check_lambda(a,b);
  set_lambda_bounds(IP,a,b)
end;
```

```
procedure lambda_bounds(var it: data_of_iteration; a,b:
 real);
begin if it.A.kind=Poisson_model_problem then set_lambda_
 bounds(it.IP,a,b)
      else determine_lambda_bounds(it.IP)
end;

function asymptotical_semiiterative_rate(var IP:
 iterationparameter): real;
begin with IP do
      asymptotical_semiiterative_rate := gmg/(gpg+sqrt
       (gpg*gpg-gmg*gmg))
end;
```

Similarly, values for the bounds γ and Γ from (22) can be obtained by

```
function check_gamma(lower,upper: real): Boolean;
procedure set_gamma_bounds(...);
procedure determine_gamma_bounds(var IP: iteration-
   parameter);
procedure gamma_bounds(var it: data_of_iteration;
   lower,upper: real);
```

The procedures set the flag `IP.Nr:=0` in order to start the semi-iteration. The semi-iterative method follows the representation (16b–c). The parameter σ_m corresponds to the component `IP.sigma`. The previous iterate y^{m-2} can be stored on `IP.previous^`. After finishing the semi-iteration, this storage may be released again by calling «`procedure release_iteration-parameter (var IP: iterationparameter)`». The last parameter required in procedure `semi_iteration` is the underlying basic iteration. Examples for semi-iterative methods are described in §7.4.

`procedure semi_iteration` (7.3.26)

```
procedure semi_iteration(var neu:gridfunction;
                    var A: data_of_discretisation;
                    var x,b:gridfunction;
                    var IP: iterationparameter;
                    procedure basisiteration
                              (var neu:gridfunction;
                               var A: data_of_discretisation;
                               var x,b: gridfunction;
                               var IP: iterationparameter));
var y: gridfunction;
begin with IP do with A do
  if Nr<0 then writeln('Semi-iteration parameters are not yet
   defined!') else
```

```
  begin if Nr=0 then
  begin if previous=nil then new(previous); previous^:=x;
   sigma:=1;
        basisiteration(y,A,x,b,IP);
        factor_x_vector_plus_factor_x_vector(nx,ny,neu,
         2/gpg,y,1-2/gpg,x)
  end else
  begin sigma:=4/(4-sqr(gmg/gpg)*sigma); basisiteration
   (y,A,x,b,IP);
  factor_x_vector_plus_factor_x_vector(nx,ny,y,2/gpg,y,
   1-2/gpg,x);
  factor_x_vector_plus_factor_x_vector(nx,ny,y,sigma,y,
   1-sigma,previous^);
  previous^:=x; neu:=y
  end;
  transfer_boundary_values(nx,ny,x,neu); Nr:=Nr+1
end end;
```

Analogously to the frame program from §3.5.5, the iteration data can be collected in the variable `it`. The procedure `semiiteration_1` uses this variable and performs one semi-iteration step.

```
procedure semiiteration_1(var it: data_of_iteration;
                procedure basisiteration
                              (var new: gridfunction;
                               var A: data_of_discretisation;
                               var x,b: gridfunction;
                               var IP: iterationparameter));
begin with It do begin semi_iteration(x,A,x,b,IP,
 basisiteration)
end end;
```

A complete frame program similar to (3.5.9) can read as follows:

Frame program for calling a semi-iterative method (7.3.27)

```
program Semiiteration;
var it: data_of_iteration; i,itnr: integer; v: data_for_
 comparison;
    l2: history_of_iteration;
{right_hand_side, zerofunction, exact_solution, boundary_
 value as in (3.5.9)}
{Further, "Euclidean_norm", "lex_SSOR", etc. are to be
 defined}
```

```
begin initialise_IT(it); initialise_comparison(v);
 repeat release_IT(it); release_comparison(v);
 define_problem(it,boundary_value,right_hand_side);
 define_SSOR_parameter(it); writeln('SSOR parameter ω=',
  it.IP.omega);
 define_SSOR_semiiteration_parameters(it);
 define_starting_iterate(it,zerofunction); writeln
  ('Starting value defined.');
 define_comparison_solution(v,it.A,exact_solution);
 comparison_with_exact_solution(l2,v,it,Euclidean_norm);
 write('--> Number of iterations ='); readln(itnr);
 for i:=1 to itnr do
 begin
  semiiteration_1(it,lex_SSOR);
  comparison_with_exact_solution(l2,v,it,Euclidean_norm);
  writeln('Iteration Nr.',it.IP.Nr,'Euclidean norm:',
   l2.value[l2.last])
 end;
 writeln('*** Run completed ***'); {results can be stored as
  in (3.5.9)}
 write('Should the run be repeated?')
 until not yes_no
end.
```

7.3.11 Amount of Work of the Semi-Iterative Method

Besides the call of the basic iteration Φ, the implementation (26) (for $m \geqslant 2$) requires six operations per grid point. This leads to

$$\textit{semi-iterative amount of work}(\Phi) \leqslant \textit{amount of work}(\Phi) + 6n \qquad (7.3.28a)$$

(cf. §3.3 and §4.6). Hence, the *cost factor* amounts to

$$C_{\Phi,\text{semiiterative}} = C_\Phi + \frac{6}{C_A}, \qquad (7.3.28b)$$

where C_A is defined in §3.3 as the number of nonzero elements of A.

Replacing in (3.3.4a) the convergence rate by the asymptotical value c from (25), one obtains the *effective amount of work*

$$\mathit{Eff}_{\text{semiiterative}}(\Phi) = -\left(C_\Phi + \frac{6}{C_A}\right)\Big/\log c. \qquad (7.3.28c)$$

If $\gamma/\Gamma \ll 1$ holds as in the examples discussed in §7.4, one can exploit the asymptotical behaviour $\log c = -2\sqrt{\gamma/\Gamma} + O(\gamma/\Gamma)$:

$$\mathit{Eff}_{\text{semiiterative}}(\Phi) \approx \left(\frac{C_\Phi}{2} + \frac{3}{C_A}\right)\sqrt{\frac{\Gamma}{\gamma}}. \qquad (7.3.28d)$$

Exercise 7.3.12. Assume that the iteration matrix of Φ fulfils $\sigma(M) \subset [a,b]$ with $b = 1 - O(h^{-\varkappa})$, $\varkappa > 0$. Prove the following comparison of the iterative and semi-iterative amount of work:

$$Eff_{\text{semiiterative}}(\Phi) \approx \left(C_\Phi + \frac{6}{C_A}\right)\sqrt{Eff(\Phi)/[(1-a)C_\Phi]}. \tag{7.3.29}$$

7.4 Application to Iterations Discussed Above

7.4.1 Preliminaries

The essential condition for the applicability of the Chebyshev method is that the spectrum $\sigma(M)$ is real. This excludes the SOR method. Even no semi-iterative variant based on other sets σ_M can be successfully applied to the SOR method with $\omega \geqslant \omega_{\text{opt}}$ (cf. §7.3.6)). The reason lies in statement (e) of Theorem 5.6.5: For $\omega \geqslant \omega_{\text{opt}}$ all eigenvalues $\lambda \in \sigma(M_\omega^{\text{SOR}})$ are situated on the boundary of the complex circle $|\zeta| = \omega - 1$, for which no convergence acceleration is possible, as stated in Theorem 3.9.

If A is positive definite, the following already mentioned iterations lead to a real spectrum: Richardson, (block-)Jacobi, and (block-)SSOR method. Numerical results for these choices of basic iterations will be presented for the Poisson-model problem in the following sections.

Besides the iterations named above, we have constructed in §4.3 their damped variants. However, for a discussion of semi-iterative methods, the damped variants are without any interest, as stated in

Lemma 7.4.1 *Let the iteration Φ have a real spectrum $\sigma(M)$. Then Φ and the corresponding damped iterations Φ_ϑ for $\vartheta > 0$ generate the same semi-iterative results y^m.*

Proof. By (1.9a, b), the semi-iterate y^m generated by Φ has the representation $y^m = x^0 + p_m(M)(x^0 - x)$. The damped iteration has the iteration matrix $M_\vartheta = I - \vartheta W^{-1}A = I - W_\vartheta^{-1}A$ with $W_\vartheta := W/\vartheta$. For W_ϑ, inequality (3.23a) can be written as $\sigma(W_\vartheta^{-1}A) \subset [\gamma', \Gamma']$ with $\gamma' := \vartheta\gamma$ and $\Gamma' := \vartheta\Gamma$. The right-hand side in (3.24) is invariant against the replacement of γ, Γ, W by γ', Γ', W_ϑ. Hence, (3.24) proves $p_m(M_\vartheta) = p_m(M)$. Therefore, the iterates $y_\vartheta^m = x^0 + p_m(M_\vartheta)(x^0 - x)$ of Φ_ϑ coincide with those of Φ. □

For the sake of simplicity, the PASCAL procedures from §7.3.10 make use of the basic iteration in the previously programmed form. Another conceivable alternative would be as follows: The procedure `residual` (cf. §4.3.1.2) for computing $b - Ax$ already exists. It would be sufficient to represent the respective iteration Φ by its matrix $N = W^{-1}$ in the sense that a subroutine for computing of $r \mapsto W^{-1}r$ is supplied. The representation of the semi-iterative method by (3.16c) shows that $N(b - Ay) = W^{-1}(b - Ay)$ is the core of the method.

7.4.2 Semi-Iterative Richardson Method

According to Lemma 1, we may fix the damping factor of Richardson's method (4.3.3) by $\Theta = 1$: $x^{m+1} = x^m - (Ax - b)$. Then the matrix of the third normal form is $W = I$ and condition (3.23a) becomes $\sigma(A) \subset [\gamma, \Gamma]$. An immediate consequence is

Remark 7.4.2. (a) The Chebyshev method is applicable if A has only positive eigenvalues. For the estimation of γ and Γ from (3.22/23a) one has to use the respective bounds for the extreme eigenvalues of A.
(b) In particular, the assumptions are satisfied if A is positive definite. In this case, one has to choose $\gamma \leqslant 1/\|A^{-1}\|_2$ and $\Gamma \geqslant \|A\|_2$.

For the Poisson-model problem, one obtains

$$\gamma = \lambda_{\min} = 8h^{-2}\sin^2(\pi h/2), \qquad \Gamma = \lambda_{\max} = 8h^{-2}\cos^2(\pi h/2),$$

according to (4.1.1.b, c). Inserting these values into the asymptotical convergence rate (3.25), we arrive at

$$\lim_{m\to\infty} (1/C_m)^{1/m} = c = \cos(\pi h)/(1 + \sin(\pi h)) = 1 - \pi h + O(h^2).$$

For $h = 1/16$ and $h = 1/32$, one obtains $c = 0.82$ and $c = 0.906$. The numerical results from Tables 1–2 show that the reduction factor approximates the convergence rate for sufficiently large m only. The ratios

$$\varrho_m := \|y^m - x\|_2/\|y^{m-1} - x\|_2, \qquad \hat{\varrho}_m := (\|y^m - x\|_2/\|y^0 - x\|_2)^{1/m}$$

tend to c from above.

Table 7.4.1 Semi-iterative Richardson method, $h = 1/16$

m	$\|y^m - x\|_2$	ϱ_m	$\hat{\varrho}_m$
1	$6.44_{10}-1$	$9.09_{10}-1$	$9.09_{10}-1$
10	$2.44_{10}-1$	$8.91_{10}-1$	$8.99_{10}-1$
20	$6.35_{10}-2$	$8.59_{10}-1$	$8.86_{10}-1$
30	$1.29_{10}-2$	$8.48_{10}-1$	$8.75_{10}-1$
40	$2.36_{10}-3$	$8.41_{10}-1$	$8.67_{10}-1$
50	$4.07_{10}-4$	$8.36_{10}-1$	$8.61_{10}-1$
60	$6.75_{10}-5$	$8.34_{10}-1$	$8.57_{10}-1$
70	$1.08_{10}-5$	$8.32_{10}-1$	$8.53_{10}-1$
80	$1.72_{10}-6$	$8.31_{10}-1$	$8.50_{10}-1$
90	$2.67_{10}-7$	$8.29_{10}-1$	$8.48_{10}-1$
100	$4.11_{10}-8$	$8.28_{10}-1$	$8.46_{10}-1$

Table 7.4.2 Semi-iterative Richardson method, $h = 1/32$

m	$\|y^m - x\|_2$	ϱ_m	$\hat{\varrho}_m$
1	$7.14_{10}-1$	$9.54_{10}-1$	$9.54_{10}-1$
10	$4.47_{10}-1$	$9.48_{10}-1$	$9.49_{10}-1$
30	$1.40_{10}-1$	$9.36_{10}-1$	$9.45_{10}-1$
50	$3.21_{10}-2$	$9.24_{10}-1$	$9.38_{10}-1$
70	$6.26_{10}-3$	$9.19_{10}-1$	$9.33_{10}-1$
80	$2.66_{10}-3$	$9.17_{10}-1$	$9.31_{10}-1$
100	$4.65_{10}-4$	$9.15_{10}-1$	$9.28_{10}-1$
120	$7.80_{10}-5$	$9.13_{10}-1$	$9.26_{10}-1$
130	$3.15_{10}-5$	$9.13_{10}-1$	$9.25_{10}-1$
140	$1.27_{10}-5$	$9.12_{10}-1$	$9.24_{10}-1$
150	$5.09_{10}-6$	$9.12_{10}-1$	$9.23_{10}-1$

The following PASCAL procedures determine γ and Γ for the Richardson method and store these values in `it.IP`:

```
function upper_gamma_Richardson(var it: data_of_iteration):
 real;
begin upper_gamma_Richardson:=it.IP.theta*maximal_ev(it.A)
end;

function lower_gamma_Richardson(var it: data_of_iteration):
 real;
begin lower_gamma_Richardson:=it.IP.theta*minimal_ev(it.A)
end;

procedure define_Richardson_semiiteration_parameters
                                    (var it: data_of_iteration);
begin
 gamma_bounds(it,lower_gamma_Richardson(it),
                 upper_gamma_Richardson(it))
end;
```

7.4.3 Semi-Iterative Jacobi and Block-Jacobi Method

The procedures corresponding to the Jacobi method are

```
function upper_gamma_Jacobi(var it: data_of_iteration): real;
begin upper_gamma_Jacobi:=maximal_ev(it.A)/it.A.S[0,0] end;

function lower_gamma_Jacobi(var it: data_of_iteration):
 real;
```

```
begin lower_gamma_Jacobi:=minimal_ev(it.A)/it.A.S[0,0] end;

function lower_gamma_column_Jacobi(var it: data_of_
 iteration): real;
begin lower_gamma_column_Jacobi:=
                1-(1-2*sqr(sin(pi/(2*it.A.nx))))/
                (1+2*sqr(sin(pi/(2*it.A.ny))))
end;

function upper_gamma_column_Jacobi(var it: data_of_
 iteration): real;
begin upper_gamma_column_Jacobi:=2_lower_gamma_column_
 Jacobi(it)
end;
```

They are necessary for defining the semi-iteration parameters by

```
procedure define_Jacobi_semiiteration_
   parameters (...);
procedure define_column_Jacobi_semiiteration_
   parameters (...);
```

Numerical examples are unnecessary, since in the Poisson-model case, the Jacobi method coincides with a damped Richardson method and hence, according to Lemma 1, reproduces the results from Tables 1–2.

For the determination of the *lower bound* a of the spectrum $\sigma(M^{\mathrm{Jac}})$, Lemma 5.2.1 proves that $a = -b$ for a particular case.

Lemma 7.4.3. *If* $\{A, D\}$ *is weakly* 2-*cyclic* (*cf. Definition* 5.1.2), M^{Jac} *has a symmetric spectrum*: $\sigma(M^{\mathrm{Jac}}) = -\sigma(M^{\mathrm{Jac}})$. *The smallest enclosing interval is* $[a, b] = [-\varrho(M^{\mathrm{Jac}}), \varrho(M^{\mathrm{Jac}})]$.

A comparison with the SOR method is possible. In the weakly 2-*cyclic case,* (3.17b) *is applicable because of Lemma* 3 *and yields the asymptotical semi-iterative convergence rate*

$$\beta/[1 + \sqrt{1 - \beta^2}], \qquad \beta := \varrho(M^{\mathrm{Jac}}).$$

This size coincides with the square root of the optimal SOR convergence rate $\omega_{\mathrm{opt}} - 1$; *hence,* the semi-iterative Jacobi iteration is half as fast as the SOR method. *The order improvement by an optimal* ω-*choice in the SOR case and the order improvement by the Chebyshev method* (*cf. Theorem* 3.8b) *lead to very similar results.*

The block variants of the Jacobi method converge faster than the pointwise version. Correspondingly, the semi-iterative column-Jacobi results from Tables 3–4 are also better than those in Tables 1–2. The factors should tend to the asymptotical value 0.7565 for $h = 1/16$ and 0.8702 for $h = 1/32$.

7.4.4 Semi-Iterative SSOR and Block-SSOR Method

As already mentioned in §7.4.1, the Gauß-Seidel and SOR methods are not suited for semi-iterative purposes, since, in general, the spectrum is not real. A remedy is offered by the symmetric Gauß-Seidel and SSOR method. In Theorem 4.8.11 it was stated that the spectrum of the SSOR method is real for Hermitian matrices A. Theorem 4.8.14 gave an upper bound for the spectral radius $\varrho(M_\omega^{\mathrm{SSOR}})$. Hence, under conditions (4.8.18a, b), the spectrum can be enclosed by the interval $[a, b]$ with

$$a = 0, \quad b = 1 - 2\Omega \Big/ \left[\frac{\Omega^2}{\gamma} + \Omega + \frac{\Gamma}{4}\right], \quad \text{where } \Omega := \frac{2 - \omega}{2\omega}, \quad 0 < \omega < 2. \tag{7.4.1}$$

Here, Γ is defined by (4.8.18b). Corollary 4.4.25 helps to determine Γ. For the Poisson-model problem, Lemma 4.7.7 yields the value $\Gamma = 2$. Inequality (4.8.18a) states that γ coincides with the equally denoted bound in inequality (3.23a′) applied to the (block-)Jacobi method. In the Poisson-model case, $\gamma = 2\sin^2(\pi h/2)$ holds.

Table 7.4.3 Semi-iterative column-Jacobi iteration, $h = 1/16$

m	$\Vert y^m - x\Vert_2$	ϱ_m	$\hat{\varrho}_m$
1	$6.09_{10}-1$	$8.60_{10}-1$	$8.60_{10}-1$
20	$1.62_{10}-2$	$7.95_{10}-1$	$8.27_{10}-1$
40	$1.19_{10}-4$	$7.75_{10}-1$	$8.04_{10}-1$
60	$6.68_{10}-7$	$7.69_{10}-1$	$7.93_{10}-1$
80	$3.33_{10}-9$	$7.65_{10}-1$	$7.86_{10}-1$
90	$2.1_{10}-10$	$7.55_{10}-1$	$7.84_{10}-1$

Table 7.4.4 Semi-iterative column-Jacobi iteration, $h = 1/32$

m	$\Vert y^m - x\Vert_2$	ϱ_m	$\hat{\varrho}_m$
1	$6.94_{10}-1$	$9.28_{10}-1$	$9.28_{10}-1$
20	$1.53_{10}-1$	$9.12_{10}-1$	$9.23_{10}-1$
40	$1.84_{10}-2$	$8.92_{10}-1$	$9.11_{10}-1$
60	$1.70_{10}-3$	$8.85_{10}-1$	$9.03_{10}-1$
80	$1.40_{10}-4$	$8.81_{10}-1$	$8.98_{10}-1$
100	$1.08_{10}-5$	$8.78_{10}-1$	$8.94_{10}-1$

Theorem 7.4.4. *Let $A = D - E - E^H > 0$ and γ, Γ satisfy the assumptions* (4.8.18a, b). *Furthermore, assume* $0 < \omega \leqslant 2/(\Gamma + 1)$. *Then*

$$a = \left(\frac{1-\xi}{1+\xi}\right)^2 \quad \text{with } \xi := \frac{2-\omega}{\Gamma\omega} \tag{7.4.2}$$

is a lower bound for the spectrum $\sigma(M_\omega^{\mathrm{SSOR}})$.

Proof. Using the parameter Ω from (4.4.32c), we can rewrite $W_\omega^{\mathrm{SSOR}} = \left(\frac{1}{\omega}D - E\right)\left[\left(\frac{2}{\omega} - 1\right)D\right]^{-1}\left(\frac{1}{\omega}D - E\right)^H$ as $W_\omega^{\mathrm{SSOR}} = [\Omega D + \Delta](2\Omega D)^{-1} \cdot [\Omega D + \Delta]^H$ with $\Delta := \frac{1}{2}D - E$. Defining $X := \Omega D + (1-\alpha)\Delta$ for some real α, we have $[\Omega D + \Delta] = X + \alpha\Delta$. The expansion of $[X + \alpha\Delta](2\Omega D)^{-1}[X + \alpha\Delta]^H$ yields

$$W_\omega^{\mathrm{SSOR}} = \frac{1}{2\Omega}XD^{-1}X^H + \frac{\alpha}{2}A + \frac{1}{2\Omega}(2\alpha - \alpha^2)\Delta D^{-1}\Delta^H$$

because of $\Delta + \Delta^H = A$. The factor $(2\alpha - \alpha^2)$ is negative for $\alpha \geqslant 2$. Hence, (4.8.18b) can be applied. Together with $XD^{-1}X^H \geqslant 0$, one obtains

$$W_\omega^{\mathrm{SSOR}} \geqslant g(\alpha)A \quad \text{with } g(\alpha) := \frac{\alpha}{2}\left[1 + \frac{\Gamma}{2}\frac{2-\alpha}{\Omega}\right] \quad \text{for } \alpha \geqslant 2.$$

The assumption $\omega \leqslant 2/(\Gamma + 1)$ implies $\alpha_0 := 1 + 2\Omega/\Gamma \geqslant 2$. Remark 4.8.3c with $1 - a = 1/g(\alpha_0)$ yields the value (2) for a. □

The statement is less interesting, since (because of $\Gamma = 2$ for the Poisson-model case) Theorem 4 applies only for the strong underrelaxation $\omega \leqslant 2/3$. The definition of the semi-iteration parameters reads as follows:

```
function SSOR_contraction_number(omega,small_gamma,
                                    capital_gamma: real): real;
begin omega:=(2-omega)/(2*omega);
      SSOR_contraction_number:=
      1-2*omega/(capital_gamma/4+omega*(1+omega/small_gamma))
end;

function lambda_max_SSOR(var it: data_of_iteration): real;
begin lambda_max_SSOR:=
      SSOR_contraction_number(it.IP.omega,lower_gamma_
       Jacobi(it),2)
end;

function lambda_min_SSOR(var it: data_of_iteration): real;
 var l: real;
```

```
begin l:=it.IP.omega;
  if l=0 then l:=1 else
  begin l:=(2-l)/(2*l); if l>1 then l:=sqr((l-1)/(l+1)) else
    l:=0
  end;
  lambda_min_SSOR:=l
end;

function lambda_max_column_SSOR(var it: data_of_iteration):
  real;
begin lambda_max_column_SSOR:=
  SSOR_contraction_number(it.IP.omega,lower_gamma_column_
    Jacobi(it),2)
end;

procedure define_SSOR_semiiteration_parameters(var it: data_
  of_iteration);
begin lambda_bounds(it,lambda_min_SSOR(it),lambda_max_
  SSOR(it))
end;

procedure define_column_SSOR_semiiteration_parameters
                                    (var it: data_of_iteration);
begin
    lambda_bounds(it,lambda_min_SSOR(it),lambda_max_column_
      SSOR(it))
end;
```

The frame program for the semi-iterative lexicographical SSOR iteration can be found in (3.27) or [Prog].

There are two possibilities for improving (halving) the convergence order. First, this can be achieved by the optimal choice of ω in the SOR and SSOR method (cf. Remarks 4.4.26 and 4.8.15). Second, the semi-iterative method leads to halving of the order compared with the basic iteration. In the case of SSOR as the basic iteration, both techniques can be applied simultaneously. First, the optimal ω' is chosen as in (4.4.33b) as a relaxation parameter of the SSOR iteration. The thus defined $\Phi_{\omega'}^{\mathrm{SSOR}}$ is chosen as the basic iteration of the Chebyshev method. Together, one succeeds in a quartering of the order. In the Poisson-model case, one obtains the asymptotical convergence rate $1 - O(h^{1/2})$.

The bound b from (1)becomes minimal for $\omega' = 2/(1 + \sqrt{\gamma\Gamma})$:

$$b = \frac{\sqrt{\Gamma} - \sqrt{\gamma}}{\sqrt{\Gamma} + \sqrt{\gamma}} = \frac{1 - \sqrt{\gamma/\Gamma}}{1 + \sqrt{\gamma/\Gamma}}. \tag{7.4.3a}$$

Insertion of this value into (3.17c) yields the asymptotical convergence rate

$$\lim_{m\to\infty} (1/C_m)^{1/m} = c = \frac{1 - \sqrt{1-b}}{1 + \sqrt{1-b}} \quad \text{with } b \text{ from (3a).} \tag{7.4.3b}$$

The condition number $\varkappa = \varkappa((W^{\mathrm{SSOR}})^{-1}A)$ from (3.23c′) equals $\frac{1}{2}(1 + \sqrt{\Gamma/\gamma})$. Using the inequality $\gamma \geqslant 1/\varkappa(A)$ from Exercise 4.4.15c, one ends up with the result

$$\varkappa((W^{\mathrm{SSOR}})^{-1}A) \leqslant \tfrac{1}{2}(1 + \sqrt{\Gamma\varkappa(A)}). \tag{7.4.3c}$$

For the values γ and Γ from Lemma 4.7.7 (Poisson-model case), the convergence rate (3b) asymptotically equals the value

$$c = 1 - Ch^{1/2} + O(h) \quad \text{with } C = 2\sqrt{\pi}. \tag{7.4.4}$$

The results from Table 5 refer to the parameters

$$h = 1/32, \quad \omega = 1.8455, \quad a = 0, \quad b = 0.878. \tag{7.4.5}$$

In §4.8.6 the ω-value proved to be optimal (note that ω' is optimal only for the bound in (4.8.18c)). We learned from Table 4.8.2 that $b = 0.878$ is an upper bound of the convergence rate. From (3b) with $b = 0.878$, one calculates the rate $c = 0.482$, which is numerically well confirmed (cf. Table 5). From $C_\Phi^{\mathrm{SSOR}} = 2 + 6/C_A = 3.2$ (according to Remark 4.8.12 and because of $C_A = 5$ for five-point formulae), one obtains the effective amount of work.

$$\mathit{Eff}_{\mathrm{semiiterative}}(\Phi^{\mathrm{SSOR}}) = -3.2/\log c = 4.38 \tag{7.4.6}$$

for the semi-iterative SSOR method with $h = 1/32$, which can be compared, e.g., with $\mathit{Eff}(\Phi^{\mathrm{SOR}}) = 7.05$ from Example 3.3.2.

If we use the values ω' from (4.4.33b), Eq. (3b) yields the asymptotical convergence rates c reported in Table 6. These values might give an impression of the asymptotic $c = 1 - O(h^{1/2})$.

Table 7.4.5 Semi-iterative lexicographical SSOR for parameters (5). For ϱ_m and $\hat{\varrho}_m$ compare Table 1

m	$\lVert y^m - x \rVert_2$	ϱ_m	$\hat{\varrho}_m$
1	$4.673_{10}-1$	$6.24_{10}-1$	$6.24_{10}-1$
2	$2.761_{10}-1$	$5.90_{10}-1$	$6.07_{10}-1$
3	$1.359_{10}-1$	$4.92_{10}-1$	$5.66_{10}-1$
4	$7.681_{10}-2$	$5.65_{10}-1$	$5.66_{10}-1$
5	$3.801_{10}-2$	$4.94_{10}-1$	$5.51_{10}-1$
20	$2.080_{10}-6$	$5.08_{10}-1$	$5.27_{10}-1$
21	$1.007_{10}-6$	$4.84_{10}-1$	$5.25_{10}-1$
22	$5.195_{10}-7$	$5.15_{10}-1$	$5.24_{10}-1$
23	$2.541_{10}-7$	$4.89_{10}-1$	$5.23_{10}-1$
29	$3.395_{10}-9$	$4.82_{10}-1$	$5.15_{10}-1$
30	$1.628_{10}-9$	$4.79_{10}-1$	$5.14_{10}-1$

7.5 Method of Alternating Directions (ADI)

«*A*lternating-*d*irection *i*mplicit iterative method» or ADI was first described in 1955 by Peaceman–Rachford [1] in connection with parabolic differential equations.

7.5.1 Application to the Model Problem

For the model problem from §1.2, the matrix A can be split into

$$A = B + C, \qquad \text{where} \tag{7.5.1a}$$

$$(Bu)(x, y) = h^{-2}[-u(x - h, y) + 2u(x, y) - u(x + h, y)], \tag{7.5.1b}$$

$$(Cu)(x, y) = h^{-2}[-u(x, y - h) + 2u(x, y) - u(x, y + h)] \tag{7.5.1c}$$

for $(x, y) \in \Omega_h$ are the second differences of u with respect to the x and y direction. If we choose the rows (x direction) of Ω_h as blocks, $B + 2h^{-2}I$ represents the block-diagonal. Similarly, $C + 2h^{-2}I$ is the block-diagonal of A, if the columns (y direction) are chosen as blocks.

Remark 7.5.1. For A, B, and C from (1a, b, c), the statements (2a, b) hold:

$$B \text{ and } C \text{ are positive definite,} \tag{7.5.2a}$$

$$A, B, C \text{ are pairwise commutative.} \tag{7.5.2b}$$

The last statement is equivalent to

$$A, B, C \text{ can be simultaneously transformed to diagonal form.} \tag{7.5.2b'}$$

Table 7.4.6 Optimal ω' and asymptotical rate c for $h = 1/N$

N	ω'	c
2	0.8284	0.0470
4	1.1329	0.1467
8	1.4386	0.2727
16	1.6721	0.4059
32	1.8212	0.5315
64	1.9064	0.6408
128	1.9520	0.7305
256	1.9757	0.8010
512	1.9878	0.8549
1028	1.9939	0.8953
5000	1.9987	0.9511
10000	1.9993	0.9651

Proof. In Lemma 4.7.5, we analysed the block-diagonal of A (with respect to the row-block structure). Because of the $x - y$ symmetry, the same result holds for the column-block structure. Therefore, the spectrum of $B + 2h^{-2}I$ and $C + 2h^{-2}I$ equals $\{h^{-2}[2 + 4\sin^2(jh\pi/2)]: 1 \leqslant j \leqslant N - 1\}$, i.e., $4h^{-2}\sin^2(jh\pi/2)$ are the eigenvalues of B and C. Since these values are positive, (2a) is proved. By Lemma 4.7.5, the eigenvectors e^{ij} of A (cf. Lemma 4.1.2) are also the eigenvectors of $B + 2h^{-2}I$, $C + 2h^{-2}I$ and hence of B, C. This proves (2b′) and (2b). □

The first half-step of the ADI method corresponds to the splitting

$$A = W - R \quad \text{with } W = \omega I + B,\ R = \omega I - C \tag{7.5.3a}$$

and reads

$$x^{m+1/2} := \Phi_\omega^B(x^m, b) := (\omega I + B)^{-1}(b + \omega x^m - Cx^m), \tag{7.5.4a}$$

where ω is a (real) parameter. Interchanging the rôles of B and C, i.e., *alternating the directions*, we generate the splitting (3b) of the second half-step (4b):

$$A = W - R \quad \text{with } W = \omega I + C, \quad R = \omega I - B, \tag{7.5.3b}$$

$$x^{m+1} := \Phi_\omega^C(x^{m+1/2}, b) := (\omega I + C)^{-1}(b + \omega x^{m+1/2} - Bx^{m+1/2}). \tag{7.5.4b}$$

Remark 7.5.2. The single half-steps (4a, b) resemble the block-Jacobi method. For $\omega = 2h^{-2}$, iteration (4a) represents the row and (4b) the column-block-Jacobi method. Because of (2a) the matrices $\omega I + B$ and $\omega I + C$ are positive definite and therefore regular for $\omega \geqslant 0$; hence, the steps (4a, b) are well-defined. Since, furthermore, $\omega I + B$ and $\omega I + C$ are *tridiagonal matrices*, the solution of $(\omega I + B)z = c$ or $(\omega I + C)z = c$ required in (4a, b) is easy to perform.

The complete ADI method $x^m \mapsto x^{m+1}$ is the product iteration

$$\Phi_\omega^{\mathrm{ADI}} := \Phi_\omega^C \Phi_\omega^B. \tag{7.5.4c}$$

7.5.2 General Representation

In the general case, we start from a splitting (1a): $A = B + C$ and assume (2a) in a weakened form: One of the matrices B or C may be only positive *semi-definite*. Without loss of generality, this might be C:

$$B \text{ positive definite}, \quad C \text{ positive semi-definite}. \tag{7.5.5a}$$

Therefore, for

$$\omega > 0 \tag{7.5.5b}$$

the matrices $\omega I + B$ and $\omega I + C$ are positive definite and, in particular, regular. Hence, the ADI iteration (4c) can be defined by means of (4a, b). To ensure practicability, we assume (5c):

$$\text{equations with } \omega I + B \text{ or } \omega I + C \text{ are easy to solve}. \tag{7.5.5c}$$

Theorem 7.5.3 (convergence). (a) *The iteration matrix of the ADI method is*

$$M_\omega^{\mathrm{ADI}} = (\omega I + C)^{-1}(\omega I - B)(\omega I + B)^{-1}(\omega I - C). \tag{7.5.6a}$$

(b) *Under the assumptions* (5a, b), *the ADI iteration converges.*

Proof. M_ω^{ADI} is the product of the iteration matrices $(\omega I + C)^{-1}(\omega I - B)$ and $(\omega I + B)^{-1}(\omega I - C)$ of the respective half-steps Φ_ω^C and Φ_ω^B (cf. §3.2.7). Lemma 2.4.16 allows the cyclic permutation of the factors in the argument of the spectral radius:

$$\begin{aligned}\varrho(M_\omega^{\mathrm{ADI}}) &= \varrho((\omega I - B)(\omega I + B)^{-1}(\omega I - C)(\omega I + C)^{-1})\\ &\leqslant \|(\omega I - B)(\omega I + B)^{-1}(\omega I - C)(\omega I + C)^{-1}\|_2\\ &\leqslant \|(\omega I - B)(\omega I + B)^{-1}\|_2 \|(\omega I - C)(\omega I + C)^{-1}\|_2. \end{aligned}\tag{7.5.6b}$$

As B is Hermitian, $B_\omega := (\omega I - B)(\omega I + B)^{-1}$ is too. In particular, it is a normal matrix, implying that $\varrho(B_\omega) = \|B_\omega\|_2$ (cf. Theorem 2.9.5). Therefore, (6b) becomes

$$\varrho(M_\omega^{\mathrm{ADI}}) \leqslant \varrho(B_\omega)\varrho(C_\omega), \tag{7.5.6c}$$

because analogous considerations apply to the second matrix $C_\omega := (\omega I - C)(\omega I + C)^{-1}$. By Remark 2.4.11b, the spectrum of B_ω equals

$$\sigma(B_\omega) = \left\{\frac{\omega - \beta}{\omega + \beta}: \beta \in \sigma(B)\right\}, \quad \varrho(B_\omega) = \max\left\{\left|\frac{\omega - \beta}{\omega + \beta}\right|: \beta \in \sigma(B)\right\}. \tag{7.5.6d}$$

By assumption (5a), β is positive. This fact implies that $|\omega - \beta| < |\omega + \beta|$ for all $\omega > 0$. This proves $\varrho(B_\omega) < 1$. Since C is only positive semi-definite, a similar argument leads to $\varrho(C_\omega) \leqslant 1$. (6c) proves $\varrho(M_\omega^{\mathrm{ADI}}) < 1$. □

Exercise 7.5.4. Formulate a convergence statement for the case of normal matrices B and C. For that purpose, prove $\varrho(B_\omega) < 1$ and $\varrho(C_\omega) \leqslant 1$ for the factors in (6c) by showing that (3a, b) are regular splittings. Which restrictions are needed for ω in the model case?

In the following, we have to determine the optimal value ω_{opt} of the ADI method. Here, we restrict ourselves to the minimisation of $\varrho(B_\omega)$. If, as for the model problem, $\varrho(C_\omega) = \varrho(B_\omega)$ holds, this problem is equivalent to the minimisation of the bound $\varrho(B_\omega)\varrho(C_\omega)$ in (6c).

The extreme eigenvalues of B (or their bounds) are assumed to be

$$0 < \beta_{\min} \leqslant \beta_{\max} \quad \text{with } \sigma(B) \subset [\beta_{\min}, \beta_{\max}]. \tag{7.5.7a}$$

As we have seen in the proof of Remark 1, in the model case the eigenvalues of B are $4h^{-2}\sin^2(jh\pi/2)$ for $1 \leqslant j \leqslant N - 1$. This implies that

$$\beta_{\min} = 4h^{-2}\sin^2(h\pi/2), \qquad \beta_{\max} = 4h^{-2}\cos^2(h\pi/2). \tag{7.5.7b}$$

For any $\beta \in [\beta_{\min}, \beta_{\max}]$ and therefore for any $\beta \in \sigma(B)$, we have

$$\left|\frac{\omega - \beta}{\omega + \beta}\right| \leqslant \max\left\{\left|\frac{\omega - \beta_{\min}}{\omega + \beta_{\min}}\right|, \left|\frac{\omega - \beta_{\max}}{\omega + \beta_{\max}}\right|\right\} \qquad (\omega > 0), \tag{7.5.7c}$$

since $|\omega - \beta|/|\omega + \beta|$ as a function of β is decreasing in $[0, \omega]$ and increasing in $[\omega, \infty)$. To minimise the right-hand side in (7c), one has to determine ω from $\left|\frac{\omega - \beta_{\min}}{\omega + \beta_{\min}}\right| = \left|\frac{\omega - \beta_{\max}}{\omega + \beta_{\max}}\right|$. The result is given by

$$\omega_{\mathrm{opt}} = \sqrt{\beta_{\min}\beta_{\max}}. \tag{7.5.7d}$$

Inserting this value into (6d), one obtains

$$\varrho(B_{\omega_{\mathrm{opt}}}) = (\sqrt{\beta_{\max}} - \sqrt{\beta_{\min}})/(\sqrt{\beta_{\max}} + \sqrt{\beta_{\min}}). \tag{7.5.7e}$$

Exercise 7.5.5. Prove for the model problem: (a) Equations (8a–c) hold:

$$\omega_{\mathrm{opt}} = 2h^{-2}\sin h\pi, \tag{7.5.8a}$$

$$\varrho(B_{\omega_{\mathrm{opt}}}) = [\cos(\tfrac{1}{2}\pi h) - \sin(\tfrac{1}{2}\pi h)]/[\cos(\tfrac{1}{2}\pi h) + \sin(\tfrac{1}{2}\pi h)], \tag{7.5.8b}$$

$$\varrho(M_{\omega_{\mathrm{opt}}}^{\mathrm{ADI}}) = [1 - \sin(\pi h)]/[1 + \sin(\pi h)]. \tag{7.5.8c}$$

(b) The convergence speed (8c) coincides exactly with the optimal convergence rate (5.6.8) of the SOR method.

If we replace the definiteness in assumption (5a) by the M-matrix property, a convergence proof becomes much more difficult. A general convergence result of this kind (also for instationary ADI methods) is due to Alefeld [1]. Here, we call the method *stationary* if ω is constant during the iteration and *instationary* if it varies (as, e.g., is assumed throughout the following section).

7.5.3 ADI Method in the Commutative Case

In addition to the assumptions (5a–c), we require that

$$BC = CB. \tag{7.5.9a}$$

The commutativity is equivalent to the simultaneous diagonalisability:

$$Q^H B Q = D_B = \operatorname{diag}\{\beta_\alpha: \alpha \in I\}, \qquad Q^H C Q = D_C = \operatorname{diag}\{\gamma_\alpha: \alpha \in I\} \tag{7.5.9b}$$

(cf. Theorem 2.8.9), which here can be achieved by a unitary transformation Q, since B and C are Hermitian. (9b) implies that B_ω, C_ω, and the iteration matrix M_ω^{ADI} built from these matrices can also be transformed by Q to diagonal form (cf. (6a)):

$$Q^H M_\omega^{\mathrm{ADI}} Q = \operatorname{diag}\left\{\frac{\omega - \gamma_\alpha}{\omega + \gamma_\alpha}\frac{\omega - \beta_\alpha}{\omega + \beta_\alpha}: \alpha \in I\right\}. \tag{7.5.9c}$$

In the following, we apply the ADI method with *varying parameters* $\omega = \omega_m$:

$$y^{m+1} = \Phi^{\mathrm{ADI}}_{\omega_m}(y^m, b) \qquad (m \in \mathbb{N}). \tag{7.5.10}$$

Exercise 7.5.6. Let x be the solution of $Ax = b$. Prove that the error $\eta^m = y^m - x$ has the representation

$$\eta^m = M^{\mathrm{ADI}}_{\omega_m} \cdot \ldots \cdot M^{\mathrm{ADI}}_{\omega_2} M^{\mathrm{ADI}}_{\omega_1} \eta^0. \tag{7.5.11}$$

We would like to choose the parameters $\omega_1, \omega_2, \ldots, \omega_m \geqslant 0$ so that the spectral norm of this matrix becomes as small as possible:

$$\|M^{\mathrm{ADI}}_{\omega_m} \cdot \ldots \cdot M^{\mathrm{ADI}}_{\omega_1}\|_2 = \min. \tag{7.5.12a}$$

Multiplications by unitary matrices do not change the spectral norm:

$$\|Q^H M^{\mathrm{ADI}}_{\omega_m} \cdot \ldots \cdot M^{\mathrm{ADI}}_{\omega_1} Q\|_2 = \|Q^H M^{\mathrm{ADI}}_{\omega_m} Q \cdot \ldots \cdot Q^H M^{\mathrm{ADI}}_{\omega_2} Q Q^H M^{\mathrm{ADI}}_{\omega_1} Q\|_2.$$

Together with (9c), we obtain

$$\left\| \prod_{i=1}^m \mathrm{diag}\left\{\frac{\omega_i - \gamma_\alpha}{\omega_i + \gamma_\alpha} \frac{\omega_i - \beta_\alpha}{\omega_i + \beta_\alpha} : \alpha \in I\right\} \right\|_2 = \left\| \mathrm{diag}\left\{\prod_{i=1}^m \frac{\omega_i - \gamma_\alpha}{\omega_i + \gamma_\alpha} \frac{\omega_i - \beta_\alpha}{\omega_i + \beta_\alpha} : \alpha \in I\right\} \right\|_2$$
$$= \max_{\alpha \in I} \left| \prod_{i=1}^m \frac{\omega_i - \gamma_\alpha}{\omega_i + \gamma_\alpha} \frac{\omega_i - \beta_\alpha}{\omega_i + \beta_\alpha} \right|.$$

Hence, the minimisation problem (12a) is equivalent to

$$\max_{\alpha \in I} \left| \prod_{i=1}^m \frac{\omega_i - \gamma_\alpha}{\omega_i + \gamma_\alpha} \frac{\omega_i - \beta_\alpha}{\omega_i + \beta_\alpha} \right| = \min. \tag{7.5.12b}$$

Remark 7.5.7. From $m \geqslant n := \# I$, as in Theorem 3.1, one finds parameters ω_i, bringing the left-hand side in (12b) to the minimum 0. For this purpose, one may choose ω_i as the eigenvalues $\{\gamma_\alpha : \alpha \in I\}$ as well as $\{\beta_\alpha : \alpha \in I\}$.

Since γ_α or β_α, in general, are not known, we change over to optimisation over a larger set $[a, b]$ containing the spectra of B and C, as we did in the third minimisation problem (3.9):

$$0 < a \leqslant \gamma_\alpha, \beta_\alpha \leqslant b \quad \text{for all } \alpha \in I. \tag{7.5.13}$$

Then, the minimisation problem takes the following form. Let

$$r_m(\zeta) := \prod_{i=1}^m \frac{\omega_i - \zeta}{\omega_i + \zeta} \tag{7.5.14}$$

be the rational function with a numerator and denominator of degree m replacing the previous polynomials. Substituting the discrete eigenvalues in (12b) by the interval $[a, b]$, one arrives at the problem

$$\begin{aligned} &\text{determine parameters } \{\omega_i : 1 \leqslant i \leqslant m\} \text{ such that} \\ &\max\{|r_m(\beta) r_m(\gamma)| : a \leqslant \beta, \gamma \leqslant b\} = \min. \end{aligned} \tag{7.5.15a}$$

Because of $\max|r_m(\beta)r_m(\gamma)| = \max|r_m(\beta)|\max|r_m(\gamma)|$, one may optimise each factor separately. Hence, problem (15a) simplifies to

$$\begin{aligned}&\text{determine parameters } \{\omega_i\colon 1 \leqslant i \leqslant m\} \text{ such that}\\ &\max\{|r_m(\zeta)|\colon a \leqslant \zeta \leqslant b\} = \min.\end{aligned} \tag{7.5.15b}$$

The following results are due to Wachspress (from the years 1957 and 1962). We omit their proofs, since the derivation of Eqs. (16a–c) is presented in detail in the book by Varga [2, pages 224–225].

Theorem 7.5.8. (a) *For any* $m \in \mathbb{N}$, *the problem* (15b) *has a unique solution* $\{\omega_1, \ldots, \omega_m\}$. *The parameters* ω_i *are disjointly situated in* (a, b).
(b) *The increasingly ordered parameters* $\omega_1 < \omega_2 < \cdots < \omega_m$ *satisfy*

$$\omega_{m+1-i} = ab/\omega_i \quad \textit{for } 1 \leqslant i \leqslant m. \tag{7.5.16a}$$

(c) *Denote the parameters* $\omega_1 < \omega_2 < \cdots < \omega_m$ *belonging to* $m \in \mathbb{N}$ *and the interval* $[a, b]$ *(with* $0 < a < b$*) by* $\omega_i(a, b, m)$ $(1 \leqslant i \leqslant m)$. *Then we have*

$$\omega_{2m+1-i}(a, b, 2m) = \omega_i\left(\sqrt{ab}, \frac{a+b}{2}, m\right) + \sqrt{\omega_i\left(\sqrt{ab}, \frac{a+b}{2}, m\right)^2 - ab}$$
$$\textit{for } i = 1, \ldots, m. \tag{7.5.16b}$$

(d) *The minimised quantities* $\delta_m := \max\{|r_m(\zeta)|\colon a \leqslant \zeta \leqslant b\}$ *for* $m = 2^p$ *are*

$$\delta_m = (\sqrt{b_p} - \sqrt{a_p})/(\sqrt{b_p} + \sqrt{a_p}), \tag{7.5.16c}$$

where $a_0 = a$, $b_0 = b$, $a_{i+1} = \sqrt{a_i b_i}$, $b_{i+1} = \frac{1}{2}(a_i + b_i)$ $(0 \leqslant i \leqslant p - 1)$.

The determination of the ADI parameters ω_i *is very easy for binary powers* $m = 2^p$. *For* $p = 0$ *(i.e.,* $m = 1$*), one concludes from* (16a) *that*

$$\omega_1(a, b, 1) = \sqrt{ab}, \tag{7.5.16d}$$

repeating the result from (7d). *As soon as the parameters for* $m = 2^{p-1}$ *are known, those for* $2m = 2^p$ *can be obtained from formula* (16b), *with index* $2m + 1 - i$ *replaced by* $2^{p-1} + 1, \ldots, 2^p$. *The* ω_i *for* $1 \leqslant i \leqslant m$ *follow from* (16a). *The algorithm can read as follows*:

```
procedure ADI_parameter(var omega: ADIparameters; a,b: real;
 m: integer);
var i: integer; ab,w: real;
begin ab:=a*b; if m=1 then omega[1]:=sqrt(ab) else
    begin m:=m div 2; ADI_parameter(omega,sqrt(ab),
        (a+b)/2,m);
        for i:=m downto 1 do
        begin w:=sqr(omega[i])-ab; if w<0 then w:=0 else
         w:=sqrt(w);
              omega[2*i]:=omega[i]+w; omega[2*i-1]:=
               ab/omega[2*i]
end end end;
```

Evidently, one may apply the calculated parameters ω_i in a *cyclic* manner: $\omega_{i+km} := \omega_i$ $(1 \leqslant i \leqslant m,\ k > 0)$. Different from the case in §7.3.7, the cyclic ADI process does not lead to stability problems.

δ_m from (16c) is the bound for $r_m(B_\omega)$ and $r_m(C_\omega)$. Therefore, the asymptotical rate equals $\varrho_m := \delta_m^{2/m}$. One recognises from (16c) that ϱ_m depends only on the ratio a/b, which in the model case has the size $O(h^2)$. The recursion $a_{i+1} = \sqrt{a_i b_i}$, $b_{i+1} = \frac{1}{2}(a_i + b_i)$ leads to

Remark 7.5.9. Let $a/b = O(h^\varkappa)$. For the optimal parameter choice, the cyclic ADI method with m parameters has the order $\varkappa/m$: $\varrho_m = 1 - O(h^{\varkappa/2m}) = 1 - C_m h^{\varkappa/2m} + O(h^{\varkappa/m})$ in the commutative case.

Hence, the instationary ADI method permits not only the halving of order (for the case $m = 1$, compare also Exercise 5b), but any arbitrarily small (and hence favourable) order can be reached with sufficiently large m. However, we will see in §7.5.6 that the simple conclusion of choosing a possibly large number m leads to practical difficulties.

The construction of the ω_i's in Theorem 8b is restricted to $m = 2^p$. For other m, a representation of ω_i is, in principle, possible but requires elliptic integrals (cf. Jordan in Wachspress [1], Samarskij–Nikolaev [1, page 276]). Lebedev [1] was the frist to suggest that the solution of the approximation problem (15b) could be reformulated into another problem for rational functions that had already been solved in 1877 by Zolotarev. In this connection, we refer to the review paper of Todd [1] concerning the «legacy of Zolotarev». The approximation problems appearing here also play an important rôle in the iterative solution of the *matrix equation* $AX - XB = C$ (A, B, C given, X unknown; cf. Starke [1]). Concerning parameter determination in the case of nonsymmetric matrices B and C, refer to Starke–Niethammer [1].

Even when the asymptotical convergence rates ϱ_m in Remark 9 and the following Table 1 look quite favourable, the effective amount of work is less favourable because of the relatively expensive iteration (4a, b) (cf. Remark 10). Furthermore, the assumption of commutativity (9a) is seldom satisfied in practice. As soon as it becomes violated, one is not able to achieve good convergence acceleration.

Table 7.5.1 Asymptotical convergence rates ϱ_m for cycle length m

m	$h = 1/32$	$h = 1/64$	$h = 1/128$
1	0.8215	0.9065	0.9521
2	0.5231	0.6373	0.7291
4	0.3735	0.4607	0.5365
8	0.3141	0.3874	0.4513
16	0.2880	0.3553	0.4139

7.5.4 ADI Method and Semi-Iterative Methods

After choosing the Richardson method as the basic iteration, the half-steps (4a, b) have the representation (1b):

$$y^{m+1/2} = \Theta_{m+1/2}(M_1^{\mathrm{Rich}} y^m + N_1^{\mathrm{Rich}} b) + (1 - \Theta_{m+1/2}) y^m,$$
$$y^{m+1} = \Theta_{m+1}(M_1^{\mathrm{Rich}} y^{m+1/2} + N_1^{\mathrm{Rich}} b) + (1 - \Theta_{m+1}) y^{m+1/2}$$

with $M_1^{\mathrm{Rich}} = I - A$ and $N_1^{\mathrm{Rich}} = I$, if one defines

$$\Theta_{m+1/2} = (\omega I + B)^{-1}, \qquad \Theta_{m+1} = (\omega I + C)^{-1}. \tag{7.5.17}$$

(17) corresponds to the second formulation. If, as in the case of §7.5.3, B and C commute with A, we obtain the first formulation (1.6): $y^m = \sum \alpha_{mj} x^j$, where x^j are the Richardson iterates and α_{mj} some *matrices* commuting with A. In this sense, one might view the ADI method as a semi-iterative one.

On the other hand, the ADI method can function as a basic iteration of the Chebyshev method, as shown in

Exercise 7.5.10. Assume that B, C, and ω satisfy (5a, b) and (9). Prove: (a) The matrix of the third normal form of $\Phi_\omega^{\mathrm{ADI}}$ is

$$W_\omega = \frac{1}{2\omega}(\omega I + C)(\omega I + B) \qquad (\textit{hint}: (3.2.20c)).$$

(b) $\Phi_\omega^{\mathrm{ADI}}$ is a symmetric iteration.

Table 7.5.2 ADI results for the model problem with 4 parameters ω_i and $h = 1/128$

m	value in the middle	$\|e^m\|_2$	$\frac{\|e^m\|_2}{\|e^{m-1}\|_2}$	$\left(\frac{\|e^m\|_2}{\|e^0\|_2}\right)^{1/m}$
1	$-3.209132566_{10}-2$	$5.01883_{10}-1$	$6.44595_{10}-1$	$6.44595_{10}-1$
2	$-3.428610244_{10}-2$	$4.61552_{10}-1$	$9.19641_{10}-1$	$7.69933_{10}-1$
3	$3.534506991_{10}-1$	$5.97883_{10}-2$	$1.29537_{10}-1$	$4.25044_{10}-1$
4	$3.538351873_{10}-1$	$5.49257_{10}-2$	$9.18670_{10}-1$	$5.15365_{10}-1$
5	$4.031600829_{10}-1$	$3.87484_{10}-2$	$7.05469_{10}-1$	$5.48767_{10}-1$
6	$4.063222547_{10}-1$	$3.67588_{10}-2$	$9.48655_{10}-1$	$6.01184_{10}-1$
7	$4.976831847_{10}-1$	$4.46545_{10}-3$	$1.21480_{10}-1$	$4.78402_{10}-1$
8	$4.976688610_{10}-1$	$4.25818_{10}-3$	$9.53584_{10}-1$	$5.21481_{10}-1$
9	$4.961625617_{10}-1$	$3.10151_{10}-3$	$7.28366_{10}-1$	$5.41205_{10}-1$
10	$4.961175712_{10}-1$	$2.96750_{10}-3$	$9.56792_{10}-1$	$5.72938_{10}-1$
11	$4.990489164_{10}-1$	$3.49870_{10}-4$	$1.17901_{10}-1$	$4.96238_{10}-1$
12	$4.990525844_{10}-1$	$3.36761_{10}-4$	$9.62530_{10}-1$	$5.24405_{10}-1$
13	$4.993912351_{10}-1$	$2.51027_{10}-4$	$7.45416_{10}-1$	$5.38785_{10}-1$
14	$4.994041549_{10}-1$	$2.41296_{10}-4$	$9.61235_{10}-1$	$5.61531_{10}-1$
15	$4.999776724_{10}-1$	$2.78508_{10}-5$	$1.15422_{10}-1$	$5.05322_{10}-1$
16	$4.999776614_{10}-1$	$2.69345_{10}-5$	$9.67099_{10}-1$	$5.26244_{10}-1$

(c) Products $\Phi := \Phi^{\text{ADI}}_{\omega_1} \circ \Phi^{\text{ADI}}_{\omega_2} \circ \cdots \circ \Phi^{\text{ADI}}_{\omega_m}$ with positive ω_j form a symmetric iteration. *Hint*: Write the iteration matrix of Φ as $M = I - NA$, shown that $N > 0$, and use $W = N^{-1}$.
(d) In the stationary case, choose ω according to (16d). Determine the bounds in $\gamma W \leqslant A \leqslant \Gamma W$. What is the optimal damping factor for $\Phi^{\text{ADI}}_{\omega}$ (cf. also Exercise 8.3.1)?

7.5.5 PASCAL Procedures

The parameters ω_i for the cyclic ADI iteration are computed by `define_ADI_parameter` and stored in `it.IP.cycle^`. The procedure `ADI_halfstep` corresponds to (4b) with C from (1c). The first half-step (4a) is equivalent to an application of `ADI_halfstep` to the reflected grid function (executed by `XY_reflexion`).

```
procedure define_ADI_parameter(var it: data_of_iteration);
var p: integer; a,b: real;
begin if it.IP.cycle=nil then new(it.IP.cycle);
 with it.A do with it.IP.cycle^do
 begin writeln; writeln('*** Choice of the ADI parameters:');
       if kind=Poisson_model_problem then
       begin a:=4*sqr(sin(pi/(2*maximum(nx,ny)))); b:=4-a
        end else
       begin repeat write('--> lower bound a ='); readln(a)
        until a>0;
             repeat write('--> upper bound b ='); readln(b)
              until b>=a
       end;
       writeln('Cycle length is computed as 2 to the power
        p.');
       repeat write('--> p ='); readln(p);
              length:=1; for p:=p downto 1 do length:=
               2*length;
              if length>ADImax then writeln('exponent too
               large')
       until length<=ADImax;
       ADI_parameter(om,a,b,length); actual:=1
end end;

procedure XY_reflexion(var x: gridfunction;
                       var A: data_of_discretisation;
                        reflect_A: Boolean);
```

```
var i,j: integer; u: real;
begin with A do
    begin for i:=0 to maximum(nx,ny) do for j:=0 to i-1 do
        begin u:=x[i,j]; x[i,j]:=x[j,i]; x[j,i]:=u end;
        if reflect_A then
        begin i:=nx; nx:=ny; ny:=i;
            if T<>nil then T^.decomposition_computed:=false;
            for i:=-1 to 1 do for j:=-1 to i-1 do
            begin u:=S[i,j]; S[i,j]:=S[j,i]; S[j,i]:=u
end end end end;

procedure ADI_halfstep(var A: data_of_discretisation;
                       var x,b: gridfunction; diagonal:
                        real);
var d: real; i,j: integer; v,z: column;
begin with A do
 begin v:=x[0]; d:=S[0,0]-diagonal;
  define_tridiag(A,S[0,-1],diagonal,S[0,1],true);
  if kind=ninepoint-formula then
     message('ADI not implemented for 9-point formulae');
  for i:=1 to nx-1 do
  begin if kind=Poisson_model_problem then
    for j:=1 to ny-1 do z[j]:=b[i,j]-d*x[i,j]+[i-1,j]+
     x[i+1,j] else
    for j:=1 to ny-1 do z[j]:=b[i,j]-d*x[i,j]-S[-1,0]*
     x[i-1,j]-S[1,0]*x[i+1,j];
    x[i-1]:=v; v[0]:=x[i,0]; v[ny]:=x[i,ny]; solve_
     tridiag(A,v,z)
  end;
  x[nx-1]:=v
end end;

procedure ADI_method(var new: gridfunction; var A: data_of_
                      discretisation;
                     var x,b: gridfunction; var IP:
                      iterationparameter);
var dc: real; label 1;
begin if IP.cycle=nil then 1: message('ADI parameters not yet
 defined!')
      else with IP.cycle^do with A do
      begin if (length<1) or (actual<1) then goto 1;
            while actual>length do actual:=actual-length;
            dc:=-S[-1,0]-S[1,0]; {diagonal of the matrix C}
            new:=x; XY_reflexion(new,A,false); XY_reflexion
             (b,A,true);
            ADI_halfstep(A,new,b,dc+om[actual]);
```

```
        XY_reflexion(new,A,false); XY_reflexion
         (b,A,true);
        ADI_halfstep(A,new,b,S[0,0]-dc+om[actual]);
        actual:=actual+1
end end;
```

7.5.6 Amount of Work and Numerical Examples

In the following, we consider the case of a general five-point formula ($C_A = 5$). The amount of work for solving the equations with the tridiagonal matrices $\omega I + C$, $\omega I + B$ amounts to $5n$ operations. The evaluation of $b + \omega x - Cx$ and $b + \omega x - Bx$ requires $6n$ operations each. Because of $C_A = 5$, this leads to

$$C_\Phi^{\mathrm{ADI}} = \frac{22}{5}; \tag{7.5.18}$$

in the Poisson-model case, $C_\Phi^{\mathrm{ADI}} = 4$.

The asymptotical rates $\varrho_m = \delta_m^{2/m}$ attainable by (16c) are reported in Table 1. One observes that for small step sizes, good rates are also achieved. The concrete results for $h = 1/128$ with $m = 4$ different parameters from Table 2 confirm that the factor 0.5365 from Table 1 is reached. The convergence behaves regularly only modulo m. Each second ratio $\|e^m\|_2/\|e^{m-1}\|_2$ is ≈ 1. However, since one cannot achieve accuracy $\|e^k\|_2 \approx \delta_m^k \|e^0\|_2$ with less than m iteration steps, the following dilemma arises:
(i) To exploit the good (asymptotical) convegence rate δ_m for large m, one must perform at least m iterations.
(ii) On the other hand, one would like to stop the iteration, e.g., as soon as the error becomes $\|e^m\|_2 \approx 1/1000$ (cf. Remark 3.3.4). The better the convergence rate, the fewer iterations one is willing to perform.

In the example of Table 2, about eight steps would be sufficient. Hence, one could still enlarge the cycle length from 4 to 8 (the result is $\|e^8\|_2 = 1.05_{10} - 3$); a further increase to 16 or more parameters would not help. The last two columns in Table 2 correspond to $\varrho_{m,m-1}$ and $\varrho_{m,0}$.

Remark 7.5.11. Good convergence rates are combined with a relatively high cost factor $C_\Phi^{\mathrm{ADI}} = 4$ in thePoisson-model case. For the example from Table 2, the effective amount of work equals $Eff(\Phi^{\mathrm{ADI}}) = -4/\log 0.5365 = 6.42$. For $h = 1/32$ and four parameters, we obtain $Eff(\Phi^{\mathrm{ADI}}) = 4.06$ (for comparison: $Eff(\Phi^{\mathrm{SOR}}) = 7.05$ (cf. Example 3.3.2) and $Eff(\Phi^{\mathrm{SSOR}}_{\mathrm{semiit.}}) = 4.38$ (cf. (4.6)).

8
Transformations, Secondary Iterations, Incomplete Triangular Decompositions

To describe the variety of iterative methods applied nowadays, one needs further techniques producing new iterations from already known ones or introducing new features. Techniques of the former kind are the transformations described in §§8.1–3 and the composed methods from §8.4 generated by a secondary iteration, whereas the ILU decomposition from §8.5 is of the latter kind, since it produces a new method.

8.1 Generation of Iterations by Transformations

8.1.1 Already Discussed Techniques for Generating Iterations

Section 4.2 introduced the (additive) *splitting technique*: The matrix A is split into

$$A = W - R \qquad (W \text{ regular}). \tag{8.1.1}$$

This leads to the iteration

$$x^{m+1} = x^m - W^{-1}(Ax^m - b). \tag{8.1.2}$$

The obvious choices for W are (i) $W = \operatorname{diag}\{A\}$ ($\rightarrow$ Jacobi method), (ii) $W = D - E$ lower triangular matrix ($\rightarrow$ Gauß-Seidel method), and (iii) $W = \Theta I$ ($\rightarrow$ Richardson iteration). The choice $W = (D - E)D^{-1}(D - F)$, however, is by no means evident. For generating the hereby defined symmetric Gauß-Seidel iteration, we have applied a second technique: the *construction of a symmetric iteration* $\Phi^*\Phi$ by means of the adjoint iteration Φ^*. This involves the construction of a *product iteration*, which also belongs to the catalogue of generating techniques. A further technique is the *damping of an iteration*, which corresponds to the replacement of W in (2) by $W_\Theta = \frac{1}{\Theta} W$.

8.1.2 Left Transformation

The (square) system of equations

$$Ax = b \tag{8.1.3}$$

can be transferred by multiplication from the left by the regular matrix T_ℓ into the equivalent system

$$T_\ell A x = T_\ell b. \tag{8.1.4}$$

We may regard (4) as a new system of equations

$$\hat{A}x = \hat{b} \quad \text{with } \hat{A} := T_\ell A,\, \hat{b} := T_\ell b, \tag{8.1.4'}$$

on which we now base the various iterative methods. Here, two approaches are imaginable. The first one is to compute $\hat{A}$ and $\hat{b}$ from $T_\ell A$ and $T_\ell b$ and to work afterwards only with these quantities. If, for instance, T_ℓ is a diagonal matrix, the transition to (4) means that the equations in (3) are suitably scaled. However, if T_ℓ is not of a diagonal structure, one has to advise against this naive approach, since for sparse A, the sparsity structure of $\hat{A} = T_\ell A$ is, in general, enlarged or $\hat{A}$ even becomes a full matrix. This complicates the programming and storage effort. However, even if A is already full, the matrix-matrix-multiplication $T_\ell A$ requires a large amount of work.

Instead, we describe in the following a second mode of performance. So far, we have applied the iteration $\Phi = \Phi(x, b)$ to a fixed system matrix A. In the case of varying matrices A, one can define an iterative method $\Phi(x, A, b)$ {more precisely, its third normal form (2)} by a prescription $A \mapsto W(A)$. Hence, (2) becomes

$$x^{m+1} = x^m - W(A)^{-1}(Ax^m - b). \tag{8.1.5}$$

For instance, Jacobi's method is defined by $A \mapsto W(A) := \operatorname{diag}\{A\}$ for all A with $a_{\alpha\alpha} \neq 0$ ($\alpha \in I$). Applying the prescription (5) to $(\hat{A}, \hat{b})$ instead of (A, b), we obtain

$$x^{m+1} = x^m - W(\hat{A})^{-1}(\hat{A}x^m - \hat{b}). \tag{8.1.6}$$

The definition of $\hat{A}$, $\hat{b}$ allows the representation $\hat{A}x^m - \hat{b} = T_\ell(Ax^m - b)$. Hence, (6) can be rewritten as

$$x^{m+1} = x^m - W(T_\ell A)^{-1} T_\ell (Ax^m - b). \tag{8.1.7}$$

Remark 8.1.1. To perform the iteration (7) one needs

(i) the computation of the defect $Ax - b$ (in the *former* quantities A, b),

(ii) the multiplication of T_ℓ by a vector,

(iii) a method for solving $W(T_\ell A)\delta = \xi$ (ξ given).

Note that we do *not* need the matrix $\hat{A} = T_\ell A$ when we proceed according to Remark 1. Indirectly, $\hat{A}$ appears in $W(T_\ell A)$ only. If, for instance, the Jacobi method is the underlying iteration Φ, one has to evaluate only the diagonal entries of $\hat{A} = T_\ell A$ for $W(T_\ell A) := \operatorname{diag}\{T_\ell A\}$.

Defining

$$\hat{W}(A) := T_\ell^{-1} W(T_\ell A), \tag{8.1.8}$$

one generates a new iterative method $\hat{\Phi}$ by means of $A \mapsto \hat{W}(A)$ that is identical to (7):

$$x^{m+1} = x^m - \hat{W}(A)^{-1}(Ax^m - b). \tag{8.1.7'}$$

This shows that the left transformation T_ℓ by means of (8) is able to generate a new iteration $\hat{\Phi}$ from Φ, which we will denote by

$$\hat{\Phi} = \Phi \circ T_\ell, \tag{8.1.9}$$

in order to express that we apply the iteration Φ after the transformation by T_ℓ.

With this notation, Remark 4.3.2 can be reformulated as

$$\Phi = \Phi_1^{\mathrm{Rich}} \circ W_\Phi^{-1} \qquad (W_\Phi \text{ matrix of } \Phi),$$

i.e., one regains any iteration Φ with regular W_Φ from Richardson's iteration together with the left transformation by the matrix W_Φ of the third normal form of Φ.

One advantage of generating iterations by transformations is that no new convergence analysis is necessary.

Remark 8.1.2. Let $\hat{\Phi} = \Phi \circ T_\ell$. The convergence properties of $\hat{\Phi}$ (applied to the matrix A) are identical to those of Φ applied to $T_\ell A$. Care is advisable only for the interpretation of convergence statements with respect to a norm depending on A (e.g., the energy norm).

For illustration purposes, we choose the left transformation $T_\ell = A^H$. Then, (7) becomes

$$x^{m+1} = x^m - W(A^H A)^{-1} A^H (Ax^m - b). \tag{8.1.10a}$$

Since $A^H A$ is positive definite for regular A, almost all methods mentioned above can be applied to $A^H A$. As an example, the Richardson iteration with the optimal damping factor

$$\Theta_{\mathrm{opt}} = 2/(\Gamma + \gamma) \quad \text{with } \Gamma := \lambda_{\max}(A^H A) = \|A\|_2^2,\ \gamma := \lambda_{\min}(A^H A) = \|A^{-1}\|_2^{-2}$$

may be chosen (cf. Theorem 4.4.3):

$$x^{m+1} = x^m - \Theta_{\mathrm{opt}} A^H (Ax^m - b). \tag{8.1.10b}$$

For the new method (10b), we draw the following conclusion from the convergence properties of the Richardson method (cf. §4.4.1).

Remark 8.1.3. The «squared Richardson iteration» $\Phi^{\mathrm{sqRich}}_{\Theta_{\mathrm{opt}}} := \Phi^{\mathrm{Rich}}_{\Theta_{\mathrm{opt}}} \circ A^H$ defined by (10b) converges for all regular matrices A with the rate

$$\varrho(I - \Theta_{\mathrm{opt}} A^H A) = (\Gamma - \gamma)/(\Gamma + \gamma) = (\mathrm{cond}_2(A)^2 - 1)/(\mathrm{cond}_2(A)^2 + 1).$$

Exercise 8.1.4. Prove that (a) $(\Phi \circ T_1) \circ T_2 = \Phi \circ (T_1 T_2)$.
(b) $\hat{\Phi} = \Phi \circ T$ is equivalent to $\Phi = \hat{\Phi} \circ T^{-1}$.

8.1.3 Right Transformation

The unknown vector $x \in \mathbb{R}^I$ can be substituted by

$$x = T_r \hat{x}, \tag{8.1.11}$$

where T_r is a regular matrix. Insertion into (3) yields the right-sided transformed equation

$$A T_r \hat{x} = b. \tag{8.1.12}$$

It would be a naive approach to compute the matrix $\hat{A}$ from

$$\hat{A}\hat{x} = b \quad \text{with } \hat{A} := A T_r \tag{8.1.13}$$

explicitly in a first step and then to apply iterative methods directly to (13), in order to obtain $x = T_r \hat{x}$ finally from the (approximation of the) solution $\hat{x}$ by means of (11).

Let Φ be an iterative method with the matrix $W = W(A)$. When applied to the system (13), the iteration Φ can be rewritten as

$$\hat{x}^{m+1} = \hat{x}^m - W(AT_r)^{-1}(AT_r\hat{x}^m - b).$$

Introducing

$$x^m := T_r \hat{x}^m,$$

we obtain the iteration

$$x^{m+1} = x^m - T_r W(AT_r)^{-1}(Ax^m - b). \tag{8.1.14}$$

This is a newly generated iteration $\hat{\Phi}$ for solving the original equation (3) with the characterising matrix

$$\hat{W}(A) := W(AT_r) T_r^{-1}. \tag{8.1.15}$$

In analogy to (9), $\hat{\Phi}$ is denoted by

$$\hat{\Phi} = T_r \circ \Phi. \tag{8.1.16}$$

Exercise 8.1.5. Prove that $T_2 \circ (T_1 \circ \Phi) = (T_2 T_1) \circ \Phi$.

Remark 8.1.6. The convergence rate of $\hat{\Phi} = T_r \circ \Phi$ (applied to the matrix A) is identical to the convergence rate of Φ applied to AT_r. The convergence properties of Φ referring to a norm of $e^m = x^m - x$ carry over to the corresponding properties of $\hat{\Phi}$ with respect to the norm of $T_r^{-1}e^m = T_r^{-1}(x^m - x)$.

The analogue to (10a) is the right transformation $T_r = A^H$, leading to

$$x^{m+1} = x^m - A^H W(AA^H)^{-1}(Ax^m - b). \tag{8.1.17}$$

Choosing the Richardson method as the basic iteration, (17) generates again the method (10b), because $\sigma(AA^H) = \sigma(A^H A)$ (cf. Theorem 2.4.6).

8.1.4 Two-Sided Transformation

Applying transformations by T_ℓ from the left and T_r from the right, we obtain the two-sided transformed iteration $\hat{\Phi} = T_r \circ \Phi \circ T_\ell$ generated by

$$\hat{W}(A) := T_\ell^{-1} W(T_\ell A T_r) T_r^{-1}, \tag{8.1.18a}$$

$$x^{m+1} = x^m - T_r W(T_\ell A T_r)^{-1} T_\ell (Ax^m - b). \tag{8.1.18b}$$

Exercise 8.1.7. Prove that $(T_r \circ \Phi) \circ T_\ell = T_r \circ (\Phi \circ T_\ell)$.

Concerning the convergence properties of $T_r \circ \Phi \circ T_\ell$, the same statements apply as for $T_r \circ \Phi$ in Remark 6.

If an iteration Φ satisfies $A = A^H$ and $W = W(A) > 0$, it is symmetric (cf. §4.8.1). In general, the matrices $\hat{W}(A) := T_\ell^{-1} W(T_\ell A)$ from (8) or $\hat{W}(A) := W(AT_r)T_r^{-1}$ from (15) arising from a transformation do not satisfy $\hat{W} > 0$ because of missing symmetry. A remedy is provided by

Remark 8.1.8. Assume that A is Hermitian (or positive definite) and that the left and right transformations satisfy $T_r = T_\ell^H$. Then $\hat{W}(A)$ from (18a) becomes $T_\ell^{-1} W(\hat{A}) T_\ell^{-H}$ with the Hermitian (or positive definite) matrix $\hat{A} := T_\ell A T_\ell^H$. If $W(\hat{A}) > 0$, the two-sided transformed iteration $T_\ell^H \circ \Phi \circ T_\ell$ is a synmmetric iteration.

Usually, we require that the iterations generated by the transformations are feasible ones. However, occasionally, one needs transformations only for the purpose of a suitable theoretical representation. If one views iteration (2) as Richardson's method for $\hat{A} := W^{-1}A$, the fact that, in general, $\hat{A}$ is no longer Hermitian will be unfavourable in later applications. Here, the substitution (11) with $T_r := W^{-1/2}$ helps.

Remark 8.1.9. Let $A = A^H$. The iteration (2) with $W > 0$ yielding the sequence $\{x^m\}$ is equivalent to the iteration

$$\hat{x}^{m+1} = \hat{x}^m - W^{-1/2}(AW^{-1/2}\hat{x}^m - b) \tag{8.1.19}$$

producing $\hat{x}^m := W^{1/2}x^m$. (19) is the Richardson iteration applied to $\hat{A}\hat{x} = \hat{b} := W^{-1/2}b$ with the Hermitian matrix $\hat{A} := W^{-1/2}AW^{-1/2}$.

Since, in general, $W^{-1/2}$ cannot be computed practically, the representation (19) make sense only for a theoretical presentation.

Exercise 8.1.10. Prove that any factorisation of W into $W = V^H V$ can be used for the formulation of $\hat{x}^{m+1} = \hat{x}^m - V^{-H}(AV^{-1}\hat{x}^m - b)$ with $\hat{x}^m := Vx^m$ and leads to a Hermitian matrix $\hat{A} := V^{-H}AV^{-1}$, provided that $A = A^H$. In particular, V can be defined by means of the Cholesky decomposition. The iterates $x^m := V^{-1}\hat{x}^m$ do not depend on the choice of V.

8.2 Kaczmarz Iteration

8.2.1 Original Formulation

In 1937 Kaczmarz [1] described a method for which he could prove convergence for all regular matrices A. In the original formulation, the projections

$$x \mapsto P_i(x, b) := x - \boldsymbol{a}_i\langle Ax - b, \boldsymbol{e}_i\rangle / \langle \boldsymbol{a}_i, \boldsymbol{a}_i\rangle \qquad (1 \leqslant i \leqslant n) \tag{8.2.1a}$$

onto the hyperplanes $\{x \in \mathbb{K}^I : \langle Ax - b, \boldsymbol{e}_i\rangle = 0\}$ are used, where $\boldsymbol{e}_i$ is the ith unit vector and $\boldsymbol{a}_i = A^H \boldsymbol{e}_i$ the transposed ith row of the matrix A. Using $\langle A\boldsymbol{a}_i, \boldsymbol{e}_i\rangle = \langle \boldsymbol{a}_i, A^H\boldsymbol{e}_i\rangle = \langle \boldsymbol{a}_i, \boldsymbol{a}_i\rangle$, one learns that $x' = P_i(x, b)$ lies in the plane $\langle Ax - b, \boldsymbol{e}_i\rangle = 0$. The complete iteration is the product of all projections in the succession $i = 1, \ldots, n$:

$$\Phi^{\mathrm{Kacz}} := P_n \circ P_{n-1} \circ \cdots \circ P_1. \tag{8.2.1b}$$

We note that Kaczmarz' method admits interesting applications to overdetermined systems of equations (cf. Maess [2], Tanabe [1]). The Kaczmarz method is also applied to problems from tomography and is then termed algebraical reconstruction technique (ART; cf. Natterer [1, §V.3–4]).

8.2.2 Interpretation as Gauß-Seidel Method

The right transformation $T_r = A^H$ generates the system

$$AA^H\hat{x} = b \tag{8.2.2a}$$

for $\hat{x}$ with $x = A^H\hat{x}$. Using the $\hat{x}$ variables, we can rewrite the projection (1a) as $\hat{x} = A^{-H}x \mapsto \hat{P}_i(\hat{x}, b)$ with

$$\hat{P}_i(\hat{x}, b) := A^{-H}P_i(A^H\hat{x}, b) = \hat{x} - \boldsymbol{e}_i\langle AA^H\hat{x} - b, \boldsymbol{e}_i\rangle / \langle \boldsymbol{a}_i, \boldsymbol{a}_i\rangle, \tag{8.2.2b}$$

since $A^{-H}\boldsymbol{a}_i = A^{-H}A^H\boldsymbol{e}_i = \boldsymbol{e}_i$. $\hat{P}_i$ is the projection onto $\langle AA^H\hat{x} - b, \boldsymbol{e}_i\rangle = 0$, i.e., $\hat{x} \mapsto \hat{x}' := \hat{P}_i(\hat{x}, b)$ represents the solution of the scalar equation $(AA^H\hat{x})_i = b_i$ with respect to $\hat{x}_i$. Therefore, the performance of $\hat{x} \mapsto \hat{P}_i(\hat{x}, b)$ for $i = 1, \ldots, n$ executes the Gauß-Seidel method for the system (2a). The denominators $\langle \boldsymbol{a}_i, \boldsymbol{a}_i\rangle = \langle AA^H\boldsymbol{e}_i, \boldsymbol{e}_i\rangle$ represent the diagonal entries of AA^H. Hence, the following lemma is proved.

Lemma 8.2.1. *The Kaczmarz method for solving $Ax = b$ coincides with the Gauß-Seidel iteration for $AA^H\hat{x} = b$:*

$$\Phi^{\mathrm{Kacz}} = A^H \circ \Phi^{\mathrm{GS}}. \tag{8.2.3}$$

Since AA^H is positive definite for regular A, the theorem of Ostrowski (Theorem 4.4.21) *yields convergence.*

Theorem 8.2.2. *The Kaczmarz method converges for all regular A.*

To obtain a quantitative statement, one has to determine and estimate the constants γ and Γ from (4.4.32a, b) for the decomposition $AA^H = D - E - F$.

Exercise 8.2.3. Instead of the right transformation, one may also choose a left one by $T_\ell = A^H$. Prove that one step of the Gauß-Seidel iteration applied to $A^HAx = A^Hb$ has the form

$$\text{for } i := 1 \text{ to } n \text{ do} \quad x := x - \boldsymbol{e}_i\langle Ax - b, \boldsymbol{a}_i\rangle / \langle \boldsymbol{a}_i, \boldsymbol{a}_i\rangle, \tag{8.2.4}$$

where $\boldsymbol{a}_i$ different from (1a) represents the ith *column* of A: $\boldsymbol{a}_i := A\boldsymbol{e}_i$.

8.2.3 Pascal Procedures and Numerical Examples

The Kaczmarz iteration is implemented according to the representation (1a, b):

```
function in_the_grid(i,j: integer; var A: data_of_
 discretisation): Boolean;
begin in_the_grid:=(i>0) and (i<A.nx) and (j>0) and (j<A.ny)
end;

function interior_point(i,j: integer; var A: data_of_
 discretisation): Boolean;
begin interior_point:=(i>1) and (i<A.nx-1) and (j>1) and
 (j<A.ny-1)
end;

procedure Kaczmarz_iteration(var new: gridfunction;
                             var A: data_of_discretisation;
                             var x,b: gridfunction; var IP:
                              iterationparameter);
var i,j,is,js: integer; d,d0: real;
function a2(i,j: integer): real; var d: real; is,js: integer;
begin d:=0; with A do for is:=-1 to 1 do for js:=-1 to 1 do
  if in_the_grid(i+is,j+js,A) then d:=d+sqr(S[is,js]);
  a2:-d
end;
```

```
begin d0:=a2(2,2); new:=x;
    with A do for j:=1 to ny-1 do for i:=1 to nx-1 do
    begin d:=b[i,j]-S[-1,-1]*new[i-1,j-1]-S[0,-1]*new[i,j-1]
                   -S[1,-1]*new[i+1,j-1]-S[-1,0]*new[i-1,j]-
                    S[1,0]*new[i+1,j]
                   -S[-1,1]*new[i-1,j+1]-S[0,1]*new[i,j+1]-
                    S[1,1]*new[i+1,j+1]
                   -S[0,0]*new[i,j];
    if interior_point(i,j,A) then d:=d/d0 else d:=d/a2(i,j);
    for is:=-1 to 1 do for js:=-1 to 1 do if in_the_grid
     (i+is,j+js,A) then
    new[i+is,j+js]:=new[i+is,j+js]+S[-is,-js]*d
end end;
```

The convergence rates for the Poisson-model case are given in Tables 1a, b. The ratios $\varrho_{m,m-1}$ behave like $1-(\alpha h)^4$ with $2.5 \leqslant \alpha \leqslant 3$. The convergence order $\varkappa = 4$ makes the Kaczmarz method very unattractive.

8.3 Preconditioning

8.3.1 Meaning of «Preconditioning»

In §7 we saw that in the case of symmetric iterations, the convergence speed depends only on the condition number $\varkappa = \varkappa(W^{-1}A)$. Hence, one can try to choose the left transformation with $T_\ell = W^{-1}$ to satisfy the following conditions:

$$\hat{A} := T_\ell A = W^{-1}A \quad \text{has a positive spectrum,} \tag{8.3.1a}$$

$$\varkappa(W^{-1}A) \quad \text{is as small as possible.} \tag{8.3.1b}$$

Table 8.2.1a $h = 1/8$ **Table 8.2.1b** $h = 1/16$
Kaczmarz results for the Poisson-model case

m	$\|e^m\|_2$	$\varrho_{m,m-1}$	m	$\|e^m\|_2$	$\varrho_{m,m-1}$
10	0.465	0.984542	10	0.624	0.993655
30	0.371	0.990501	30	0.575	0.996921
50	0.309	0.991074	50	0.564	0.997806
70	0.258	0.990876	70	0.525	0.998272
80	0.235	0.990790	80	0.517	0.998434
90	0.241	0.990728	90	0.509	0.998564
100	0.195	0.990687	100	0.502	0.998671

Often, the matrix W is called the «preconditioning matrix», «preconditioning», or «preconditioner». The names express the fact that according to (1b) the transformation by W^{-1} should improve the condition number $\varkappa(A)$, which corresponds to the trivial choice $W = I$. Concerning the applicability of the semi-iterative method to the basic iteration

$$\Phi_W(x, b) := x - W^{-1}(Ax - b) \tag{8.3.1c}$$

the requirement (1a) is not imperative, since the interval $[\gamma, \Gamma]$ with $\gamma = \lambda_{\min}(W^{-1}A)$ and $\Gamma = \lambda_{\max}(W^{-1}A)$ may also be replaced by an ellipse (cf. §7.3.6).

The requirement (1b) is by no means restricted to semi-iterative methods. Having constructed the matrix W with (1a) for an iterative method, one can, in a second step, pass over to the damped method with $M_\Theta = I - W_\Theta^{-1}A$, $W_\Theta := \frac{1}{\Theta}W$. The application of Theorem 4.4.3 to $\hat{A} := W^{-1}A$ yields the following statement.

Exercise 8.3.1. Assume that $W^{-1}A$ has a real spectrum. Prove that (a) The damping factor $\Theta = 2/[\lambda_{\min}(W^{-1}A) + \lambda_{\max}(W^{-1}A)]$ yields the optimal convergence rate

$$\varrho(M_\Theta) = \frac{\varkappa - 1}{\varkappa + 1}, \tag{8.3.2a}$$

where $\varkappa = \varkappa(W_\Theta^{-1}A) = \varkappa(W^{-1}A)$ is not dependent on $\Theta \neq 0$.
(b) Let $M = I - W^{-1}A$ be the iteration matrix of a convergent symmetric method. Then

$$\varkappa(W^{-1}A) \leqslant (1 + \varrho(M))/(1 - \varrho(M)), \tag{8.3.2b}$$

where the equality sign is attained if the iteration is optimally damped.

Because of $\varkappa = \varkappa(W^{-1}A)$, the discovery of a left transformation with (1b) is the essential step for the construction of an iteration. Here, the condition (1a) can also be generalised (cf. Theorem 4.4.8 applied to $\hat{A}$ instead of A).

The connection of the term «preconditioning» to the *left* transformation is not the only possible one. The *right* transformation $T_r = W^{-1}$ in connection to the Richardson method leads to the same iteration (1c) (cf. (1.14)).

In the literature, the term «preconditioning» is used almost exclusively in connection with semi-iterative methods (in particular, conjugate gradient methods; cf. §9). This might lead to the impression that «preconditioning» is a special technique for semi-iterations only. This is not the case: Good (i.e., fast convergent) symmetric iterative methods (1c) produce a good preconditioner as shown by (2b). Vice versa, a preconditioning W with suitable damping yields the convergence rate (2a), which becomes better the smaller $\varkappa$ is, i.e., the better we are able to precondition. The fact that the semi-iteration based on a method Φ_W converges faster than the (stationary) iteration (1c) is a property of the Chebyshev method and not at all a consequence of preconditioning.

In the sense explained above, the search for a symmetric iteration being as fast as possible and the search for a good preconditioner for a semi-iterative method are completely identical. There is only one exception: $\varkappa = \varkappa(W^{-1}A)$ is not the only criterion in the case of the conjugate gradient methods discussed in §9 (see the end of §9.4.3). Hence, there may be preconditioners W for which a conjugate gradient method is advantageous without yielding necessarily fast iterative methods (1c).

8.3.2 Examples

We can cite the matrices W of the already described symmetric iterations as examples of preconditioners for positive definite matrices $A = D - E - F$:

$$W = D := \operatorname{diag}\{A\} \qquad \text{(Jacobi)}, \tag{8.3.3a}$$

$$W = (D - E)D^{-1}(D - F) \qquad \text{(SSOR)}. \tag{8.3.3b}$$

Here, the methods can be understood pointwise as well as blockwise. Since the choice «W = diagonal matrix» is especially simple and furthermore computable in parallel, one might ask whether the Jacobi method with $D := \operatorname{diag}\{A\}$ represents the optimal diagonal preconditioning. The answer is given by Theorems 2 and 3: $D := \operatorname{diag}\{A\}$ is optimal in the 2-cyclic case, whereas D is close to the optimum in the general case.

Theorem 8.3.2 (Forsythe–Strauss [1]). *Assume that A is positive definite with $D := \operatorname{diag}\{A\}$ and $A - D$ is weakly 2-cyclic. Then $D = \operatorname{diag}\{A\}$ is the best diagonal preconditioner: $\varkappa(D^{-1}A) \leqslant \varkappa(\Delta^{-1}A)$ for all diagonal matrices Δ.*

Theorem 8.3.3 (van der Sluis [1]). *Assume that A is positive definite with $D := \operatorname{diag}\{A\}$ and each row of A does not contain more than C_A nonzero entries (cf. (3.3.1)). Then $\varkappa(D^{-1}A) \leqslant C_A\varkappa(\Delta^{-1}A)$ holds for all diagonal matrices Δ.*

According to (7.4.3c), SSOR preconditioning improves the condition number from $\varkappa(A)$ to $\frac{1}{2}(1 + \sqrt{\Gamma\varkappa(A)})$.

If A is indefinite, more precisely, if

$$A = A^H, \quad \sigma(A) \subset [-b', -a'] \cup [a, b] \quad (0 < a \leqslant b, 0 < a' \leqslant b'), \tag{8.3.4}$$

all previous methods apart from Kaczmarz' method of §8.2 are divergent. Matrices with the property (4) arise, for instance, from the discretisation of the *Helmholtz equation* $-\Delta u - cu = f$ with positive c, leading to $A = B - cI$ (here, B is the matrix of the Poisson-model problem, which has been called A before). If $0 < \lambda_1 \leqslant \lambda_2 \leqslant \cdots \leqslant \lambda_n$ are the eigenvalues of B and $\lambda_k < c < \lambda_{k+1}$

holds for some k, assumption (4) is satisfied with $a = \lambda_{k+1} - c$ and $b = \lambda_n - c$, $a' = c - \lambda_k$, $b' = c - \lambda_1$. In the case of indefinite matrices, (1a) becomes the essential part of the requirement for preconditioning.

The Kaczmarz method offers the preconditioning $W = A^{-H}$ leading to a positive spectrum (cf. (1a)) but, in general, deteriorating the condition number: $\varkappa(A^H A) = \mathrm{cond}_2(A)^2$. If the intervals $[-b', -a']$ and $[a, b]$ are symmetric: $a = a'$, $b = b'$, preconditioning $W = A^{-1}$ is, in spite of $\varkappa(A^2) = \varkappa(A)^2$, not worse than the best semi-iterative method based on the Richardson iteration, as shown in Exercise 4.

Exercise 8.3.4. (a) Let p_m be the optimal polynomial solving the problem (7.3.9) for $\sigma_M = [-b, -a] \cup [a, b]$ with $0 < a \leqslant b$. Prove that if m is even, $p_m(\lambda) = \hat{p}_{m/2}(\lambda^2)$ holds, where $\hat{p}_{m/2}$ is the optimal polynomial belonging to $\hat{\sigma}_M = [a^2, b^2]$ (cf. (7.3.12b)).
(b) What is the transformation $T_\ell = A - \vartheta I$ corresponding to the more general case of $b - a = b' - a'$, mapping $\sigma_M = [-b, -a] \cup [a, b]$ onto an interval $[\hat{a}, \hat{b}]$ with $0 < \hat{a} < \hat{b}$ and minimal $\hat{b}/\hat{a}$.

The left transformation by polynomials in A is relatively simple and even executable in parallel; therefore, it is of practical interest for the case (4). A discussion of such preconditioners can be found, e.g., in the paper by Ashby–Manteuffel–Saylor [1]. The next exercise explains why left transformations with polynomials in A are uninteresting for *positive definite* A.

Exercise 8.3.5. Let A be positive definite. (a) The transformation $W^{-1} = T_\ell = q_{m-1}(A)$ by means of the optimal polynomial q_{m-1} of degree $m - 1$ with respect to condition (1a, b): $\varkappa(q_{m-1}(A)A) = \min$ is related to the optimal polynomial p_m from (7.3.12b) by $p_m(\lambda) = 1 - \lambda q_{m-1}(\lambda)$. (b) The Chebyshev method for the problem preconditioned by $q_{m-1}(A)$ is not as effective as the Chebyshev method for the nonconditioned problem (Richardson iteration). *Hint*: m steps without preconditioning are as expensive as one step with preconditioning.

So far, we have restricted our discussion to the iterations described previously. In §8.5 we will introduce the ILU decompositions, which are well-suited as preconditioners. The multi-grid method described in §10 yields a further possibility for preconditioning (cf. 10.8.3).

Remark 8.3.6. The previously discussed preconditionings may be termed *algebraical preconditionings*, because they are derived from the matrix of the linear system. An alternative is *problem-oriented preconditioning*. If the system $Ax = b$ describes a problem similar to $By = c$ (e.g., because $Ax = b$ and $By = c$ are discretising related boundary value problems), we may expect that $\varkappa(B^{-1}A) \ll \varkappa(A)$, i.e., that $W := B$ is a good preconditioner.

So far, we have only discussed the possibility of preconditioning the linear system. Sometimes, one can already reformulate the boundary value problem before it is discretised in order to obtain a discrete system that is easier to solve. For an example, refer to Axelsson–Eijkhout–Polman–Vassilevski [1]. There, the convection-diffusion problem $-\varepsilon\Delta u + \langle v, \operatorname{grad} u\rangle = f$ (cf. Hackbusch [15, §10.2]) is discussed, which usually leads to an unsymmetric matrix.

8.3.3 Rules of Calculation for Condition Numbers

We recall (2.10.7/8): The number $\operatorname{cond}_2(A) = \|A\|_2 \|A^{-1}\|_2$ is defined by the spectral norm, whereas the definition of $\varkappa(A) = \varrho(A)\varrho(A^{-1})$ refers to the spectral radius.

Exercise 8.3.7. Let the matrices A, B, C be regular. Prove:

$$\varkappa(A) = \varkappa(A^{-1}), \quad \operatorname{cond}_2(A) = \operatorname{cond}_2(A^{-1}), \tag{8.3.5a}$$

$$\varkappa(A) = \varkappa(\lambda A), \quad \operatorname{cond}_2(A) = \operatorname{cond}_2(\lambda A) \quad \text{for all } \lambda \in \mathbb{C}\backslash\{0\}, \tag{8.3.5b}$$

$$\varkappa(A) = \operatorname{cond}_2(A) \quad \text{for normal matrices } A, \tag{8.3.5c}$$

$$\operatorname{cond}_2(AB) \leqslant \operatorname{cond}_2(A)\operatorname{cond}_2(B), \tag{8.3.5d}$$

$$\operatorname{cond}_2(C^{-1}A) \leqslant \operatorname{cond}_2(C^{-1}B)\operatorname{cond}_2(B^{-1}A), \tag{8.3.5e}$$

$$\varkappa(B^{-1}A) = \operatorname{cond}_2(B^{-1}A) \qquad \text{for } A, B > 0. \tag{8.3.5f}$$

Lemma 8.3.8. *Let A and B be positive definite. Then $\varkappa(B^{-1}A)$ can be represented as*

$$\varkappa(B^{-1}A) = \bar{\alpha}/\underline{\alpha}, \tag{8.3.6a}$$

where $\underline{\alpha}$ and $\bar{\alpha}$ are the best bounds in the inequality

$$\underline{\alpha}B \leqslant A \leqslant \bar{\alpha}B \quad \textit{with } \underline{\alpha} > 0. \tag{8.3.6b}$$

Vice versa, the inequality (6c) *follows from* (6b):

$$\varkappa(B^{-1}A) \leqslant \bar{\alpha}/\underline{\alpha}. \tag{8.3.6c}$$

Proof. The best bounds in (6b) are the extreme eigenvalues of $B^{-1/2}AB^{-1/2}$ and $B^{-1}A$. Hence, (6a) follows from (2.10.9). Compare also Lemma 7.3.11. □

Exercise 8.3.9. Prove: (6b) is equivalent to (6b′) or (6b″):

$$\frac{1}{\underline{\alpha}}A \leqslant B \leqslant \frac{1}{\bar{\alpha}}A \qquad \text{with } \underline{\alpha} > 0, \tag{8.3.6b′}$$

$$\underline{\alpha}A^{-1} \leqslant B^{-1} \leqslant \bar{\alpha}A^{-1} \quad \text{with } \underline{\alpha} > 0. \tag{8.3.6b″}$$

The inequalities (5e) and (5f) yield the next lemma:

Lemma 8.3.10. *Let A, B, C be positive definite. Then*

$$\varkappa(C^{-1}A) \leqslant \varkappa(C^{-1}B)\varkappa(B^{-1}A). \tag{8.3.7}$$

The interpretation of (5e) *and* (7) *concerning the preconditioning technique is as follows: If B is a good preconditioner for A and C is a good preconditioner for B, then C also represents a good preconditioning of A.*

Definition 8.3.11. Let $H \subset (0, \infty)$ be an index set with $0 \in \overline{H}$ (e.g., H: set of all grid sizes). If $\{A_h\}_{h \in H}$ and $\{B_h\}_{h \in H}$ are two families of regular matrices, then $\{A_h\}_{h \in H}$ and $\{B_h\}_{h \in H}$ are called *spectrally equivalent* if there is a constant C independent of $h \in H$ such that

$$\mathrm{cond}_2(B_h^{-1}A_h) \leqslant C \quad \text{for all } h \in H. \tag{8.3.8}$$

(5a) shows symmetry and (5e) transitivity; hence, the property «spectrally equivalent» defines an equivalence relation. For families of positive definite matrices, inequality (8) can be written as

$$\varkappa(B_h^{-1}A_h) \leqslant C \quad \text{for all } h \in H. \tag{8.3.8'}$$

The spectral equivalence could also be defined by means of other norms or even by (8′). Concerning the introduction of the term «spectrally equivalent» we refer to the early papers of D'Yakonov [1] and Gunn [1].

8.4 Secondary Iterations

8.4.1 Examples of Secondary Iterations

The differrential equation

$$-\Delta u + u_{xy} + \alpha u_x = f \quad \text{in } \Omega \tag{8.4.1a}$$

with the boundary condition (1.2.1b): $u = 0$ on Γ can be discretised, e.g., by the *seven-point formula*

$$\frac{1}{2}h^{-2}\begin{bmatrix} -1 & -1 & 0 \\ -1-\alpha h & 6 & -1+\alpha h \\ 0 & -1 & -1 \end{bmatrix} u = f, \tag{8.4.1b}$$

which abbreviates the equations

$$\begin{aligned} \frac{1}{2}h^{-2}[&6u(x,y) - u(x-h,y+h) - u(x,y+h) - u(x,y-h) \\ &- u(x+h,y-h) - (1+\alpha h)u(x-h,y) \\ &- (1-\alpha h)h(x+h,y)] = f \end{aligned} \tag{8.4.1b'}$$

for $(x, y) \in \Omega_h$ (cf. §2.1.3 and Hackbusch [15, §5.1.4]). As long as $|\alpha h| \leqslant 1$, the arising matrix A is an M-matrix. However, note that A is not symmetric, unless $\alpha = 0$.

According to Remark 3.6, one can choose the matrix

$$B \mathrel{\hat{=}} h^{-2} \begin{bmatrix} & -1 & \\ -1 & 4 & -1 \\ & -1 & \end{bmatrix}$$

corresponding to the Poisson-model problem as preconditioning.

Lemma 8.4.1. *Let A be the matrix of system* (1b′) *and B the matrix of the Poisson-model problem. Then* $\operatorname{cond}_2(B^{-1}A) \leqslant C$ *holds for all* $h \leqslant 1/|\alpha|$, *i.e., $A = A_h$ and $B = B_h$ are spectrally equivalent.*

Proof. (i) Since the preconditioning chosen is problem-oriented, the proof is also problem-oriented. Since the proof would become too complicated, we demonstrate the spectral equivalence with respect to the energy norm $\|x\|_B = \|B^{1/2}x\|_2 = \langle Bx, x\rangle^{1/2}$:

$$\operatorname{cond}_B(B^{-1}A) \leqslant C \quad \text{for all } h \leqslant 1/|\alpha|. \tag{8.4.2a}$$

$\|\cdot\|_B$ is equivalent to the H_h^1 norm, which is denoted in Hackbusch [15, §9.2] by $\|\cdot\|_1$. The *dual norm* $\|x\|_{-1} := \sup\{|\langle x, y\rangle|/\|y\|_B \colon y \neq 0\}$ is

$$\|x\|_{-1} = \|x\|_{B^{-1}} = \|B^{-1/2}x\|_2 = \langle B^{-1}x, x\rangle^{1/2}.$$

(ii) From Hackbusch [15, Exerc. 9.2.4/6], one obtains the «H_h^1 coercivity»

$$\langle Ax, x\rangle \geqslant C_E\|x\|_B^2 - C_K\|x\|_2^2. \tag{8.4.2b}$$

From the M-matrix property of A one concludes that $\|A^{-1}\|_\infty \leqslant C'$ and $\|A^{-T}\|_\infty \leqslant C'$, and hence, $\|A^{-1}\|_2 = \varrho(A^{-T}A^{-1}) \leqslant C'^2$ independently of h (cf. Hackbusch [15, Theorems 4.3.16 and 5.1.22]). The ℓ_2 stability $\|A^{-1}\|_2 \leqslant C$ and the H_h^1 coercivity yield the «H_h^1 regularity»

$$\|A^{-1}\|_{1\leftarrow -1} := \sup\{\|A^{-1}x\|_B/\|x\|_{B^{-1}} \colon x \neq 0\} \leqslant \text{const} \tag{8.4.2c}$$

for all $h \leqslant 1/|\alpha|$ (cf. Hackbusch [15, Theorem 9.2.3]). The substitution $x = By$ together with $\|x\|_{B^{-1}} = \|y\|_B$ shows

$$\|A^{-1}B\|_B \leqslant \text{const} \quad \text{for all } h \leqslant 1/|\alpha|. \tag{8.4.2c′}$$

The reverse inequality $\|A\|_{-1\leftarrow 1} \leqslant$ const is easy to prove and can be rewritten as

$$\|B^{-1}A\|_B \leqslant \text{const} \quad \text{for all } h \leqslant 1/|\alpha|. \tag{8.4.2d}$$

(2c′) and (2d) yield (2a).
(iii) By the discrete «H_h^2 regularity» (cf. Hackbusch [15, §9.2.4]), one can even prove $\operatorname{cond}_2(B^{-1}A) \leqslant C$ for all $h \leqslant 1/|\alpha|$. □

Applying Theorem 4.4.8 to $B^{-1/2}AB^{-1/2}$ instead of A, we obtain

Lemma 8.4.2. *Let A be decomposed into $A_0 + iA_1$ ($A_0 = A_0^H$ symmetric, $iA_1 = iA_1^H$ skew-symmetric part). Let B be positive definite and assume that*

$$\lambda B \leqslant A_0 \leqslant \Lambda B, \quad \lambda > 0, \quad -\sigma B \leqslant A_1 \leqslant \sigma B, \quad 0 < \Theta < 2\lambda/[\lambda\Lambda + \sigma^2]. \tag{8.4.3a}$$

Then the iteration

$$x^{m+1} = x^m - B^{-1}(Ax^m - b) \tag{8.4.3b}$$

converges monotonically with respect to the $\|\cdot\|_B$ norm:

$$\|M_\Theta\|_B = \|I - \Theta B^{-1/2} A B^{-1/2}\|_2$$
$$\leqslant \tfrac{1}{2}\Theta(\Lambda - \lambda) + \sqrt{[1 - \tfrac{1}{2}\Theta(\Lambda + \lambda)]^2 + \Theta^2\sigma^2} < 1. \tag{8.4.3c}$$

The optimal Θ is described in (4.4.11c).

Remark 8.4.3. According to Lemma 1, the inequalities (3a) hold with h-independent λ, Λ, σ for problem (1b). Hence, for sufficiently small Θ, the iteration has a convergence rate independent of h, i.e., the convergence order has the optimal value $\varkappa = 0$. Different from the previous iterations, the convergence does not deteriorate as h tends to zero.

The good convergence properties of iteration (3b) are offset by the difficulty in performing the mapping $c \mapsto B^{-1}c$ required for a solution of the system $B\delta = c$. In the given case, this would, in principle, be possible, because there are direct solvers for the Poisson model (see the end of §1.3) and the fast FFT techniques can be applied also. The solvers, however, do not work for domains other than rectangles.

One remedy is the approximate solution of the mapping $c \mapsto B^{-1}c$ (i.e., of the equation $B\delta = c$) by some iterative technique. The iteration for solving the auxiliary problem $B\delta = c$ is called the *secondary iteration* and leads to the following *composed iteration*:

composed iteration Φ_k:	(8.4.4)
$x^m \mapsto c := Ax^m - b,$	(8.4.4a)
perform the secondary iteration for solving $B\delta = c$:	(8.4.4b)
set the starting iterate $\delta^0 := 0,$	(8.4.4b$_1$)
perform the (semi-)iteration $\delta^0 \mapsto \delta^1 \mapsto \cdots \mapsto \delta^k.$	(8.4.4b$_2$)
$x^{m+1} := x^m - \delta^k.$	(8.4.4c)

The number k of inner iteration steps may depend on m or be constant. The larger k is, the better the sequences x^m from (3b) and (4a–c) coincide. Vice versa, one would like to choose k as small as possible, since the amount of work per (outer) iteration step increases with k.

In the case of the block variants of the different iterative methods, we have already solved systems of equations for each block. Since in the model cases tridiagonal blocks always appear, the exact solution is easily and cheaply computable (only for parallel computations is another procedure imaginable). This is different for three-dimensional problems (e.g., for the Poisson equation $-u_{xx} - u_{yy} - u_{zz} = f$ in the cube $\Omega = (0, 1)^3$).

Remark 8.4.4. In the two-dimensional case, the rows and columns are the natural block structures. In the three-dimensional case, one has a doubly nested block structure: The grid Ω_h consists of blocks formed by planes (either the x-y, x-z, or y-z plane). The blocks corresponding to the planes have the structure of the two-dimensional problems treated thus far and decompose again into rows or columns (in the x, y, or z direction) as sub-blocks. If we choose the planes as blocks for the block-Jacobi, block-Gauß-Seidel method, or another block variant, the block equations are of the same form as the previous, two-dimensional model problems. Performing a secondary iteration for each of the blocks is one obvious approach.

At first view, the secondary iteration recommended in Remark 4 seems to lead to an expensive algorithm. The following exercise, however, shows that the blocks, e.g., arising from the block-Jacobi method are strictly diagonally dominant and have an h-independent condition number.

Exercise 8.4.5. Prove (a) In three dimensions, the discretisation corresponding to the five-point formula reads

$$h^{-2}[6u(x,y,z) - u(x-h,y,z) - u(x+h,y,z) - u(x,y-h,z) \\ - u(x,y+h,z) - u(x,y,z-h) - u(x,y,z+h)] = f(x,y,z)$$

and leads to diagonal blocks with the five-point star

$$D = h^{-2}\begin{bmatrix} 0 & -1 & 0 \\ -1 & 6 & -1 \\ 0 & -1 & 0 \end{bmatrix}.$$

(b) The matrices D from (a) have a spectrum contained in $[2h^{-2}, 10h^{-2}]$.
(c) What is the optimal damping of the Jacobi method for the matrix D? Prove the h-independent rate 2/3 for the optimally damped Jacobi method.

8.4.2 Convergence Analysis in the General Case

In the following, we denote the matrix of the third normal form of $\Phi = \Phi_A$ by B (cf. (3b)) instead of W. The iteration matrix of

$$x^{m+1} = x^m - B^{-1}(Ax^m - b) \tag{8.4.5a}$$

is

$$M_A = I - B^{-1}A. \tag{8.4.5b}$$

For solving the auxiliary equation $B\delta = c$, we apply the secondary iteration Φ_B:

$$\delta^{m+1} = \delta^m - C^{-1}(B\delta^m - c) = M_B\delta^m + N_B c \tag{8.4.6}$$

with the iteration matrix $M_B = I - C^{-1}B$.

In the following, we will always apply the secondary iteration in (4b) with a *constant* value k, since otherwise we would have to describe a suitable stopping criterion.

Lemma 8.4.6. *Let Φ_A be the iteration for solving $Ax = b$ by* (5a), *while Φ_B is a linear iteration for solving $B\delta = c$. The composed iteration Φ_k defined for fixed $k \geqslant 0$ by* (4a–c) *is a linear and consistent iteration $\Phi = \Phi_k$ for the solution of $Ax = b$. Its iteration matrix is*

$$M_k = I - \sum_{q=0}^{k-1} M_B^q N_B A \qquad (M_B, N_B \textit{ from } (6)). \tag{8.4.7a}$$

If, in addition, Φ_B is consistent, (7a) *simplifies to*

$$M_k = M_A + M_B^k B^{-1} A \qquad (M_A \textit{ from } (5b)). \tag{8.4.7b}$$

The matrix of the second normal form (3.2.4) *is*

$$N_k = (I - M_B^k)B^{-1}. \tag{8.4.7c}$$

If M_B has an eigenvalue λ with $\lambda^k = 1$, the iteration Φ_k diverges; otherwise, the matrix of the third normal form (3.2.5) *can be written as*

$$W_k = B(I - M_B^k)^{-1}. \tag{8.4.7d}$$

Proof. According to Theorem 3.2.5, the iterate δ^k from (4b$_2$) has the representation $\delta^k = \sum_{q=0}^{k-1} M_B^q N_B c$ (note that $\delta^0 = 0$ and $c = Ax^m - b$). This proves (7a). The consistency of Φ_B implies $N_B = (I - M_B)B^{-1}$ (cf. (3.2.3″)). (7b) and (7c) can be concluded from $\sum_{q=0}^{k-1} M_B^q(I - M_B) = I - M_B^k$ and (5b). $W_k = N_k^{-1}$ holds for invertible N_k and indicates (7d). □

The representation (7b) permits interpretation of the iteration matrix M_k as a perturbation of the iteration matrix M_A. The contraction number of Φ_k can be estimated as follows.

Lemma 8.4.7. *Let the iterations Φ_A and Φ_B from Lemma 6 be linear and consistent. The contraction numbers of Φ_k with respect to the spectral norm and, if B or A are positive definite, with respect to the norms $\|x\|_B = \|B^{1/2}x\|_2$ and $\|x\|_A = \|A^{1/2}x\|_2$ are*

$$\|M_k\|_2 \leqslant \|M_A\|_2 + \|M_B\|_2^k \|B^{-1}A\|_2, \tag{8.4.8a}$$

$$\|M_k\|_B \leqslant \|M_A\|_B + \|M_B\|_B^k \|B^{-1/2}AB^{-1/2}\|_2 \quad (\text{if } B > 0), \tag{8.4.8b}$$

$$\|M_k\|_A \leqslant \|M_A\|_A + \|M_B\|_A^k \|A^{1/2}B^{-1}A^{1/2}\|_2 \quad (\text{if } A > 0). \tag{8.4.8c}$$

Knowledge of the spectral radius $\varrho(M_B)$ is not sufficient for analysis of the secondary iteration, because the spectral radius describes the convergence only asymptotically, whereas here we need precise upper bounds after the fixed number of k iteration steps. At best, the contraction number of Φ_B may be replaced by the numerical radius $r(M_B)$.

Exercise 8.4.8. Prove: (a) Let Φ_A and Φ_B be linear and consistent. Then

$$r(M_k) \leqslant r(M_A) + 2r(M_B)^k \|B^{-1}A\|_2. \tag{8.4.8d}$$

(b) The factor $\|B^{-1}A\|_2$ in (8a, d) is bounded by

$$1 - \|M_A\|_2 \leqslant \|B^{-1}A\|_2 \leqslant 1 + \|M_A\|_2. \tag{8.4.8e}$$

The conclusions that one can draw from (8a–d) are the subject of

Remark 8.4.9. (a) Assume that one of the expressions $\|M_A\|_2$, $\|M_A\|_B$, $\|M_A\|_A$, $r(M_A)$ together with the corresponding quantity $\|M_B\|_2$, $\|M_B\|_B$, $\|M_B\|_A$, $r(M_B)$ is smaller than 1. Then the composed method Φ_k converges for sufficiently large k.
(b) One should choose k sufficiently large, so that the right-hand side of (8a) has a size comparable with $\|M_A\|_2$, e.g., $\frac{1}{2}(1 + \|M_A\|_2)$ (similarly for (8b–d)). If $\|M_A\|_2 \leqslant \zeta < 1$ (ζ independent of h) and $\|M_B\|_2 = 1 - O(h^\beta)$ ($\beta > 0$), inequality $\|M_k\|_2 \leqslant (1 + \zeta)/2$ can be achieved with $k = O(h^{-\beta})$. In this case, the effective amount of work for Φ_k is also of the order $\mathit{Eff}(\Phi_k) = O(h^{-\beta})$. If, however, $\|M_A\|_2 = 1 - O(h^\alpha)$ ($\alpha > 0$), inequality (8a) admits only the unfavourable estimate $\mathit{Eff}(\Phi_k) = O(h^{-\alpha-\beta})$.

In particular, (8a–d) yields no statement that could guarantee the convergence of Φ_k for small k. Since, according to (8e), the factor $\|B^{-1}A\|_2$ attains at best the value ≈ 1, one needs at least $k = O(h^{-\beta})$ iterations to make the right-hand side of (8a) smaller than 1.

Since Φ_k is again a linear and consistent iteration, Φ_k may be used as a basic iteration, e.g., of a semi-iteration. Another situation arises, if k is not fixed but determined by means of some stopping criterion or a semi-iteration is applied as a secondary process. In these cases, Φ_k is nonlinear; hence, the suitability of Φ_k as the basic iteration of a semi-iterative method is questionable. For a discussion of this problem, we refer to Golub–Overton [1] and Axelsson–Vassilevski [2].

8.4.3 Analysis in the Symmetric Case

In the following, let Φ_A and Φ_B be symmetric iterations, i.e., the matrices in (5a) and (6) satisfy

$$A = A^H, \qquad B > 0, \qquad C > 0. \tag{8.4.9a}$$

Lemma 8.4.10. *Let Φ_A and Φ_B be symmetric. A necessary condition for the convergence of Φ_B is*

$$0 < B < 2C. \tag{8.4.9b}$$

Under this assumption, the composed iteration Φ_k defined in (4) *is also symmetric for all $k \in \mathbb{N}$.*

Proof. According to (7c), Φ_k has the first normal form

$$x^{m+1} = M_k x^m + N_k b \quad \text{with } N_k = (I - M_B^k)B^{-1}. \tag{8.4.10a}$$

We have to show $W_k > 0$ for the matrix of the third normal form of Φ_k:

$$W_k(x^m - x^{m+1}) = Ax^m - b. \tag{8.4.10b}$$

By Remark 4.8.3a, (9b) is equivalent to the convergence of Φ_B. $\varrho(M_B) < 1$ implies that $I - M_B^k$ is regular; hence, the matrix $W_k = N_k^{-1} = B(I - M_B^k)^{-1}$ exists. The representation $M_B = I - C^{-1}B = B^{-1/2}(I - B^{1/2}C^{-1}B^{1/2})B^{1/2}$ proves the symmetry $W_k = W_k^H$:

$$W_k = B^{1/2}[I - (I - B^{1/2}C^{-1}B^{1/2})^k]^{-1}B^{1/2}. \tag{8.4.10c}$$

Since $\varrho(M_B) < 1$ implies $\varrho(I - B^{1/2}C^{-1}B^{1/2}) < 1$, this inequality together with $I - (I - B^{1/2}C^{-1}B^{1/2})^k > 0$ (because of $k > 0$) yields the positive definiteness

$$W_k > 0. \tag{8.4.10d}$$

According to (4.8.1a, b), Φ_k is a symmetric iteration. □

It is not true that the convergence of Φ_A and Φ_B implies that of Φ_k, but convergence can always be achieved for a suitable damping. In the following, we assume the inequalities

$$\gamma B \leqslant A \leqslant \Gamma B \quad \text{with } 0 < \gamma \leqslant \Gamma, \tag{8.4.11a}$$

$$\delta C \leqslant B \leqslant \Delta C \quad \text{with } 0 < \delta \leqslant \Delta. \tag{8.4.11b}$$

The spectrum of $B^{1/2}C^{-1}B^{1/2}$ lies in $[\delta, \Delta]$ (cf. (3.6b″)), i.e., $\sigma(I - B^{1/2}C^{-1}B^{1/2}) \subset [1 - \Delta, 1 - \delta]$. The spectrum of $I - (I - B^{1/2}C^{-1}B^{1/2})^k$ is contained in the interval $[\underline{\beta}, \overline{\beta}]$ with

$$\underline{\beta} := \left\{ \begin{array}{ll} 1 - (1 - \delta)^k & \text{for odd } k \\ 1 - \max\{(1 - \Delta)^k, (1 - \delta)^k\} & \text{for even } k \end{array} \right\}, \tag{8.4.11c$_1$}$$

$$\overline{\beta} := \left\{ \begin{array}{ll} 1 - (1 - \Delta)^k & \text{for odd } k \text{ or } \Delta < 1 \\ 1 & \text{for even } k \text{ and } \Delta \geqslant 1 \end{array} \right\}. \tag{8.4.11c$_2$}$$

(10c) proves that $\underline{\beta} W_k \leqslant B \leqslant \overline{\beta} W_k$. By means of (11a) one obtains

Lemma 8.4.11. *The inclusions* (11a, b) *prove* (12) *for W_k from* (10b):

$$\gamma_k W_k \leqslant A \leqslant \Gamma_k W_k \quad \text{with } \gamma_k := \gamma\underline{\beta},\ \Gamma_k := \Gamma\overline{\beta} \qquad (\underline{\beta}, \overline{\beta} \text{ from (11c)}). \tag{8.4.12}$$

Let δ, Δ, γ, Γ be the optimal bounds in (11a, b). *Then* (13) *holds*:

$$\varkappa(W_k^{-1}A) = \frac{\Gamma_k}{\gamma_k} = \frac{\Gamma}{\gamma}\frac{\bar{\beta}}{\underline{\beta}} = \frac{\bar{\beta}}{\underline{\beta}}\varkappa(B^{-1}A). \tag{8.4.13}$$

Analysing the iteration Φ_B separately, we obtain the optimal damping parameter

$$\Theta_B = 2/(\delta + \Delta) \qquad (cf.\ (4.4.5)). \tag{8.4.14a}$$

The matrix of the third normal form of the damped *iteration Φ_{B,Θ_B} is $\Theta_B^{-1}C$ instead of C and leads to the bounds $\delta\Theta_B$ and $\Delta\Theta_B$ instead of δ and Δ. This scaling changes the ratio $\bar{\beta}/\underline{\beta}$. The next exercise discusses the factor Θ_B minimising the condition number* (13).

Exercise 8.4.12. Prove that (a) for even k, the parameter Θ_B from (14a) yields the optimal condition number (13). For odd k, however, the minimum of $\varkappa(W_k^{-1}A)$ is attained by a value of Θ_B in the open interval

$$1/\Delta < \Theta_B < 2/(\delta + \Delta). \tag{8.4.14b}$$

(b) For $k = 1$, $\varkappa(W_1^{-1}A) = \varkappa(B^{-1}A)\varkappa(C^{-1}B)$ holds independently of Θ_B.
(c) For $k = 3$, the optimal value is $\Theta_B = 3/[\Delta + \delta + \sqrt{\Delta(\Delta - \delta) + \delta^2}]$.

Since in the case of an odd k, the optimal Θ_B is not explicitly described, in the following we will always use (14a).

8.4.4 Estimate of the Amount of Work

An important question concerns the number k of secondary iterations for which one obtains an effective amount of work as favourable as possible. A trivial statement is given in

Remark 8.4.13. The effective amount of work $Eff(\Phi_k)$ is minimal for finite k, because $Eff(\Phi_k) = O(k)$ for $k \to \infty$.

Proof. $2C_A + 1$ operations are needed for (4a) and (4c) (concerning C_A, see §3.3.1). Let C_B be the amount of work for one secondary iteration step. Then the cost factor for Φ_k amounts to

$$C_k = C' + kC'' \quad \text{with } C' := 2 + 1/C_A,\ C'' := C_B/C_A. \tag{8.4.15}$$

C_k increases for $k \to \infty$ as $O(k)$, while at best the convergence rate of Φ_k tends to that of Φ_A. □

We assume that $\varkappa := \varkappa(C^{-1}B) = \Delta/\delta \gg 1$ for the condition number corresponding to the method Φ_B. Furthermore, let Φ_B already be optimally damped, i.e., $\Theta_B = 2/(\delta + \Delta) = 1$ (cf. (14a)). Then

$$-(1 - \Delta) = 1 - \delta = (\varkappa - 1)/(\varkappa + 1) = 1 - \frac{1}{\varkappa} + O(\varkappa^{-2})$$

proves

$$(1 - \delta)^k = 1 - \frac{k}{\varkappa} + O((k/\varkappa)^2), \qquad (1 - \Delta)^k = (-1)^k(1 - \delta)^k. \tag{8.4.16a}$$

From $(11c_{1,2})$, one obtains the following expansion for $k \leqslant \varkappa$:

$$\frac{\bar{\beta}}{\underline{\beta}} = \begin{cases} \varkappa/k + O(1) & \text{for odd } k, \\ \varkappa/2k + O(1) & \text{for even } k. \end{cases} \tag{8.4.16b}$$

First, we consider the case in which Φ_k serves as the (stationary) iterative method. Then the convergence rate

$$\varrho(\Phi_k) = \frac{\varkappa(W_k^{-1}A) - 1}{\varkappa(W_k^{-1}A) + 1} \approx 1 - 2/\varkappa(W_k^{-1}A) = 1 - 2\frac{\gamma}{\Gamma}\frac{\underline{\beta}}{\bar{\beta}} \approx 1 - 2\alpha k \tag{8.4.16c}$$

can be shown for optimal damping (cf. (3.2a)) with $\alpha = \gamma/(\varkappa\Gamma)$ for odd and $\alpha = 2\gamma/(\varkappa\Gamma)$ for even k. From $-\log\varrho(\Phi_k) \approx 2\alpha k$ and (15) we obtain

$$Eff(\Phi_k) \approx \frac{C' + kC''}{2\alpha k} = \left(\frac{1}{k}C' + C''\right)\Big/(2\alpha). \tag{8.4.16d}$$

Remark 8.4.14. The effective amount of work of the iterative method Φ_k initially decreases with k until for $k \approx \varkappa$ the asymptotical representations (16b, c) lose their validity. Because of the better value $\bar{\beta}/\underline{\beta}$ for even k, one should prefer even numbers k.

A different situation arises when the (symmetric) iteration Φ_k is used as the basic iteration of the Chebyshev method, since then the asymptotical rate is given by (7.3.25) instead of (16c). According to (7.3.28d), the expansion

$$\begin{aligned} Eff_{\text{semiiterative}}(\Phi_k) &\approx \left[\frac{1}{2}(C' + kC'') + \frac{3}{C_A}\right]\left(\frac{\Gamma}{\gamma}\frac{\bar{\beta}}{\underline{\beta}}\right)^{1/2} \\ &\approx \left[\frac{1}{2}(C' + kC'') + \frac{3}{C_A}\right]\Big/\sqrt{\alpha k} \end{aligned} \tag{8.4.16e}$$

holds with the same α as in (16c).

Remark 8.4.15. The semi-iterative effective amount of work (16e) becomes minimal for the even number k next to the value k_0 from (16f):

$$k_0 = \left(\frac{C'}{2} + \frac{3}{C_A}\right)\Big/\frac{C''}{2} = \left(C' + \frac{6}{C_A}\right)\Big/C'' = \left(2 + \frac{7}{C_A}\right)\Big/C''. \tag{8.4.16f}$$

Since $k_0 < 3$ is realistic, $k = 2$ is the optimum.

8.4.5 Pascal Procedures

The composed iteration (4) damped by $\Theta =$ `IP.theta` can be called by the procedure `composed_iteration`, where the parameter `k` denotes the number of secondary iteration steps.

```
procedure k_steps(k: integer; var A: data_of_discretisation;
    var x,b: gridfunction; var IP: iterationparameter;
    procedure iteration(var new: gridfunction; var A: data_
                         of_discretisation;
                        var x,b: gridfunction; var IP:
                         iterationparameter));
begin zero_gridfunction(A.nx,A.ny,x); IP.Nr:=0;
      for k:=k downto 1 do iteration(x,A,x,b,IP)
end;

procedure composed_iteration(var primary, secondary: data_of_
  iteration;
          k: integer;
          procedure secondaryiteration(var new: gridfunction;
          var A: data_of_discretisation;
          var x,b: gridfunction; var IP: iterationparameter));
begin with primary do residual(secondary.b,A,x,b);
      with primary do with A do if kind=Poisson_model_
       problem then
          factor_x_vector(nx,ny,secondary.b,1/h2,secondary.b);
      with secondary do with A do
            if kind=Poisson_model_problem then factor_x_vector
             (nx,ny,b,h2,b);
      with secondary do k_steps(k,A,x,b,IP,
       secondaryiteration);
      with primary do
           vector_plus_factor_x_vector(A.nx,A.ny,x,x,IP.theta,
             secondary.x)
end;
```

8.4.6 Numerical Examples

We solve the introductory example (1b) for $\alpha = 1$ by choosing the matrix B as that of the Poisson-model problem. The solution of the auxiliary equation $B\delta = r$ is approximated by the ADI method with the cycle length $4 = 2^p$ with $p = 2$, where $k = 4$ is also chosen in ($4b_2$). Note that the ADI method is not directly applicable to the original problem, since A is not a five-point matrix and, furthermore, not symmetric. The composed method Φ_4 is performed without damping (`theta` = 1) for $h = 1/32$, $1/64$. The exact solution is again $u = x^2 + y^2$ if we choose $f = 2x - 4$ in (1a):

```
function right_hand_side(x,y: real): real;

begin right_hand_side:=-4+2*x end;
```

The main part of the PASCAL program for this problem reads:

```
program secondary_iteration;
var it,its: data_of_iteration; i,is,itnr: integer;
{Insert the necessary procedures as, e.g., right_hand_side,
 etc.}
begin initialise_IT(it); initialise_IT(its);
    writeln('Define primary problem:');
    define_problem(it,boundary_value,right_hand_side);
    determine_theta(it.IP);
    writeln('Define secondary problem:');
    define_problem(its,zerofunction,zerofunction);
    define_ADI_parameter(its);
    define_starting_iterate(it,zerofunction); writeln
     ('Starting value defined.');
    write('--> Number of primary iterations =');
     readln(itnr);
    is:=its.IP.cycle^.length; writeln('Number of secondary
     ADI iterations =',is);
    for i:=1 to itnr do
    begin composed_iteration(it,its,is,ADI_method);
          it.IP.nr:=it.IP.nr+1; {comparison with exact
           solution, output}
end end;
```

Table 8.4.1 Φ_4 for $h = 1/32$

m	$\|x^m - x\|_2$	$\varrho_{m,m-1}$
1	$3.06_{10}-2$	$4.093_{10}-2$
2	$3.79_{10}-3$	$1.239_{10}-1$
3	$9.80_{10}-4$	$2.582_{10}-1$
4	$3.40_{10}-4$	$3.473_{10}-1$
5	$1.33_{10}-4$	$3.927_{10}-1$
6	$5.90_{10}-5$	$4.415_{10}-1$
7	$2.61_{10}-5$	$4.427_{10}-1$
8	$1.19_{10}-5$	$4.557_{10}-1$
9	$5.53_{10}-6$	$4.640_{10}-1$
10	$2.56_{10}-6$	$4.640_{10}-1$

Table 8.4.2 Φ_4 for $h = 1/64$

m	$\|x^m - x\|_2$	$\varrho_{m,m-1}$
1	$2.78_{10}-2$	$3.628_{10}-2$
2	$4.36_{10}-3$	$1.565_{10}-1$
3	$1.57_{10}-3$	$3.611_{10}-1$
4	$6.68_{10}-4$	$4.243_{10}-1$
5	$2.90_{10}-4$	$4.340_{10}-1$
6	$1.32_{10}-4$	$4.576_{10}-1$
7	$5.99_{10}-5$	$4.512_{10}-1$
8	$2.73_{10}-5$	$4.568_{10}-1$
9	$1.25_{10}-5$	$4.566_{10}-1$
10	$5.73_{10}-6$	$4.581_{10}-1$

The results for the different step widths show a rate of ≈ 0.46. Note, however, that one Φ_k step consists of $k = 4$ ADI steps. Therefore, $0.46^{1/4} = 0.82$ gives a better idea of the rate. The reader may determine the effective amount of work. Tables 1 and 2 contain the Euclidean norm of the errors $\|x^m - x\|_2$ (m: number of the outer iterations Φ_k for $k = 4$) and their ratios $\varrho_{m,m-1} = \|x^m - x\|_2 / \|x^{m-1} - x\|_2$.

8.5 Incomplete Triangular Decompositions

8.5.1 Introduction and ILU Iteration

In the following, the index set I is ordered. Here, the standard choice is lexicographical ordering. ILU stands for «*i*ncomplete *LU* decomposition», where L and U refer to the lower and upper triangular matrices, respectively. Triangular matrices have already appeared in the additive decomposition $A = D - E - F = D(I - U - L)$ on which the Gauß-Seidel method is based. However, the matrices used in this chapter correspond to the multiplicative decomposition $A = LU$. By Remark 1.3.8, the so-called LU decomposition $A = LU$ has proved to be *inappropriate* for sparse matrices, since the factors L and U contain many more nonzero entries than the original matrix A. The computation of the LU decomposition is completely identical to Gauß elimination: U is the upper triangular matrix remaining after the elimination of the entries below the diagonal, whereas L contains the elimination factors $L_{ji} = a_{ji}^{(i)} / a_{ii}^{(i)}$ $(j \geqslant i)$ (cf. Stoer [1, §4.1], Stoer–Bulirsch [1, §4.1]). Instead of computing L and U by means of the Gauß elimination, one may determine the $n^2 + n$ unknown entries L_{ji}, U_{ij} $(j \geqslant i)$ directly from the normalisation condition

$$L_{ii} = 1 \qquad (1 \leqslant i \leqslant n) \tag{8.5.1a}$$

and the n^2 equations $A = LU$:

$$\sum_{j=1}^{n} L_{ij} U_{jk} = A_{ik} \qquad (1 \leqslant i, k \leqslant n). \tag{8.5.1b}$$

The incomplete LU decomposition is based on the idea of not eliminating all matrix entries of A in order to avoid the fill-in of the matrix during the elimination process. Since, after an incomplete elimination, entries remain the lower triangular part, an exact solution of the system is not possible. Instead, one has to work iteratively. The previous equality $A = LU$ holds up to a remainder R:

$$A = LU - R \tag{8.5.2}$$

For the exact description of the ILU process, one has to choose a subset $E \subset I \times I$ of the product of the ordered index set $I = \{1, 2, \ldots, n\}$: The elimination is to be performed only for the pairs $(i,j) \in E$. We always require

$$(i,i) \in E \quad \text{for all } i \in I. \tag{8.5.3a}$$

In general, one should choose E large enough, so that the graph $G(A)$ of A is contained in E (cf. Definition 6.2.1):

$$G(A) \subset E. \tag{8.5.3b}$$

E is called the *pattern* of the ILU decomposition. Through the definition of the triangular matrices, we have

$$L_{ij} = U_{ji} = 0 \quad \text{for } 1 \leqslant i < j \leqslant n. \tag{8.5.4a}$$

In order to construct *sparse* matrices L and U, nonzero entries are allowed only at positions of the pattern E; otherwise, we require

$$L_{ij} = U_{ij} = 0 \quad \text{for } (i,j) \notin E. \tag{8.5.4b}$$

Exercise 8.5.1. The number of matrix entries of L, U not yet determined by (1a), (4a), or (4b) equals $\#E$.

In analogy to (1b), we pose $\#E$ equations for the same number of unknowns:

$$\sum_{j=1}^{n} L_{ij} U_{jk} = A_{ik} \quad \text{for } (i,k) \in E, \tag{8.5.4c}$$

where the remainder $R = LU - A$ is obtained from (4d, e):

$$R_{ik} = 0 \qquad\qquad \text{for } (i,k) \in E, \tag{8.5.4d}$$

$$R_{ik} = \sum_{j=1}^{n} L_{ij} U_{jk} - A_{ik} \quad \textit{for } (i,k) \notin E. \tag{8.5.4e}$$

Under assumption (3b), the term A_{ik} may be omitted because of $A_{ik} = 0$.

The ILU factors satisfying (1a) and (4a–c) can, e.g., be constructed by the following algorithm:

```
L := 0;     U := 0;
for i := 1 to n do
begin L_ii := 1;
   for k := 1 to i − 1 do if (i, k) ∈ E then L_ik := (A_ik − Σ' L_ij U_jk)/U_kk;
                                                                    (8.5.5a)
   for k := 1 to i do if (k, i) ∈ E then U_ki := A_ki − Σ'' L_kj U_ji       (8.5.5b)
end;
```

The sums $\sum'$ and $\sum''$ are taken over all j with $j \neq k$. Since all indices with vanishing terms can be omitted, we may write:

$$\sum{}' = \sum_{j \in I} j < k,\ (i,j) \in E,\ (j,k) \in E \quad \text{and} \quad \sum{}'' = \sum_{j \in I} j < k,\ (k,j) \in E,\ (j,i) \in E. \tag{8.5.5c}$$

The definition of L_{ik} in (5a) is obtained from (4c). To prove (5b), interchange i and k in (4c). One verifies that only those components of L and U are involved in the right-hand sides of (5a, b) that are already computed. Remark 2 will enable a simplification of the algorithm.

Remark 8.5.2. The definitions $D := \operatorname{diag}\{U\}$, $U' := U - D$, $L' := (L - I)D$ lead to a strictly lower triangular matrix L' and a strictly upper triangular matrix U'. Eq. (2) rewritten with the new quantities reads

$$A = (D + L')D^{-1}(D + U') - R. \tag{8.5.6}$$

The quantities D, L', and U' are obtainable directly from the algorithm

$D := 0; \quad L' := 0; \quad U' := 0;$
for $i := 1$ to n do
begin
for $k := 1$ to $i - 1$ do if $(i, k) \in E$ then $L'_{ik} := A_{ik} - \sum' L'_{ij} D_{jj}^{-1} U'_{jk}$; (8.5.5a′)
for $k := 1$ to $i - 1$ do if $(k, i) \in E$ then $U'_{ki} := A_{ki} - \sum'' L'_{kj} D_{jj}^{-1} U'_{ji}$; (8.5.5b′)
$D_{ii} := A_{ii} - \sum'' L'_{ij} D_{jj}^{-1} U'_{ji}$; (8.5.5c′)
end;

Remark 8.5.3. (a) If A is Hermitian, (5a′–c′) immediately implies the symmetries $L' = U'^H$, $D = D^H$.
(b) The *incomplete Cholesky decomposition* $A = L''L''^H - R$ for Hermitian matrices A follows from (6) with $L'' := (D + L')D^{-1/2}$.

Tacitly, we have assumed that the quantities U_{kk} (pivot entries) in (5a) and D_{jj} in (5a′) do not vanish and that, in the case of Remark 3b, even $D_{jj} > 0$ holds. Concerning these conditions, we refer to §8.5.5.

Exercise 8.5.4. *Complete* LU decompositions are characterised by $R = 0$ in (2). Prove: (a) $R = 0$ holds for the cases (i) $E = I \times I$ and (ii) $E = \{(i,j): |i - j| \leqslant w\}$ for band matrices of band width $w \geqslant 0$. (b) $D = \operatorname{diag}\{A\}$ and $L' = U' = 0$ hold for the diagonal pattern $E = \{(i, i): i \in I\}$.

The additive decomposition $A = W - R$ of A given by (2) or (6) defines the corresponding *ILU iteration*:

$$W(x^m - x^{m+1}) = Ax^m - b \quad \text{with} \tag{8.5.7a}$$

$$W = LU \quad \text{or} \quad W = (D + L')D^{-1}(D + U'), \text{ respectively.} \tag{8.5.7b}$$

The other matrices of the first and second normal forms are

$$M = NR \quad \text{with} \quad N = U^{-1}L^{-1} \quad \text{or} \quad N = (D + U')^{-1}D(D + L')^{-1}. \tag{8.5.7c}$$

Remark 8.5.5. One should either store A and the factors L, U (or D, L', U', respectively) or L, U (or D, L', U', respectively) and R. In the latter case, the representation (8) can be exploited instead of (7a):

$$Wx^{m+1} = b + Rx^m. \tag{8.5.8}$$

8.5.2 Incomplete Decomposition with Respect to a Star Pattern

For the description of the pattern E, one should not use the ordered indices $1, \ldots, n$. If, for instance, one tries to describe the graph $G(A)$ in such a manner, this is still possible for the model problem with lexicographical ordering: $G(A)$ consists of $(i, i \pm 1)$ (if i and $i \pm 1$ belong to the same row) and $(i, i \pm (N-1))$. However, as soon as the square from Fig. 1.2.1a–c is replaced by another domain with a varying number of points per row (e.g., an L-shaped domain), a systematic description of the set $G(A)$ becomes rather difficult.

An alternative is «star notation», which has already been used in §3.1.2 for the brief definition of matrices. In the following, we use the so-called *star patterns*. The entry $*$ in the examples

$$\begin{bmatrix} * & * & * \\ * & * & * \\ * & * & * \end{bmatrix}, \quad \begin{bmatrix} * & * & \\ * & * & * \\ & * & * \end{bmatrix}, \quad \begin{bmatrix} & \cdot & \cdot & \cdot & * \\ & \cdot & \cdot & \cdot & \\ * & \cdot & \cdot & \cdot & \end{bmatrix}$$

refers to an element in the set E. If, for instance, $*$ is the right neighbour of the midpoint, this means that for all $\alpha \in I$ having a right neighbour $\beta \in I$, the pair (α, β) belongs to E. Nonmarked positions or the sign «$\cdot$» signifies that the corresponding pairs (α, β) do not belong to E. For example, the minimal set $E = \{(i, i): i \in I\}$ is characterised by $[*]$.

8.5.3 Application to General Five-Point Formulae

Algorithm (5a′, b′) should be regarded more as a definition than a method for the practical computation of the matrices D, L', U'. For the example of a general five-point formula A, we demonstrate how to derive the proper description for the computation. For the sake of convenience, we assume that the coefficients are constant:

$$A = \begin{bmatrix} & -e & \\ -a & d & -b \\ & -c & \end{bmatrix} \quad \text{(cf. (3.5.5))}. \tag{8.5.9a}$$

To ensure that A is an M-matrix, we assume

$$a, b, c, e \geqslant 0, \qquad d \geqslant a + b + c + e. \tag{8.5.9b}$$

The smallest pattern satisfying (3b) is

$$E = G(A), \quad \text{i.e.,} \quad E = [*\overset{*}{\underset{*}{*}}*] \qquad (\text{«five-point pattern»}). \quad (8.5.10a)$$

The strictly triangular matrix L' has the pattern $[*\,\dot{\underset{*}{\cdot}}\,\cdot]$, since the $*$-marked positions are the only matrix entries corresponding to the pattern E and located below the diagonal. If we assume lexicographical numbering, the marked positions in $[*\,\dot{\underset{*}{\cdot}}\,\cdot]$ have a smaller index than the midpoint. Correspondingly, U' has the pattern $[\cdot\,\overset{*}{\dot{\cdot}}\,*]$.

In (5a′, b′), we replace the indices i, j, $k \in \{1, \ldots, n\}$ by α, β, $\gamma \in I = \Omega_h$ and, subsequently, we identify $\alpha = (x, y) = (k_\alpha h, \ell_\alpha h) \in \Omega_h$ with the pair (k_α, ℓ_α), where now $1 \leqslant k_\alpha$, $\ell_\alpha \leqslant N - 1$ holds (cf. (1.2.3)). First, one has to discuss the sum $\sum'$ in (5a′). $L'_{\alpha\gamma} \neq 0$ can be true only for $\gamma = (k_\gamma, \ell_\gamma) = (k_\alpha - 1, \ell_\alpha)$ or $\gamma = (k_\alpha, \ell_\alpha - 1)$, whereas $U'_{\gamma\beta} \neq 0$ leads to $\beta = (k_\gamma + 1, \ell_\gamma)$ and $\beta = (k_\gamma, \ell_r + 1)$. Hence, $L'_{\alpha\gamma} D^{-1}_{\gamma\gamma} U'_{\gamma\beta} \neq 0$ requires $\beta = \alpha$ or $\beta = (k_\alpha + 1, \ell_\alpha - 1)$. Both possibilities contradict $\alpha \neq \beta$ {in (5a′), written as $k \leqslant i - 1$} and $(\alpha, \beta) \in E$. Therefore, $\sum'$ is an empty sum and (5a′) reduces to $L'_{\alpha\beta} = A_{\alpha\beta}$ for $\alpha > \beta$, $(\alpha, \beta) \in E$. Hence, L' is the constant two-point star

$$L' = \begin{bmatrix} & 0 & \\ -a & 0 & 0 \\ & -c & \end{bmatrix}. \qquad (8.5.10b)$$

Similarly, one obtains

$$U' = \begin{bmatrix} & -e & \\ 0 & 0 & -b \\ & 0 & \end{bmatrix}. \qquad (8.5.10c)$$

Only for $\alpha = \beta$, the sum $\sum''$ in (5c′) is not empty and contains the two indices $\gamma = (i_\alpha - 1, j_\alpha)$ and $\gamma = (i_\alpha, j_\alpha - 1)$. We abbreviate the diagonal entry $D_{\alpha\alpha}$ by $d_\alpha = D_{i\alpha, j\alpha}$. Because of $A_{\alpha\alpha} = d$ and the already known values in (10b, c), definition (5c′) can be rewritten as

$$d_{i,j} = d - ab/d_{i-1,j} - ce/d_{i,j-1} \qquad (1 \leqslant i, j \leqslant N - 1), \qquad (8.5.10d)$$

where the terms with $j - 1 = 0$ or $i - 1 = 0$ are to be ignored. In particular, one obtains $d_{11} = d$ for the first grid point. For the five-point formula (9a), the double loop in (5a′ c′) is reduced to a simplc loop over all $(i,j) \in \Omega_h$.

It is also possible to determine the remainder matrix R. Equations (4d, e) become

$$R_{\alpha\beta} = 0 \quad \text{for } (\alpha, \beta) \in E, \qquad R_{\alpha\beta} = (L'D^{-1}U')_{\alpha\beta} \quad \text{for } (\alpha, \beta) \notin E. \quad (8.5.10e)$$

One verifies that R has two (variable) coefficients per row:

$$R = \begin{bmatrix} r_{ij} & \cdot & \cdot \\ \cdot & \cdot & \cdot \\ \cdot & \cdot & s_{ij} \end{bmatrix} \quad \text{with } r_{ij} = ae/d_{i-1,j}, \quad s_{ij} = cb/d_{i,j-1}, \quad (8.5.10f)$$

where $r_{ij} = 0$ holds for $i = 1$ and $s_{ij} = 0$ is to be inserted for $j = 1$.

Remark 8.5.6. The ILU decomposition of a five-point formula with constant or variable coefficients requires $6n$ operations for computing the d_{ij}'s in (10d). The solution of $W\delta = (D + L')D^{-1}(D + U')\delta = r$ takes $10n$ operations; hence, because of the additional $10n$ operations for computing $Ax^m - b$, one ILU iteration step (7a) requires, in total, $20n$ operations. Note that the d_{ij} values from (10d) have to be determined only once. An alternative is the determination of R by an additional $4n$ operations. Afterwards, the iteration (8) requires only $14n$ operations. Together with $C_A = 5$, the following cost factors result:

$$C_\Phi^{\mathrm{ILU}} = 4 \quad \text{or} \quad C_\Phi^{\mathrm{ILU}} = 2.8, \text{ respectively} \qquad (E \text{ from (10a)}). \tag{8.5.10g}$$

8.5.4 Modified ILU Decompositions

On may pose the question of whether or not completely ignoring the matrix entries α_{ij} for $(i,j) \notin E$ is the optimal approach. We recall the Gauß-Seidel method, where the matrix $W = D - E$ is changed into $W = \frac{1}{\omega} D - E$ for the SOR method. Hence, overrelaxation, which in general leads to convergence improvement, corresponds to a diminishing of the diagonal in W. The modification next presented, which we introduce following Wittum [5], also leads to a diminishing or enlargement of the diagonal corresponding to the choice of ω.

Let $\mathbb{1}$ be the vector $(1)_{\alpha \in I}$ consisting only of 1 components. Gustafsson [1] has proposed replacing the equation $R_{ii} = 0$ [i.e., (4d) for $i = k$] by

$$A\mathbb{1} = W\mathbb{1}, \qquad \text{i.e., } R\mathbb{1} = 0. \tag{8.5.11}$$

One may view $\mathbb{1}$ as «test vector». By (11) W is gauged in such a way that A and W coincide with respect to their application to $\mathbb{1}$. We generalise the condition $R\mathbb{1} = 0$ by

$$R_{ii} = \omega \sum_{j \neq i} R_{ij} \qquad (\omega \in \mathbb{R}) \tag{8.5.12}$$

and denote the thus determined decomposition as *ILU_ω decomposition* (its existence is not yet claimed).

Remark 8.5.7. (a) For $\omega = 0$, Eq. (12) coincides with (4d) for $i = k$: $R_{ii} = 0$. Hence, the previous ILU decomposition is the ILU_0 decomposition.
(b) For $\omega = -1$, conditions (11) and (12) are identical, i.e., ILU_{-1} decomposition describes the modification by Gustafsson [1].

In the case of the five-point formula (9a) and the five-point pattern (10a), L' and U' are still obtainable from (10b, c), whereas the recursion (10d) for the entries d_{ij} of D becomes

$$d_{ij} := d + (\omega e - b)a/d_{i-1,j} + (\omega b - e)c/d_{i,j-1} \tag{8.5.13}$$

(terms with $i - 1 = 0$ and $j - 1 = 0$ are again to be ignored).

8.5.5 On the Existence and Stability of the ILU Decomposition

In this section the inequalities $A \leqslant B$ are used in the sense of *elementwise* inequalities $A_{\alpha\beta} \leqslant B_{\alpha\beta}$ ($\alpha, \beta \in I$) as in §6.

It is known that the (complete) LU decomposition exists if and only if all principal submatrices $(a_{ij})_{1 \leqslant i, j \leqslant k}$ are regular for $1 \leqslant k \leqslant n$. However, even if the decomposition $A = LU$ exists, it can be useless, since the solution process of the equations $Ly = b$, $Ux = y$ may be unstable. Choose, e.g., $A = LU$ with $L = \mathrm{tridiag}\{\alpha, 1, 0\}$ for $\alpha < 1$, $U = L^T$ and investigate the error propagation (cf. Elman [1]). The criterion mentioned above is satisfied for positive definite matrices. However, there are positive definite matrices permitting no ILU decomposition (because of $U_{kk} = 0$ in (5a)). The paper of Meijerink–van der Vorst [1] contains the first part of the following criterion, while the second part is due to Manteuffel [2].

Criterion 8.5.8. Let $E \subset I \times I$ satisfy (3a). (a) M-matrices A permit an ILU decomposition $A = W - R$ with W from (7b), which, furthermore, represents regular splitting in the sense of Definition 6.5.1.
(b) H-matrices A with a positive diagonal D corresponding to the M-matrix $\hat{A} := |D| - |A - D|$ (cf. Definition 6.6.7) have an ILU decomposition $(D + L')D^{-1}(D + U')$ with $0 \leqslant \hat{D} \leqslant D$, $\hat{L}'\hat{D}^{-1} \leqslant -|L'D^{-1}| \leqslant 0$, and $\hat{D}^{-1}\hat{U}' \leqslant -|D^{-1}U'| \leqslant 0$, where $\hat{D}$, $\hat{L}'$, $\hat{U}'$ stem from the ILU decomposition of $\hat{A}$.

Meijerink–van der Vorst [1] prove part (a) by interpreting the ILU decomposition as a sequence of Gauß elimination steps, which conserve the M-matrix property (cf. Lemma 6.4.19). We give another proof that directly refers to the defining equations (4c) and requires weaker assumptions.

Given an arbitrary matrix X, we denote by X_E the matrix

$$(X_E)_{\alpha\beta} := \begin{Bmatrix} X_{\alpha\beta} & \text{if } (\alpha, \beta) \in E \\ 0 & \text{otherwise} \end{Bmatrix}, \tag{8.5.14a}$$

i.e., X_E is the restriction of X to the pattern E. The matrices denoted in the following by the letters D, L, and U with different indices should always be of a diagonal structure or strictly lower or upper triangular structure, respectively. Note that a triple $\{D, L, U\}$ is uniquely defined by the sum $C = D + L + U$. In order to express the single components of this triple, we write $C = \mathrm{diag}\{C\} + L(C) + U(C)$.

In the following, it is not necessarily assumed that $A_{\alpha\beta} \leqslant 0$ holds for $\alpha \neq \beta$, as it is necessary for M-matrices. We define

$$(A_-)_{\alpha\beta} := \begin{Bmatrix} A_{\alpha\beta} & \text{if } \alpha = \beta \text{ or } A_{\alpha\beta} \leqslant 0 \\ 0 & \text{otherwise} \end{Bmatrix}. \tag{8.5.14b}$$

The matrix A is assumed to fulfil the following conditions:

$$A_{\alpha\beta} \leqslant (L(A_-)_E \operatorname{diag}\{A\}^{-1} U(A_-)_E)_{\alpha\beta} \quad \text{for all } \alpha \neq \beta, (\alpha, \beta) \in E, \tag{8.5.15a}$$

A has a *complete LU* decomposition

$$A = (\underline{D} + \underline{L})\underline{D}^{-1}(\underline{D} + \underline{U}) = \underline{D} + \underline{L} + \underline{U} + \underline{L}\underline{D}^{-1}\underline{U} \quad \text{with}$$

$$\underline{D} \geqslant 0, \underline{L} \leqslant 0, \underline{U} \leqslant 0. \tag{8.5.15b}$$

Remark 8.5.9. All M-matrices A satisfy the assumptions (15a, b). (15b) implies the inverse positivity of A: $A^{-1} \geqslant 0$. (15a) is always satisfied if A fulfils the sign condition $A_{\alpha\beta} \leqslant 0$ ($\alpha \neq \beta$) *inside* of the pattern E.

Proof. Since the complete LU decomposition is generated by Gauß elimination, the inequalities in (15b) follow from Lemma 6.4.19. Vice versa, the inequalities in (15b) imply $(\underline{D} + \underline{L})^{-1} \geqslant 0$, $\underline{D}^{-1} \geqslant 0$ $(\underline{D} + \underline{U})^{-1} \geqslant 0$, from which $A^{-1} \geqslant 0$ can be concluded. If A is an M-matrix and therefore $A_{\alpha\beta} \leqslant 0$ for $\alpha \neq \beta$, $(A_-)_{\alpha\beta} \leqslant 0 \leqslant (L(A_-)_E \operatorname{diag}\{A\}^{-1} U(A_-)_E)_{\alpha\beta}$ follows. □

Theorem 8.5.10. *Assume that $E \subset I \times I$ satisfies* (3a) *and the matrix A fulfils* (15a, b). *Then A permits an ILU decomposition $A = W - R$ with W from* (7b). *$A = W - R$ is a regular splitting if $A_{\alpha\beta} \leqslant 0$ for $(\alpha, \beta) \notin E$* ((3b) *is sufficient). The inclusions* (16) *hold with $\underline{D}$, $\underline{L}$, $\underline{U}$ from* (15b):

$$(\underline{D} + \underline{L} + \underline{U})_E \leqslant D + L' + U' \leqslant (A_-)_E \tag{8.5.16}$$

Proof. The conditions (4d) can be written as $R_E = 0$. Inserting $R = D + L' + U' + L'D^{-1}U' - A$, we obtain $(D + L' + U' + L'D^{-1}U' - A)_E = 0$ or

$$(D + L' + U')_E = (A - L'D^{-1}U')_E. \tag{8.5.17}$$

Using the mapping

$$\Phi(C) := (A - L(C) \operatorname{diag}\{C\}^{-1} U(C))_E, \tag{8.5.18a}$$

we may write the defining equation (17) as a *fixed-point equation*

$$D + L' + U' = \Phi(D + L' + U'). \tag{8.5.17'}$$

Assume the monotonicity properties

$$C_1 \leqslant C_2, \quad \operatorname{diag}\{C_1\} \geqslant 0, \quad L(C_2) + U(C_2) \leqslant 0 \Rightarrow \Phi(C_1) \leqslant \Phi(C_2). \tag{8.5.18b}$$

(15b) implies $A = \underline{D} + \underline{L} + \underline{U} + \underline{L}\underline{D}^{-1}\underline{U}$. We set

$$A_0 := \underline{D} + \underline{L} + \underline{U} \quad \text{and} \quad A^0 := (A_-)_E. \tag{8.5.18c}$$

$\underline{L}\underline{D}^{-1}\underline{U} \geqslant 0$ yields $A_0 = (A_0)_- \leqslant ((A_0)_-)_E \leqslant ((A_0 + \underline{L}\underline{D}^{-1}\underline{U})_-)_E = (A_-)_E = A^0$:

$$A_0 \leqslant A^0. \tag{8.5.18d}$$

Next, we show that

$$A_0 \leqslant \Phi(A_0) \quad \text{and} \quad \Phi(A^0) \leqslant A^0. \tag{8.5.18e}$$

$\Phi(A_0) = (A - \underline{L}\underline{D}^{-1}\underline{U})_E = (\underline{D} + \underline{L} + \underline{U})_E = (A_0)_E \geqslant A_0$ holds because of $\underline{L}$, $\underline{U} \leqslant 0$. The second inequality in (18e) is identical to (15a). Φ defines the following *fixed-point iteration*:

$$A_{m+1} := \Phi(A_m), \qquad A^{m+1} := \Phi(A^m). \tag{8.5.18f}$$

The monotonicity (18b) and inequalities (18d, e) lead to

$$A_0 \leqslant A_1 \leqslant \cdots \leqslant A_m \leqslant \cdots \leqslant A^m \leqslant \cdots \leqslant A^1 \leqslant A^0 \tag{8.5.18g}$$

(cf. Theorem 6.6.6). Hence, both sequences must converge to a limit $C = D + L' + U'$ satisfying the fixed-point equation (17′). (16) follows from (18c) and $A_0 \leqslant D + L' + U' \leqslant A^0$.

$W^{-1} = (D + U')^{-1}D(D + L')^{-1} \geqslant 0$ is a consequence of the inequalities $D \geqslant 0$ and L', $U' \leqslant 0$. The remainder R vanishes on E: $R_E = 0$; otherwise, $R_{\alpha\beta} = (L'D^{-1}U' - A)_{\alpha\beta}$ holds. $A_{\alpha\beta} \leqslant 0$ for the indices $(\alpha, \beta) \notin E$ implies $R_{\alpha\beta} \geqslant (L'D^{-1}U')_{\alpha\beta} \geqslant 0$. Hence, the splitting $A = W - R$ is regular. □

The stability of the ILU decomposition is expressed in (16) by the estimate of the diagonal D from below by $\underline{D}$. Concerning the problem that the solution of the systems $(D + L)x = b$ or $(D + U)x = b$ may lead to instabilities, we refer to Elman [1], where ILU decompositions for nonsymmetric matrices A are discussed.

To generalise Theorem 10 to the ILU_ω decomposition with $\omega \neq 0$, one may write the equations $R_{ij} = 0$ for $i \neq j$, $(i,j) \in E$, and (12) as

$$R_E - \omega\,\mathrm{diag}\{R_{E'}\mathbb{1}\} = 0 \quad \text{with } R = D + L + U + LD^{-1}U - A,$$
$$E' = I \times I \backslash E, \tag{8.5.19a}$$

where for a given vector $v = (v_1, \ldots, v_n)^T$, we denote the diagonal matrix $\mathrm{diag}\{v_1, \ldots, v_n\}$ by $\mathrm{diag}\{v\}$. Carrying over the proof technique, we are led to the fixed-point equation $C = \Phi_\omega(C)$ with

$$\Phi_\omega(C) := \Phi(C) - \omega\,\mathrm{diag}\{(A - L(C)\,\mathrm{diag}\{C\}^{-1}U(C))_{E'}\mathbb{1}\}, \tag{8.5.19b}$$

where Φ stems from (18a). However, in general, Φ_ω does not have the desired properties. The monotonicity corresponding to (18b) may be violated for $\omega > 0$, whereas for $\omega < 0$, it may happen that no A_0 exists with $\Phi_\omega(A_0) \geqslant A_0$ (and hence, no solution exists).

For a precise discussion, we study the five-point formula (9a) with the five-point pattern (10a). Since L' and U' are already uniquely determined (cf. (10b, c)), the fixed-point equation reduces to the determination of D as a solution of

$$D = \Phi_\omega(D) := \mathrm{diag}\{A - L'D^{-1}U'\} - \omega\,\mathrm{diag}\{(A - L'D^{-1}U')_{E'}\mathbb{1}\}$$
$$= \mathrm{diag}\{d + (\omega a - c)e/D_{i-1,j} + (\omega c - a)b/D_{i,j-1}\} \tag{8.5.20a}$$

(cf. (13)). For the analysis of this equation, we investigate the one-dimensional fixed-point equation

$$\delta = \varphi_\omega(\delta) := d + [(\omega e - b)a + (\omega b - e)c]/\delta. \tag{8.5.20b}$$

A discussion of the function φ_ω, which is left to the reader, shows: (i) The fixed-point equation (20b) is solvable if and only if

$$4\gamma < d^2 \quad \text{for} \quad \gamma := ce + ab - \omega(ae + cb). \tag{8.5.20c}$$

(ii) If (20c) is satisfied, the solutions of (20b) are

$$\delta_\pm = \tfrac{1}{2}(d \pm \sqrt{d^2 - 4\gamma}). \tag{8.5.20d}$$

(iii) δ_+ is the stable fixed point, because (20e) leads to (20f):

$$\varphi_\omega(\delta) < \delta \quad \text{for } \delta > \delta_+, \qquad \varphi_\omega(\delta) > \delta \quad \text{for } \delta_- < \delta < \delta_+, \tag{8.5.20e}$$

$$\lim \delta_m = \delta_+ \quad \text{for } \delta_0 > \delta_-, \delta_{m+1} := \varphi_\omega(\delta_m). \tag{8.5.20f}$$

(iv) On the other hand, starting iterates $\delta_0 < \delta_-$ generate sequences $\{\delta_m\}$ containing at least one element $\delta_m \leqslant 0$.

Exercise 8.5.11. Let A from (9a) be diagonally dominant and symmetric:

$$a = b \geqslant 0, \quad c = e \geqslant 0, \ \sigma := a + c > 0, \quad d = 2\sigma + \varepsilon \quad \text{with } \varepsilon \geqslant 0. \tag{8.5.21a}$$

Prove that for $\omega = -1$, the value δ_+ is obtained from (20d) with $\gamma = \sigma^2$. For small ε, this value has the expansion

$$\delta_+ = \sigma + \sqrt{\varepsilon\sigma} + O(\varepsilon). \tag{8.5.21b}$$

Assuming (21a), one obtains for $\omega = 0$

$$\delta_+ = a + c + \sqrt{2ac} + O(\sqrt{|\varepsilon|}). \tag{8.5.21c}$$

Theorem 8.5.12. *Let $\omega \in [-1, \omega^*]$, where $\omega^* := \min\{c/a, a/c\}$. Assume that the matrix A from* (9a) *satisfies* (21a). *Then the ILU_ω decomposition exists. Furthermore, the entries of the diagonal D are enclosed by*

$$\delta_+ = \tfrac{1}{2}(d + \sqrt{d^2 - 4(c^2 + a^2 - 2\omega ac)}) < d_{ij} \leqslant d \quad \textit{for } (i,j) \in I. \tag{8.5.22}$$

The fixed-point iteration (20a) *with starting iterate $D^0 := \mathrm{diag}\{d\mathbb{1}\}$ converges from above to D.*

Proof. (20c) is satisfied for $\omega \geqslant -1$, while for $\omega \leqslant \omega^*$, Φ_ω is monotone. One verifies $D_0 \leqslant \Phi_\omega(D_0)$ and $\Phi_\omega(D^0) \leqslant D^0$ for $D_0 := \mathrm{diag}\{\delta_+\mathbb{1}\}$ and $D^0 := \mathrm{diag}\{d\mathbb{1}\}$. Hence, we can draw the same conclusions as in the proof for Theorem 9. □

8.5.6 Properties of the ILU Decomposition

An immediate consequence of Theorem 6.5.2 is the following convergence statement:

Theorem 8.5.13. *If A is an M-matrix or if according to Theorem 10 $A = W - R$ describes a regular splitting, the ILU iteration* (7a, b) *converges with the convergence rate* $\varrho(A^{-1}R)/(1 + \varrho(A^{-1}R))$.

In the standard case, one may assume $\|R\| = O(\|A\|)$, *so that* $\varrho(A^{-1}R) \leqslant \|A^{-1}\|\,\|R\| \leqslant C\|A^{-1}\|\,\|A\| = C\,\mathrm{cond}(A) \gg 1$ *leads to the convergence rate* $(1 + 1/\varrho(A^{-1}R))^{-1} \approx 1 - O(1/\mathrm{cond}(A))$. *Hence, the ILU decomposition has the same order as the Jacobi or Gauß-Seidel iteration. A better result can be derived for the modified* ILU_{-1} *decomposition (cf.* (11) *or* (12) *with* $\omega = -1$). *We prepare its analysis with the following lemma (cf. Wittum* [4]).

Lemma 8.5.14. *Assume* (23a), *where* A, D_A, *and* D *are positive definite*:

$$A = D_A - L - L^H, \qquad W = (D + L')D^{-1}(D + L'^H). \tag{8.5.23a}$$

The spectrum $\sigma(W^{-1}A)$ *is contained in* $[0, \Gamma]$ *if*

$$\left(2 - \frac{1}{\Gamma}\right)D - D_A + L + L^H + L' + L'^H \quad \textit{is positive semi-definite.} \tag{8.5.23b}$$

Proof. We write $D + L'$ as $\frac{1}{\Gamma}D + C$ with $C := \left(1 - \frac{1}{\Gamma}\right)D + L'$. From

$$\begin{aligned}\Gamma W - A &= \left(\frac{1}{\Gamma}D + C\right)\left(\frac{1}{\Gamma}D\right)^{-1}\left(\frac{1}{\Gamma}D + C\right)^{H} - A \\ &\geqslant \frac{1}{\Gamma}D + C + C^H - A \\ &= \left(2 - \frac{1}{\Gamma}\right)D - D_A + L + L^H + L' + L'^H \geqslant 0\end{aligned}$$

with "$\geqslant$" in the sense of positive semi-definiteness, it follows that $\sigma(W^{-1}A) \subset [0, \Gamma]$. □

Theorem 8.5.15. *Let* $-1 \leqslant \omega \leqslant \omega^*$ *(cf. Theorem* 12). *The five-point formula* (9a) *and the five-point pattern* (10a) *are assumed to satisfy* (21a). *Then the inequality*

$$\gamma W \leqslant A \leqslant \Gamma W \quad \textit{with } \gamma = 1\Big/\left[1 + (1 + \omega)\frac{2ac}{\delta_+\lambda_{\min}}\right],\ \Gamma = \frac{\delta_+}{2\delta_+ - d}, \tag{8.5.24}$$

holds with "$\leqslant$" *in the sense of positive (semi-)definiteness, where* δ_+ *is defined in* (22) *and* $\lambda_{\min} = \varepsilon + 4(a + c)\sin^2\frac{\pi h}{2}$ *is the smallest eigenvalue of* A. *In particular,* (25) *holds*:

$$\gamma = 1, \quad \Gamma = \frac{1}{2}\sqrt{\frac{\sigma}{\varepsilon}} - \frac{1}{4} + O\left(\sqrt{\frac{\varepsilon}{\sigma}}\right) \quad \textit{for } \omega = -1. \tag{8.5.25}$$

Proof. (i) (23b) becomes $\left(2 - \frac{1}{\Gamma}\right)D - D_A \geqslant 0$, since by (10b, c), $L = -L'$ holds in Lemma 14. Thanks to $D_A = dI$ and $\delta_+ I \leqslant D$ (cf. (22)), Γ with $\left(2 - \frac{1}{\Gamma}\right)\delta_+ = d$ is sufficient for (23b). Solving for Γ, we obtain $\Gamma = \delta_+/(2\delta_+ - d)$.

(ii) The entries α_{ij}, β_{ij} of R (cf. (10f)) are bounded from above by $2ac/\delta_+$. According to (12), the diagonal entries of R equal $\omega(\alpha_{ij} + \beta_{ij})$. The eigenvalues of R lie in the Gershgorin circles around $\omega(\alpha_{ij} + \beta_{ij})$ with radius $\alpha_{ij} + \beta_{ij}$ (cf. Hackbusch [15, Criterion 4.3.4]) and, hence, they are bounded by $(1 + \omega)(\alpha_{ij} + \beta_{ij}) \leqslant 2(1 + \omega)ac/\delta_+$, implying $R \leqslant [2(1 + \omega)ac/\delta_+]I$. From $\lambda_{\min} I \leqslant A$, one deduces $R \leqslant \varrho A$ with $\varrho := 2(1 + \omega)ac/(\delta_+\lambda_{\min})$. $A = W - R \geqslant W - \varrho A$ yields $W \leqslant (1 + \varrho)A$. Hence, $\gamma = 1/(1 + \varrho)$ leads to the representation of γ.

(iii) For $\omega = -1$, insert the representation (21b) into (24). □

Remark 8.5.16. (a) Replacing the Poisson equation $-\Delta u = f$ by the Helmholtz equation $-\Delta u + \varepsilon u = f$ with $\varepsilon > 0$, one obtains the coefficients $a = b = c = e = -h^{-2}$, $d = 4h^{-2} + \varepsilon$ in (21a). (25) yields the bound and condition number $\Gamma = \Gamma/\gamma = h^{-1}/\sqrt{2\varepsilon} + O(1)$.
(b) Let $\omega = -1$. The (modified) ILU_{-1} iteration *damped* by $\Theta_{\mathrm{opt}} = 2/(\gamma + \Gamma) = 2\sqrt{2\varepsilon}h + O(h^2)$ has the convergence speed

$$\varrho(M_{\Theta_{\mathrm{opt}}}^{\mathrm{ILU}}) \leqslant (\Gamma - 1)/(\Gamma + 1) \approx 1 - 2/\Gamma \approx 1 - 2\sqrt{2\varepsilon}h. \tag{8.5.26}$$

Hence, similarly to the SSOR method with an optimal relaxation parameter ω_{SSOR}, it is of first order as long as $\varepsilon > 0$.

Proof. cf. Exercise 3.1a. □

The fact that in Theorem 15 and Remark 16b we require strong diagonal dominance $\varepsilon > 0$, does not restrict at all the applicability of the ILU_{-1} decomposition, as shown by the following remark.

Remark 8.5.17 (enlargement of the diagonal). Let $A = A_\varepsilon$ be a matrix satisfying (21a) only with $\varepsilon > -4(a + c)\sin^2\frac{\pi h}{2}$ (i.e., $\lambda_{\min} > 0$) instead of $\varepsilon > 0$. Then the ILU_{-1} decomposition is to be applied to the matrix $A_\eta := A + (\eta - \varepsilon)I$ with $\eta > 0$: $A_\eta = W_\eta - R_\eta$ in order to reestablish diagonal dominance $d > 2\sigma$. W_η can be viewed as the ILU decomposition of $A = A_\varepsilon$ with the remainder $R = W_\eta - A = R_\eta - (\eta - \varepsilon)I$. Remark 16 yields the condition number $\varkappa(W_\eta^{-1}A_\eta)$. Let $\lambda = \lambda_{\min}$ and $\Lambda = \Lambda_{\max}$ be the extreme eigenvalues of A. Because of

$$\varkappa(A_\eta^{-1}A) = \varkappa(A_\eta^{-1}A_\varepsilon) = \frac{\Lambda(\lambda + \eta - \varepsilon)}{\lambda(\Lambda + \eta - \varepsilon)} \approx 1 + \frac{\eta - \varepsilon}{\lambda_{\min}}, \tag{8.5.27a}$$

Lemma 3.10 shows that

$$\varkappa(W_\eta^{-1}A_\varepsilon) \lessapprox h^{-1}\left(1+\frac{\eta-\varepsilon}{\lambda_{\min}}\right)\Big/\sqrt{2\eta}. \tag{8.5.27b}$$

Exercise 8.4.18. Prove that the right-hand side of (27b) becomes minimal for $\eta = 4(a+c)\sin^2\dfrac{\pi h}{2}$.

Exercise 8.5.19. Prove that the ILU decomposition coincides with the exact LU decomposition, if A has one of the tridiagonal patterns $\left[\begin{smallmatrix} & \cdot & \\ * & * & * \\ & \cdot & \end{smallmatrix}\right]$ or $\left[\begin{smallmatrix} & * & \\ \cdot & * & \cdot \\ & * & \end{smallmatrix}\right]$. Then, the ILU iteration solves $Ax = b$ directly.

8.5.7 ILU Decompositions Corresponding to Other Patterns

Condition (3b): $E \supsetneqq G(A)$ is a minimum requirement to construct new methods. When choosing a pattern E larger than $G(A)$, one should add those positions where $R = 0$ is violated: According to (10f), these are the positions $\left[\begin{smallmatrix} * & \cdot & \cdot \\ \cdot & \cdot & \cdot \\ \cdot & \cdot & * \end{smallmatrix}\right]$. Adding $\left[\begin{smallmatrix} * & \cdot & \cdot \\ \cdot & \cdot & \cdot \\ \cdot & \cdot & * \end{smallmatrix}\right]$ to the five-point pattern, one obtains

$$E = \begin{bmatrix} * & * & \\ * & * & * \\ & * & * \end{bmatrix} \qquad \text{(«seven-point pattern»)}. \tag{8.5.28}$$

Now, the lower triangular matrix L' and upper triangular matrix U' have the form

$$L' = -\begin{bmatrix} 0 & 0 & \\ a_{ij} & 0 & 0 \\ & c & f_{ij} \end{bmatrix}, \qquad U' = -\begin{bmatrix} g_{ij} & e & \\ 0 & 0 & b_{ij} \\ & 0 & 0 \end{bmatrix}, \tag{8.5.29}$$

whose coefficients result from the recursions

$$\begin{aligned} d_{ij} = d - ec/d_{i,j-1} &+ a_{ij}(\omega g_{i-1,j} - b_{i-1,j})/d_{i-1,j} \\ &+ f_{ij}(\omega b_{i+1,j-1} - g_{i+1,j-1})/d_{i+1,j-1}, \end{aligned} \tag{8.5.30a}$$

$$a_{ij} = a + g_{i,j-1}c/d_{i,j-1}, \qquad b_{ij} = b + ef_{ij}/d_{i+1,j-1}, \tag{8.5.30b}$$

$$f_{ij} = b_{i,j-1}c/d_{i,j-1}, \qquad g_{ij} = a_{ij}e/d_{i-1,j} \tag{8.5.30c}$$

for $1 \leqslant i, j \leqslant N-1$, where all terms with indices $i-1=0$, $j-1=0$, or $i+1=N$ have to be ignored. This seven-point ILU decomposition has properties similar to those of the five-point version in Theorem 15 (cf. Gustafsson [1], Axelsson–Barker [1]).

Exercise 8.5.20. Prove: (a) For $-1 \leqslant \omega \leqslant 0$, the fixed-point iteration (19b) converges for the starting iterate $C = A$ to values satisfying the inequali-

ties $a_{ij} \leqslant \alpha := a/\Delta$, $b_{ij} \leqslant \beta := b/\Delta$, $f_{ij} \leqslant \beta c/\delta$, $g_{ij} \leqslant \alpha e/\delta$, $d_{ij} \geqslant \delta$ with $\Delta := 1 - ec/\delta^2$, where δ is the maximal solution of the fixed-point equation $\delta = \varphi(\delta) := d - [ec + ab(1 + ec/\delta^2)/\Delta^2 - \omega(\alpha^2 e + \beta^2 c)/\delta]/\delta$.
(b) For the next considerations, assume symmetry $a = b$, $c = e$ as well as diagonal dominance $d = 2(a + c) + \varepsilon$ with $\varepsilon \geqslant 0$. Furthermore, choose $\omega = -1$ (i.e., modified ILU). Prove that the equation $\delta = \varphi(\delta)$ can be brought into the form $2a + \varepsilon = a(\xi + \xi^{-1})$ with $\xi := a\delta/(\delta - c)^2$. Hence, the solution is $\delta = c + a/(2\xi) + [ac/\xi + a^2/(4\xi^2)]^{1/2}$ with $\xi = 1 + \varepsilon/(2a) + [\varepsilon/a + \varepsilon^2/(4a^2)]^{1/2}$.
(c) For $\varepsilon \geqslant 0$, a solution $\delta = \delta_0 + C\sqrt{\varepsilon} + O(\varepsilon)$ exists.
(d) δ solves the equation $(\delta - \gamma - e - \beta)^2 = \varepsilon\delta$.
(e) The weak diagonal dominance, which is sufficient for (23b), leads to the condition $2\varphi + 2|a - \alpha| \leqslant (2 - 1/\Gamma)\delta - d$. Show that $\Gamma = \delta/(2\sqrt{\varepsilon\delta} - \varepsilon)$.
(f) As in (25), the estimate $\gamma W \leqslant A \leqslant \Gamma W$ holds with $\gamma = 1$.

For the construction of ILU decompositions with a general k-point pattern, note that the amount of computational work increases more than linearly with the number k of pattern entries.

8.5.8 Approximative ILU Decompositions

The ILU decompositions, as defined in (10d) or (13), are strictly sequential algorithms. The same statement holds for the solution of the systems $(D + L)x = b$ and $(D + U)x = b$, that arise during the solution of $W\delta = r$. This is a disadvantage when using vector computers. For this case, the treatment of the systems is discussed by van der Vorst [2] (cf. also Ortega [1, §3.4]). Here, we discuss the computation of the ILU decomposition. The fixed-point iteration (18f) from the proof of Theorem 9 is also suited to numerical computations. The upper starting iterate $A^0 = (A_-)_E$ (in general, $A^0 = A$) is available (in contrast to A_0), so that the iterates $A^{m+1} = \Phi(A^m)$ are computable.

Remark 8.5.21. The evaluation of the function Φ from (18a) can be performed in parallel for all coefficients $\Phi(C)_{\alpha\beta}$, $(\alpha, \beta) \in E$.

The equations (17'): $C = \Phi(C)$ or, more precisely, the recursions (13) and (30a–c) represent simple systems of equations for the unknowns d_{ij} (and possibly $a_{ij}, b_{ij}, f_{ij}, g_{ij}$), which can be solved by back substitutions. Independently of the starting iterate, the values for (i,j) with $\max\{i,j\} \leqslant m$ are *exact* after m iteration steps. If A and therefore also the starting iterate A^0 (cf. (18c)) have constant coefficients, the mth iterate A^m has identical constant coefficients for all positions (i,j) with $\min\{i,j\} \geqslant m$ [at positions with $\min\{i,j\} < m$, other values may result, since in (13) or (30a–c) some terms may be absent because of $i - 1 = 0$ or $j - 1 = 0$]. Since the coefficients of A^m coincide for

$\min\{i,j\} \geqslant m$, one need not calculate all of them. This consideration leads us to the *truncated ILU version* introduced by Wittum [1] for constant coefficients:

> Compute d_{ij} (and possibly a_{ij}, b_{ij}, f_{ij}, g_{ij}) from (13) or (30a–c) for all i, j with $\max\{i,j\} = k$ for $k = 1, 2, \ldots, m$ and continue these values constantly by means of

$$d_{ij} := d_{\min\{i,m\},\min\{j,m\}} \qquad \text{for } \max\{i,j\} > m \tag{8.5.31}$$

(analogously for a_{ij}, b_{ij}, f_{ij}, g_{ij}). The amount of computational work is $O(m^2)$ independent of the dimension n of the matrix. The same statement holds for the storage requirement. The truncated ILU decomposition is not only a good substitute for the standard ILU decomposition but also has favourable stability properties (cf. Wittum–Liebau [1]).

8.5.9 Blockwise ILU Decompositions

Choosing the row or column variables as blocks, A has a block structure with tridiagonal diagonal-blocks. In the decomposition ansatz (6): $A = (D + L')D^{-1}(D + U') - R = D + L' + U' + L'D^{-1}U' - R$, we may also require that D be a block-diagonal matrix with a tridiagonal block-pattern and that L' and U' are strictly (lower/upper) block-triangular matrices. The algorithm is similar to (5a′, b′) (cf. (10.9.2a–c)). With the increased amount of computational work, one gains, in general, more robust convergence properties. Block-ILU decompositions were introduced in the early 1980s (cf. §8.5.12).

8.5.10 Pascal Procedures

The ILU iteration uses three parameters: ω for the decomposition (13) or (30a–c), the damping parameter Θ for the ILU iteration (7a), and the quantity `diag` that is added to the diagonal entry to improve diagonal dominance (cf. Remark 17). The introduction of Θ is necessary, since otherwise, the ILU iteration would diverge for $\omega = -1$. The cited parameters can be defined by means of `determine_ILU_parameters`. The ILU iterations for the five- and seven-point patterns are performed by `ILU_5` and `ILU_7`.

```
procedure determine_ILU_parameters(var IP:
  iterationparameter);
begin writeln; writeln('*** Input of the ILU parameters:')
      determine_omega(IP); determine_theta(IP);
      write('--> The diagonal is to be enlarged by diag =');
        readln(IP.diag)
end;
```

```
procedure compute_ILU5_decomposition(var A: data_of_
                                         discretisation;
                                        var IP:
                                         iterationparameter);
var d,d0: real; i,j: integer;
begin with A do
  if kind>fivepoint_formula then message('ILU5 not
   implemented!') else
  if ILUD=nil then with IP do
  begin new(ILUD); d0:=diag; if kind=Poisson_model_problem
   then d0:=d0*h2;
        for j:=1 to ny-1 do for i:=1 to nx-1 do
        begin d:=S[0,0]+d0;
  if j>1 then d:=d+(omega*S[1,0]-S[0,1])*S[0,-1]/ILUD^[i,j-1];
  if i>1 then d:=d+(omega*S[0,1]-S[1,0])*S[-1,0]/ILUD^[i-1,j];
  ILUD^[i,j]:=d
end end end;

procedure solve_ILU5_decomposition(var x:gridfunction;
                  var A: data_of_discretisation; var IP:
                   iterationparameter);
var i,j: integer;
begin with A do
    begin if ILUD=nil then compute_ILU5_decomposition(A,IP);
          zero_boundary_values(nx,ny,x);
          for j:=1 to ny-1 do for i:=1 to nx-1 do
          x[i,j]:=(x[i,j]-S[-1,0]*x[i-1,j]-S[0,-1]*x[i,j-1])/
           ILUD^[i,j];
          for j:=ny-1 downto 1 do for i:=nx-1 downto 1 do
          x[i,j]:=x[i,j]-(S[1,0]*x[i+1,j]+S[0,1]*x[i,j+1])/
           ILUD^[i,j]
end end;

procedure ILU_5(var new:gridfunction; var A: data_of_
                 discretisation;
                var x,b: gridfunction; var IP:
                 iterationparameter);
var r: gridfunction;
begin residual(r,A,x,b); solve_ILU5_decomposition(r,A,IP);
      vector_plus_factor_x_vector(A.nx,A.ny,new,x,
       IP.theta,r);
      transfer_boundary_values(nx,ny,x,new)
end;

procedure compute_ILU7_decomposition(var A: data_of_
                                         discretisation;
                                        var IP:
                                         iterationparameter);
```

```
var c,d,d0: real; i,j: integer;
begin with A do
    if kind>fivepoint_formula then message('ILU7 not
     implemented') else
    begin if ILUD=nil then new(ILUD); if ILU7=nil then with
     IP do
       begin new(ILU7); d0:=diag;
           if kind=Poisson_model_problem then d0:=d0*h2;
           with ILU7^ do for j:=1 to ny-1 do for i:=1 to
            nx-1 do
           begin d:=S[0,0]+d0; A[i,j]:=-S[-1,0]; B[i,j]:
            =-S[1,0];
               if j>1 then
           begin c:=-S[0,-1]/ILUD^[i,j-1]; d:=d+S[0,1]*c;
               A[i,j]:=A[i,j]+G[i,j-1]*c; F[i,j]:=
                B[i,j-1]*c;
               if i+1<nx then
               begin c:=F[i,j]/ILUD^[i+1,j-1];
                     d:=d+c*(omega*B[i+1,j-1]-G[i+1,j-1]);
                     B[i,j]:=B[i,j]-S[0,1]*c
               end end else F[i,j]:=0;
               if i>1 then
               begin c:=A[i,j]/ILUD^[i-1,j];
                      d:=d+c*(omega*
                      G[i-1,j]-B[i-1,j]);
                     G[i,j]:=-c*S[0,1]
               end else G[i,j]:=0;
               ILUD^[i,j]:=d
end end end end;

procedure solve_ILU7_decomposition(var x: gridfunction;
                  var A: data_of_discretisation;
                  var IP: iterationparameter);
var i,j: integer;
begin with A do
    begin if ILU7=nil then compute_ILU7_decomposition(A,IP);
        zero_boundary_values(nx,ny,x);
        with ILU7^ do
        begin for j:=1 to ny-1 do for i:=1 to nx-1 do
            x[i,j]:=(x[i,j]+A[i,j]*x[i-1,j]+F[i,j]*x[i+1,j-1]
                   -S[0,-1]*x[i,j-1])/ILUD^[i,j];
            for j:=ny-1 downto 1 do for i:=nx-1 downto 1 do
            x[i,j]:=x[i,j]-(S[0,1]*x[i,j+1]-B[i,j]*x[i+1,j]
                           -G[i,j]*x[i-1,j+1])/ILUD^[i,j]
end end end;
```

```
procedure ILU_7(var new: gridfunction; var A: data_of_
                 discretisation;
                var x,b: gridfunction; var IP:
                 iterationparameter);
var r: gridfunction;
begin residual(r,A,x,b); solve_ILU7_decomposition(r,A,IP);
      vector_plus_factor_x_vector(A.nx,A.ny,new,x,
       IP.theta,r);
      transfer_boundary_values(nx,ny,x,new)
end;
```

8.5.11 Numerical Examples

Table 1 shows the errors $\|x^m - x\|_2$ after $m = 20$ iterations and the convergence factors for different ILU variants. The step size is $h = 1/32$. For $\omega = 0$ and $\omega = 1$, we choose `diag:=0` for the modified method ($\omega = -1$), `diag:=5` is chosen.

Table 8.5.1 Results of the ILU iteration in the Poisson-model case

version	ω	Θ	$\|x^{20} - x\|_2$	$\dfrac{\|x^{20} - x\|_2}{\|x^{19} - x\|_2}$
ILU_5	0	1.66	$1.617_{10} - 1$	$9.455_{10} - 1$
ILU_5	-1	0.25	$1.628_{10} - 3$	$7.666_{10} - 1$
ILU_5	1	1.9	$2.349_{10} - 1$	$9.617_{10} - 1$
ILU_7	0	1	$8.904_{10} - 2$	$9.185_{10} - 1$
ILU_7	0	1.66	$2.690_{10} - 2$	$8.646_{10} - 1$
ILU_7	-1	0.4	$4.722_{10} - 5$	$6.254_{10} - 1$

Exercise 8.5.22. Count the arithmetical operations (separately for the decompositions and the solution phase) and compare `ILU_5` and `ILU_7` with regard to the effective amount of work.

8.5.12 Comments

ILU decompositions were first mentioned in 1960 by Varga [1, §6] and Buleev [1]. The first precise analysis was due to Meijerink–van der Vorst [1]. Here, we also mention Jennings–Malik [1]. ILU methods have proved to be very robust. This means that good convergence properties are not restricted to the Poisson-model problem, but hold for a large class of problems. Since the existence of an ILU decomposition is not always ensured, there are many stabilisation variants. For literature concerning the ILU method, we refer the interested reader to Axelsson–Barker [1] (cf. also Beauwens [1]).

Because of the improved condition number Γ/γ (cf. (25)) the modified version ($\omega = -1$) of Gustafsson [1] is the preferred basis for applications of the conjugate gradient technique (cf. §9) to ILU iterations. Because of the consistency condition $R\mathbb{1} = 0$, this version is also called an ILU decomposition of the *first* order. A special decomposition for the Poisson-model problem of second order was described by Stone [1]; however, because of other disadvantages, first-order variants are preferred.

The first publication of a blockwise ILU method occurred in 1981 with Kettler [1], who referred to a preprint by Meijerink [1] published in a paper two years later. Additional early papers were written by Axelsson–Brinkkemper–Il'in [1] (1984 with a preprint in 1983) and Concus–Golub–Meurant [1] (1985 with a preprint in 1982).

In the literature, the distinction between SSOR and ILU methods is not very sharp. The SSOR method for $A = D + L' + U'$ corresponds to an ILU decomposition $W = (D + L')D^{-1}(D + U')$ with the remainder $R = W - A = L'D^{-1}U'$. R does not satisfy condition (4d); however, this condition is already weakened by (12) and the addition of a diagonal part (cf. Remark 17). Vice versa, generalised SSOR methods have been introduced in which $D = \operatorname{diag}\{A\}$ is replaced by another diagonal (cf. Axelsson–Barker [1]). The ILU iteration based on a five-point pattern also falls into this category.

In the literature, one finds a lot of abbreviations for different ILU variants. «IC» refers to the symmetric «incomplete Cholesky» variant of the ILU decomposition. Additional numbers like «(5)» or «(7)» denote the respective five- or seven-point pattern. In other papers, «(0)» indicates the pattern $E = G(A)$, whereas «(1)» means the pattern which has been enlarged by one level, etc. The supplement «Tr» characterised a truncated version (31). The letter «M» stands for the modified method with $\omega = -1$, whereas «B» may indicate a block variant. If the block corresponds to a grid line (row or column), sometimes the symbol «L» is used. Since such abbreviations are not understandable to the nonexperts, we recommend avoiding them.

8.6 A Superfluous Term: Time-Stepping Methods

The term of a time-stepping method is used, in particular, in the engineering community. The function $x(t)$, $0 \leqslant t \leqslant \infty$, is introduced as a solution of the system of ordinary differential equations

$$\frac{d}{dt}x(t) = b - Ax \quad \text{with the initial value } x(0) = x^0. \tag{8.6.1}$$

If x^0 does not belong to a particular subspace (cf. U in Remark 3.2.13b) and A is positive definite (or $\mathrm{Re}(\lambda) > 0$ for all eigenvalues $\lambda \in \sigma(A)$), $x(t)$ converges for $t \to \infty$ to the solution $x^* := A^{-1}b$, which now is interpreted as the *stationary solution* of (1). The time-stepping method tries to discretise the differential equation and to approximate $x(t)$ for large $t = t_m$. One Euler step with the time step $\Delta t := t_{m+1} - t_m$ reads

$$x(t_{m+1}) \approx x^{m+1} = x^m - \Delta t(Ax^m - b) \tag{8.6.2}$$

(cf. Stoer–Bulirsch [2, §7.2]). For a fixed (variable) step size Δt, (2) describes the stationary (instationary) Richardson method. Often Runge–Kutta-like methods are proposed. For example, the Heun method reads

$$x' := Ax^m - b, \quad x(t_{m-1}) \approx x^{m+1} = x^m - \Delta t[\alpha x' + \beta(Ax' - b)] \tag{8.6.3}$$

with $\alpha = \beta = \frac{1}{2}$. But the coefficients α, β (in the true Runge–Kutta case there are 4 coefficients) are chosen in order to attain an optimal convergence behaviour $x^m \to x^*$. The produced methods (as, e.g., (3)) are the semi-iterative variants of the Richardson iteration already described in §7.3.7 (with cycle length = number of coefficients in (3)).

In the language of the ordinary differential equations one explains the unfavourably slow convergence of the arising Richardson variants by means of the *stiffness* of the system. When preconditioning is introduced to speed up the convergence: $x^{m+1} = x^m - \Delta t\, W^{-1}(Ax^m - b)$, this is called a *quasi*-time-stepping method, which however does not longer approximate the equation (1) but only the same stationary solution x^*.

This shows that any iteration (by means of its W) as well as all semi-iterative variants (including the gradient and cg-method for §9) can be interpreted as time-stepping method, so that this term is of no significance. Moreover, it leads the reader to misunderstandings, since often one finds papers not expressing clearly whether really the differential equation (1) is to be solved by discretisations of the form (2), (3) or if only the stationary value x^* is to be approximated.

9
Conjugate Gradient Methods

9.1 Linear Systems of Equations as Minimisation Problem

9.1.1 Minimisation Problem

In the following, $A \in \mathbb{R}^{I \times I}$ and $b \in \mathbb{R}^I$ are real. We consider a system

$$Ax = b \tag{9.1.1}$$

and assume that

$$A \text{ is positive definite.} \tag{9.1.2}$$

System (1) is associated with the function

$$F(x) := \tfrac{1}{2}\langle Ax, x\rangle - \langle b, x\rangle. \tag{9.1.3}$$

The derivative (gradient) of F is $F'(x) = \frac{1}{2}(A + A^T)x - b$. Since $A = A^T$ by assumption (2), the derivative equals

$$F'(x) = \operatorname{grad} F(x) = Ax - b. \tag{9.1.4}$$

A necessary condition for a minimum of F is the vanishing of the gradient: $Ax = b$. Since the Hesse matrix $F''(x) = (Fx_ix_j)_{i,j \in I} = A$ is positive definite, the solution of $Ax = b$ (in the following denoted by $x = x^*$) leads, in fact, to a minimum. This proves

Lemma 9.1.1. *Let $A \in \mathbb{R}^{I \times I}$ be positive definite. The solution of the system $Ax = b$ is equivalent to the solution of the minimisation problem*

$$F(x) = \min. \tag{9.1.5}$$

A second proof of Lemma 1 *results from the representation*

$$F(x) = F(x^*) + \tfrac{1}{2}\langle A(x - x^*), x - x^*\rangle \quad \text{with } x^* := A^{-1}b. \tag{9.1.6}$$

(6) *proves* $F(x) > F(x^*)$ *for* $x \neq x^*$, *i.e.*, $x^* = A^{-1}b$ *is the unique minimum of* F. *The representation* (6) *is a particular case of the following expansion of* F *around an arbitrary value* $\tilde{x} \in \mathbb{R}^I$:

$$F(x) = F(\tilde{x}) + \langle A\tilde{x} - b, x - \tilde{x}\rangle + \tfrac{1}{2}\langle A(x - \tilde{x}), x - \tilde{x}\rangle. \tag{9.1.7}$$

Proof. Expanding the scalar products proves the coincidence with (3). □

9.1.2 Search Directions

In the following, the minimisation of F with respect to a particular direction $p \in \mathbb{R}^I \backslash \{0\}$ plays a central rôle. Optimisation over all $x \in \mathbb{R}^I$ is replaced by the one-dimensional minimisation problem (8a, b):

$$f(\lambda) = \min_{\lambda \in \mathbb{R}} \quad \text{for the function} \tag{9.1.8a}$$

$$f(\lambda) := F(x + \lambda p) \qquad (x, p \in \mathbb{R}^I \text{ fixed}). \tag{9.1.8b}$$

Replacing the variables x and $\tilde{x}$ in (7) by $x + \lambda p$ and x, one obtains

$$f(\lambda) = F(x) + \lambda\langle Ax - b, p\rangle + \lambda^2\tfrac{1}{2}\langle Ap, p\rangle. \tag{9.1.8c}$$

$p \neq 0$ implies $\langle Ap, p\rangle > 0$ (cf. (2)); hence, the minimum of the parabola f can be determined from $f'(\lambda) = 0$.

Lemma 9.1.2. *Assume* $p \neq 0$ *and* (2): $A > 0$. *The unique minimum of problem* (8a, b) *is attained with*

$$\lambda = \lambda_{\text{opt}}(x, p) := \frac{\langle r, p\rangle}{\langle Ap, p\rangle}, \tag{9.1.9a}$$

where

$$r := b - Ax. \tag{9.1.9b}$$

In the following, the letter r always denoted the residual (*residue*) $b - Ax$. *It is the negative defect* $Ax - b$ *and also the negative gradient* $F' = Ax - b$.

Concerning an optimal search direction, $p = x^* - x$ (*or a multiple* $\neq 0$) *is evidently optimal, because* $f(\lambda_{\text{opt}}) = F(x^*)$ *yields the global minimum. However, since* $p = x^* - x$ *requires knowledge of the solution, another proposal is needed. Let p be normalised by* $\|p\|_2 = 1$. *The directional derivative* $f'(0) = -\langle r, p\rangle = \langle \text{grad}\, F(x), p\rangle$ *at* $\lambda = 0$ *is maximal for the gradient direction* $p = -r/\|r\|_2$ *and minimal for the reverse direction* $p = r/\|r\|_2$, grad $F(x) = -r$ *is the direction of the steepest ascent, while the residual r is the* direction of the steepest descent. *This consideration shows the optimality of* $p = r$ *from a local* *point of view. For $p = r$ the expression* (9a) *becomes*

$$\lambda = \lambda_{\text{opt}}(x, r) = \frac{\|r\|_2^2}{\langle Ar, r\rangle} \quad \textit{for } r = b - Ax \neq 0. \tag{9.1.9c}$$

The case

$$\lambda_{\text{opt}}(x, 0) := 0 \tag{9.1.9d}$$

is added by formal reasons only. As soon as $r = 0$ occurs, x is already the exact solution x^*.

9.1.3 Other Quadratic Functionals

The function F from (3) is not the only quadratic function with $x^* := A^{-1}b$ as the minimising argument.

Lemma 9.1.3. (a) *Any quadratic form with a unique minimum at $x^* = A^{-1}b$ has the form*

$$\begin{aligned} F(x) &= \tfrac{1}{2}\langle HA(x - x^*), A(x - x^*)\rangle + c \\ &= \tfrac{1}{2}\langle H(Ax - b), Ax - b\rangle + c \end{aligned} \tag{9.1.10a}$$

with an arbitrary constant c and

$$H \textit{ positive definite.} \tag{9.1.10b}$$

Here, in contrast to (2), *A may be any regular matrix.*
(b) *To ensure that the calculation of* $\operatorname{grad} F(x) = A^H HA(x - x^*) = A^H Hr$ *from the residual $r = Ax - b$ is practical, the matrix H must be such that the matrix-vector multiplication $r \mapsto A^H Hr$ is feasible.*
(c) *Under assumption* (2), *$H := A^{-1}$ and $c := -\frac{1}{2}\langle b, x^*\rangle$ may be chosen. Then F from* (10a) *coincides with F from* (3).

Proof of (c). By (2), $H = A^{-1}$ satisfies (10b) (cf. Lemma 2.10.4b). A comparison of (10a) and (6) shows $c = F(x^*) = \frac{1}{2}\langle Ax^*, x^*\rangle - \langle b, x^*\rangle = \frac{1}{2}\langle b, x^*\rangle$. □

Remark 9.1.4. Let A be positive definite. The (energy) scalar product $\langle\cdot,\cdot\rangle_A$ and (energy) norm $\|\cdot\|_A$ are defined by (11a):

$$\langle x, y\rangle_A := \langle Ax, y\rangle, \qquad \|x\|_A := \|A^{1/2}x\|_2 = \sqrt{\langle x, x\rangle_A}. \tag{9.1.11a}$$

The minimisation of F from (3) is equivalent to the minimisation problem

$$\|x - x^*\|_A = \min. \tag{9.1.11b}$$

Proof. (11b) may be replaced by $\|x - x^*\|_A^2 = \min$. The identity

$$\|x - x^*\|_A^2 = 2[F(x) - F(x^*)] \tag{9.1.11c}$$

(cf. (6)) completes the proof. □

Remark 9.1.5. (a) For the choice $H = I$ and $c = 0$, Eq. (10a) becomes $F(x) = \frac{1}{2}\|Ax - b\|_2^2$ and describes the «least-squares minimisation».
(b) For $H = A^{-H}A^{-1}$ and $c = 0$, the identity $F(x) = \frac{1}{2}\|x - x^*\|_2^2$ holds.
(c) For a positive definite K, the minimisation of the norm $\|x - x^*\|_K^2 = \|K^{1/2}(x - x^*)\|_2^2$ corresponds to problem (10a) with $H = \frac{1}{2}A^{-H}KA^{-1}$, $c = 0$. According to Lemma 3b, multiplication by KA^{-1} must be feasible.

Remark 9.1.6. Any iteration converging (weakly) monotonically with respect to the norm $\|\cdot\|_A$ leads to a descent sequence $F(x^0) \geqslant F(x^1) \geqslant \cdots$.

9.1.4 Complex Case

In the complex case of $A \in \mathbb{C}^{I \times I}$ and $b \in \mathbb{C}^I$, the function F can again be defined by (10a, b), provided that c from (10a) is real. Definition (3) cannot be generalised without change, since only real functions F can be minimised and, in general, F is not real because of the term $\langle b, x\rangle$. One has to replace F from (3) by

$$F(x) := \tfrac{1}{2}\langle Ax, x\rangle - \operatorname{Re}\langle b, x\rangle \quad \text{for } x \in \mathbb{C}^I. \tag{9.1.12a}$$

Exercise 9.1.7. Assume (2) and let F be defined by (12a). Prove that (a) F is real, $\operatorname{Re}\langle b, x^*\rangle = \langle b, x^*\rangle$ for $x^* = A^{-1}b$. (b) (12b, c) hold:

$$F(x) = \tfrac{1}{2}(\langle A(x - x^*), x - x^*\rangle - \langle b, x^*\rangle), \tag{9.1.12b}$$

$$F(x) = F(\tilde{x}) + \operatorname{Re}\langle A\tilde{x} - b, x - \tilde{x}\rangle + \tfrac{1}{2}\langle A(x - \tilde{x}), x - \tilde{x}\rangle. \tag{9.1.12c}$$

(c) The minimum of $f(\lambda) = F(x + \lambda p)$ over $\lambda \in \mathbb{C}$ with F from (12a) is attained at the value $\lambda_{\text{opt}}(x, p)$ from (9a), which in general is complex.

9.2 Gradient Method

9.2.1 Construction

Generally speaking, the gradient method is an algorithm for solving a minimisation problem $F(x) = \min$ with a differentiable function $F\colon \mathbb{R}^I \to \mathbb{R}$ (cf., e.g., Kosmol [1, §4]). In the following, we apply the gradient method only to the quadratic function F from (1.3) or (1.10a).

The gradient method minimises F iteratively in the direction of the steepest descent:

$$x^0\text{: arbitrary starting iterate,} \tag{9.2.1a}$$

$$\text{iteration } m = 0, 1, \ldots:$$

$$r^m := b - Ax^m, \tag{9.2.1b}$$

$$x^{m+1} := x^m + \lambda_{\text{opt}}(x^m, r^m)r^m. \tag{9.2.1c}$$

The representation $r^{m+1} = b - Ax^{m+1} = b - A(x^m + \lambda_{\mathrm{opt}} r^m) = r^m - \lambda_{\mathrm{opt}} A r^m$ allows the following update of the residual:

start:

$$x^0 \text{ arbitrary}, \quad r^0 := b - Ax^0, \tag{9.2.2a}$$

iteration $m = 0, 1, \ldots$:

$$x^{m+1} := x^m + \lambda_{\mathrm{opt}}(x^m, r^m) r^m, \tag{9.2.2b}$$

$$r^{m+1} := r^m - \lambda_{\mathrm{opt}}(x^m, r^m) A r^m \tag{9.2.2c}$$

with $\lambda_{\mathrm{opt}}(x^m, r^m)$ from (1.9c, d). The advantage of (2c) over (1b) is the fact that the product Ar^m was already calculated in (1.9c) during the λ_{opt} determination.

9.2.2 Properties of the Gradient Method

Remark 9.2.1. Assume (1.2). (a) In contrast to the previous methods, the iteration $x^m \mapsto \Phi(x^m, b)$ defined in (2a–c) is *not linear.*
(b) $\Phi(\cdot, \cdot)$ is continuous with respect to both of its arguments.
(c) The gradient method is consistent and convergent.

Proof. (a) $\lambda_{\mathrm{opt}}(x^m, r^m)$ is a nonconstant function of x^m and b. Hence, $\Phi(x, b) = x + \lambda_{\mathrm{opt}}(x, b - Ax)(b - Ax)$ is not linear.
(c) The convergence will be proved in Theorem 3. If x^* is a solution of $Ax = b$, the residual r vanishes. Together with (1.9d), one concludes $\Phi(x^*, b) = x^*$, i.e., Φ is consistent. □

Although the gradient method is not linear, it can be interpreted as semi-iterative method applied to a linear basic iteration.

Remark 9.2.2. The sequence $\{x^m\}$ of the gradient method (2a–c) is identical to the sequence $\{y^m\}$ of the semi-iterative Richardson method

$$y^{m+1} = y^m - \Theta_{m+1}(Ay^m - b) = \Phi^{\mathrm{Rich}}_{\Theta_{m+1}}(y^m, b) \tag{9.2.3a}$$

(cf. (7.2.1b)), if one chooses $y^0 = x^0$ and fixes the factors Θ_{m+1} by

$$\Theta_{m+1} := \lambda_{\mathrm{opt}}(x^m, b - Ax^m). \tag{9.2.3b}$$

Theorem 9.2.3. *Let A be positive definite and denote the extreme eigenvalues of A by $\lambda = \lambda_{\min}(A)$ and $\Lambda = \lambda_{\max}(A)$. F is defined according to* (1.3). *Then for any starting iterate x^0, the sequence $\{x^m\}$ of the gradient method converges to the solution $x^* = A^{-1}b$ and satisfies the error estimates*

$$F(x^m) - F(x^*) \leqslant \left(\frac{\Lambda - \lambda}{\Lambda + \lambda}\right)^{2m} [F(x^0) - F(x^*)], \tag{9.2.4a}$$

$$\|x^m - x^*\|_A \leqslant \left(\frac{\Lambda - \lambda}{\Lambda + \lambda}\right)^{m} \|x^0 - x^*\|_A. \tag{9.2.4b}$$

Proof. (i) By (1.11c) the estimates (4a, b) are equivalent.
(ii) For proving (4b), it suffices to consider the case $m = 1$. The Richardson iteration

$$x^1_{\text{Rich}} = x^0 - \Theta_{\text{Rich}}(Ax^0 - b) \quad \text{with } \Theta_{\text{Rich}} = 2/(\Lambda + \lambda)$$

yields the error $e^1_{\text{Rich}} = Me^0$. The iteration matrix $M = M^{\text{Rich}}_{\Theta_{\text{Rich}}} = I - \Theta_{\text{Rich}} A$ has the norm $\|M\|_2 \leqslant \eta$, where

$$\eta = \frac{\Lambda - \lambda}{\Lambda + \lambda} \tag{9.2.4c}$$

(cf. Theorem 4.4.3). Since M commutes with A and $A^{1/2}$, we have

$$\tilde{e}^1_{\text{Rich}} = M\tilde{e}^0 \quad \text{for } \tilde{e}^1_{\text{Rich}} := A^{1/2}e^1_{\text{Rich}},\ \tilde{e}^0 := A^{1/2}e^0.$$

By $\|\tilde{e}^0\|_2 = \|e^0\|_A$ and $\|\tilde{e}^1_{\text{Rich}}\|_2 = \|e^1_{\text{Rich}}\|_A$, we can estimate e^1_{Rich} by

$$\|e^1_{\text{Rich}}\|_A = \|\tilde{e}^1_{\text{Rich}}\|_2 \leqslant \|M\|_2 \|\tilde{e}^0\|_2 = \eta \|e^0\|_A.$$

Both x^1_{Rich} and x^1 are of the form $x^0 + \Theta r^0$. Since the iterate x^1 of the gradient method minimises the error $\|x^1 - x^*\|_A$ (cf. Remark 1.4), the assertion follows for $m = 1$: $\|x^1 - x^*\|_A \leqslant \|e^1_{\text{Rich}}\|_A \leqslant \eta \|e^0\|_A$. □

Corollary 9.2.4. (a) *The factor* η *from* (4c) *is the minimal one in* (4a, b).
(b) *The asymptotical convergence rate of the gradient method is* η.
(c) η *depends only on the condition number* $\varkappa = \varkappa(A) = \text{cond}_2(A) = \Lambda/\lambda$:

$$\eta = \frac{\varkappa - 1}{\varkappa + 1}. \tag{9.2.5}$$

Proof. Let v_1 and v_2 with $\|v_1\|_2 = \lambda$ and $\|v_2\|_2 = \Lambda$ be the eigenvectors corresponding to Λ and λ. For $x^0 := x^* + e^0$ with $e^0 := v_1 \pm v_2$, one obtains $e^1 = \eta(v_1 \mp v_2)$ and $e^2 = \eta^2(v_1 \pm v_2) = \eta^2 e^0$. $\|e^2\|_A / \|e^0\|_A = \eta^2$ proves part (a). Analogously, $e^{2k} = \eta^{2k} e^0$ shows (b). □

Remark 9.2.5. (a) Let $\langle e^0, v_i \rangle \neq 0$ hold for the eigenvectors v_1, v_2 of A corresponding to Λ and λ. Then $\varrho_{m+1,m} := \|x^{m+1} - x^*\|_A / \|x^m - x^*\|_A$ converges to $\eta = (\varkappa - 1)/(\varkappa + 1)$ from (5).
(b) Using $\varrho(M^{\text{Rich}}_\Theta) = 1 - \Theta\lambda$ (e.g., for $\Theta = 1/\|A\|_\infty$), we can approximate λ from the convergence behaviour of the Richardson method. The approximation of η by $\Lambda/\lambda = \varkappa = (1 + \eta)/(1 - \eta)$ allows us to determine the other extreme eigenvalue Λ.

9.2.3 Numerical Examples

At first view, the gradient method seems to surpass the semi-iterative method, because in the latter case the parameters Θ_k are to be chosen *a priori* (cf. (3a)), whereas the gradient method determines these values *a posteriori* in an optimal way. However, while the Chebyshev method leads to an order improvement, Corollary 4a yields the convergence rate η from (4c), which is as slow as the stationary Richardson method with $\Theta = \Theta_{\text{opt}}$ (cf. Theorem 4.4.3).

In the model case, λ and Λ from (4.1.1b, c) are known and lead to

$$\eta = \frac{\cos^2 \pi h/2 - \sin^2 \pi h/2}{\cos^2 \pi h/2 + \sin^2 \pi h/2} = \cos \pi h = 1 - \tfrac{1}{2}\pi^2 h^2 + O(h^4).$$

The low convergence speed of the gradient method is confirmed by the following numerical example (Poisson-model problem (3.4.1a, b)). Table 1 contains the results for the step size $h = 1/32$ and the starting iterate $x^0 = 0$. The factors $\|x^{m+1} - x^*\|_A / \|x^m - x^*\|_A$ from the last column of Table 1 clearly approximate the asymptotical convergence rate $\eta = \cos \pi/32 = 0.9951847$. Even after 300 iterations, the value $u_{16,16}$ at the midpoint is wrong by 50%: 0.2778 instead of 0.5. The error measured in the scaled energy norm $h^2 \|x^m - x^*\|_A$ deviates very little from the maximum norm $\|e^m\|_\infty$. However, the error with respect to the energy norm $\|\cdot\|_A$ decreases uniformly, whereas the ratios of $\|e^m\|_\infty$ oscillate. By $\eta = \varrho(M^{\text{Jac}})$, the results from Table 1 and Table 4.7.1 prove to be very similar.

Table 9.2.1 Results of the gradient method for $h = 1/32$

m	value in the middle	$\|e^m\|_A / \|e^{m-1}\|_A$
1	$-1.86560_{10}-3$	
2	$-3.52293_{10}-3$	0.844824
3	$-4.84034_{10}-3$	0.907804
4	$-5.97611_{10}-3$	0.935293
5	$-7.10198_{10}-3$	0.946906
6	$-8.16295_{10}-3$	0.953838
7	$-9.23998_{10}-3$	0.958895
8	$-1.02699_{10}-2$	0.962711
9	$-1.13230_{10}-2$	0.965778
10	$-1.23360_{10}-2$	0.968271
100	$-1.89771_{10}-2$	0.993444
110	$-5.13520_{10}-3$	0.993749
120	$1.01805_{10}-2$	0.993990
200	$1.45146_{10}-1$	0.994852
250	$2.18301_{10}-1$	0.995024
295	$2.73556_{10}-1$	0.995100
296	$2.73548_{10}-1$	0.995102
297	$2.75710_{10}-1$	0.995103
298	$2.75702_{10}-1$	0.995104
299	$2.77844_{10}-1$	0.995105
300	$2.77836_{10}-1$	0.995106

9.2.4 Gradient Method Based on Other Iterations

By Remark 2, the gradient method is a particular semi-iterative method with Richardson's iteration as the basic iteration. From the analysis of semi-iterative methods, we know that other basic iterations Φ may better suit semi-iterations because of a more favourable condition number $\varkappa(W^{-1}A)$ (W: matrix of the third normal form of Φ). This suggests replacing Richardson's iteration by another one (e.g., the SSOR iteration; cf. §7.4.2). For this purpose, the matrix A is to be replaced formally by $\hat{A} := W^{-1}A$, because the Richardson method for the left-transformed (preconditioned) system $\hat{A}x = \hat{b} := W^{-1}b$ is equivalent to Φ (cf. Remark 4.3.2).

Let A and W be positive definite. Since, in general, the matrix $\hat{A} = W^{-1}A$ is no longer symmetric, $\hat{A}$ does not satisfy the assumption (1.2), which is necessary for the applicability of the gradient method. A remedy is offered by Remark 8.1.9: The iteration $\check{\Phi}$ defined by

$$\check{x}^{m+1} = \check{x}^m - W^{-1/2}(AW^{-1/2}\check{x}^m - b) = \check{x}^m - (\check{A}\check{x}^m - \check{b}), \tag{9.2.6a}$$

$$\check{A} := W^{-1/2}AW^{-1/2}, \qquad \check{b} := W^{-1/2}b, \tag{9.2.6b}$$

is equivalent to the basic iteration $\Phi(x^m, b) = x^m - W^{-1}(Ax^m - b)$ via $\check{x}^m = W^{1/2}x^m$ (but not feasible) and represents the Richardson iteration for the system $\check{A}\check{x} = \check{b}$ with a *positive definite* matrix $\check{A}$. Therefore, the gradient method is to be applied not to F from (1.3), but

$$\check{F}(\check{x}) := \tfrac{1}{2}\langle \check{A}\check{x}, \check{x}\rangle - \langle \check{b}, \check{x}\rangle. \tag{9.2.6c}$$

Its negative gradient is the new residual

$$\check{r} := \check{b} - \check{A}\check{x} = W^{-1/2}r \qquad (r = b - Ax). \tag{9.2.6d}$$

The gradient method (2b, c) associated with $\check{A}$ reads

$$\check{x}^{m+1} := \check{x}^m + \check{\lambda}_{\text{opt}}(\check{x}^m, \check{r}^m)\check{r}^m,$$

$$\check{r}^{m+1} := \check{r}^m - \check{\lambda}_{\text{opt}}(\check{x}^m, \check{r}^m)\check{A}\check{r}^m$$

where

$$\check{\lambda}_{\text{opt}}(\check{x}^m, \check{r}^m) = \|\check{r}^m\|_2^2/\langle \check{A}\check{r}^m, \check{r}^m\rangle.$$

Inserting $\check{A} = W^{-1/2}AW^{-1/2}$, $\check{x}^m = W^{1/2}x^m$, $\check{r}^m = W^{-1/2}r^m$, and solving the defining equations for x^{m+1} and r^{m+1}, we obtain the following algorithm for the iterates $\{x^m\}$:

$$x^{m+1} := x^m + \check{\lambda}_{\text{opt}} W^{-1}r^m, \tag{9.2.7a}$$

$$r^{m+1} := r^m - \check{\lambda}_{\text{opt}} AW^{-1}r^m \quad \text{with} \tag{9.2.7b}$$

$$\check{\lambda}_{\text{opt}} := \langle W^{-1}r^m, r^m\rangle/\langle AW^{-1}r^m, W^{-1}r^m\rangle. \tag{9.2.7c}$$

The quantities $W^{-1/2}$ and $W^{1/2}$ no longer appear in (7a–c), so that (7a–c) is a practical algorithm. We call (7a–c) the *gradient method applied to the basic iteration* $\Phi(x,b) = x - W^{-1}(Ax - b)$. Usually, the term «*preconditioned gradient method*» is used, which, more precisely, should be named «*method of the preconditioned gradients*» (cf. Remark 7b). In analogy to Remark 2, Eq. (7a) proves

Remark 9.2.6. The sequence $\{x^m\}$ of the gradient method (7a–c) applied to the symmetric iteration Φ is identical to the sequence $\{y^m\}$ of the semi-iterative method $y^{m+1} = y^m - \Theta_{m+1} W^{-1}(Ay^m - b) = \Theta_{m+1}\Phi(y^m, b) + (1 - \Theta_{m+1})y^m$ with Φ as the basic iteration, when the factors Θ_{m+1} are defined by λ_{opt} from (7c).

The amount of work needed in (7a–c) can be reduced by introducing $q^m := W^{-1}r^m$ and $a^m := Aq^m$. Note that q^m and a^m need not be saved for the next iteration step.

start:

$$x^0 \text{ arbitrary}, \quad r^0 := b - Ax^0, \quad q^0 := W^{-1}r^0, \tag{9.2.8a}$$

iteration $m = 0, 1, \ldots$:

$$q^m := W^{-1}r^m, \quad a^m := Aq^m, \tag{9.2.8b}$$

$$\lambda_{\text{opt}} = \lambda_{\text{opt}}(x^m, q^m) = \langle q^m, r^m\rangle / \langle a^m, q^m\rangle, \tag{9.2.8c}$$

$$x^{m+1} := x^m + \lambda_{\text{opt}} q^m, \tag{9.2.8d}$$

$$r^{m+1} := r^m - \lambda_{\text{opt}} a^m. \tag{9.2.8e}$$

Remark 9.2.7. (a) The representation (8a–e) shows that for each iteration step only one multiplication by W^{-1} and one by A are necessary.
(b) The optimal factor λ_{opt} from (8c) regarded as a function of (x^m, q^m) coincides with $\lambda_{\text{opt}}(x, p)$ from (1.9a).

Remark 7b enables the following interpretation: While the method (2a–c) takes the (negative) gradient r^m as the search direction, this is replaced in (8a–e) by the «preconditioned» gradient $q = W^{-1}r$. In this sense, (8a–e) represent the *method of the preconditioned gradients.*

The convergence of the method (8a–c) follows by applying the convergence statement of Theorem 3 to the transformed problem (6c): $\breve{F}(\breve{x}) = \min$. First, we obtain an error estimate for $\breve{x}^m$ with respect to the corresponding $\breve{A}$ norm $\|\cdot\|_{\breve{A}}$. Because of

$$\begin{aligned}\|\breve{x}^m - \breve{x}^*\|_{\breve{A}}^2 &= \langle \breve{A}(\breve{x}^m - \breve{x}^*), \breve{x}^m - \breve{x}^*\rangle \\ &= \langle A(x^m - x^*), x^m - x^*\rangle \\ &= \|x^m - x^*\|_A^2, \qquad (\breve{x}^* = W^{1/2}x^*)\end{aligned}$$

the $\breve{A}$ estimates of $\breve{x}^m - \breve{x}^*$ carry over to the A norm of the error $x^m - x^*$:

Theorem 9.2.8 (convergence). *Let A and W be positive definite. If*

$$\gamma W \leqslant A \leqslant \Gamma W \qquad \text{(cf. (7.3.23a'))}, \tag{9.2.9a}$$

the iterates from (8a–e) *satisfy the error estimate*

$$\|x^m - x^*\|_A \leqslant \left(\frac{\Gamma - \gamma}{\Gamma + \gamma}\right)^m \|x^0 - x^*\|_A. \tag{9.2.9b}$$

Remark 9.2.9. Under an assumption analogous to that in Remark 5a, we conclude for (8a–e) that the convergence factors converge to $\eta = (\varkappa - 1)/(\varkappa + 1)$ with $\varkappa := \varkappa(W^{-1}A) = \Gamma/\gamma$ (here, γ and Γ are the *optimal* bounds in (9a)). Therefore, the gradient method (8a–e) can be used to determine the condition number Γ/γ. A procedure determining the optimal damping parameter Θ of an iteration Φ may read as follows:

```
procedure determine_optimal_theta (var IP:
 iterationparameter);
var eta,rho: real;
begin writeln;
     writeln('*** The optimal damping factor theta is to be
      determined:');
     write('--> convergence rate of the gradient method=');
      readln(eta);
     write('--> convergence rate of the iteration=');
      readln(rho);
     writeln('Is the eigenvalue dominating the rate
      positive?');
     if yes_no then IP.theta:=(1-eta)/(1-rho) else
      IP.theta:=(1+eta)/(1+rho)
end;
```

Different from the presentations of other authors, we regard the gradient method as a general technique that can be applied to all symmetric iterations with $A > 0$, just as the Chebyshev method requires specification of the basic iteration.

Theorem 9.2.10. *Let Φ be a symmetric iteration and assume that $A > 0$. The gradient method applied to Φ converges as fast as the optimally damped iteration Φ_Θ, $\Theta = 2/(\gamma + \Gamma)$. However, explicit knowledge of the optimal bounds γ and Γ from* (9a) *is not necessary.*

Proof. Compare the results from Exercise 8.3.1 and (9b). □

In (6a) we have interpreted the iteration Φ as Richardson's iteration with the symmetric matrix $\bar{A}$. This is not the only possibility. Φ is also equivalent to

$$\bar{x}^{m+1} := \bar{x}^m - (\bar{A}\bar{x}^m - \bar{b}) \quad \text{with} \tag{9.2.10a}$$

$$\bar{A} := A^{1/2}W^{-1}A^{1/2} > 0, \qquad \bar{b} := A^{1/2}W^{-1}b, \qquad \bar{x}^m := A^{1/2}x^m. \tag{9.2.10b}$$

Exercise 9.2.11. Prove the following: (a) The application of the gradient method to the minimisation of

$$\bar{F}(\bar{x}) := \tfrac{1}{2}\langle \bar{A}\bar{x}, \bar{x}\rangle - \langle \bar{b}, \bar{x}\rangle \tag{9.2.10c}$$

yields (11a–c) after a reformulation by mean of the x quantities:

$$\text{start: } x^0 \text{ arbitrary}, \quad q^0 := W^{-1}(b - Ax^0), \tag{9.2.11a}$$

$$x^{m+1} := x^m + \bar{\lambda}q^m \quad \text{with } \bar{\lambda} := \langle Aq^m, q^m\rangle / \langle W^{-1}Aq^m, Aq^m\rangle, \tag{9.2.11b}$$

$$q^{m+1} := q^m - \bar{\lambda}W^{-1}Aq^m. \tag{9.2.11c}$$

(b) The methods (8a–e) and (11a–c) are different.
(c) Let γ and Γ be the bounds from (9a). The error estimate (10d) holds:

$$\|W^{-1/2}A(x^m - x^*)\|_2 \leqslant \left(\frac{\Gamma - \gamma}{\Gamma + \gamma}\right)^m \|W^{-1/2}A(x^0 - x^*)\|_2. \tag{9.2.10d}$$

(d) Interpret the factor $\bar{\lambda}$ from (11b) as the optimal factor for the search direction q^m with respect to the minimisation of $F(x)$ from (1.10a) with $H = W^{-1}$.

9.2.5 Pascal Procedures and Numerical Examples

During the iteration, the actual residual r^m is stored in `IP.Res^`. r^0 is computed by `start_gradient_method`. The procedure `gradient_method` performing (8a–e) has a parameter `basisiteration` specifying the underlying iteration Φ. The choice `Richardson_iteration` yields the simple version (2a–c). In order to calculate the product $W^{-1}y = Ny$ (N: matrix of the second normal form) by `N_y`, the iteration $\Phi(x, b) = Mx + Nb$ is performed with $x = 0$ and $b = y$. This approach is not the cheapest one, but it exempts us from writing new procedures for all W_Φ. A separate procedure `A_x` performs the multiplication Ax (alternative: compute Ax from $r = b - Ax$ with $b := 0$ by means of `residual`). A single step of the gradient method with the parameter `IT` is performed by the procedure `gradient_method_1`.

```
function Eucl_scalarproduct(var x,y: gridfunction;
                            var A: data_of_discretisation):
                             real;
```

```
var i,j: integer; e: real;
begin e:=0; for i:=1 to A.nx-1 do for j:=1 to A.ny-1 do e:=e+
 x[i,j]*y[i,j];
      Eucl_scalarproduct:=e*A.h2
end;

function Euclidean_norm (var x: gridfunction;
                         var A: data_of_discretisation):
                          real;
begin Euclidean_norm:=sqrt(Eucl_scalarproduct(x,x,A))
end;

function lambda_opt(var r,p,Ap: gridfunction;
                    var A: data_of_discretisation): real;
var s: real;
begin s:=Eucl_scalarproduct(Ap,p,A); if s=0 then lambda_
 opt:=0 else
      lambda_opt:=Eucl_scalarproduct(r,p,A)/s
end;

procedure A_x(var Ax: gridfunction;
              var A: data_of_discretisation; var x:
               gridfunction);
var i,j: integer; v,z: column;
begin with A do begin zero_boundary_values(nx,ny,x); v:=x[0];
        for i:=1 to nx-1 do
        begin case kind of
Poisson_model_problem:
        for j:=1 to ny-1 do z[j]:=4*x[i,j]-x[i,j-1]-x[i,j+1]-
         x[i-1,j]-x[i+1,j];
fivepoint_formula:
      for j:=1 to ny-1 do
      z[j]:=S[-1,0]*x[i-1,j]+S[1,0]*x[i+1,j]+S[0,-1]*x[i,j-1]
           +S[0,1]*x[i,j+1]+S[0,0]*x[i,j];
ninepoint_formula:
      for j:=1 to ny-1 do
      z[j]:=S[-1,-1]*x[i-1,j-1]+S[0,-1]*x[i,j-1]+S[1,-1]*
            x[i+1,j-1]
            +S[-1,0]*x[i-1,j]+S[0,0]*x[i,j]+S[1,0]*x[i+1,j]
            +S[-1,1]*x[i-1,j+1]+S[0,1]*x[i,j+1]+S[1,1]*
            x[i+1,j+1]
      end {case};
      Ax[i-1]:=v; v:=z
   end;
   Ax[nx-1]:=v; zero_boundary_values(nx,ny,Ax)
end end;
```

```
function A_scalarproduct(var x,y: gridfunction;
                          var A: data_of_discretisation):
                           real;
var Ax: gridfunction;
begin A_x(Ax,A,x); A_scalarproduct:=Eucl_scalarproduct
 (Ax,y,A)
end;
function A_norm(var x: gridfunction; var A: data_of_
 discretisation): real;
begin A_norm:=sqrt(A_scalarproduct(x,x,A)) end; {energy norm}

procedure zero_gridfunction(nx,ny: integer; var x:
 gridfunction);
var i,j: integer;
begin for j:=0 to ny do for i:=0 to nx do x[i,j]:=0
end;

procedure N_y(var Ny,y: gridfunction; var A: data_of_
               discretisation;
              var IP: iterationparameter;
              procedure basisiteration(var new: gridfunction;
                         var A: data_of_discretisation;
                         var x,b: gridfunction;
                         var IP: iterationparameter));
begin zero_gridfunction(A.nx,A.ny,Ny); basisiteration
 (Ny,A,Ny,y,IP)
end;

procedure start_gradient_method(var A: data_of_
 discretisation;
                          var x,b: gridfunction;
                          var IP: iterationparameter);
begin with IP do begin Nr:=0; if Res=nil then new(Res);
 residual(Res^,A,x,b)
end end;

procedure gradient_method(var new: gridfunction;
                  var A: data_of_discretisation; var x,b:
                   gridfunction;
                  var IP: iterationparameter;
                  procedure basisiteration(var new:
                   gridfunction;
                      var A: data_of_discretisation;
                      var x,b: gridfunction;
                      var IP: iterationparameter));
```

```
var q,Aq: gridfunction; lambda: real;
begin if IP.Res=nil then start_gradient_method(A,x,b,IP);
    with A do with IP do
    begin N_y(q,Res^,A,IP,basisiteration); A_x(Aq,A,q);
          lambda:=lambda_opt(Res^,q,Aq,A);
          vector_plus_factor_x_vector(nx,ny,new,x,lambda,q);
          transfer_boundary_values(nx,ny,x,new);
          vector_plus_factor_x_vector(nx,ny,Res^,Res^,
           -lambda,Aq)
end end;

procedure gradient_method_1(var it: data_of_iteration;
               procedure basisiteration(var new: gridfunction;
                         var A: data_of_discretisation;
                         var x,b: gridfunction;
                         var IP: iterationparameter));
var number: integer;
begin with it do with IP do
    begin number:=Nr; if Nr=0 then start_gradient_
           method(A,x,b,IP);
          gradient_method(x,A,x,b,IP,basisiteration); Nr:=
           number+1
end end;
```

The frame program calling the gradient method applied to the SSOR iteration can be organised as follows:

```
program gradient_method;
var it: data_of_iteration; i,itnr: integer; v: data_for_
 comparison;
    AN: history_of_iteration; {insert right_hand_side, etc.}
begin initialise_IT(it); initialise_comparison(v);
 repeat release_IT(it); release_comparison(v);
  define_problem(it,boundary_value,right_hand_side);
  writeln('Probl. defined');
  define_optimal_SSOR_parameter(it);
  define_starting_iterate(it,zerofunction);
  writeln('Starting value defined.');
  define_comparison_solution(v,it.A,exact_solution);
  comparison_with_exact_solution(AN,v,it,A_norm);
  writeln('iteration nr.' ,IT.IP.Nr,' A-norm: ',AN.value[0]);
  write('--> Number of iterations='); readln(itnr);
  for i:=1 to itnr do
```

```
  begin gradient_method_1(it,lex_SSOR);
        comparison_with_exact_solution(AN,v,it,A_norm);
        writeln('iteration nr. ',it.IP.Nr, ' A norm: ',
                AN.value[AN.last])
  end {here, the results may be stored}
 until not yes_no {termination?}
end.
```

As an example, the SSOR iteration is used as a basic iteration in the Poisson-model case. As in Table 4.8.1, we choose the relaxation parameter $\omega = 1.82126912$ for the step size $h = 1/32$. The results given in Table 2 suggest the convergence rate $\eta \approx 0.769$. From (5) one concludes the condition number $\Gamma/\gamma = \varkappa = (1+\eta)/(1-\eta) = 7.66$. According to Tables 4.8.1/2, the convergence rate of the SSOR iteration equals 0.8796. From $\varrho(M^{\mathrm{SSOR}}) = 1 - \lambda$, one deduces $\lambda = 0.1204$, leading to $\Gamma = 7.66$ and $\gamma = 0.922$. Hence, $\Theta = 2/(\gamma + \Gamma) \approx 1.92$ is the optimal damping or (more precisely) extrapolation factor for $\Phi^{\mathrm{SSOR}}_{\omega=1.82}$ in the Poisson-model case with $h = 1/32$.

Table 9.2.2 Gradient method for SSOR iteration, $h = 1/32$

m	value in the middle	$\frac{\lVert e^m \rVert_A}{\lVert e^{m-1} \rVert_A}$
1	0.2851075107	0.457624
2	0.9245177570	0.519182
3	0.1780816984	0.588553
4	0.2274720552	0.645408
5	0.2956906889	0.685785
10	0.4381492069	0.757746
20	0.4954559469	0.767196
30	0.4996724015	0.768216
40	0.4999764630	0.768512
50	0.4999983084	0.768691
60	0.4999998782	0.768827
70	0.4999999912	0.768935

9.3 The Method of the Conjugate Directions

9.3.1 Optimality with Respect to a Direction

The slowness of the gradient method was demonstrated in Theorem 2.3 by means of the two-dimensional subspace spanned by the two extreme eigenvectors. Therefore, a system of two equations is able illustrate this situation. The matrix $A = \operatorname{diag}\{\lambda_1, \lambda_2\}$ with $0 < \lambda_1 \leqslant \lambda_2$ has the condition $\operatorname{cond}_2(A) = \lambda_2/\lambda_1$. The corresponding function F from (1.3) leads to ellipses as level curves $N_c := \{x \in \mathbb{R}^2 : F(x) = c\}$, where $c \in \mathbb{R}$. In the two-dimensional case, the gradient method can be illustrated graphically as follows: The point $x^m [x^{m+1}]$ lies on the ellipse $N^{(m)} := N_c$ with $c = F(x^m)$ [or $N^{(m+1)} := N_c$ with $c = F(x^{m+1})$, respectively]. The straight line $x^m x^{m+1}$ is vertical to $N^{(m)}$ and tangential to $N^{(m+1)}$. Therefore, succeeding straight lines (i.e., the corrections $x^{m+1} - x^m$) form right angles. Fig. 1 shows the case of an elongated ellipse, where the iteration path forms a zigzag line. This illustrates that the approximation to the centre requires many iteration steps. Note that the ellipses are more elongated the larger the condition is. In the case of a circle ($\lambda_1 = \lambda_2$), the first correction would yield the exact solution x^*.

From the fact that the corrections $x^{m+3} - x^{m+2}$ and $x^{m+1} - x^m$ are parallel, one understands that the iterate x^{m+2} must be corrected in exactly the same direction in which x^m had been corrected previously. Hence, x^{m+2} has lost the property of x^{m+1} being optimal with respect to the direction $x^{m+1} - x^m$. We define:

$$x \text{ is } \textit{optimal with respect to a direction } p \neq 0 \text{ if} \tag{9.3.1a}$$

$$F(x) \leqslant F(x + \lambda p) \quad \text{for all } \lambda \in \mathbb{K}. \tag{9.3.1b}$$

Lemma 9.3.1. *The optimality of x with respect to p is equivalent to*

$$p \perp r := b - Ax. \tag{9.3.1c}$$

Proof. A necessary condition for $f(\lambda) = F(x + \lambda p)$ from (1.8c) to be minimal at $\lambda = 0$ is $\langle Ax - b, p \rangle = -\langle r, p \rangle = 0$. As (1.8c) is restricted to $\mathbb{K} = \mathbb{R}$, use (1.12c) for the complex case $\mathbb{K} = \mathbb{C}$. ☐

Exercise 9.3.2. x is called *optimal with respect to a subspace U* if $F(x) \leqslant F(x + \xi)$ for all $\xi \in U$. Prove that x is optimal with respect to U if

$$r = b - Ax \perp U. \tag{9.3.1d}$$

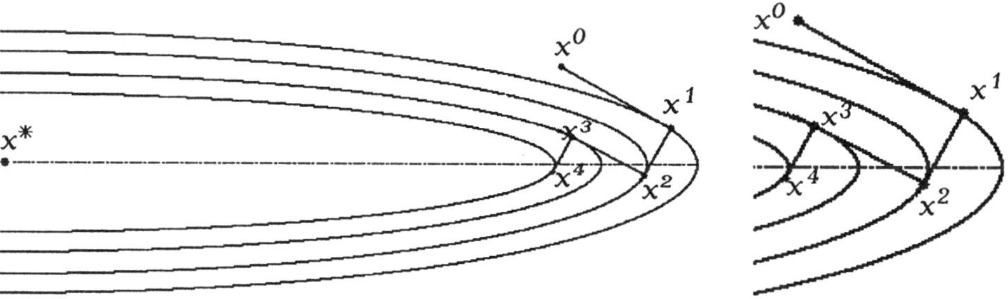

Fig. 9.3.1 The iterates x^m on the level curves of F

Remark 9.3.3. The iterates x^m of the gradient method satisfy (2a, b):

$$x^{m+1} \ (m \geqslant 0) \text{ is optimal with respect to } r^m = b - Ax^m, \tag{9.3.2a}$$

$$r^{m+1} \perp r^m. \tag{9.3.2b}$$

Proof. By Lemma 1, (2a) and (2b) are equivalent. $r^m \perp r^{m+1} = r^m - \lambda_{\text{opt}}(x^m, r^m)Ar^m$ follows from definition (1.9c, d) of λ_{opt}. □

The dilemma of the gradient method can be stated in the following form: The relation $r^{m+1} \perp r^m$ is not transitive, i.e., $r^m \perp r^{m+1}$ and $r^{m+1} \perp r^{m+2}$ do *not* imply $r^m \perp r^{m+2}$. Therefore, in general, x^{m+2} has lost its optimality with respect to r^m.

9.3.2 Conjugate Directions

The change of x into $x' := x + q$ $(q \neq 0)$ transforms the residual $r = b - Ax$ of x into the residual

$$r' = b - Ax' = b - A(x + q) = b - Ax - Aq = r - Aq \tag{9.3.3a}$$

of x'. Let x be optimal with respect to the direction p:

$$r \perp p. \tag{9.3.3b}$$

The new value x' remains optimal with respect to p if and only if $r' \perp p$, i.e., $Aq \perp p$, because the latter property is equivalent to $-\langle Aq, p\rangle = \langle r - Aq, p\rangle = \langle r', p\rangle = 0$. This proves

Lemma 9.3.4. *The optimality of x with respect to $p \neq 0$ implies the optimality of $x' = x + q$ with respect to the same $p \neq 0$ if and only if*

$$Aq \perp p. \tag{9.3.3c}$$

Vectors p, q with the property (3c) are called *conjugate*. The term «conjugate» can also be replaced with «*A-orthogonal*», abbreviated as

$$q \perp_A p, \tag{9.3.3c'}$$

where $\perp_A$ denotes orthogonality with respect to the scalar product $\langle \cdot, \cdot \rangle_A$ from (1.11a).

Condition (3c) leads us to the following *method of conjugate directions*:

start: x^0 arbitrary, $r^0 := b - Ax^0$,

Loop for $m = 0, 1, \ldots, n-1$: $(n := \#I)$

Choose a direction $p^m \neq 0$ that is conjugate to all preceding directions p^ℓ $(\ell < m)$, (9.3.4a)

$$x^{m+1} := x^m + \lambda_{\text{opt}}(x^m, p^m)p^m \quad \text{with} \tag{9.3.4b}$$

$$\lambda_{\text{opt}}(x^m, p^m) := \langle r^m, p^m\rangle / \langle Ap^m, p^m\rangle, \tag{9.3.4c}$$

$$r^{m+1} := r^m - \lambda_{\text{opt}}(x^m, p^m)Ap^m. \tag{9.3.4d}$$

The step (4b, c) shows that x^{m+1} is optimal with respect to the direction p^m: $F(x^{m+1}) = \min\{F(x^m + \lambda p^m): \lambda \in \mathbb{K}\}$ or

$$r^{m+1} \perp p^m. \tag{9.3.4e}$$

Definition (4d) is equivalent to

$$r^{m+1} := b - Ax^{m+1}. \tag{9.3.4d'}$$

The properties of this method are collected in

Theorem 9.3.5. (a) *The directions $\{p^m: 0 \leqslant m \leqslant n-1\}$ form the basis for pairwise conjugate vectors, i.e., an A-orthogonal basis.*
(b) *The algorithm ends at $m = n-1$ with the exact solution $x^{m+1} = x^n = x^*$.*
(c) *The iterate x^m is optimal with respect to all directions $p^0, p^1, \ldots, p^{m-1}$, i.e., it is optimal with respect to the subspace U_{m-1}, where*

$$U_\ell := \operatorname{span}\{p^0, \ldots, p^\ell\}. \tag{9.3.5a}$$

The residuals r^m satisfy

$$r^m \perp p^\ell \qquad (0 \leqslant \ell \leqslant m-1), \tag{9.3.5b}$$

$$r^m \perp U_\ell \qquad (0 \leqslant \ell \leqslant m-1). \tag{9.3.5c}$$

(d) *The error $e^m = x^m - x^*$ fulfils the conditions*

$$e^m \perp_A p^\ell \ (0 \leqslant \ell \leqslant m-1), \text{ i.e., } e^m \text{ is conjugate to } p^\ell. \tag{9.3.5d}$$

(e) *x^m is the solution of the minimisation problem*

$$F(x^m) = \min\left\{F(\xi): \xi = x^0 + \sum_{\ell=0}^{m-1} \lambda_\ell p^\ell, \lambda_\ell \in \mathbb{K}\right\} = \min_{\xi - x^0 \in U_\ell} F(\xi), \tag{9.3.5e}$$

where the minimum in (5e) is attained for $\lambda_\ell = \lambda_{\text{opt}}(x^\ell, p^\ell)$.

Proof of (a). First, we state that division by $\langle Ap^m, p^m\rangle$ in (4c) is well-defined because of $p^m \neq 0$, as long as an additional conjugate direction exists, i.e., as long as $m < n$. As soon as $m = n-1$ is reached, $p^0, \ldots, p^{n-1}$ span the whole space $\mathbb{K}^I$ and the process cannot be continued.
(c) The statement (5b) is true for $m = 0$, since $\{\ell: 0 \leqslant \ell \leqslant m-1\}$ is the empty set. Suppose that (5b) holds for m. By Lemma 1, x^m is optimal with respect to all directions p^ℓ $(0 \leqslant \ell \leqslant m-1)$. According to Lemma 4, this property is transmitted to x^{m+1} because of $p^m \perp_A p^\ell$ $(0 \leqslant \ell \leqslant m-1)$; hence, $r^{m+1} \perp p^\ell$ for all $0 \leqslant \ell \leqslant m-1$. The missing condition $r^{m+1} \perp p^m$ follows from (4e).
(d) (5d) follows from (5b), as $Ae^m = A(x^m - x^*) = Ax^m - b = -r^m$.
(b) (5c) proves that $r^n \perp U_{n-1}$. Since $U_{n-1} = \mathbb{K}^I$ (cf. (a)), $r^n = 0$ follows, i.e., $x^n = x^*$.
(e) Inserting Eqs. (4b) one into another, one obtains

$$x^m = x^0 + \sum_{\ell=0}^{m-1} a_\ell p^\ell \quad \text{with } a_\ell = \lambda_{\text{opt}}(x^\ell, p^\ell). \tag{9.3.5f}$$

From (1.12c) with $\tilde{x} := x^m$, $x := \xi$ and from $r^m \perp U_{m-1}$, one deduces that $F(\xi) - F(x^m) = \mathrm{Re}\langle r^m, \sum(\lambda_\ell - a_\ell)p^\ell\rangle + \frac{1}{2}\langle A(\xi - x^m, \xi - x^m\rangle = \frac{1}{2}\|\xi - x^m\|_A^2 \geqslant 0$ with an equality sign only for $\xi = x^m$, i.e., for $\lambda_\ell = a_\ell$. This proves (5e). □

The method of conjugate directions is not interesting in practice, unless directions p^m in (4a) are suitably selected. If, for instance, one chooses a *fixed* conjugate system $\{p^0, \ldots, p^{n-1}\}$, the starting value $x^0 := x^* - p^{n-1}$ with the residual $r^0 = Ap^{n-1}$ leads to a sequence $x^0 = x^1 = \cdots = x^{n-1}$ that only in the last step changes to the exact solution $x^n = x^*$. This explains, why, in general, no convergence estimate as in (2.4b) can be given.

9.4 Conjugate Gradient Method (cg Method)

9.4.1 First Formulation

In the following, the gradient method and method of conjugate directions will be combined. In order not to lose optimality with respect to the previous search directions, we permit only conjugate directions. The residuals (negative gradients) are used to determine the search direction p^m in (3.4a).

Having already constructed p^0, p^1, ..., p^{m-1} (all $\neq 0$), one can orthogonalise r^m (with respect to the scalar product $\langle \cdot, \cdot \rangle_A$, Remark 2.7.3):

$$p^m := r^m - \sum_{\ell=0}^{m-1} \frac{\langle Ar^m, p^\ell\rangle}{\langle Ap^\ell, p^\ell\rangle} p^\ell. \tag{9.4.1a}$$

Because of the empty sum for $m = 0$, the construction starts with

$$p^0 := r^0. \tag{9.4.1b}$$

Remark 9.4.1. (a) p^m from (1a) is conjugate to all p^ℓ, $0 \leqslant \ell \leqslant m - 1$.
(b) The directions span the subspace

$$U_m := \mathrm{span}\{p^0, \ldots, p^m\} = \mathrm{span}\{p^0, p^1, \ldots, p^{m-1}, r^m\}. \tag{9.4.2a}$$

(c) Having constructed x^m and its residual r^m by means of the method of conjugate directions, then r^m and p^m can vanish only simultaneously. This means that either $x^m = x^*$ is the exact solution or $p^m \neq 0$ holds.
(d) The residual is orthogonal to the preceding subspaces:

$$r^m \perp U_\ell \quad \text{for } \ell < m. \tag{9.4.2b}$$

Proof of (a). By construction (1a), $\langle Ap^m, p^j\rangle = 0$ holds for $j < m$.
(b) The assertion follows directly from (1a).
(d) Repetition of (3.5c) from Theorem 3.5.
(c) By (1a), $p^m = 0$ follows from $r^m = 0$. Assume the case $p^m = 0$. (1a) shows that $r^m \in U_{m-1}$. On the other hand, $r^m \perp U_{m-1}$ holds (cf. (2b)). Both together imply that $r^m = 0$. □

A first provisional representation of the *conjugate gradient method* reads as follows:

$$x^0 \text{ arbitrary}, \quad r^0 := b - Ax^0, \tag{9.4.3a}$$

loop for $m = 0, 1, \ldots, n-1$: $\quad (n := \#I)$

$$\begin{array}{l}\text{stop if } r^m = 0;\\ \text{otherwise, compute } p^m \text{ from } r^m \text{ according to (1a, b).}\end{array} \tag{9.4.3b}$$

$$x^{m+1} := x^m + \lambda_{\text{opt}}(x^m, p^m)p^m \quad \text{with } \lambda_{\text{opt}} \text{ from (3.4c)}, \tag{9.4.3c}$$

$$r^{m+1} := r^m - \lambda_{\text{opt}}(x^m, p^m)Ap^m. \tag{9.4.3d}$$

The properties of this method are summarised in

Theorem 9.4.2. (a) *The loop* (3b–d) *terminates at least for* $m = n$ *with* $x^m = x^*$. *In the following, let* $m_0 \leqslant n$ *be the first index with* $x^{m_0} = x^*$.
(b) *The iterates* x^m $(0 \leqslant m \leqslant m_0)$ *can be characterised by each of the following minimisation problems*:

$$F(x^m) = \min\left\{F\left(x^0 + \sum_{\ell=0}^{m-1} \lambda_\ell p^\ell\right): \lambda_0, \ldots, \lambda_{m-1} \in \mathbb{K}\right\}, \tag{9.4.4a}$$

$$F(x^m) = \min\left\{F\left(x^0 + \sum_{\ell=0}^{m-1} \mu_\ell r^\ell\right): \mu_0, \ldots, \mu_{m-1} \in \mathbb{K}\right\}, \tag{9.4.4b}$$

$$F(x^m) = \min\{F(x^0 + p_{m-1}(A)r^0): p_{m-1} \text{ polynomial of degree } \leqslant m-1\}. \tag{9.4.4c}$$

(c) *The following subspaces coincide for all* $0 \leqslant m \leqslant m_0$:

$$\begin{aligned} U_m &:= \operatorname{span}\{p^0, \ldots, p^m\} \\ &= \operatorname{span}\{r^0, \ldots, r^m\} \\ &= \operatorname{span}\{r^0, Ar^0, \ldots, A^m r^0\}. \end{aligned} \tag{9.4.4d}$$

Before we prove the theorem, the next exercise shows us how to deal with the subspaces that appear.

Exercise 9.4.3. Let $U = \operatorname{span}\{u_1, \ldots, u_m\}$ be a subspace of $\mathbb{K}^I$.
(a) Prove that $\operatorname{span}\{U, x\} = \operatorname{span}\{U, y\}$, provided that $x - y \in U$.
(b) Let $A \in \mathbb{K}^{I \times I}$ be any matrix. AU abbreviates the subspace $\{Ax: x \in U\}$. Prove that $AU = \operatorname{span}\{Au_1, \ldots, Au_m\}$.

Proof of Theorem 2. (a) When $r^m = 0$ with $m < n$ occurs in (3b) for the first time, we have $m_0 := m$ and $x^m = x^*$. Otherwise, $r^m \neq 0$ holds and, by Remark 1c, $p^m \neq 0$ must also hold for all $m = 0, \ldots, n-1$. Therefore, Theorem 3.5b yields $x^n = x^*$; hence, $m = n$.

(c) (4d) is proved by induction on m. For $m = 0$, (4d) holds because of (1b): $p^0 = r^0$. Let (4d) be valid for $m - 1$. Remark 1b proves $U_m = \operatorname{span}\{p^0, \ldots, p^{m-1}, p^m\} = \operatorname{span}\{p^0, \ldots, p^{m-1}, r^m\} = \operatorname{span}\{r^0, \ldots, r^{m-1}, r^m\}$ (cf. Exercise 3a). By $r^{m-1} \in \operatorname{span}\{r^0, \ldots, r^{m-1}\} = \operatorname{span}\{r^0, \ldots, A^{m-1}r^0\}$ and

$$Ap^{m-1} \in AU_{m-1} = A\operatorname{span}\{r^0, \ldots, A^{m-1}r^0\} = \operatorname{span}\{Ar^0, \ldots, A^m r^0\}, \quad (9.4.4e)$$

one concludes from the representation $r^m = r^{m-1} - \lambda A p^{m-1}$ that $r^m \in \operatorname{span}\{r^0, \ldots, A^m r^0\}$. Together with the induction hypothesis, we obtain

$$U_m = \operatorname{span}\{r^0, \ldots, r^{m-1}, r^m\} \subset \operatorname{span}\{r^0, \ldots, A^{m-1}r^0, A^m r^0\}.$$

Since $\dim U_m = m + 1$, the subspaces must be equal. This proves (4d).

(b) Because of (4d), all minimisation problems (4a–c) are of the form

$$F(x^m) = \min\{F(x^0 + \xi) \colon \xi \in U_{m-1}\}. \quad (9.4.4f) \quad \square$$

Corollary 9.4.4. *The statements* (4a–c) *can also be expressed by means of the energy norm* $\|\cdot\|_A$:

$$\|e^m\|_A = \|x^m - x^*\|_A$$

$$= \min\left\{\left\| e^0 + \sum_{\ell=0}^{m-1} \lambda_\ell p^\ell \right\|_A : \lambda_0, \ldots, \lambda_{m-1} \in \mathbb{K}\right\}, \quad (9.4.4a')$$

$$\|e^m\|_A = \|x^m - x^*\|_A$$

$$= \min\left\{\left\| e^0 + \sum_{\ell=0}^{m-1} \mu_\ell r^\ell \right\|_A : \mu_0, \ldots, \mu_{m-1} \in \mathbb{K}\right\}, \quad (9.4.4b')$$

$$\|e^m\|_A = \min\{\|e^0 + p_{m-1}(A)r^0\|_A \colon p_{m-1} \text{ polynomial of degree } \leqslant m - 1\}. \quad (9.4.4c')$$

The proposed algorithm (3a–d) can be simplified significantly in step (3b). The computation of most of the scalar products $\langle Ar^m, p^\ell \rangle$ in (1a) can be avoided.

Lemma 9.4.5. $\langle Ar^m, p^\ell \rangle = 0$ *holds for all* $0 \leqslant \ell \leqslant m - 2$, $m \leqslant m_0$.

Proof. We have $\langle Ar^m, p^\ell \rangle = \langle r^m, Ap^\ell \rangle$ and (4e) shows $Ap^\ell \in U_{m-1}$. Therefore, the assertion follows from (2b): $r^m \perp U_{m-1}$. $\square$

Only the term for $\ell = m - 1$ remains in the sum (1a):

$$p^m := r^m - \frac{\langle Ar^m, p^{m-1} \rangle}{\langle Ap^{m-1}, p^{m-1} \rangle} p^{m-1} = r^m - \frac{\langle r^m, Ap^{m-1} \rangle}{\langle Ap^{m-1}, p^{m-1} \rangle} p^{m-1}. \quad (9.4.5)$$

The second representation in (5) has the advantage that only the product Ap^{m-1} that has already appeared [in the denominator, in λ_{opt} (cf. (3.4c)), and in (3d)] is needed.

9.4.2 cg Method (Applied to the Richardson Iteration)

Using (5), we present the cg method (3a–d) in the following form:

$$x^0 \text{ arbitrary}, \quad r^0 := b - Ax^0, \quad p^0 := r^0. \tag{9.4.6a}$$

For $m = 0, 1, \ldots, n-1$: stop, if $r^m = 0$, otherwise:

$$x^{m+1} := x^m + \lambda_{\text{opt}}(x^m, p^m)p^m \quad \text{with} \tag{9.4.6b}$$

$$\lambda_{\text{opt}}(x^m, p^m) := \langle r^m, p^m \rangle / \langle Ap^m, p^m \rangle, \tag{9.4.6c}$$

$$r^{m+1} := r^m - \lambda_{\text{opt}}(x^m, p^m)Ap^m, \tag{9.4.6d}$$

$$p^{m+1} := r^{m+1} - \frac{\langle r^{m+1}, Ap^m \rangle}{\langle Ap^m, p^m \rangle} p^m. \tag{9.4.6e}$$

Exercise 9.4.6. The following alternatives are equivalent to (6c, e):

$$\lambda_{\text{opt}}(x^m, p^m) := \|r^m\|_2^2 / \langle Ap^m, p^m \rangle, \tag{9.4.6c$'$}$$

$$p^{m+1} := r^{m+1} + \frac{\|r^{m+1}\|_2^2}{\|r^m\|_2^2} p^m. \tag{9.4.6e$'$}$$

Remark 9.4.7. One cg step $x^m \mapsto x^{m+1}$ requires one multiplication Ap^m and, in addition, only simple vector operations and scalar products. On the other hand, the storage requirement is higher: Besides x^m, r^m and p^m are also needed.

The cg method was first presented in 1952 by Stiefel [1] in a paper still worth reading. Independently, the method was described in the same year by Hestenes (cf. Hestenes [1], Hestenes–Stiefel [1]).

The cg method can be interpreted in two completely different directions:

* as a *direct* method,

* as an *iterative* method.

Formally, the cg algorithm is a direct method, because it produces the exact solution x^* after finitely many operations (at the least after n steps). For the practical performance, this is not true. Since the later and smaller residuals r^m arise from linear combinations of larger quantities, cancellation leads to error amplification, so that the vectors $\{p^0, \ldots, p^{n-1}\}$ do not form a conjugate system. This difficulty suggests a modification toward a true *cg iteration*: Every n steps (i.e., for $m = 0, n, 2n, \ldots$), one starts with the descent direction $p^n := r^n$. The arising method (here denoted as the «cyclic cg-iteration») is usually called the «*restarted cg method*».

$$\begin{aligned}&x^0\text{: arbitrary starting iterate,}\\&\textit{cyclic cg iteration:}\\&x^m\text{: as in (6b, d)}\\&r^m, p^m\text{: as in (6c–e), if } m \text{ is not a multiple of } n = \#I,\\&r^m := p^m := b - Ax^m, \text{ if } m = 0, n, 2n, \ldots\end{aligned} \tag{9.4.7}$$

We dispense with a discussion of the rounding error influence by reasons connected to the following interpretation of the cg algorithm as an iteration.

Apart from the fact that the cg method is not of the form $x^m \mapsto x^{m+1} = \Phi(x^m, b)$ (the result also depends on p^m), even in a wider sense an iteration must be an *infinite* process. If only finitely many iterates $x^0, \ldots, x^m$ can be constructed, the terms «convergence» and «asymptotical convergence rate» lose their meaning, because no limit can be formed. Nevertheless, it still makes sense to view the cg method (without the cyclic extension (7)) as a (semi-)iteration. The reason is that the cg method is of practical interest only if a sufficiently accurate iterate x^m can already be achieved for indices m that are significantly smaller than the dimension n. In the examples from §9.4.6, according to Table 1 [or Table 2], satisfactory accuracy is attained after 50 to 100 [$m = 10$] steps, whereas the dimension equals $n = 31^2 = 961$. The propagation of the cg method as an iterative method is due to Reid [1].

9.4.3 Convergence Analysis

The convergence analysis is based on the following observation corresponding to Remark 2.2 in the case of the gradient method.

Remark 9.4.8. Let $x^0, \ldots, x^m$ be the sequence of the cg iterates.
(a) For all $0 \leqslant k \leqslant m$, there are polynomials P_k of degree $\leqslant k$ (depending on the starting iterate x^0) with $P_k(1) = 1$ such that the *semi-iterative Richardson iteration* (7.2.1a, b) corresponding to this sequence of polynomials results in the same iterates $x^0, \ldots, x^m$. In particular, the following error representation holds:

$$e^k = x^k - x^* = P_k(I - A)e^0 = P_k(M_1^{\text{Rich}})e^0 \qquad (M_1^{\text{Rich}} = I - A). \tag{9.4.8a}$$

(b) The polynomials P_k and $Q_{k-1}(\xi) := [1 - P_k(1 - \xi)]/\xi$ are the respective solutions of the minimisation problems

$$\|e^k\|_A = \|P_k(I - A)e^0\|_A \leqslant \|\tilde{P}_k(I - A)e^0\|_A$$
$$\text{for all polynomials } \tilde{P}_k \text{ of degree} \leqslant k \text{ with } \tilde{P}_k(1) = 1, \tag{9.4.8b}$$

$$\|e^k\|_A = \|e^0 + Q_{k-1}(A)r^0\|_A \leqslant \|e^0 + \tilde{Q}_{k-1}(A)r^0\|_A$$
$$\text{for all polynomials } \tilde{Q}_{k-1} \text{ of degree} \leqslant k - 1. \tag{9.4.8c}$$

Proof. (a) (6b) shows that $x^k = x^0 + \sum_{\nu=0}^{k-1} \beta_\nu p^\nu$ with $\beta_\nu := \lambda_{\text{opt}}(x^\nu, p^\nu)$, i.e.,

$$x^k - x^0 \in \operatorname{span}\{p^0, \ldots, p^{k-1}\} = \operatorname{span}\{r^0, \ldots, A^{k-1}r^0\}$$

(cf. (4d)). Using $A(x^k - x^0) = (b - Ax^0) - (b - Ax^k) = r^0 - r^k$, one finds

$$r^k = r^0 + \sum_{\nu=1}^{k} \alpha_\nu A^\nu r^0, \quad \text{i.e., } r^k = R_k(A) r^0$$

with a polynomial $R_k(\xi) = \sum \alpha_\nu \xi^\nu$ of degree $\leqslant k$ with $\alpha_0 = 1$ (hence, $R_k(0) = 1$). Define $P_k(\xi) := R_k(1 - \xi)$. The new polynomial has the properties

$$P_k(1) = 1, \qquad \text{degree } P_k \leqslant k, \qquad P_k(I - A) = R_k(A), \tag{9.4.8d}$$

$$r^k = P_k(I - A) r^0. \tag{9.4.8e}$$

Using $e^k = -A^{-1}r^k$, $e^0 = -A^{-1}r^0$ and $A^{-1}P_k(I - A)A = P_k(I - A)$, we deduce Eq. (8a) from (8e).

(b) By construction, Q_{k-1} satisfies $P_k(I - A) = I - Q_{k-1}(A)A$, and hence, $e^k = P_k(I - A)e^0 = e^0 - Q_{k-1}(A)Ae^0$. $Ae^0 = -r^0$ shows the equality sign in the first part of (8c). The inequality in the second part of (8c) coincides with the characterisation (4c′) in Corollary 4. Since each polynomial $\tilde{P}_k$ of degree $\leqslant k$ satisfying $\tilde{P}_k(1) = 1$ is associated with a polynomial $\tilde{Q}_{k-1}(\xi) := [1 - \tilde{P}_k(1 - \xi)]/\xi$ of degree $\leqslant k - 1$ and $\tilde{P}_k(I - A)e^0 = e^0 + \tilde{Q}_{k-1}(A)r^0$ holds, (8a) and (8b) are equivalent. □

Remark 9.4.9. The cg iterates x^m are not the solutions of the minimisation problem posed in §7.3.1, because there minimisation is required with respect to the Euclidean norm $\|\cdot\|_2$. However, if $\|\cdot\|_2$ is replaced by $\|\cdot\|_A$, the cg method offers the possibility of solving the thus modified minimisation problem (7.3.1) without knowledge of the initial error e^0 and the spectrum of $M_1^{\text{Rich}} = I - A$ (equivalently, of the spectrum of A).

Remark 9.4.10. For any polynomial P_m of degree $\leqslant m$ with $P_m(1) = 1$ the errors $e^m = x^m - x^*$ of the cg iterates satisfy the error estimate

$$\|e^m\|_A \leqslant \max\{|P_m(1 - \lambda)| : \lambda \in \sigma(A)\} \, \|e^0\|_A. \tag{9.4.9}$$

Proof. (8b) shows that $\|e^m\|_A \leqslant \|P_m(I - A)\|_A \|e^0\|_A$. The matrix norm $\|\cdot\|_A$ has the representation $\|X\|_A = \|A^{1/2}XA^{-1/2}\|_2$ (cf. (2.6.10)). $A^{1/2}$ commutes with polynomials in A: $A^{1/2}P_m(I - A)A^{-1/2} = P_m(I - A)$. The assertion (9) follows from $\|P_m(I - A)\|_2 = \max\{|P_m(1 - \lambda)| : \lambda \in \sigma(A)\}$. □

Exercise 9.4.11. Prove by means of (9) that $x^m = x^*$ holds for some for $m \leqslant$ degree of the minimum function of A.

The following theorem shows that, as in the case of the Chebyshev method, an *order improvement* can be achieved.

Theorem 9.4.12. *Let A be positive definite with $\lambda := \lambda_{\min}(A)$, $\Lambda := \lambda_{\max}(A)$ and abbreviate the condition number by $\varkappa = \varkappa(A) = \Lambda/\lambda$. The errors e^m of the cg iterates x^m satisfy the estimate*

$$\|e^m\|_A \leqslant \frac{2(1 - 1/\varkappa)^m}{(1 + 1/\sqrt{\varkappa})^{2m} + (1 - 1/\sqrt{\varkappa})^{2m}} \|e^0\|_A$$

$$= c^m \frac{2}{1 + c^{2m}} \|e^0\|_A \quad \text{with } c := \frac{\sqrt{\varkappa} - 1}{\sqrt{\varkappa} + 1} = \frac{\sqrt{\Lambda} - \sqrt{\lambda}}{\sqrt{\Lambda} + \sqrt{\lambda}}. \tag{9.4.10}$$

Proof. Let P_m be the transformed Chebyshev polynomial (7.3.12b) belonging to $\sigma_M := [a, b] \supset \sigma(M^{\text{Rich}}) = \sigma(I - A)$ with $a = 1 - \Lambda$, $b = 1 - \lambda$. (9) and (7.3.12c) yield $\|e^m\|_A \leqslant \|e^0\|_A / C_m$. (7.3.14e/13c) prove (10). □

The error estimate (10) is an upper bound that may be too pessimistic. It is based on the Chebyshev polynomial P_m that is the optimal choice for minimising $\max\{|P_m(\xi)|: \xi \in \sigma_M\}$, but not necessarily for minimising $\max\{|P_m(\xi)|: \xi \in \sigma(M^{\text{Rich}}) = \sigma(I - A)\} = \max\{|P_m(1 - \lambda)|: \lambda \in \sigma(A)\}$. This leads to the following statement.

Remark 9.4.13. Although the asymptotical convergence rate of the gradient method depends exclusively on the condition number $\varkappa(A)$ and therefore extreme eigenvalues, the convergence of the cg method is influenced by the whole spectrum.

A simple example is given as an illustration. Assume that the inclusion $\sigma(M^{\text{Rich}}) \subset [a, b]$ with $a = 1 - \Lambda$, $b = 1 - \lambda$ can be strengthened to $\sigma(M^{\text{Rich}}) \subset \sigma_M := [a, a'] \cup [b', b]$ with $a \leqslant a' < b' \leqslant b$. Then one may find a polynomial P_m for which $\max\{|P_m(1 - \lambda)|: \lambda \in \sigma(A)\}$ is smaller than for the Chebyshev polynomial (cf. §7.3.6). Hence, P_m yields a better estimate than (10). Generally speaking, if the eigenvalues of A are not distributed uniformly over $[\lambda, \Lambda]$ (e.g., if they accumulate in smaller subintervals), the cg method converges better than estimated by (10).

Even if the eigenvalue distribution permits no better polynomial than the Chebyshev polynomial, the ratios $\|e^{m+1}\|_A / \|e^m\|_A$ improve with an increasing iteration number m and becomes smaller than $c \approx 1 - 2/\sqrt{\varkappa}$ from (10). The reason is as follows: In the case of the *gradient* method (9.2.2a–c), the error e^m converges to the subspace $V := \text{span}\{v_1, v_2\}$ spanned by the eigenvectors belonging to $\lambda := \lambda_{\min}(A)$ and $\Lambda := \lambda_{\max}(A)$ (see the proof of Corollary 2.4). For the *cg* case, this behaviour cannot occur: If the *cg* error e^m lies exactly in the subspace V, $2 = \dim V$ steps of the cg methods would suffice to obtain $e^{m+2} = 0$. It can be proved that the cg error moves toward $V^\perp$. A, however, restricted to $V^\perp$ has the spectrum $\sigma(A) \backslash \{\lambda, \Lambda\}$ and the condition number Λ_2/λ_2, where λ_2 is the second smallest and Λ_2 the second largest eigenvalue.

Hence, after a certain number of steps, the error ratios behave more like $c' \approx 1 - 2/\sqrt{\Lambda_2/\lambda_2} < c$. A precise analysis of this superconvergence result is given by van der Sluis–van der Vorst [1]. Compare also Strakoš [1].

9.4.4 cg Method Applied to Symmetric Iterations

Like the gradient method, the method of conjugate gradients can be applied to other symmetric iterations than the Richardson method. This yields the so-called «*preconditioned cg method*» (but notice that the gradients not the cg method are preconditioned). Let Φ be the symmetric iteration

$$x^{m+1} = x^m - W^{-1}(Ax^m - b), \qquad A, W \text{ positive definite.} \tag{9.4.11a}$$

As in (2.6b), we introduce $\breve{A} := W^{-1/2}AW^{-1/2}$ and $\breve{b} := W^{-1/2}b$. Algorithm (11a) is equivalent to the iteration (11b) for solving $\breve{A}\breve{x} = \breve{b}$:

$$\breve{x}^{m+1} = \breve{x}^m - (\breve{A}\breve{x}^m - \breve{b}). \tag{9.4.11b}$$

Applying the cg algorithm (6a–e) to $\breve{A}\breve{x} = \breve{b}$, we obtain

$$\breve{x}^0 = W^{1/2}x^0, \qquad \breve{r}^0 := \breve{b} - \breve{A}\breve{x}^0, \qquad \breve{p}^0 := \breve{r}^0, \tag{9.4.12a}$$

For $m = 0, 1, 2, \ldots$ (as long as $m < n$ and $\breve{r}^m \neq 0$):

$$\breve{x}^{m+1} := \breve{x}^m + \breve{\lambda}_{\text{opt}}(\breve{x}^m, \breve{p}^m)\breve{p}^m \quad \text{with} \tag{9.4.12b}$$

$$\breve{\lambda}_{\text{opt}}(\breve{x}^m, \breve{p}^m) = \langle \breve{r}^m, \breve{p}^m \rangle / \langle \breve{A}\breve{p}^m, \breve{p}^m \rangle, \tag{9.4.12c}$$

$$\breve{r}^{m+1} := \breve{r}^m - \breve{\lambda}_{\text{opt}}(\breve{x}^m, \breve{r}^m)\breve{A}\breve{p}^m \qquad (= \breve{b} - \breve{A}\breve{x}^{m+1}), \tag{9.4.12d}$$

$$\breve{p}^{m+1} := \breve{r}^{m+1} - \langle \breve{r}^{m+1}, \breve{A}\breve{p}^m \rangle / \langle \breve{A}\breve{p}^m, \breve{p}^m \rangle \breve{p}^m. \tag{9.4.12e}$$

Insert $\breve{A} = W^{-1/2}AW^{-1/2}$ and $\breve{b} = W^{-1/2}b$, define x^m and p^m by

$$\breve{x}^m = W^{1/2}x^m, \qquad \breve{p}^m = W^{1/2}p^m, \tag{9.4.12f}$$

and use $W^{-1/2}r^m = W^{-1/2}(b - Ax^m) = \breve{b} - \breve{A}\breve{x}^m = \breve{r}^m$. (12a–e) becomes

$$x^0 \text{ arbitrary}, \qquad r^0 := b - Ax^0, \qquad p^0 := W^{-1}r^0, \tag{9.4.13a}$$

$$x^{m+1} := x^m + \lambda_{\text{opt}}(x^m, p^m)p^m \quad \text{with} \tag{9.4.13b}$$

$$\lambda_{\text{opt}}(x^m, p^m) = \langle r^m, p^m \rangle / \langle Ap^m, p^m \rangle, \tag{9.4.13c}$$

$$r^{m+1} := r^m - \lambda_{\text{opt}}(x^m, p^m)Ap^m, \tag{9.4.13d}$$

$$p^{m+1} := W^{-1}r^{m+1} - \langle W^{-1}r^{m+1}, Ap^m \rangle / \langle Ap^m, p^m \rangle p^m. \tag{9.4.13e}$$

The expression (13c) coincides with the original definition (1.9a) for λ_{opt}. (13e) shows that the search directions p^m are produced from the «preconditioned» gradient $W^{-1}r^m$ by an A orthogonalisation. Exploiting the equivalent formulations (6c′, e′), one ends up with

$$\lambda_{\text{opt}}(x^m, p^m) = \langle W^{-1}r^m, r^m \rangle / \langle Ap^m, p^m \rangle, \tag{9.4.13c′}$$

$$p^{m+1} := W^{-1}r^{m+1} + \langle W^{-1}r^{m+1}, r^{m+1} \rangle / \langle W^{-1}r^m, r^m \rangle p^m. \tag{9.4.13e′}$$

If one carries along the variables x^m, p^m, r^m, and $\varrho_m := \langle W^{-1}r^m, r^m\rangle$ during the iteration, the cg algorithm takes the form (14a–e):

$$x^0 \text{ arbitrary}, \quad r^0 := b - Ax^0, \quad p^0 := W^{-1}r^0,$$

$$\varrho_0 := \langle p^0, r^0\rangle, \tag{9.4.14a}$$

iteration: For $m = 0, 1, \ldots$ (as long as $m < n$ and $r^m \neq 0$):

$$a^m := Ap^m, \qquad \lambda_{\text{opt}} := \varrho_m/\langle a^m, p^m\rangle, \tag{9.4.14b}$$

$$x^{m+1} := x^m + \lambda_{\text{opt}} p^m \tag{9.4.14c}$$

$$r^{m+1} := r^m - \lambda_{\text{opt}} a^m, \tag{9.4.14d}$$

$$q^{m+1} := W^{-1}r^{m+1}, \qquad \varrho_{m+1} := \langle q^{m+1}, r^{m+1}\rangle \tag{9.4.14e}$$

$$p^{m+1} := q^{m+1} + \frac{\varrho_{m+1}}{\varrho_m} p^m. \tag{9.4.14f}$$

The error estimate for $e^m = x^m - x^*$ follows as in §9.2.4, since inequality (10) for $\check{e}^m = \check{x}^m - \check{x}^* = W^{1/2}e^m$ carries over to e^m because of $\|\check{e}^m\|_{\check{A}} = \|e^m\|_A$. Further, notice that $\varkappa = \varkappa(\check{A}) = \varkappa(W^{-1/2}AW^{-1/2}) = \varkappa(W^{-1}A) = \Gamma/\gamma$ with Γ, γ from (15a).

Theorem 9.4.14 (error estimate). *Let Φ be a symmetric iteration. Its matrix W of the third normal form is assumed to satisfy*

$$\gamma W \leqslant A \leqslant \Gamma W \qquad (\gamma > 0 \text{ cf. (2.9a)}). \tag{9.4.15a}$$

Then the iterates x^m of the cg method (14a–f) *applied to Φ fulfil the energy norm estimate*

$$\|e^m\|_A \leqslant \frac{2(1 - 1/\varkappa)^m}{(1 + 1/\sqrt{\varkappa})^{2m} + (1 - 1/\sqrt{\varkappa})^{2m}} \|e^0\|_A$$

$$= c^m \frac{2}{1 + c^{2m}} \|e^0\|_A \quad \textit{with } \varkappa = \frac{\Gamma}{\gamma},\ c := \frac{\sqrt{\varkappa} - 1}{\sqrt{\varkappa} + 1} = \frac{\sqrt{\Gamma} - \sqrt{\gamma}}{\sqrt{\Gamma} + \sqrt{\gamma}}. \tag{9.4.15b}$$

Lemma 9.4.15. (a) *At the least for $m = n$, one obtains the exact solution: $x^m = x^*$. Let m_0 be the first index with $x^{m_0} = x^*$.*
(b) *The search directions generated by* (14a–f) *are conjugate w.r.t. A:*

$$\langle p^k, p^\ell\rangle_A = 0 \quad \textit{for } k \neq \ell. \tag{9.4.16a}$$

(c) *For all $0 \leqslant m \leqslant m_0$ the following subspaces (Krylov spaces) coincide:*

$$U_m := \operatorname{span}\{p^0, \ldots, p^m\} \tag{9.4.16b$_1$}$$

$$= \operatorname{span}\{W^{-1}r^0, \ldots, W^{-1}r^m\} \tag{9.4.16b$_2$}$$

$$= \operatorname{span}\{W^{-1}r^0, \ldots, (W^{-1}A)^m W^{-1}r^0\}. \tag{9.4.16b$_3$}$$

(d) x^m *is the minimising argument of the expressions*

$$F(x^m) = \min\left\{F\left(x^0 + \sum_{\ell=0}^{m-1} \lambda_\ell p^\ell\right): \lambda_0, \dots, \lambda_{m-1} \in \mathbb{K}\right\}, \tag{9.4.16c$_1$}$$

$$F(x^m) = \min\left\{F\left(x^0 + W^{-1}\sum_{\ell=0}^{m-1} \mu_\ell r^\ell\right): \mu_0, \dots \mu_{m-1} \in \mathbb{K}\right\}, \tag{9.4.16c$_2$}$$

$$F(x^m) = \min\{F(x^0 + p_{m-1}(W^{-1}A)W^{-1}r^0): p_{m-1} \textit{ polynomial of degree} \leqslant m-1\}. \tag{9.4.16c$_3$}$$

Proof. Part (a) is identical to Theorem 2a. Part (b) follows from (12f), $\langle \check{p}^k, \check{p}^\ell \rangle_{\check{A}} = \langle \check{A}\check{p}^k, \check{p}^\ell \rangle = \langle W^{-1/2}AW^{-1/2}W^{1/2}p^k, W^{1/2}p^\ell \rangle = \langle Ap^k, p^\ell \rangle = \langle p^k, p^\ell \rangle_A$, and the $\check{A}$ orthogonality of the search directions $\check{p}^k$. (c) and (d) are consequences of (4a–d) applied to the ˇ quantitites from (12f). □

9.4.5 Pascal Procedures

The record parameter `cg` in the variable `IP: iterationparameter` is a pointer to `cg_parameter` containing the components `r` for the residual r^m, `p` for the cg direction p^m, and `rho` for ϱ_m from (14a, e). By `start_cg_method`, the quantities from (14a) are computed. The procedure `cg_method` executes (14a–f) with Φ = `basisiteration`. The choice `Richardson_iteration` corresponds to the simple version (6a–e). For computing the product $W^{-1}y = Ny$ by `N_y` compare §9.2.5. Procedure `cg_method_1` serves to call a single cg step with the parameter `IT`. The required frame program is completely analogous to that from §9.2.5.

```
procedure start_cg_method(var A: data_of_discretisation;
                   var x,b: gridfunction; var IP:
                    iterationparameter;
                   procedure basisiteration(var neu:
                    gridfunction;
                         var A: data_of_discretisation;
                         var x,b: gridfunction; var IP:
                          iterationparameter));
begin with IP do begin Nr:=0; if cg=nil then new(cg) end;
      with IP.cg^ do begin residual(r,A,x,b); N_y(p,r,A,IP,
                              basisiteration);
                             rho:=Eucl_scalarproduct(p,r,A)
end end;

procedure cg_method(var new: gridfunction; var A: data_of_
 discretisation;
```

```
                    var x,b: gridfunction; var IP:
                     iterationparameter;
                    procedure basisiteration(var new:
                     gridfunction;
                         var A: data_of_discretisation;
                         var x,b: gridfunction; var IP:
                          iterationparameter));
var c,q: gridfunction; lambda,rhonew: real;
begin if IP.cg=nil then start_cg_method(A,x,b,IP,
 basisiteration);
    with A do with IP.cg^ do
    begin A_x(c,A,p); lambda:=Eucl_scalarproduct(c,p,A);
        if lambda=0 then message('Cancellation of cg because
         of pAp=0');
        lambda:=rho/lambda;
        vector_plus_factor_x_vector(nx,ny,new,x,
         lambda,p);                                      {(14c)}
        transfer_boundary_values(nx,ny,x,new);
        vector_plus_factor_x_vector(nx,ny,r,r,
         -lambda,c);                                     {(14d)}
        N_y(c,r,A,IP,basisiteration); rhonew:=Eucl_
         scalarproduct(c,r,A);
        if rhonew=0 then writeln('cg reaches exact
         solution!');
        if rho>0 then vector_plus_factor_x_vector(nx,ny,p,c,
         rhonew/rho,p);
        rho:=rhonew
end end;

procedure cg_method_1(var it: data_of_iteration;
procedure basisiteration(var new: gridfunction; var A: data_
                          of_discretisation;
                         var x,b: gridfunction; var IP:
                          iterationparameter));
var number: integer;
begin with it do with IP do begin number:=Nr;
       if Nr=0 then start_cg_method(A,x,b,IP,
        basisiteration);                                 {(14a)}
       cg_mthod(x,A,x,b,IP,basisiteration);
        Nr:=number+1                                    {(14b-f)}
end end;
```

9.4.6 Numerical Examples in the Model Case

As the system of equations we choose the Poisson-model problem with $h = 1/32$. The application of the cg method to the Richardson iteration (i.e., algo-

Table 9.4.1 cg results for the Richardson iteration as basic iteration, model problem for $h = 1/32$

m	value in the middle	$\|e^m\|_A/\|e^{m-1}\|_A$
1	$-1.8656097815_{10}-3$	0.670874
2	$-4.6008798010_{10}-3$	0.791286
3	$-7.3924161408_{10}-3$	0.860663
4	$-1.1116057550_{10}-2$	0.865691
10	$-4.4081878259_{10}-2$	0.917138
20	$-1.1796241337_{10}-1$	0.939358
30	$4.0673579950_{10}-1$	0.918423
40	$4.9137792828_{10}-1$	0.843496
50	$5.0013929834_{10}-1$	0.832459
60	$5.0010381735_{10}-1$	0.738779
70	$5.0001053720_{10}-1$	0.761377
80	$5.0000013936_{10}-1$	0.708295
90	$5.0000000342_{10}-1$	0.661969
100	$5.0000000001_{10}-1$	0.665531

rithm (6a–e)) yields the results given in Table 1. Due to inequality (10), the convergence factors $\|e^m\|_A/\|e^{m-1}\|_A$ measured with respect to the energy norm $\|\cdot\|_A$ should become smaller than

$$c = (\sqrt{\Lambda} - \sqrt{\lambda})/(\sqrt{\Lambda} + \sqrt{\lambda}).$$

Inserting the eigenvalues λ and Λ from (4.1.1.b, c) for $h = 1/32$, one obtains $c = 0.9063471$. For $m \geqslant 30$, the ratios clearly fall short of this value: The convergence factor decreases from 0.9 to 0.66 for $m \geqslant 90$. This «superlinear» convergence behaviour illustrates the improvement of the effective condition number during the iteration as discussed in the last paragraph of §9.4.3.

Table 2 reports the cg results for $h = 1/32$ with the SSOR and ILU iteration as the basic iteration. The (optimal) SSOR parameter is the same as for Table 9.2.2. The ILU iteration is the modified five-point version `ILU_5` with $\omega = -1$ and `diag` $= 5$ (cf. §8.5.11). The condition number of the SSOR method was determined in §9.2.5 to be $\varkappa \approx 7.66$. This results in the value $c \approx 0.47$ for c from (15b). In the SSOR case, the averaged convergence factors $[\|e^m\|_A/\|e^0\|_A]^{1/m}$ are around 0.47 until $m = 11$. Afterwards they decrease to 0.42 for $m \approx 30$. The values $u_{16,16}$ given in Table 2 show that for $m \geqslant 27$, the rounding errors acquire the upper hand. Nevertheless, the cg algorithm is stable.

The superlinear convergence behaviour mentioned in connection with Table 1 should not be overrated. Its advantage can be exploited only if m becomes sufficiently large. In the case of Table 1, this means $m \geqslant 30$; in the SSOR case of Table 2, it is $m \geqslant 17$. An inspection of the values in these tables illustrates the following dilemma:

Table 9.4.2 cg method applied to the LIU and SSOR iterations

	5-point ILU with $\omega = -1$		SSOR with $\omega = 1.8212691200$	
m	$u_{16,16}$	$\|e^m\|_A/\|e^{m-1}\|_A$	$u_{16,16}$	$\|e^m\|_A/\|e^{m-1}\|_A$
1	$2.262513522_{10}-1$	$1.56365_{10}-1$	$2.851075107_{10}-2$	$4.57624_{10}-1$
2	$5.320480495_{10}-1$	$4.46360_{10}-1$	$1.146321025_{10}-1$	$3.07093_{10}-1$
3	$4.582969109_{10}-1$	$4.65620_{10}-1$	$2.093879771_{10}-1$	$5.99140_{10}-1$
4	$4.818928890_{10}-1$	$4.59572_{10}-1$	$3.500438579_{10}-1$	$5.30214_{10}-1$
5	$4.827955876_{10}-1$	$4.90598_{10}-1$	$4.301535841_{10}-1$	$4.91911_{10}-1$
10	$4.999129317_{10}-1$	$3.80570_{10}-1$	$4.992951874_{10}-1$	$4.64830_{10}-1$
11	$5.000044282_{10}-1$	$3.58332_{10}-1$	$4.998541213_{10}-1$	$4.65082_{10}-1$
12	$4.999850353_{10}-1$	$4.29905_{10}-1$	$4.999456258_{10}-1$	$3.94760_{10}-1$
20	$5.000000033_{10}-1$	$3.42381_{10}-1$	$5.000000087_{10}-1$	$3.20139_{10}-1$
21	$5.000000026_{10}-1$	$3.88711_{10}-1$	$5.000000020_{10}-1$	$4.87606_{10}-1$
22	$5.000000008_{10}-1$	$4.05064_{10}-1$	$5.000000055_{10}-1$	$4.05755_{10}-1$
23	$5.000000002_{10}-1$	$3.13452_{10}-1$	$5.000000041_{10}-1$	$4.08013_{10}-1$
24	$5.000000000_{10}-1$	$3.55741_{10}-1$	$5.000000000_{10}-1$	$3.32715_{10}-1$
25	$5.000000000_{10}-1$	$4.51311_{10}-1$	$5.000000005_{10}-1$	$4.32772_{10}-1$
26	$5.000000000_{10}-1$	$5.57156_{10}-1$	$5.000000001_{10}-1$	$3.34264_{10}-1$
27	$5.000000000_{10}-1$	$5.17255_{10}-1$	$5.000000001_{10}-1$	$3.66209_{10}-1$
28	$5.000000000_{10}-1$	$8.02069_{10}-1$	$5.000000000_{10}-1$	$3.65471_{10}-1$
29	$5.000000000_{10}-1$	$9.69482_{10}-1$	$5.000000000_{10}-1$	$4.87797_{10}-1$
30	$5.000000000_{10}-1$	$1.00102_{10}+0$	$5.000000000_{10}-1$	$7.76690_{10}-1$

(i) Either the iteration is fast (as in Table 2). Then one would like to stop the iteration before reaching the critical value of m.

(ii) Or the iteration is slow (as in Table 1). Then one would prefer to reject this iteration.

9.4.7 Amount of Work of the cg Method

One iteration step (14b–f) requires on evaluation of $p \mapsto Ap$ and $r \mapsto W^{-1}r$, three vector additions, three multiplications of a vector by a scalar number, and two scalar products. This adds up to

$$\textit{cg amount of work } (\Phi) = C(A) + C(W) + 8n \tag{9.4.17a}$$

operations for the cg method applied to Φ, where

$$C(A)\text{: work for } p \mapsto Ap, \qquad C(W)\text{: work for } r \mapsto W^{-1}r. \tag{9.4.17b}$$

Performing the Φ-iteration step in the form $\Phi(x, b) = x - W^{-1}(Ax - b)$, we need $C(A) + C(W) + 2n$ operations:

$$\textit{cg amount of work } (\Phi) = \textit{amount of work } (\Phi) + 6n. \tag{9.4.17c}$$

Hence, as in the semi-iterative case (cf. §7.3.11), the cost factor equals

$$C_{\Phi,\text{cg method}} = C_\Phi + 6/C_A. \tag{9.4.17d}$$

According to the discussion of convergence behaviour from above, we choose $c = (\sqrt{\Gamma} - \sqrt{\gamma})/(\sqrt{\Gamma} + \sqrt{\gamma})$ from (15b) as the asymptotical rate on which we base the *effective amount of work*:

$$Eff_{\text{cg}}(\Phi) = -(C_\Phi + 6/C_A)\log(\sqrt{\Gamma} - \sqrt{\gamma})/(\sqrt{\Gamma} + \sqrt{\gamma}). \tag{9.4.17e}$$

Remark 9.4.16. Even if these numbers coincide exactly with those obtained in §7.3.11 for the Chebyshev method, one has to emphasize one important advantage of the cg method: The eigenvalue bounds γ and Γ may be unknown to the user. Vice versa, the effectivity of the Chebyshev method deteriorates if too pessimistic γ, Γ bounds are inserted.

9.4.8 Suitability for Secondary Iterations

In §8.4 we discussed composed iterations arising from $x \mapsto x - B^{-1}(Ax - b)$ by replacing the exact solution of $B\delta = c$ by the approximation by means of a secondary iteration. Now we can start with $\delta^0 = 0$ and perform m steps of the cg algorithm. Positive and negative comments concerning this approach are given in

Lemma 9.4.17. *Let A, B be positive definite. $\Phi_A(x,b) = x - B^{-1}(Ax - b)$ is the primary iteration. For solving $B\delta = c$, the cg method based on an iteration $\Phi_B(\delta, c) = c - C^{-1}(B\delta - c)$ with a starting iterate $\delta^0 = 0$ is inserted as the secondary solver. The number k of cg steps is chosen such that $2c^k \leqslant \varepsilon$ holds with $c = (\sqrt{\Delta} - \sqrt{\delta})/(\sqrt{\Delta} + \sqrt{\delta})$, $0 < \delta C \leqslant B \leqslant \Delta C$. The composed iteration Φ_k is no longer linear, but it still can be written in the form*

$$\Phi_k(x,b) = M_k(Ax - b)x + N_k(Ax - b)b \tag{9.4.18a}$$

and has the contraction number (18b) *with respect to the energy norm*:

$$\|M_k(Ax - b)\|_A \leqslant \|M_A\|_A + \varepsilon\|A^{1/2}B^{-1}A^{1/2}\|_2 \qquad (M_A = I - B^{-1}A). \tag{9.4.18b}$$

Before proving the lemma, we comment on (18b): If, as in §8.4.1, B is a preconditioner with $\varkappa(B^{-1}A) = \|A^{1/2}B^{-1}A^{1/2}\|_2 = O(1)$, the right-hand side in (18b) is bounded by $\|M_A\|_A + C\varepsilon$. For example, one should choose ε such that $\|M_A\|_A + C\varepsilon \leqslant \frac{1}{2}(1 + \|M_A\|_A) < 1$.

Proof of the Lemma. The right-hand side c in $B\delta = c$ is the defect $c = Ax^m - b$ (cf. (8.4.4a)). Because of $\delta^0 = 0$, the error estimate (15b) yields the B-energy norm $\|\delta^k - \delta\|_B \leqslant \varepsilon\|\delta^0 - \delta\|_B = \varepsilon\|\delta\|_B$, $\delta := B^{-1}c$. From

$$\|\delta\|_B - B^{1/2}\delta\|_2 = \|B^{-1/2}A(x^m - x^*)\|_2 \leqslant \|B^{-1/2}AB^{-1/2}\|_2\|x^m - x^*\|_B,$$

one deduces

$$\begin{aligned}\|x^{m+1} - x^*\|_B &= \|x^m - \delta^k - x^*\|_B \\ &\leqslant \|x^m - \delta - x^*\|_B + \|\delta^k - \delta\|_B \\ &= \|\Phi_A(x^m, b) - x^*\|_B + \|\delta^k - \delta\|_B \\ &\leqslant \|M_A\|_B \|x^m - x^*\|_B + \varepsilon\|\delta\|_B \\ &\leqslant [\|M_A\|_B + \varepsilon\|B^{-1/2}AB^{-1/2}\|_2]\,\|x^m - x^*\|_B.\end{aligned}$$

The identity $\|B^{-1/2}AB^{-1/2}\|_2 = \|A^{1/2}B^{-1/2}\|_2^2 = \|A^{1/2}B^{-1}A^{1/2}\|_2$ (cf. (2.9.4a)) proves the contraction number (18b). The definition of $M_k(Ax - b)$ and $N_k(Ax - b)$ in (18a) is obvious. Since the cg method is nonlinear (analogous to Remark 2.1a), Φ_k is also nonlinear. □

Remark 9.4.18. The composed iteration Φ_k defined in Lemma 17 is not well-suited to be the basic iteration for the Chebyshev or cg method, because the matrix $W_k(d) = A(I - M_k(d))$, $d = Ax - b$, of the third normal form of Φ_k depends on the value of the iterates x^m. Concerning these problems, compare Golub–Overton [1] and Axelsson–Vassilevski [2].

9.5 Generalisations

9.5.1 Formulation of the cg Method with a More General Bilinear Form

As a preparation for the next section we generalise the cg algorithm (5.14a–f) by replacing the Euclidean scalar product $\langle\cdot,\cdot\rangle$ by a bilinear form (or a respective sesquilinear form in the complex case $\mathbb{K} = \mathbb{C}$). The mapping $x, y \mapsto (x, y) \in \mathbb{K}$ is called a *bilinear form* ($\mathbb{K} = \mathbb{R}$) or *sesquilinear form* ($\mathbb{K} = \mathbb{C}$) if (2.7.1b, b′) are satisfied.

Exercise 9.5.1. Let $\langle\cdot,\cdot\rangle$ be the Euclidean scalar product and prove:
(a) Any bilinear or sesquilinear form is associated with a matrix B with

$$(x, y) = \langle Bx, y\rangle \quad \text{for all } x, y \in \mathbb{K}^I. \tag{9.5.1a}$$

(b) For each $B \in K^{I\times I}$, (1a) defines a sesquilinear form.
(c) If and only if $BAW^{-1} = A^H W^{-H} B$, we have

$$(AW^{-1}x, y) = (x, W^{-1}Ay) \quad \text{for all } x, y \in \mathbb{K}^I. \tag{9.5.1b}$$

In the following, we require no assumptions except the regularity of W. After replacing $\langle\cdot,\cdot\rangle$ by $(\cdot,\cdot)$, algorithm (4.14a–f) reads as follows:

$$\textit{start}: x^0 \text{ arbitrary}, \quad r^0 := b - Ax^0, \quad p^0 := W^{-1}r^0, \tag{9.5.2a}$$

iteration for $m = 0, 1, \ldots,$ as long as $(Ap^m, p^m) \neq 0$:

$$x^{m+1} := x^m + \lambda p^m \quad \text{with } \lambda := (r^m, p^m)/(Ap^m, p^m), \tag{9.5.2b}$$

$$r^{m+1} := r^m - \lambda A p^m, \tag{9.5.2c}$$

$$p^{m+1} := W^{-1}r^{m+1} - \omega p^m \quad \text{with } \omega := \frac{(AW^{-1}r^{m+1}, p^m)}{(Ap^m, p^m)}. \tag{9.5.2d}$$

Remark 9.5.2. Algorithm (2a–d) can be performed if and only if W is regular and the arising search directions p^m satisfy the condition $(Ap^m, p^m) \neq 0$. The latter inequality does not follow from $p^m \neq 0$.

Most of the properties from Lemma 4.15 can be maintained.

Theorem 9.5.3. *Assume that algorithm* (2) *can be performed for all* $0 \leqslant m \leqslant m_0$, *i.e., assume*

$$(Ap^m, p^m) \neq 0 \quad \textit{for all } 0 \leqslant m \leqslant m_0. \tag{9.5.3}$$

Furthermore, (1b) *must hold. Then we have for all* $0 \leqslant m \leqslant m_0$

$$(Ap^m, p^\ell) = 0 \quad \textit{for } \ell < m, \tag{9.5.4a}$$

$$(r^m, W^{-1}r^\ell) = (r^m, p^\ell) = 0 \quad \textit{for } \ell < m, \tag{9.5.4b}$$

$$r^m = b - Ax^m, \tag{9.5.4c}$$

$$U_m := \operatorname{span}\{p^0, \ldots, p^m\} \textit{ has dimension } m + 1, \tag{9.5.4d}$$

$$U_m = \operatorname{span}\{W^{-1}r^0, \ldots, W^{-1}r^m\}$$
$$= \operatorname{span}\{(W^{-1}A)^\nu W^{-1}r^0 \colon 0 \leqslant \nu \leqslant m\}, \tag{9.5.4e}$$

$$(r^\ell, p^\ell) = (r^\ell, W^{-1}r^\ell) \neq 0 \quad \textit{for } \ell < m_0. \tag{9.5.4f}$$

The proof is very similar to that from §9.4. However, since that proof was given in part in both Sections 9.3 and 9.4, we repeat it here. First, we comment on the interpretation of the results of Theorem 3. (4a) can be read as «p^m is conjugate to p^ℓ». However, note that this statement is not symmetric, since (4a) is restricted to $\ell < m$ only. An analogous statement holds for (4b).

Proof. (i) Obviously, the representation (4c) follows from (2a, c). The other statements are shown by induction on m_0, where $m_0 = 0$ is a trivial case. Let (4a–e) be valid for $m \leqslant m_0$. The coincidence of the subspaces from (4d, e) follows as for the corresponding statement (4.4d).
(ii) By the definition of λ in (2b), (r^{m+1}, p^ℓ) vanishes for $\ell = m$. For $\ell < m$, we obtain from (2c): $(r^{m+1}, p^\ell) = (r^m, p^\ell) - \lambda(Ap^m, p^\ell) = 0$ because of (4a, b). Thanks to (4e), $(r^{m+1}, p^\ell) = 0$ $(\ell \leqslant m)$ also implies $(r^{m+1}, W^{-1}r^\ell) = 0$.

(iii) The product (Ap^{m+1}, p^ℓ) vanishes for $\ell = m$ because of the definition of ω in (2d). For $\ell < m$, insert (2d):

$$(Ap^{m+1}, p^\ell) = (AW^{-1}r^{m+1}, p^\ell) - \omega(Ap^m, p^\ell).$$

(4a) yields $(Ap^m, p^\ell) = 0$. (1b) allows the reformulation $(AW^{-1}r^{m+1}, p^\ell) = (r^{m+1}, W^{-1}Ap^\ell)$. Since $p^\ell \in U_{m-1}$ and thus $W^{-1}Ap^\ell \in U_m$ (cf. (4e)), property (4b) provided in (ii) (with m replaced by $m+1$) shows that $(AW^{-1}r^{m+1}, p^\ell) = 0$. Hence, $(Ap^{m+1}, p^\ell) = 0$ holds for all $\ell \leqslant m$.
(iv) Assume $v := \sum_{\ell=0}^{m+1} \alpha_\ell p^\ell = 0$ in contradiction to (4d). Part (iii) proves $0 = (Ap^k, v) = \bar{\alpha}_k$ for $k = m+1$. Induction shows that $\alpha_k = 0$ for all $0 \leqslant k \leqslant m+1$; hence, $v = 0$. Therefore, the search directions p^ℓ, $0 \leqslant \ell \leqslant m+1$, are linear independent. The assumption $(r^m, p^m) = 0$ leads to $r^{m+1} = r^m$. This would be a contradiction to $\dim U_{m+1} = \dim\{W^{-1}r^0, \ldots, W^{-1}r^{m+1}\} = m+2$, from which $(r^m, p^m) \neq 0$ follows. From $p^m - W^{-1}r^m \in U_{m-1}$, one deduces the identity $(r^m, p^m) = (r^m, W^{-1}r^m)$, completing the proof of (4f). □

The replacement of (r^m, p^m) in (2b) by $(r^m, W^{-1}r^m)$ is already mentioned in (4f). The statement analogous to (4.6e′) is the subject of

Exercise 9.5.4. Prove that the following definition is equivalent to (2d):

$$p^{m+1} := W^{-1}r^{m+1} + (r^{m+1}, W^{-1}r^{m+1})/(r^m, W^{-1}r^m)p^m. \tag{9.5.2d′}$$

9.5.2 Method of Conjugate Residuals

By Exercise 1c, assumption (1b) is equivalent to the condition $BAW^{-1} = A^H W^{-H} B$ for the matrix B generating the bilinear form (1a). Sufficient conditions are:

$$AW^{-1} = A^H W^{-H}, \qquad B \text{ commutes with } AW^{-1}. \tag{9.5.5a}$$

Evidently, the last condition holds for $B = I$. However, this case is uninteresting, as it leads back to algorithm (4.14a–f). Another possibility is

$$B = AW^{-1} = A^H W^{-H}. \tag{9.5.5b}$$

The choice (5b) yields the bilinear form

$$(x, y) = \langle Bx, y\rangle = \langle AW^{-1}x, y\rangle = \langle A^H W^{-H}x, y\rangle = \langle x, W^{-1}Ay\rangle. \tag{9.5.5c}$$

Algorithm (2a–d) with B from (5b) becomes

start: x^0 arbitrary, $r^0 := b - Ax^0$, $p^0 := W^{-1}r^0$; (9.5.6a)

iteration: for $m = 0, 1, \ldots,$ as long as $\langle Ap^m, W^{-1}Ap^m\rangle \neq 0$:

$$x^{m+1} := x^m + \lambda p^m, \qquad r^{m+1} := r^m - \lambda Ap^m \quad \text{with} \tag{9.5.6b}$$

$$\lambda := \frac{\langle r^m, W^{-1}Ap^m\rangle}{\langle Ap^m, W^{-1}Ap^m\rangle} = \frac{\langle AW^{-1}r^m, W^{-1}r^m\rangle}{\langle Ap^m, W^{-1}Ap^m\rangle}, \tag{9.5.6c}$$

$$p^{m+1} := W^{-1}r^{m+1} - \omega p^m \quad \text{with} \tag{9.5.6d}$$

$$\omega := \frac{\langle AW^{-1}r^{m+1}, W^{-1}Ap^m\rangle}{\langle Ap^m, W^{-1}Ap^m\rangle} = -\frac{\langle AW^{-1}r^{m+1}, W^{-1}r^{m+1}\rangle}{\langle AW^{-1}r^m, W^{-1}r^m\rangle}. \tag{9.5.6e}$$

For $W = I$, this method is equivalent to the *method of the conjugate residuals* (CR) of Stiefel [2]. If $A > 0$ and $W > 0$, one also obtains (6a–e) directly from the standard cg method (4.6a–e) as stated in

Exercise 9.5.5. Apply (4.6a–e) to the equivalent system $\bar{A}\bar{x} = \bar{b}$ with $\bar{A}$ and $\bar{b}$ from (2.10b) and prove that after reformulating the algorithm by the quantities x and $p = A^{-1/2}\bar{p}$, one obtains (6a–e).

Exercise 5 allows an immediate transfer of the error statement (4.10) with $\bar{A} = A^{1/2}W^{-1}A^{1/2}$ (instead of A), which in analogy to (2.10d) involves the $\|\cdot\|_{AW^{-1}A}$ norm.

Theorem 9.5.6. *Let Φ, γ, Γ, $\varkappa$, and c be as in Theorem* 4.14. *The errors $e^m = x^m - x^*$ of the iterates from* (6a–e) *satisfy the estimate*

$$\|W^{-1/2}A(x^m - x^*)\|_2 = \|W^{-1/2}(Ax^m - b)\|_2 \leqslant \frac{2c^m}{1 + c^{2m}}\|W^{-1/2}(Ax^0 - b)\|_2. \tag{9.5.7}$$

In the case of $W = I$, (7) represents an estimate of the residuals r^m.

Lemma 9.5.7. *Let $A = A^H$, $W > 0$. U_m is the subspace from* (4d). *Then the iterate x^m from* (6a–e) *minimises the norm*

$$\|W^{-1/2}r^m\|_2 = \min\{\|W^{-1/2}A(x - x^*)\|_2 \colon x - x^0 \in U_{m-1}\}. \tag{9.5.8}$$

Proof. Because of $AW^{-1}A > 0$, $\{\langle A(x - x^*), W^{-1}A(x - x^*)\rangle \colon x - x^0 \in U_{m-1}\}$ attains its minimum at $x = x^m$ if and only if the gradient $AW^{-1}A(x^m - x^*) = -AW^{-1}r^m$ is orthogonal to U_{m-1}. Since $p^0, \ldots, p^{m-1}$ span the subspace U_{m-1}, it suffices to prove $\langle AW^{-1}r^m, p^\ell\rangle = 0$ for $0 \leqslant \ell \leqslant m - 1$. By (5c), this condition reads $(r^m, p^\ell) = 0$ and holds because of (4b). □

The assumptions of Lemma 7 also apply to an *indefinite* matrix A (i.e., A has both positive *and* negative eigenvalues). In this situation, the bilinear form (5c) is not a scalar product; however, the denominator $\langle Ap^m, W^{-1}Ap^m\rangle$ represents the scalar product belonging to the $\|\cdot\|_{AW^{-1}A}$ norm. The denominator in (6c) vanishes only if $p^m = 0$. Despite that, the method (6a–e) is not suited for solving the equation $Ax = b$, as the following counterexample will show.

For the sake of simplicity, assume $W = I$. If A is indefinite, there are eigenvalues $\lambda_1 < 0$, $\lambda_2 > 0$ with corresponding eigenvectors v_1, v_2. For $v := \sqrt{\lambda_2}\|v_2\|_2 v_1 + \sqrt{-\lambda_1}\|v_1\|_2 v_2 \neq 0$ one verifies that $\langle v, Av\rangle = 0$. We choose a starting iterate x^0 with $r^0 = v$. Because of $p^0 = r^0$ and $\langle r^0, Ap^0\rangle = 0$, $\lambda = 0$ in (6c) leads to $x^1 = x^0$ and $r^1 = r^0$. (6d, e) yield the search direction $p^1 = r^1 - \omega p^0 = r^0 - 1 \cdot r^0 = 0$; hence, the algorithm (6a–e) terminates for $m = 1$ because of $\langle Ap^m, W^{-1}Ap^m\rangle = \langle 0,0\rangle = 0$ without having reached the exact solution. The reason of this failure is that the subspaces $\operatorname{span}\{r^0, r^1\} = \operatorname{span}\{r^0\}$ and $\operatorname{span}\{r^0, Ar^0\}$ differ. This fact suggests that the subspace $\operatorname{span}\{r^0, \ldots, A^m r^0\}$ or, more generally, $\operatorname{span}\{W^{-1}r^0, \ldots, (W^{-1}A)^m W^{-1}r^0\}$ is better suited than $\operatorname{span}\{r^0, \ldots, r^m\}$. $\operatorname{Span}\{q, Xq, \ldots, X^{m-1}q\}$ is called the *Krylov space* (of dimension m) generated by q (and the matrix X).

Even if $\langle Ar^m, r^m\rangle = 0$ does not occur during the calculations, it may happen that r^m is «almost» contained in U_{m-1}, leading to the instability of the algorithm. One remedy is the construction of the search directions p^m by another recursion, which first is explained for the standard cg variant before considering the indefinite case in §9.5.4.

9.5.3 Three-Term Recursion for p^m

Inserting definition (4.14d): $r^{m+1} := r^m - \lambda A p^m$ into (4.14f), one obtains $p^{m+1} := W^{-1}r^m - \lambda W^{-1}Ap^m + \text{const}\, p^m$. Since the scaling of the search direction is irrelevant, we may replace p^{m+1} by $-p^{m+1}/\lambda$. Because of $W^{-1}r^m$ and $p^m \in U_m$, we arrive at the following ansatz:

$$p^{m+1} := W^{-1}Ap^m - \sum_{\mu=0}^{m} \alpha_{\mu,m+1} p^{m-\mu}. \tag{9.5.9a}$$

The condition (4.16a): $\langle Ap^{m+1}, p^m\rangle = 0$ determines the coefficients

$$\alpha_{0,m+1} = \langle AW^{-1}Ap^m, p^m\rangle / \langle Ap^m, p^m\rangle, \tag{9.5.9b}$$

because $\langle Ap^{m-\mu}, p^m\rangle = 0$ for $\mu > 0$. Similarly,

$$\alpha_{1,m+1} = \langle AW^{-1}Ap^m, p^{m-1}\rangle / \langle Ap^{m-1}, p^{m-1}\rangle. \tag{9.5.9c}$$

Lemma 9.5.8. *Assume $A = A^H$, $W = W^H$. In (9a), $\alpha_{\mu,m+1} = 0$ holds for $\mu \geqslant 2$.*

Proof. The condition $\langle Ap^{m+1}, p^{m-\mu}\rangle = 0$ yields the equation

$$\langle AW^{-1}Ap^m, p^{m-\mu}\rangle = \alpha_{\mu,m}\langle Ap^{m-\mu}, p^{m-\mu}\rangle. \tag{9.5.9d}$$

The assertion follows from

$$\begin{aligned}\langle AW^{-1}Ap^m, p^{m-\mu}\rangle &= \langle Ap^m, W^{-1}Ap^{m-\mu}\rangle \\ &= \left\langle Ap^m, p^{m+1-\mu} + \sum_{\nu\geqslant 0} \alpha_{\nu,m+1-\mu} p^{m-\mu-\nu}\right\rangle \\ &= 0.\end{aligned}$$

□

Thanks to Lemma 8, p^{m+1} can be calculated from the three-term recursion

$$p^{m+1} := W^{-1}Ap^m - \alpha_0 p^m - \alpha_1 p^{m-1} \quad \text{with} \tag{9.5.10a}$$

$$\alpha_0 = \frac{\langle AW^{-1}Ap^m, p^m\rangle}{\langle Ap^m, p^m\rangle}, \qquad \alpha_1 = \frac{\langle AW^{-1}Ap^m, p^{m-1}\rangle}{\langle Ap^{m-1}, p^{m-1}\rangle}, \tag{9.5.10b}$$

where the last term is absent for $m = 0$ (formally, one may set $\alpha_1 = 0$, $p^{-1} = 0$). The arising cg algorithm is described by means of the bilinear form $(\cdot,\cdot)$. The particular choice $(\cdot,\cdot) = \langle\cdot,\cdot\rangle$ leads to an algorithm equivalent to (4.14a–f).

$$\textit{start}: x^0 \text{ arbitrary}, \quad r^0 := b - Ax^0, \quad p^{-1} := 0,$$

$$p^0 := W^{-1}r^0; \tag{9.5.11a}$$

iteration for $m = 0, 1, \ldots,$ as long as $(Ap^m, p^m) \neq 0$:

$$x^{m+1} := x^m + \lambda p^m, \qquad r^{m+1} := r^m - \lambda Ap^m \quad \text{with} \tag{9.5.11b}$$

$$\lambda := (r^m, p^m)/(Ap^m, p^m), \tag{9.5.11c}$$

$$p^{m+1} := W^{-1}Ap^m - \alpha_0 p^m - \alpha_1 p^{m-1} \quad \text{with} \tag{9.5.11d}$$

$$\alpha_0 = \frac{(AW^{-1}Ap^m, p^m)}{(Ap^m, p^m)}, \qquad \alpha_1 = \frac{(AW^{-1}Ap^m, p^{m-1})}{(Ap^{m-1}, p^{m-1})} \tag{9.5.11e}$$

where again $\alpha_1 := 0$ is chosen for $m = 0$.

Theorem 9.5.9. *Assume* (1b). *Let m_0 be the maximal index such that the directions generated in* (11) *satisfy $(Ap^m, p^m) \neq 0$ for all $0 \leqslant m \leqslant m_0$.*
(a) *The statements* (4a–d) *are valid for the quantities x^m, r^m, p^m $(0 \leqslant m \leqslant m_0)$ from* (11a–e). *Statement* (4e) *is replaced by*

$$\begin{aligned} U_m &:= \operatorname{span}\{p^0, \ldots, p^m\} \\ &= \operatorname{span}\{(W^{-1}A)^\nu W^{-1}r^0 \colon 0 \leqslant \nu \leqslant m\} \\ &\supset \operatorname{span}\{W^{-1}r^0, \ldots, W^{-1}r^m\} \qquad \textit{for } 0 \leqslant m \leqslant m_0. \end{aligned} \tag{9.5.12a}$$

More precisely, we have

$$W^{-1}r^m \in \operatorname{span}\{p^m, p^{m-1}\} \qquad \textit{for } 0 \leqslant m \leqslant m_0. \tag{9.5.12b}$$

(b) *As long as algorithm* (2a–d) *does not terminate,* (2a–d) *and* (11a–e) *produce the same iterates x^m, whereas the search directions may differ by a factor $\neq 0$.*
(c) *If the iteration* (11b–e) *terminates because of $p^m = 0$, the iterate x^m already represents the exact solution.*

Proof. (i) The properties (4a–d) are shown as in Theorem 3. For the proof of (12b), apply induction on m. The assertion is trivial for $m = 0$. Assume (12b) for $m - 1$. The definition of r^m in (11b) shows that $W^{-1}r^m = W^{-1}r^{m-1} - \lambda W^{-1}Ap^{m-1}$. The induction hypothesis together with (11d) yields $W^{-1}r^m \in \operatorname{span}\{p^m, p^{m-1}, p^{m-2}\}$; hence,

$$W^{-1}r^m = \beta_0 p^m + \beta_1 p^{m-1} + \beta_2 p^{m-2}.$$

(4a) permits the representation $\beta_2 = (AW^{-1}r^m, p^{m-2})/(Ap^{m-2}, p^{m-2})$. Thanks to assumption (1b), $(AW^{-1}r^m, p^{m-2}) = (r^m, W^{-1}Ap^{m-2})$ holds. From $AW^{-1}p^{m-2} \in U_{m-1}$ and $(r^m, y) = 0$ for all $y \in U_{m-1}$ (cf. (4b)) one concludes that $\beta_2 = 0$.

(ii) As long as the algorithm (2a–d) does not terminate, the generated p^m span the same Krylov space $U_m := \mathrm{span}\{(W^{-1}A)^\nu W^{-1}r^0 : 0 \leqslant \nu \leqslant m\}$. Since vectors p^m with the property (4a): $(Ap^m, p^\ell) = 0$ for $m > \ell$ and $p^0 = W^{-1}r^0$ differ only by a contant ($\neq 0$, because $(Ap^m, p^m) \neq 0$), (2b) and (6b) must lead to the same x^m.
(iii) Assume $p^m = 0$. This implies the coincidence $U_m = U_{m-1}$. Statement (12b) becomes $W^{-1}r^m \in \mathrm{span}\{p^{m-1}\}$; hence, $r^m = 0$ is equivalent to $(AW^{-1}r^m, p^{m-1}) = 0$. (1b) allows the reformulation $(AW^{-1}r^m, p^{m-1}) = (r^m, W^{-1}Ap^{m-1})$. $W^{-1}Ap^{m-1} \in U_m = U_{m-1}$ and $(r^m, y) = 0$ for all $y \in U_{m-1}$ (cf. (4b)) prove the assertion. □

9.5.4 Stabilised Method of the Conjugate Residuals

We emphasize that, in general, the iteration (11a–e) may terminate because of $(Ap^m, p^m) = 0$ without ensuring $p^m = 0$. In order to guarantee that in the case of termination, the algorithm yields the exact solution, the bilinear form must be chosen so that the implication $(Az, z) = 0 \Rightarrow z = 0$ holds.

Define the bilinear form by (1a) with B from (5b). Then, the termination criterion $(Ap^m, p^m) = 0$ becomes $\langle Ap^m, W^{-1}Ap^m \rangle = 0$. Under the assumption $W > 0$ (e.g., $W = I$) this implies $p^m = 0$; hence, according to Theorem 9c, the exact solution $x^m = x^*$ is reached in the case of termination. For B from (5b), algorithm (11a–e) takes the following form:

$$\textit{start}: x^0 \text{ arbitrary}, \quad r^0 := b - Ax^0, \quad p^{-1} := 0,$$

$$p^0 := W^{-1}r^0; \tag{9.5.13a}$$

iteration for $m = 0, 1, \ldots$, as long as $\langle Ap^m, W^{-1}Ap^m \rangle \neq 0$:

$$x^{m+1} := x^m + \lambda p^m, \qquad r^{m+1} := r^m - \lambda Ap^m \quad \text{with} \tag{9.5.13b}$$

$$\lambda := \langle r^m, W^{-1}Ap^m \rangle / \langle Ap^m, W^{-1}Ap^m \rangle, \tag{9.5.13c}$$

$$p^{m+1} := W^{-1}Ap^m - \alpha_0 p^m - \alpha_1 p^{m-1} \quad \text{with} \tag{9.5.13d}$$

$$\alpha_0 = \frac{\langle AW^{-1}Ap^m, W^{-1}Ap^m \rangle}{\langle Ap^m, W^{-1}Ap^m \rangle}, \qquad \alpha_1 = \frac{\langle AW^{-1}Ap^m, W^{-1}Ap^{m-1} \rangle}{\langle Ap^{m-1}, W^{-1}Ap^{m-1} \rangle}. \tag{9.5.13e}$$

Exercise 9.5.10. By $AW^{-1}Ap^m$ appearing in (13e), algorithm (13a–e) seems to cost two multiplications by the matrix A per iteration step. Rewrite algorithm (13a–e) so that an additional recursion for $a^m := Ap^m$ is used and only one A-multiplication per iteration is required.

9.5.5 Convergence Results for Indefinite Matrices A

Lemma 7 carries over with the same proof.

Lemma 9.5.11. *Let* $A = A^H$ *and* $W > 0$. *Let* U_m *be the subspace from* (12a). *The iterate* x^m *of algorithm* (13a–e) *minimises the norm* (8):

$$\| W^{-1/2} r^m \|_2 = \min\{ \| W^{-1/12} A(x - x^*) \|_2 : x - x^0 \in U_{m-1} \}. \tag{9.5.8}$$

The algorithm (13a–e) of the conjugate residuals is interesting for *indefinite* matrices because of

Remark 9.5.12. Let W be positive definite. The assumptions of Lemma 11 permit indefinite Hermitian matrices A. Conditions (5a, b) are satisfied for a regular matrix $A = A^H$ that may be indefinite. $\langle Ap^m, W^{-1}Ap^m \rangle = 0$ implies $p^m = 0$.

The error estimate (7) from Theorem 6 cannot be transferred directly to indefinite matrices, because the spectrum of $W^{-1}A$ no longer lies in the positive part. In the general case, the resulting convergence speed is slower than in the positive definite case.

Theorem 9.5.13. *Assume* $W > 0$, $A = A^H$ *regular, and* $\varkappa = \varkappa(W^{-1}A)$. *Then the iterates* x^m *of algorithm* (13a–e) *satisfy the error estimate*

$$\| W^{-1/2}A(x^m - x^*)\|_2 \leqslant \frac{2c^\mu}{1 + c^{2\mu}} \| W^{-1/2}(Ax^0 - b)\|_2 \tag{9.5.14}$$

with $c := (\varkappa - 1)/(\varkappa + 1)$ *and* $\frac{m}{2} - 1 < \mu \leqslant \frac{1}{2}m$, *where* $\mu \in \mathbb{Z}$. *Hence, the asymptotical convergence rate amounts to* $\sqrt{c} = 1 - 1/\varkappa + O(\varkappa^{-2})$.

Proof. For odd m, we exploit the (weakly) monotone convergence $\| W^{-1/2}Ae^m\|_2 \leqslant \| W^{-1/2}Ae^{m+1}\|_2$ following from (8). Therefore, consider an even $m = 2\mu$. In analogy to Remark 4.10, one finds

$$\| W^{-1/2}Ae^m\|_2 \leqslant \max\{|P_m(1 - \lambda)|: \lambda \in \sigma(W^{-1}A)\} \| W^{-1/2}Ae^0\|_2 \tag{9.5.15}$$

for any polynomial P_m of degree $\leqslant m$ with $P_m(1) = 1$. Let p_μ be a polynomial of degree $\leqslant \mu = m/2$ with $p_\mu(1) = 1$. $P_m(\xi) := p_\mu(\xi(2 - \xi))$ is of degree $\leqslant m$ and satisfies $P_m(1) = 1$. Evidently, $P_m(1 - \lambda) = p_\mu(1 - \lambda^2)$, from which

$$\| W^{-1/2}Ae^m\|_2 \leqslant \max\{|p_\mu(1 - \lambda^2)|: \lambda \in \sigma(W^{-1}A)\} \| W^{-1/2}Ae^0\|_2$$

follows. If $\lambda \in \sigma(W^{-1}A)$, we have $|\lambda| \in [\gamma, \Gamma]$ and $\lambda^2 \in [\gamma^2, \Gamma^2]$, where

$$\gamma := 1/\varrho(A^{-1}W) = \min\{|\lambda|: \lambda \in \sigma(W^{-1}A)\}, \qquad \Gamma \in \varrho(W^{-1}A).$$

Since $[\gamma^2, \Gamma^2]$ is in the positive part, the Chebyshev polynomial (7.3.12b) yields the following estimate with $c = (\Gamma - \gamma)/(\Gamma + \gamma) = (\varkappa - 1)/(\varkappa + 1)$:

$$\max\{|p_\mu(1 - \lambda^2)|: \lambda \in \sigma(W^{-1}A)\} \leqslant \max\{|p_\mu(1 - \xi)|: \gamma^2 \leqslant \xi \leqslant \Gamma^2\}$$
$$\leqslant \frac{2c^\mu}{1 + c^{2\mu}}. \qquad \square$$

Often, a *milder form of indefiniteness* occurs. If, for instance, the Helmholtz equation $-\Delta u - cu = f$ with $c > 0$ is discretised, A has the eigenvalues λ_μ^h, where

$$\lambda_\mu^h = \lambda_{\mu,0}^h - c,\ \lambda_{\mu,0}^h > 0 \text{ eigenvalues of the Poisson-model case,}$$
$$1 \leqslant \mu \leqslant n = n_h.$$

Since for $h \to 0$, the discrete eigenvalues λ_μ^h tend to values λ_μ that cannot accumulate (cf. Hackbusch [15, §11]), the following properties are satisfied:

The number k of negative eigenvalues is bounded for $h \to 0$. (9.5.16a)

For all $h > 0$, the nonpositive eigenvalues belong to $[-c_1, -c_0]$ with $0 < c_0 \leqslant c_1$. (9.5.16b)

The positive eigenvalues are in $[\gamma, \Gamma]$ with $0 < \gamma \leqslant \Gamma$. (9.5.16c)

Let $k = k_h$ be the number of negative eigenvalues λ_μ^h, $1 \leqslant \mu \leqslant k$. Define

$$\pi_h(1 - \xi) = \prod_{\ell=1}^{k} (1 - \xi/\lambda_\mu^h).$$

Let p_μ be the Chebyshev polynomial (7.3.12b) of degree $\mu := m - k$ for $a = 1 - \gamma$ and $b = 1 - \Gamma$. The product $P_m(\xi) := \pi_h(\xi)p_\mu(\xi)$ is of degree m with $P_m(1) = 1$. Since $P_m(1 - \lambda) = 0$ for the negative eigenvalues $\lambda \in \sigma(W^{-1}A)$, the factor on the right-hand side in (15) reduces to

$$\max\{|P_m(1 - \lambda)|: \lambda \in [\gamma, \Gamma]\} \leqslant \max\{|\pi_h(1 - \lambda)|: \lambda \in [\gamma, \Gamma]\} \frac{2c^\mu}{1 + c^{2\mu}}$$

with $c := \dfrac{\sqrt{\Gamma} - \sqrt{\gamma}}{\sqrt{\Gamma} + \sqrt{\gamma}}$. $|\pi_h(1 - \lambda)|$ can be estimated by $(1 + \Gamma/c_0)^k$ (cf. (16b)). The mth root of the bound $2(1 + \Gamma/c_0)^k c^\mu/(1 + c^{2\mu})$ tends to c. Hence, the asymptotical convergence rate is not influenced by the negative eigenvalues. This proves

Theorem 9.5.14. *Assume* $A = A^H$, $W > 0$ *and let the eigenvalues of* $W^{-1}A$ *satisfy* (16a–c). *Replace the condition number* $\varkappa(W^{-1}A)$ *by the possibly better number* $\varkappa := \Gamma/\gamma$ (γ, Γ *from* (16c)). *Then the error estimate for the algorithm* (13a–e) *of the conjugate residuals reads*

$$\|W^{-1/2}A(x^m - x^*)\|_2 \leqslant 2\left(\frac{1 + \Gamma/c_0}{c}\right)^k c^m \|W^{-1/2}(Ax^0 - b)\|_2 \quad (9.5.17)$$

with $c := \dfrac{\sqrt{\varkappa} - 1}{\sqrt{\varkappa} + 1} = \dfrac{\sqrt{\Gamma} - \sqrt{\gamma}}{\sqrt{\Gamma} + \sqrt{\gamma}}$ *being the asymptotical convergence rate.*

An alternative to the method (13a–e) of conjugate residuals is the application of the standard cg method to the Kaczmarz iteration (cf. §9.5.9). Then the convergence speed is as slow as in Theorem 13. In the situation (16a–c), however, the convergence rate would not improve.

9.5.6 PASCAL Procedures

The following procedures have a structure analogous to those of §9.4.5.

```
procedure start_CR_method(var A: data_of_discretisation;
          var x,b: gridfunction; var IP: iterationparameter;
          procedure basisiteration(var neu: gridfunction;
                var A: data_of_discretisation;
                var x,b: gridfunction; var IP:
                 iterationparameter));
begin with IP do begin Nr:=0; if cg=nil then new(cg) end;
    with IP.cg^ do
    begin if CR=nil then new(CR); rho:=0;
          residual(r,A,x,b); N_y(p,r,A,IP,basisiteration);
           A_x(CR^.a,A,p)
end end;

procedure CR_method(var new: gridfunction;
                    var A: data_of_discretisation; var x,b:
                     gridfunction;
                    var IP: iterationparameter;
                    procedure basisiteration(var new:
                     gridfunction;
                        var A: data_of_discretisation;
                        var x,b: gridfunction; var IP:
                         iterationparameter));
var c,aw: gridfunction; alpha0,alpha1,lambda,rho_old: real;
 label 1;
begin if IP.cg=nil then 1: start_CR_method(A,x,b,IP,
 basisiteration)
    else if IP.cg^.CR=nil then goto 1;
    with A do with IP.cg^ do
    begin rho_old:=rho;
       N_y(c,CR^.a,A,IP,basisiteration); rho:=Eucl_
        scalarproduct(CR^.a,c,A)
       if rho=0 then begin writeln('CR reaches exact
        solution!'); rho:=1
       end;
       lambda:=Eucl_scalarproduct(r,c,A)/rho;
       vector_plus_factor_x_vector(nx,ny,new,x,lambda,p);
       transfer_boundary_values(nx,ny,x,new);
       vector_plus_factor_x_vector(nx,ny,r,r,-lambda,CR^.a);
       A_x(aw,A,c); alpha0:=Eucl_scalarproduct(aw,c,A)/rho;
       if rho_old<>0 then alpha1:=Eucl_scalarproduct
        (aw,CR^.w,A)/rho_old
```

```
        else alpha1:=0;
        CR^.w:=c; if alpha1<>0 then
        vector_plus_factor_x_vector(nx,ny,c,c,-alpha1,
         CR^.p_old);
        CR^.p_old:=p; vector_plus_factor_x_vector(nx,ny,p,c,
         -alpha0,p);
        if alpha1<>0 then
        vector_plus_factor_x_vector(nx,ny,aw,aw,-alpha1,
         CR^.a_old);
        CR^.a_old:=CR^.a;
        vector_plus_factor_x_vector(nx,ny,CR^.a,aw,-alpha0,
         CR^.a)
end end;

procedure CR_method_1(var it: data_of_iteration;
                      procedure basisiteration(var new:
                       gridfunction;
                              var A: data_of_discretisation;
                              var x,b: gridfunction;
                              var IP: iterationparameter));
var number: integer;
begin with it do with IP do
    begin number:=Nr;
          if Nr=0 then start_CR_method(A,x,b,IP,
           basisiteration);
          CR_method(x,A,x,b,IP,basisiteration); Nr:=number+1
end end;
```

9.5.7 Numerical Examples

For reasons of comparison, we first test the positive definite Poisson-model problem with $h = 1/32$. We apply `CR_method` to the ILU iteration (five-point pattern) with the same parameters as in Table 4.2. The results given in Table 1 are similar to those of the standard cg method from Table 4.2.

Next, we choose the discrete Helmholtz equation $-\Delta u - 50u = f$ as an indefinite example. The matrix A is the Poisson-model matrix minus $50I$. It has three negative eigenvalues $\lambda_1 = -30.277$, $\lambda_2 = \lambda_3 = -0.7866$, while $\lambda_4 = 28.7$ is the smallest positive eigenvalue. For the modified ILU decomposition the diagonal must be enlarged by 50: `diag=55`. The results of Table 2 show that the reduction factor moves toward the asymptotical convergence rate and has a size similar to the positive definite case of Table 1. The stagnation for $m \geqslant 33$ is due to rounding errors. In both of the examples, the algorithm behaves stably.

Table 9.5.1 The CR method (13) for the Poisson-model problem applied to the ILU iteration

5-point ILU with $\omega = -1$		
m	$u_{16,16}$	$\lVert e^m \rVert_A / \lVert e^{m-1} \rVert_A$
1	0.2222124445	$1.57356_{10}-1$
2	0.4269164370	$5.37790_{10}-1$
3	0.4510237348	$4.39627_{10}-1$
4	0.4759275765	$4.38732_{10}-1$
5	0.4813602944	$4.86199_{10}-1$
10	0.4998558015	$3.84330_{10}-1$
20	0.5000000047	$3.38399_{10}-1$
21	0.5000000029	$3.89211_{10}-1$
22	0.5000000012	$4.07384_{10}-1$
23	0.5000000003	$3.17164_{10}-1$
24	0.5000000002	$4.42606_{10}-1$
25	0.5000000000	$6.91876_{10}-1$
26	0.5000000000	$7.68926_{10}-1$
27	0.5000000000	$9.55932_{10}-1$
28	0.5000000000	$1.02596_{10}+0$
29	0.5000000000	$1.03161_{10}+0$
30	0.5000000000	$1.03076_{10}+0$

9.5.8 Method of Orthogonal Directions

The cg method (4.6a–e) minimises the error $\lVert e^m \rVert_A = \lVert A^{1/2} e^m \rVert_2$ with respect to the energy norm over the Krylov space $U_m := \operatorname{span}\{A^\nu r^0 : 0 \leqslant \nu \leqslant m\}$. The method of conjugate residuals (with $W = I$) minimises the residual $\lVert r^m \rVert_2 = \lVert A e^m \rVert_2$ over the same space. A more natural norm would be the $\lVert \cdot \rVert_2$ norm. Then the search directions p^m would be orthogonal in the usual sense. Algorithm (2a–f) yields a method with these properties if one chooses the bilinear form $(x, y) := \langle A^{-1} x, y \rangle$. The expressions (Ap^m, p^m) would simplify to $\langle p^m, p^m \rangle$. This approach, however, fails because of the product $(r^m, p^m) = \langle A^{-1} r^m, p^m \rangle$, whose computation is no longer practical. A remedy is the reformulation $(r^m, p^m) = \langle r^m, A^{-1} p^m \rangle$, provided that p^m has the form $A\hat{p}^m$ with computable $\hat{p}^m$. This can be achieved by replacing the Krylov space $\operatorname{span}\{A^\nu r^0 : 0 \leqslant \nu \leqslant m\}$ by $\operatorname{span}\{A^{\nu+1} r^0 : 0 \leqslant \nu \leqslant m\}$. The arising algorithm (18) is due to Fridman[1] (1963) and is called the *method of orthogonal directions* (OD), since the search directions form an orthogonal system if $W = I$:

$$\textit{start}: x^0 \text{ arbitrary}, \quad r^0 := b - Ax^0, \quad q^{-1} := r^0,$$

$$q^0 := AW^{-1} q^{-1}; \tag{9.5.18a}$$

Table 9.5.2 The CR method (13) for an indefinite problem based on the ILU iteration

m	$u_{16,16}$	$\frac{\Vert e^m \Vert_2}{\Vert e^{m-1} \Vert_2}$
1	−1.129805206	1.36998
2	0.5616735534	0.41788
3	0.9170148791	0.77945
4	0.7375934000	0.78685
5	0.6675855715	0.88467
6	0.5834957931	0.95835
7	0.5440078825	0.99228
8	0.5222771713	1.00338
9	0.5099064832	1.00768
10	0.5053055088	1.00956
11	0.5029466483	1.01213
12	0.5020970259	1.01621
13	0.5015223028	1.01159
14	0.5015388760	1.00335
15	0.5017304154	0.97466
16	0.5019212154	0.86920
17	0.5018558645	0.56086
18	0.5017568935	0.32511
19	0.5008067252	0.27287
20	0.5003741869	0.19130
21	0.5003841894	0.35875
30	0.4999998664	0.30740
31	0.4999999694	0.40910
32	0.4999999838	0.69469
33	0.4999999962	0.90769
34	0.4999999984	0.97862
35	0.4999999986	0.98121
50	0.4999999994	0.99385

iteration for $m = 0, 1, \ldots,$ as long as $q^m \neq 0$:

$$x^{m+1} := x^m + \lambda p^m, \quad r^{m+1} := r^m - \lambda A p^m \quad \text{with } p^m := W^{-1} q^m, \tag{9.5.18b}$$

$$\lambda := \langle r^m, p^{m-1} \rangle / \varrho_m, \qquad \varrho_m := \langle q^m, p^m \rangle, \tag{9.5.18c}$$

$$q^{m+1} := A p^m - \alpha_0 q^m - \alpha_1 q^{m-1} \quad \text{with} \tag{9.5.18d}$$

$$\alpha_0 := \langle A p^m, p^m \rangle / \varrho_m, \qquad \alpha_1 := \langle A p^m, p^{m-1} \rangle / \varrho_{m-1}, \tag{9.5.18e}$$

where $\alpha_1 := 0$ is defined for $m = 0$. The method (18a–e), which is contained in the collection [Prog] under the name `OD_method`, is *unstable* as one observes from the results in Tables 3 and 4. A stabilisation is given by Stoer [2, (3.16)] (in [Prog] named `stabilised_OD_method`). On the other hand, one can do without it if only few iteration steps are to be performed.

The proof of the following theorem is left to the reader.

Theorem 9.5.15. *Assume that* $W > 0$, $A = A^H$. *Let* m_0 *be the largest index with* $q^m \neq 0$, $q^{m_0+1} = 0$ *implies* $x^{m_0+1} = x^*$. *For all* $0 \leqslant m \leqslant m_0$, (19a–c) *hold*:

$$\langle W^{-1}q^k, q^\ell \rangle = 0, \qquad \langle W^{-1}q^k, q^k \rangle \neq 0 \quad \text{for } 0 \leqslant k \neq \ell \leqslant m_0, \tag{9.5.19a}$$

$$r^m \perp A^{-1}U_{m-1} \quad \text{with } U_m := \operatorname{span}\{(AW^{-1})^\nu r^0 : 1 \leqslant \nu \leqslant m+1\}, \tag{9.5.19b}$$

$$U_m = \operatorname{span}\{q^0, \dots, q^m\}. \tag{9.5.19c}$$

There is an indirect connection between cg variants and the *Lanczos method*, which serves to approximate matrix eigenvalues by transforming the matrix into a similar tridiagonal matrix (cf. Golub–van Loan [1, §9] and Parlett [1, §§12–13]). The Lanczos method is also used to construct or

Table 9.5.3 Method (18a–e) for ILU iteration; problem as in Table 1

m	$u_{16,16}$	$\|e^m\|_2$	$\frac{\|e^m\|_2}{\|e^{m-1}\|_2}$	$\left\{\frac{\|e^m\|_2}{\|e^0\|_2}\right\}^{1/m}$
1	$-4.574928591_{10}-2$	$2.93956_{10}-1$	$3.92836_{10}-1$	$3.92836_{10}-1$
10	$4.989708480_{10}-1$	$5.16807_{10}-4$	$3.74216_{10}-1$	$4.82976_{10}-1$
11	$5.002475524_{10}-1$	$1.91138_{10}-4$	$3.69844_{10}-1$	$4.71399_{10}-1$
15	$5.000015863_{10}-1$	$5.80274_{10}-6$	$4.67856_{10}-1$	$4.56356_{10}-1$
16	$5.000085958_{10}-1$	$2.76800_{10}-6$	$4.77016_{10}-1$	$4.57621_{10}-1$
17	$4.999867395_{10}-1$	$4.98160_{10}-6$	$1.79971_{10}+0$	$4.96007_{10}-1$
18	$4.999923552_{10}-1$	$1.35781_{10}-5$	$2.72567_{10}+0$	$5.45253_{10}-1$
19	$4.999645661_{10}-1$	$3.77863_{10}-5$	$2.78287_{10}+0$	$5.94095_{10}-1$
20	$4.998667835_{10}-1$	$1.05920_{10}-4$	$2.80315_{10}+0$	$6.42016_{10}-1$
27	$5.290373141_{10}-1$	$7.10159_{10}-2$	$3.20465_{10}+0$	$9.16477_{10}-1$
30	$2.379020942_{10}+0$	$1.33836_{10}+0$	$1.77625_{10}+0$	$1.01957_{10}+0$

Table 9.5.4 OD method (18a–e) for an indefinite problem as in Table 2

m	$u_{16,16}$	$\|e^m\|_2$	$\frac{\|e^m\|_2}{\|e^{m-1}\|_2}$	$\left\{\frac{\|e^m\|_2}{\|e^0\|_2}\right\}^{1/m}$
1	$1.288025563_{10}+0$	$4.58268_{10}-1$	$6.12419_{10}-1$	$6.12419_{10}-1$
10	$5.107964511_{10}-1$	$2.08681_{10}-1$	$9.93821_{10}-1$	$8.80119_{10}-1$
20	$5.084149051_{10}-1$	$1.00523_{10}-2$	$4.81485_{10}-1$	$8.06139_{10}-1$
30	$5.000072697_{10}-1$	$5.10416_{10}-6$	$4.28537_{10}-1$	$6.72659_{10}-1$
35	$4.999124543_{10}-1$	$2.09700_{10}-4$	$2.28682_{10}+0$	$7.91591_{10}-1$
40	$4.178915511_{10}-1$	$5.64009_{10}-2$	$6.53540_{10}+0$	$9.37412_{10}-1$

stabilise cg variants. For instance, this connection is described by Paige–Saunders [1]. The method SYMMLQ defined there is a further stabilisation of the method (18a–e).

A review of the algorithms discussed above and of additional variants is given by Stoer [2]. A detailed and commented reference list up to 1976 from the field of conjugate gradient and Lanczos methods is contained in the paper of Golub–O'Leary [1].

9.5.9 Solution of Unsymmetric Systems

Some of the methods described above do not require the assumption (1.2) of positive definiteness of A and are also applicable to indefinite but still symmetric matrices. The nonsymmetric situation is more difficult. Recalling the Lanczos method and its connection to the eigenvalue problem, we mention that in the nonsymmetric case a matrix A is not transformed into tridiagonal but Hessenberg form (cf. Stoer–Bulirsch [1, §6.5]). The latter form corresponds to cg variants in which the A-orthogonalisation is performed completely as in (4.1a). Then not only the amount of computational work increases with m, but also the storage requirement is enlarged, because the former search directions are needed for orthogonalisation. The algorithms called ORTHOMIN, ORTHODIR, ORTHORES (cf. Young–Jea [1] and Hageman–Young [1, §12.3]), and GMRES (cf. Saad–Schultz [1], Walker [1]) are of this type. One tries to overcome these disadvantages partially by using not all but the last s directions for orthogonalisation (cf. Axelsson [2], Hageman–Young [1, §12.3]). It is possible to interpret GMRES as secant methods with a rank-1 update. On this basis, a more advantageous method was developed by Deuflhard–Freund–Walter [1] (contained in [Prog]). Furthermore, we refer to a discussion of cg-like methods for complex matrices by Freund [1].

Again, we recall the relation to eigenvalue problems: The orthogonality of the eigenvectors of Hermitian matrices does not carry over to the general case, but the left and right eigenvectors are orthogonal: $\langle e, e' \rangle = 0$, if $Ae = \lambda e$, $A^H e' = \bar{\mu} e'$, and $\lambda \neq \mu$. This property corresponds to the bi-cg methods in which two sequences of search directions p^m and p'^m are constructed. Partially, such methods can be interpreted as one of the methods mentioned above applied to the augmented symmetric system $\begin{bmatrix} 0 & A \\ A^H & 0 \end{bmatrix} \begin{bmatrix} x' \\ x \end{bmatrix} = \begin{bmatrix} b \\ b' \end{bmatrix}$ (b' arbitrary). Connections to the Lanczos bi-orthogonalisation (compare Gutknecht [1], [2] for a comprehensive collection of algorithms) exist. We also refer to the *squared cg method* CGS by Sonneveld–Wesseling–de Zeeuw [1]. Its stabilised version named Bi-CGSTAB is discussed by van der Vorst [3]. For a recent result we refer to Bank–Chan [1].

We can return to the standard cg method if we use the Kaczmarz iteration as the basic iteration. The left- or right-sided transformation with A^H yields the equations $A^H A x = b' := A^H b$ or $A A^H \hat{x} = b$ (cf. §8.2.2) with a positive definite matrix. The condition number is squared compared with A. However, the convergence of the cg method (4.6a–e) is characterised by $\sqrt{\varkappa} = \text{cond}(A)$ (cf. Theorem 4.12); hence, compensating the squaring of the condition number. The respective estimate for the method of conjugate residuals (applied to the Richardson iteration; cf. Theorem 13) is not better, because it is determined by $\varkappa = \varkappa(A)$.

Since matrices not positive definite require more or less involved cg variants, another remedy is worth being considered. As in §8.4, an indefinite or nonsymmetric problem can be preconditioned by a positive definite matrix B, where for solving $B\delta = c$, the standard cg method is applied as a secondary iteration.

Concus–Golub [1] and Widlund [1] describe an interesting method for general matrices A that are split into their symmetric and skew-symmetric parts: $A = A_0 + A_1$, $A_0 = \frac{1}{2}(A + A^H)$. For many applications, A_0 proves to be positive definite. Left and right transformations by $A_0^{-1/2}$ yield the matrix $A' := I - S$ with the skew-symmetric term $S := A_0^{-1/2} A_1 A_0^{-1/2}$. The eigenvalues of A' lie in a complex interval instead of a real one. For the respective cg version, one finds an error estimate with respect to the A_0 norm, depending on the condition number $\Lambda := \|A_0^{-1} A_1\|_2$ and leading to the asymptotical convergence rate $1 - O(1/\Lambda)$. In the cases of systems arising from partial differential equations, Λ is in general h-independent, leading to a convergence rate independent of the step size h. For each step of the algorithm, one system $A_0 \delta = c$ must be solved. This fact limits practicability. Under similar assumptions, the multi-grid method of the second kind even achieves convergence rate $O(h^\varkappa)$ with positive (!) exponents $\varkappa$ (cf. §10.9.1).

9.5.10 Further Comments

The scalar products and vector operations occurring in the cg method are optimally suited to vector computers. Variants of the conjugate gradient method suited for parallel computers are described by O'Leary [1–2] and Hackbusch [12].

In the last paragraph of §8.5.12 we complained about the extensive use of abbreviations. In the field of cg methods, the situation is even worse, since parts of the abbreviation refer to the underlying iteration («preconditioning») and another part indicates the cg variant.

Finally, we mention that vol. 29 (issue 4) of the journal BIT is devoted to cg methods.

10
Multi-Grid Methods

Multi-grid methods belong to the class of fastest iterations, because their convergence rate is independent of the step size h. Furthermore, their applicability does not require symmetry or positive definiteness, as, e.g., the standard conjugate gradient methods.

10.1 Introduction

10.1.1 Smoothing

Let A be the matrix of the Poisson-model problem. As the simplest example, we choose Richardson's iteration

$$x^{m+1} = \Phi_{\Theta}^{\text{Rich}}(x^m, b) = x^m - \Theta(Ax^m - b) \quad \text{with} \tag{10.1.1a}$$

$$\Theta = \tfrac{1}{8}h^2 \approx 1/\lambda_{\max} = 1/\varrho(A) \qquad \text{(cf. (4.1.1c))}. \tag{10.1.1b}$$

$\varrho(A) = \lambda_{\max}(A)$ is the eigenvalue corresponding to the eigenfunction

$$e^{\alpha\beta}(x, y) = 2h \sin \alpha x \sin \beta y \qquad (1 \leqslant \alpha, \beta \leqslant N-1, (x, y) \in \Omega_h) \tag{10.1.2a}$$

of the highest frequency $\alpha = \beta = N - 1$ (cf. §4.1). The convergence rate $\varrho(M_{\Theta}^{\text{Rich}}) = 1 - \Theta\lambda_{\min} \approx 1 - \lambda_{\min}/\lambda_{\max} \leqslant 1 - O(h^2)$ is attained by the lowest frequency $\alpha = \beta = 1$, i.e, when the error $e^m = x^m - x$ is a multiple of the eigenfunction $e^{1,1}$.

Now we change the point of view and ask: How does the iteration behave when applied to the subspace of *high* frequencies α, β? Thus far, all vectors x, b, etc. belonged to the vector space $X = \mathbb{R}^I$. For all $x \in X$, there is a representation (2b) by means of the eigenvector basis (2a):

$$x = \sum_{\alpha,\beta=1}^{N-1} \xi_{\alpha\beta} e^{\alpha\beta} \quad \text{with } \xi_{\alpha\beta} := \langle x, e^{\alpha\beta} \rangle. \tag{10.1.2b}$$

Since high frequencies α, β correspond to strong oscillations of the sine functions (2a), we define

$$X_{\text{osc}} := \text{span}\left\{ e^{\alpha\beta}\colon 1 \leqslant \alpha, \beta \leqslant N-1, \max\{\alpha,\beta\} > \frac{N}{2} \right\} \tag{10.1.2c}$$

as a subspace of the «oscillatory components». Note that at least one of the indices α, β lies in the «high-frequency» part $(N/2, N)$ of the frequency interval $[1, N-1]$. If we were able to generate an approximation x^0, whose error was in the subspace X_{osc}:

$$e^0 := x^0 - x \in X_{\text{osc}}, \quad (e^0 \text{ is the error vector not eigenvector!}), \tag{10.1.2d}$$

the simple Richardson iteration would yield fast convergence.

Lemma 10.1.1. *Assume the Poisson-model case with* (2d). *Then all succeeding errors e^m also belong to X_{osc} and satisfy the error estimate*

$$\|e^m\|_2 \leqslant \tfrac{3}{4} \|e^{m-1}\|_2, \tag{10.1.3}$$

i.e., restricted to X_{osc}, the convergence rate is h-independent.

Proof. Since the vectors (2a) are orthonormal (cf. Lemma 4.1.2), we have

$$\|x\|_2^2 = \sum_{\alpha,\beta=1}^{N-1} |\xi_{\alpha\beta}|^2 \quad \text{for } x \text{ from (2b)}. \tag{10.1.2e}$$

Because of $Me^{\alpha\beta} = (1 - \Theta\lambda_{\alpha\beta})e^{\alpha\beta}$, the application of the iteration matrix $M = I - \Theta A$ to the error e^m with coefficients $\xi_{\alpha\beta}$ yields:

$$\begin{aligned} \|e^{m+1}\|_2^2 &= \sum_{\alpha,\beta} |1 - \Theta\lambda_{\alpha\beta}|^2 |\xi_{\alpha\beta}|^2 \leqslant \max_{\alpha,\beta} |1 - \Theta\lambda_{\alpha\beta}|^2 \sum |\xi_{\alpha\beta}|^2 \\ &= \max_{\alpha,\beta} |1 - \Theta\lambda_{\alpha\beta}|^2 \|e^m\|_2^2 \end{aligned}$$

with $\lambda_{\alpha\beta} = 4h^{-2}[\sin^2(\alpha\pi h/2) + \sin^2(\beta\pi h/2)]$ (cf. (4.1.1a)). The maximum is to be taken over all α, β appearing in (2c). By symmetry, we may restrict the frequencies to $0 < \alpha < N$, $N/2 < \beta < N$. For these α, β,

$$2h^{-2} = 4h^{-2}\sin^2(\pi/4) < \lambda_{\alpha\beta} \leqslant \lambda_{N-1,N-1} < 8h^{-2}$$

holds; hence, $|1 - \Theta\lambda_{\alpha\beta}| < \frac{3}{4}$ proves assertion (3). □

The statement of the lemma is not of direct practical use, because the assumption (2d) cannot be established in practice (at least not with less work than for the exact solution of $Ax = b$). However, we can conclude the following.

Remark 10.1.2. Split the starting error e^0 into

$$e^0 = e^0_{osc} + e^0_{sm}, \quad e^0_{osc} \in X_{osc}, \quad e^0_{sm} \in X_{sm} := X^{\perp}_{osc}. \tag{10.1.4a}$$

Then after m steps of Richardson's iteration, we have

$$e^m = e^m_{osc} + e^m_{sm} \quad \text{with} \tag{10.1.4b}$$

$$e^m_{osc} = M^m e^0_{osc} \in X_{osc}, \qquad e^m_{sm} = M^m e^0_{sm} \in X_{sm}, \tag{10.1.4c}$$

$$\|e^m_{osc}\|_2 \leqslant (\tfrac{3}{4})^m \|e^0_{osc}\|_2, \tag{10.1.4d}$$

while e^m_{sm} converges only very slowly to 0. Since e^m_{osc} decreases faster than e^m_{sm}, the smooth part of e^m has increased and one may regard e^m as «smoother» than e^0.

For illustration purposes, we present the numerical results for the system

$$Ax = b \quad \text{with } A = h^{-2}\,\text{tridiag}\{-1, 2, -1\} \tag{10.1.5a}$$

of $n = N - 1$ ($N := 1/h$) equations corresponding to the one-dimensional Poisson boundary value problem

$$-u''(x) = f(x) \quad \text{for } 0 < x < 1, \qquad u(0) = u_0, u(1) = u_1. \tag{10.1.5b}$$

Figure 1 shows the (piecewise linearly connected) values e^0_i ($0 \leqslant i \leqslant N - 8$) as the first solid line. The additional Richardson iterates have errors e^m ($m =$ 1, 2, 3), which are insignificantly smaller but clearly smoother.

We will call iterative methods like the Richardson iteration (1a, b) used above *smoothing iterations* and use the symbol $\mathscr{S}$ instead of Φ.

In the following, an iteration Ψ with a complementary property is desired: Ψ should effectively reduce the smooth components in

$$X_{sm} := X^{\perp}_{osc} = \text{span}\{e^{\alpha\beta}\colon 1 \leqslant \alpha, \beta \leqslant N/2\}. \tag{10.1.6}$$

Ψ is not required to have good convergence with respect to the subspace X_{osc}. Then the product method $\Psi \circ \mathscr{S}$ would have the property that for

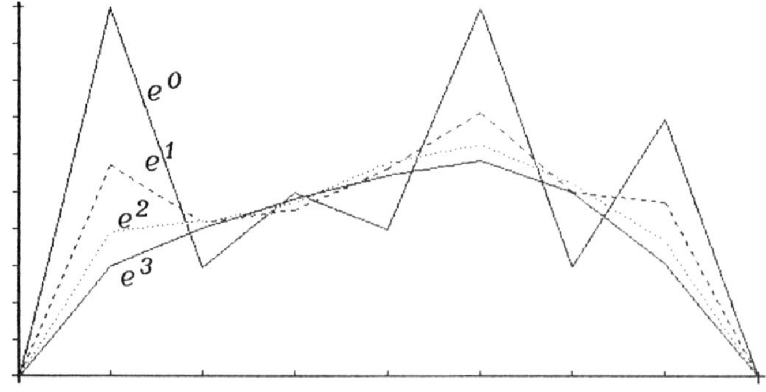

Fig. 10.1.1 Errors e^m for example (5a)

both subspaces X_{osc} and X_{sm}, one of the factors in $\Psi \circ \mathcal{S}$ would yield fast convergence.

Unfortunately, none of the methods mentioned before has this property. To construct such an iteration Ψ, we adhere to the concept that smooth grid functions can be well approximated by means of a coarser grid. After the introduction of coarser grids in §§10.1.2–4, we will return to the construction of this so-called «coarse-grid correction» Ψ in §10.1.5.

10.1.2 Hierarchy of Systems of Equations

For the followng considerations, we have to embed the problem $Ax = b$ into a family of systems. In the model case, for all step sizes $h = 1/N$, we obtain a system $Ax = b$ depending on N or h, respectively. Let

$$h_0 > h_1 > \cdots > h_{\ell-1} > h_\ell > \cdots \quad \text{with} \lim_{\ell \to \infty} h_\ell = 0 \tag{10.1.7a}$$

be a sequence of step sizes, which may be generated, e.g., by

$$h_\ell := h_0/2^\ell \qquad (\ell \geqslant 0). \tag{10.1.7b}$$

The index ℓ is the *level number*. $\ell = 0$ corresponds to the coarsest grid. In the model case, where the grid $\Omega_\ell := \Omega_{h_\ell}$ is contained in the unit square, the step size

$$h_0 = \tfrac{1}{2} \tag{10.1.7c}$$

is the coarsest one. Then $\Omega_0 = \Omega_{h_0}$ contains only *one* interior grid point.

Each step size h_ℓ (i.e., each level ℓ) corresponds to a system

$$A_\ell x_\ell = b_\ell \qquad (\ell = 0, 1, 2, \ldots) \tag{10.1.8a}$$

of equations of the dimension n_ℓ, which in the model case amounts to

$$n_\ell = (N_\ell - 1)^2 = (1/h_\ell - 1)^2. \tag{10.1.8b}$$

The family of systems (8a) for $\ell = 0, 1, 2, \ldots$ represents the *hierarchy of systems of equations*. The actual problem $Ax = b$ to be solved corresponds to one particular level $\ell = \ell_{max}$. During the solution of $A_\ell x_\ell = b_\ell$ at $\ell = \ell_{max}$, the lower levels $\ell < \ell_{max}$ are involved.

10.1.3 Prolongation

The vectors x_ℓ and b_ℓ from (8a) are elements of the vector space

$$X_\ell \cong \mathbb{R}^{n_\ell}. \tag{10.1.9}$$

To connect the different levels $\ell = 0, 1, 2, \ldots$, we introduce the *prolongation*

$$p\colon X_{\ell-1} \to X_\ell \qquad (\ell = 1, 2, \ldots), \tag{10.1.10}$$

which is assumed to be a linear and injective mapping (more precisely; a family of mappings for all $\ell \geqslant 1$) from the coarse grid into the fine one.

In the one-dimensional case (5–6), the vector x_ℓ can be considered the grid function defined on $\Omega_\ell = \{\mu h_\ell : 0 \leqslant \mu \leqslant N_\ell = 1/h_\ell\}$. We write x_ℓ as function u_ℓ connected to x_ℓ via

$$u_\ell(\mu h_\ell) = x_{\ell,\mu} \qquad (1 \leqslant \mu \leqslant N_\ell - 1), \tag{10.1.11a}$$

i.e., for all step sizes the arguments of u_ℓ belong to the interval $\Omega = (0,1)$ and are restricted to Ω_ℓ. For ease of notation, we add the boundary values

$$u_\ell(0) = u_\ell(1) = 0. \tag{10.1.11b}$$

An obvious proposal for the prolongation p is the piecewise linear interpolation between the grid points of $\Omega_{\ell-1}$:

$$(pu_{\ell-1})(\xi) := u_{\ell-1}(\xi) \qquad \text{for } \xi \in \Omega_{\ell-1} \subset \Omega_\ell, \tag{10.1.12a}$$

$$(pu_{\ell-1})(\xi) := \tfrac{1}{2}[u_{\ell-1}(\xi + h_\ell) + u_{\ell-1}(\xi - h_\ell)] \qquad \text{for } \xi \in \Omega_\ell \backslash \Omega_{\ell-1}, \tag{10.1.12b}$$

where the definition (11b) is used at $\xi = h_\ell$ and $\xi = 1 - h_\ell$. A shorter characterisation of the prolongation p is the symbol (12c):

$$p = [\tfrac{1}{2} \quad 1 \quad \tfrac{1}{2}]. \tag{10.1.12c}$$

(12c) indicates that the unit vector $x_{\ell-1} = (\ldots, 0, 1, 0, \ldots)^T$ is mapped into $x_\ell = px_{\ell-1} = (\ldots, 0, \tfrac{1}{2}, 1, \tfrac{1}{2}, 0, \ldots)^T$.

In the case of the two-dimensional Poisson equation, the vector x_ℓ is represented by the grid function u_ℓ:

$$u_\ell(\xi, \eta) = x_{\ell,ij} \quad \text{for } 1 \leqslant i, j \leqslant N_\ell - 1, (\xi, \eta) = (ih_\ell, jh_\ell) \in \Omega_\ell, \tag{10.1.13a}$$

where the boundary values are defined by

$$u_\ell(\xi, \eta) := 0 \quad \text{for } \xi = 0 \text{ or } \xi = 1 \text{ or } \eta = 0 \text{ or } \eta = 1. \tag{10.1.13b}$$

The two-dimensional generalisation of the piecewise linear interpolation (12a, b) (bilinear interpolation) reads

$$(pu_{\ell-1})(\xi, \eta) := u_{\ell-1}(\xi, \eta) \quad \text{for } (\xi, \eta) \in \Omega_{\ell-1} \subset \Omega_\ell, \tag{10.1.14a}$$

$$(pu_{\ell-1})(\xi, \eta) := \tfrac{1}{2}[u_{\ell-1}(\xi + h_\ell, \eta) + u_{\ell-1}(\xi - h_\ell, \eta)] \quad \text{for } \xi/h_\ell \text{ odd}, \eta/h_\ell \text{ even}, \tag{10.1.14b}$$

$$(pu_{\ell-1})(\xi, \eta) := \tfrac{1}{2}[u_{\ell-1}(\xi, \eta + h_\ell) + u_{\ell-1}(\xi, \eta - h_\ell)] \quad \text{for } \xi/h_\ell \text{ even}, \eta/h_\ell \text{ odd}, \tag{10.1.14c}$$

$$\begin{aligned}(pu_{\ell-1})(\xi, \eta) := \tfrac{1}{4}[&u_{\ell-1}(\xi + h_\ell, \eta + h_\ell) + u_{\ell-1}(\xi - h_\ell, \eta - h_\ell)] \\ &+ u_{\ell-1}(\xi - h_\ell, \eta + h_\ell) + u_{\ell-1}(\xi + h_\ell, \eta - h_\ell)]\end{aligned} \quad \text{for } \xi/h_\ell, \eta/h_\ell \text{ odd}. \tag{10.1.14d}$$

The abbreviation for p from (14a–d) is the star

$$p = \begin{bmatrix} 1/4 & 1/2 & 1/4 \\ 1/2 & 1 & 1/2 \\ 1/4 & 1/2 & 1/4 \end{bmatrix} \qquad \textit{(nine-point prolongation).} \tag{10.1.14e}$$

In general, the stencil

$$p = \begin{bmatrix} \pi_{-1,1} & \pi_{0,1} & \pi_{1,1} \\ \pi_{-1,0} & \pi_{0,0} & \pi_{1,0} \\ \pi_{-1,-1} & \pi_{0,-1} & \pi_{1,-1} \end{bmatrix} \tag{10.1.15a}$$

describes the mapping (15b), where the summation is taken over all i, j with $(\xi - ih_\ell, \eta - jh_\ell) \in \Omega_{\ell-1}$:

$$(pu_{\ell-1})(\xi, \eta) := \sum_{i,j} \pi_{ij} u_{\ell-1}(\xi - ih_\ell, \eta - jh_\ell) \quad \text{for } (\xi, \eta) \in \Omega_\ell. \tag{10.1.15b}$$

Other linear interpolations as well as prolongations of a higher order are discussed by Hackbusch [14, §3.4]. A so-called *matrix-dependent prolongation* is defined by (15a) with the coefficients

$$\pi_{00} := 1, \quad \pi_{\pm 1,0} := -\sum_j \alpha_{\mp 1,j} \Big/ \sum_j \alpha_{0,j}, \quad \pi_{0,\pm 1} := -\sum_i \alpha_{i,\mp 1} \Big/ \sum_i \alpha_{i,0}, \tag{10.1.16a}$$

$$(A_\ell p u_{\ell-1})(\xi, \eta) = 0 \quad \text{for } \xi/h_\ell,\ \eta/h_\ell \text{ odd}, \tag{10.1.16b}$$

where $\alpha_{i,j}$ are the coefficients of A_ℓ according to (2.1.8a, b). Condition (16b) determines $\pi_{\pm 1, \pm 1}$ (cf. Hackbusch [14, §10.3] and de Zeeuw [1]).

10.1.4 Restriction

The restriction r is a linear and surjective mapping

$$r: X_\ell \to X_{\ell-1} \qquad (\ell \geqslant 1), \tag{10.1.17}$$

which maps fine-grid functions into coarse-grid functions. If $\Omega_{\ell-1} \subset \Omega_\ell$ holds as in the model case, the simplest choice is the *trivial restriction*

$$(r_{\text{triv}} u_\ell)(\xi, \eta) = u_\ell(\xi, \eta) \quad \text{for } (\xi, \eta) \in \Omega_{\ell-1}. \tag{10.1.18}$$

However, because of certain disadvantages, we advise against its use (cf. Hackbusch [14, §3.5]). Instead, we define $(ru_\ell)(\xi, \eta)$ as the weighted mean of the neighbouring values. The stencil

$$r = \begin{bmatrix} \varrho_{-1,1} & \varrho_{0,1} & \varrho_{1,1} \\ \varrho_{-1,0} & \varrho_{0,0} & \varrho_{1,0} \\ \varrho_{-1,-1} & \varrho_{0,-1} & \varrho_{1,-1} \end{bmatrix} \tag{10.1.19a}$$

characterises the restriction

$$(ru_\ell)(\xi,\eta) = \sum_{i,j=-1}^{1} \varrho_{ij} u_\ell(\xi + ih_\ell, \eta + jh_\ell) \quad \text{for } (\xi,\eta) \in \Omega_{\ell-1}, \tag{10.1.19b}$$

The nine-point prolongation (14e) corresponds to the *nine-point restriction*

$$r = \frac{1}{4}\begin{bmatrix} 1/4 & 1/2 & 1/4 \\ 1/2 & 1 & 1/2 \\ 1/4 & 1/2 & 1/4 \end{bmatrix}, \tag{10.1.20}$$

which can be considered as adjoint to (14e), when the definition of adjoint mappings is based on the scalar products

$$\langle \cdot, \cdot \rangle = \langle \cdot, \cdot \rangle_\ell \quad \text{with } \langle u_\ell, v_\ell \rangle_\ell = h_\ell^d \sum_{\alpha \in I} u_{\ell,\alpha} \overline{v_{\ell,\alpha}} \tag{10.1.21}$$

for X_ℓ. d is the dimension of the grid $\Omega_\ell \subset \mathbb{R}^d$.

Exercise 10.1.3. Assume the two-dimensional case $d = 2$ and prove that the mapping adjoint to p defined by (15a) is r from (19a) with $\varrho_{ij} = \pi_{ij}/4$.

Having fixed the prolongation, one can always choose the adjoint mapping

$$r := p^* \tag{10.1.22}$$

as a restriction. For example, one can define the *matrix-dependent restriction* by (16a, b) and (22).

10.1.5 Coarse-Grid Correction

Let $\bar{x}_\ell$ be the result of a few steps of the smoothing iteration (1a, b). The corresponding error

$$\bar{e}_\ell := \bar{x}_\ell - x_\ell \tag{10.1.23a}$$

is the exact correction by which the solution can be computed:

$$x_\ell = \bar{x}_\ell - e_\ell \tag{10.1.23a'}$$

Since $A_\ell \bar{e}_\ell = A_\ell(\bar{x}_\ell - x_\ell) = A_\ell \bar{x}_\ell - A_\ell x_\ell = A_\ell \bar{x}_\ell - b_\ell$, $\bar{e}_\ell$ satisfies the equation

$$A_\ell \bar{e}_\ell = d_\ell \quad \text{with the } \textit{defect } d_\ell := A_\ell \bar{x}_\ell - b_\ell. \tag{10.1.23b}$$

According to the considerations of §10.1.1, $\bar{e}_\ell$ is smooth. Therefore, it should be possible to approximate $\bar{e}_\ell$ by means of the coarse grid: $\bar{e}_\ell \approx pe_{\ell-1}$. As ansatz for $e_{\ell-1}$, we take the coarse-grid equation corresponding to (23b):

$$A_{\ell-1} e_{\ell-1} = d_{\ell-1} \quad \text{with } d_{\ell-1} := rd_\ell. \tag{10.1.23c}$$

Assume that we are able to solve the coarse-grid equation (23c) exactly:

$$e_{\ell-1} = A_{\ell-1}^{-1} d_{\ell-1} \tag{10.1.23d}$$

Its image $pe_{\ell-1}$ under the prolongation p should approximate the solution $\bar{e}_\ell$ of (23b), so that the coarse-grid correction is completed by

$$x_\ell^{\text{new}} := \bar{x}_\ell - pe_{\ell-1}. \tag{10.1.23e}$$

In compact form the *coarse-grid correction* (23b–e) reads as follows:

$$\bar{x}_\ell \mapsto x_\ell^{\text{new}} := \bar{x}_\ell - pA_{\ell-1}^{-1} r(A_\ell \bar{x}_\ell - b_\ell). \tag{10.1.24}$$

If we rename $\bar{x}_\ell$ and x_ℓ^{new} by x_ℓ^m and x_ℓ^{m+1}, (24) defines an iterative method, the so-called coarse-grid correction:

$$\Phi_\ell^{\text{CGC}}(x_\ell, b_\ell) := x_\ell - pA_{\ell-1}^{-1} r(A_\ell x_\ell - b_\ell). \tag{10.1.24'}$$

Remark 10.1.4. The iteration matrix M and the matrix N of the second normal form of the coarse-grid correction are

$$M_\ell^{\text{CGC}} = I - pA_{\ell-1}^{-1} r A_\ell, \qquad N_\ell^{\text{CGC}} = pA_{\ell-1}^{-1} r. \tag{10.1.25}$$

Φ_ℓ^{CGC} (as such without smoothing) is not an interesting iteration as shown in

Remark 10.1.5. The coarse-grid correction Φ_ℓ^{CGC} is consistent, but not convergent.

Proof. The consistency is a consequence of the second normal form. $h_\ell < h_{\ell-1}$ implies $\dim X_\ell > \dim X_{\ell-1}$; hence, the kernel of r is not trivial. Let $0 \neq \xi \in \text{kernel}(r)$. Since $M_\ell^{\text{CGC}}\eta = \eta$ for $\eta := A_\ell^{-1}\xi$, $\varrho(M_\ell^{\text{CGC}}) \geqslant 1$ indicates divergence. □

For systems of equations obtained from a Galerkin discretisation (cf. Exercise 6.12 and Hackbusch [14, Note 3.6.6]), the so-called *Galerkin product* representation is valid:

$$A_{\ell-1} = rA_\ell p. \tag{10.1.26}$$

Lemma 10.1.6. *Assume* (26). *Then* $\Phi_\ell^{\text{CGC}}(\hat{x}_\ell, b_\ell) = \hat{x}_\ell$ *holds for all vectors* $\hat{x}_\ell$ *with* $\hat{e}_\ell = \hat{x}_\ell - x_\ell \in \text{range}(p)$. *In particular,* Φ_ℓ^{CGC} *is a projection.*

Proof. Use $M_\ell^{\text{CGC}} p = p - pA_{\ell-1}^{-1} r A_\ell p = p - pA_{\ell-1}^{-1} A_{\ell-1} = p - p = 0$. □

The next exercise shows that the coarse-grid equation (23c) is a reasonable ansatz for $e_{\ell-1}$.

Exercise 10.1.7. Let A_ℓ be positive definite. The best approximation of $\bar{e}_\ell \in X_\ell$ with respect to the A_ℓ norm $\|x_\ell\|_A := \langle A_\ell x_\ell, x_\ell \rangle_\ell^{1/2}$ is $pe_{\ell-1}$, where $p = r^*$ according to (22) and $e_{\ell-1}$ is the solution of (23c) with the Galerkin matrix (26).

10.2 Two-Grid Method

10.2.1 Algorithm

The smoothing iteration $\mathscr{S}_\ell$ is defined in §10.1.1 and the coarse-grid correction Φ_ℓ^{CGC} constructed in §10.1.5. The two-grid iteration is the product method

$$\Phi_\ell^{\mathrm{TGM}} := \Phi_\ell^{\mathrm{CGC}} \circ \mathscr{S}_\ell^{\nu} \qquad (\ell \geqslant 1, \nu \geqslant 1), \tag{10.2.1}$$

where ν is the number of smoothing steps. In algorithmic notation, (1) takes the form (2):

$$\textit{procedure } \Phi_\ell^{\mathrm{TGM}}(x_\ell, b_\ell); \tag{10.2.2}$$

$$\textit{begin for } i := 1 \textit{ to } \nu \textit{ do } x_\ell := \mathscr{S}_\ell(x_\ell, b_\ell); \tag{10.2.2a}$$

$$d_{\ell-1} := r(A_\ell x_\ell - b_\ell); \tag{10.2.2b}$$

$$e_{\ell-1} := A_{\ell-1}^{-1} d_{\ell-1}; \tag{10.2.2c}$$

$$x_\ell := x_\ell - p e_{\ell-1}; \tag{10.2.2d}$$

$$\Phi_\ell^{\mathrm{TGM}} := x_\ell \tag{10.2.2e}$$

end;

10.2.2 Modifications

A semi-iterative smoothing instead of (2a) will be discussed in §10.8.1. As known from Exercise 3.2.16b, $\Phi_\ell^{\mathrm{CGC}} \circ \mathscr{S}_\ell^{\nu}$ has the same convergence behaviour as

$$\Phi_\ell^{\mathrm{TGM}}(\nu_1, \nu_2) := \mathscr{S}_\ell^{\nu_2} \circ \Phi_\ell^{\mathrm{CGC}} \circ \mathscr{S}_\ell^{\nu_1} \quad \text{with } \nu = \nu_1 + \nu_2. \tag{10.2.3a}$$

In this case, ν_1 *pre-* and ν_2 *post-smoothing* steps are applied. Algorithm (1) is the special case of iteration (3a) when $\nu_1 = \nu$ and $\nu_2 = 0$. Therefore, in the sequel, we will use version (3a).

One may also use different iterations $\mathscr{S}_\ell$ and $\hat{\mathscr{S}}_\ell$ as pre- and post-smoothers:

$$\hat{\mathscr{S}}_\ell^{\nu_2} \circ \Phi_\ell^{\mathrm{CGC}} \circ \mathscr{S}_\ell^{\nu_1} \quad \text{with } \nu = \nu_1 + \nu_2. \tag{10.2.3b}$$

10.2.3 Iteration Matrix

Lemma 10.2.1. *Let $\mathscr{S}_\ell$ be a consistent iteration with iteration matrix S_ℓ. Then $\Phi_\ell^{\mathrm{TGM}(\nu_1,\nu_2)}$ is a consistent iteration with the iteration matrix*

$$M_\ell^{\mathrm{TGM}}(\nu_1, \nu_2) = S_\ell^{\nu_2}(I - pA_{\ell-1}^{-1} r A_\ell) S_\ell^{\nu_1}. \tag{10.2.4}$$

Proof. According to Exercise 3.2.16b, M_ℓ^{TGM} is the product of the iteration matrices of $\mathscr{S}_\ell^{\nu_2}$, Φ_ℓ^{CGC}, $\mathscr{S}_\ell^{\nu_1}$. (4) follows from Remark 1.4. □

10.2.4 Pascal Procedures

The list (3.5.7) contains the types `kind_of_restriction` and `kind_of_prolongation`. The following possible choices have not yet been explained: the *seven-point prolongation* and *restriction* (5) and the *five-point restriction* (half-weighting) (6), which in general should be avoided:

$$p = \begin{bmatrix} 0 & 1/2 & 1/2 \\ 1/2 & 1 & 1/2 \\ 1/2 & 1/2 & 0 \end{bmatrix}, \qquad r = \frac{1}{4}\begin{bmatrix} 0 & 1/2 & 1/2 \\ 1/2 & 1 & 1/2 \\ 1/2 & 1/2 & 0 \end{bmatrix} = p^*, \tag{10.2.5}$$

$$r = \frac{1}{2}\begin{bmatrix} 0 & 1/4 & 0 \\ 1/4 & 1 & 1/4 \\ 0 & 1/4 & 0 \end{bmatrix}, \qquad p = \begin{bmatrix} 0 & 1/2 & 0 \\ 1/2 & 1 & 1/2 \\ 0 & 1/2 & 0 \end{bmatrix}. \tag{10.2.6}$$

The *five-point prolongation* from (6) should be used only if the values at (ξ, η) for odd ξ/h_ℓ and η/h_ℓ may remain undefined because a following chequerboard Gauß-Seidel step redefines these values in any way. The choice `general_restriction` and `general_prolongation` admits an arbitrary star (1.15a) or (1.19a), respectively. The record of type `MG_data` has a component `AL` containing the `data_of_discretisation` for all levels $0 \leqslant \ell \leqslant$ `Lmax` and a component `IP` providing the iteration parameters possibly needed by $\mathscr{S}_\ell$. The component `MI` of type `MG_parameter` contains the pre- and post-smoothing numbers ν_1, ν_2 and the data R, P needed for r and p.

Below we give the procedures for determining (i) r from (1.22), (ii) p according to (1.16a, b), for performing (iii) r and (iv) p, as well as (v) for the computation of $A_{\ell-1}$ from A_ℓ according to (1.26). The procedures for performing p and r can be written in much shorter form, if only the nine-point versions are used.

```
procedure restriction_as_adjoint_prolongation(var S: star);
 var i,j: integer;
begin for i:=-1 to 1 do for j:=-1 to 1 do S[i,j]:=S[i,j]/4
end;

procedure generate_matrixdependent_prolongation_star(var PS:
 star;
                                    var A: data_of_discretisation);
var i,j: integer; h,ha,v,va: column_star; p: real;
procedure normalise(var s: column_star);
  begin s[-1]:=-abs(s[-1]); s[0]:=abs(s[0]); s[1]:=-abs(s[1]);
        if s[0]-s[-1]-s[1]<0 then s[0]:=-s[-1]-s[1]
  end;
```

```
procedure check(var s,t: column_star);
 begin p:=t[-1]+t[0]+t[1]; normalise(s); normalise(t);
      if s[0]<0.1*p*A.h2 then s:=t
 end;
begin for i:=-1 to 1 do h[i]:=0; ha:=h; v:=h; va:=h;
 {initialise by zero}
    for i:=-1 to 1 do for j:=-1 to 1 do
    begin p:=A.S[i,j]; h[i]:=h[i]+p; v[j]:=v[j]+p;
          p:=abs(p); ha[i]:=ha[i]+p; va[j]:=va[j]+p
    end; {row and column sums}
    check(v,va); check(h,ha);
    PS[0,0]:=1;
    PS[1,0]:=-h[-1]/h[0]; PS[-1,0]:=-h[1]/h[0]; {horizontal
     interpolation}
    PS[0,1]:=-v[-1]/v[0]; PS[0,-1]:=-v[1]/v[0]; {vertical
     interpolation}
    with A do
    begin PS[1,1]:=-(S[-1,-1]+S[-1,0]*PS[0,1]+S[0,-1]*
           PS[1,0])/S[0,0];
          PS[-1,-1]:=-(S[1,1]+S[1,0]*PS[0,-1]+S[0,1]*
           PS[-1,0])/S[0,0];
          PS[-1,1]:=-(S[1,-1]+S[1,0]*PS[0,1]+S[0,-1]*
           PS[-1,0])/S[0,0];
          PS[1,-1]:=-(S[-1,1]+S[-1,0]*PS[0,-1]+S[0,1]*
           PS[1,0])/S[0,0]
end end;

procedure MG_restriction(level: integer; var x_restricted, x:
 gridfunction;
           var R: data_of_restriction; var AL: hierarchy_of_
            discretisation);
var i,j,ii,jj,nx1,ny1: integer; label 1;
begin nx1:=AL[level-1].nx-1; ny1:=AL[level-1].ny-1; with R do
 case kind of
trivial_restriction:
 for i:=1 to nx1 do for j:=1 to ny1 do x_restricted[i,j]:=
  x[2*i,2*j];
fivepoint_restriction:
 for i:=1 to nx1 do
 begin ii:=2*i; for j:=1 to ny1 do
     begin jj:=2*j; x_restricted[i,j]:=
                    (4*x[ii,jj]+x[ii-1,jj]+x[ii+1,jj]+
                     x[ii,jj-1]+x[ii,jj+1])/8
 end end;
```

```
sevenpoint_restriction:
 for i:=1 to nx1 do
 begin ii:=2*i; for j:=1 to ny1 do
     begin jj:=2*j; x_restricted[i,j]:=(2*x[ii,jj]+
                    x[ii-1,jj]+x[ii+1,jj]+x[ii,jj-1]+x[ii,jj+
                     1]+x[ii-1,jj-1]+x[ii+1,jj+1])/8
 end end;
ninepoint_restriction:
 for i:=1 to nx1 do
 begin ii:=2*i; for j:=1 to ny1 do
     begin jj:=2*j; x_restricted[i,j]:=(4*x[ii,jj]+
                      2*(x[ii-1,jj]+x[ii+1,jj]+x[ii,jj-1]+
                      x[ii,jj+1])+x[ii-1,jj+1]+x[ii+1,jj-1]
                      +x[ii-1,jj-1]+x[ii+1,jj+1])/16
 end end;

general_restriction:
1: for i:=1 to nx1 do
 begin ii:=2*i; for j:=1 to ny1 do
     begin jj:=2*j;
           x_restricted[i,j]:=S[0,0]*x[ii,jj]+S[-1,0]*
           x[ii-1,jj]+S[1,0]*x[ii+1,jj]+S[0,-1]*x[ii,jj-1]
           +S[0,1]*x[ii,jj+1]+S[-1,1]*x[ii-1,jj+1]
           +S[1,-1]*x[ii+1,jj-1]+S[-1,-1]*x[ii-1,jj-1]
           +S[1,1]*x[ii+1,jj+1]
 end end;

matrixdependent_restriction:
 begin generate_matrixdependent_prolongation_
  star(S,AL[level]);
       restriction_as_adjoint_prolongation(S); goto 1
end end end; {boundary values are undefined!}

procedure MG_prolongation(level: integer; var xp,x:
 gridfunction;
         var P: data_of_prolongation; var AL: hierarchy_of_
          discretisation);
var i,j,ii,jj,nxx,nyy: integer; label 1,9;
begin nxx:=AL[level-1].nx; nyy:=AL[level-1].ny; with P do
begin for i:=nxx downto 0 do
 begin ii:=2*i; for j:=nyy downto 0 do xp[ii,2*j]:=x[i,j]
 end;
 case kind of
general_prolongation: goto 1;
```

```
matrixdependent_prolongation:
 begin generate_matrixdependent_prolongation_
  star(S,AL[level]); goto 1
 end end {case};
 for i:=0 to nxx do
 begin ii:=2*i; jj:=1; for j:=1 to nyy do
     begin xp[ii,jj]:=(xp[ii,jj-1]+xp[ii,jj+1])/2; jj:=jj+2
 end end; {linear interpolation in y direction}
 if kind=ninepoint_prolongation then
 begin for i:=1 to nxx do
   begin ii:=2*i-1; for jj:=0 to 2*nyy do xp[ii,jj]:=
    (xp[ii-1,jj]+xp[ii+1,jj])/2
   end; {linear interpolation in x direction}
   goto 9
 end;
 ii:=1; for i:=1 to nxx do
 begin jj:=0; for j:=0 to nyy do
   begin xp[ii,jj]:=(xp[ii-1,jj]+xp[ii+1,jj])/2; jj:=jj+2
   end;
   ii:=ii+2
 end {linear interpolation in x direction at coarse-grid
  points};
 if kind=fivepoint_prolongation then goto 9;
 {For the 5-point prolongation the values xp[odd,odd] remain
  undefined!}
 {It remains the case kind=sevenpoint_prolongation}
 ii:=1; for i:=1 to nxx do
 begin jj:=1; for j:=1 to nyy do
   begin xp[ii,jj]:=(xp[ii-1,jj-1]+xp[ii+1,jj+1])/2; jj:=jj+2
   end;
   ii:=ii+2
 end; {linear interpolation in diagonal direction}
 goto 9;
1: {general_prolongation}
 for i:=0 to nxx do
 begin ii:=2*i; jj:=1; for j:=1 to nyy do
     begin xp[ii,jj]:=S[0,1]*xp[ii,jj-1]+S[0,-1]*xp[ii,jj+1];
      jj:=jj+2
 end end; {continuation in y direction}
 ii:=1; for i:=1 to nxx do
 begin jj:=0; for j:=0 to nyy do
   begin xp[ii,jj]:=S[1,0]*xp[ii-1,jj]+S[-1,0]*xp[ii+1,jj];
    jj:=jj+2
   end; ii:=ii+2
```

```
  end {continuation in x direction};
  ii:=1; for i:=1 to nxx do
  begin jj:=1; for j:=1 to nyy do
    begin xp[ii,jj]:=S[1,1]*xp[ii-1,jj-1]+S[-1,-1]*
                       xp[ii+1,jj+1]
                      +S[1,-1]*xp[ii-1,jj+1]+S[-1,1]*
                       xp[ii+1,jj-1];
          jj:=jj+2
    end; ii:=ii+2
  end {continuation in points with two odd coefficients};
  if S[0,0]<>1 then for i:=0 to nxx do
  begin ii:=2*i; jj:=0; for j:=0 to nyy do
      begin xp[ii,jj]:=S[0,0]*xp[ii,jj]; jj:=jj+2
  end end; {modification at coarse-grid points}
9: end
end;

procedure Galerkin_star(var RSP: star; var fine: data_of_
  discretisation;
                        var P: data_of_prolongation; var R: data_
                         of_restriction);
var v: gridfunction; A: hierarchy_of_discretisation; i,j:
  integer;
begin with A[0] do begin nx:=4; ny:=4 end;
    A[1]:=fine; with A[1] do begin nx:=8; ny:=8 end;
    zero_gridfunction(4,4,v); v[2,2]:=1;
    MG_prolongation(1,v,v,P,A); A_x(v,A[1],v); MG_
     restriction(1,v,v,R,A);
    for i:=-1 to 1 do for j:=-1 to 1 do RSP[i,j]:=v[2-i,2-j]
end;
```

For the interactive input of p and r one may use

```
procedure determine_restriction(var R: data_of_restriction);
procedure determine_prolongation(var P: data_of_
  prolongation);
procedure determine_prolongation_and_restriction
                       (var P: data_of_prolongation; var R: data_
                        of_restriction);
```

If $A_\ell x_\ell = b_\ell$ is to be solved for $\ell = \ell_{\max}$, the matrices A_ℓ for $\ell < \ell_{\max}$ are only of an auxiliary character. Essentially, there are two possibilities for the suitable choice of the matrices A_ℓ ($\ell < \ell_{\max}$), which can be selected by means of the parameter Galerkin in MG_parameter:

A_ℓ is generated by the same discretisation as used for

$\ell = \ell_{max}$ (Galerkin:=false), (10.2.7a)

A_ℓ is generated by means of (1.26) from $A_{\ell_{max}}$

(Galerkin:=true). (10.2.7b)

The discretisation can be described by define_MG_discretisation. The coefficients $\alpha_{ij} = \alpha_{ij,0} + \alpha_{ij,1} h_\ell^{-1} + \alpha_{ij,2} h_\ell^{-2}$ of A_ℓ (cf. §2.1.3) are determined by three stars $(\alpha_{ij,k})$ $(k = 0, 1, 2)$ in order to evaluate A_ℓ separately for all h_ℓ. For the case (7b) (Galerkin:=true) mentioned above, the actual level number ℓ_{max} = actual_level can be defined by set_actual_level (cf. §10.4.2).

The two-grid iteration is not described as a procedure, because it is a particular case of the multi-grid procedure described later (cf. §10.4.2). The choice «direct:=true; lmin:=level-1» of the parameters in the multi-grid procedure yields the two-grid method. The complete frame program for the two-grid algorithm in the Poisson-model case may read as follows:

```
program two_grid_iteration;
var MGD: MG_data; level,i,itnr: integer; it: data_of_
 iteration;
{right_hand_side, zerofunction, exact_solution, boundary_
 value etc. to be inserted}
begin initialise_MG_data(MGD); initialise_it(it); with MGD do
  repeat release_MG_data(MGD); release_it(it);
   define_MG_discretisation(MGD);
   {definition of the two-grid iteration parameters}
   writeln('Should the Galerkin product (10.1.26) be used?');
   MI.Galerkin:=yes_no; MI.gamma:=1; MI.direct:=true;
   writeln('number of pre- (ny1) and post-smoothing steps
    (ny2).');
   write('--> ny1 ='); readln(MI.ny1); write('--> ny2 =');
    readln(MI.ny2);
   determine_prolongation_and_restriction(MI.P,MI.R);
   write('--> solution at level l='); readln(level); MI.lmin:=
    level-1;
   set_actual_level(level,MGD); it.A:=AL[level];
   define_data_of_system(it,boundary_value,right_hand_side);
   define_starting_iterate(it,zerofunction); writeln('start
    defined.');
   write('--> Number of iterations ='); readln(itnr);
    it.ip.Nr:=0;
```

```
  comparison_with_exact_solution(12,v,it,Euclidean_norm);
  for i:=1 to itnr do
  begin MG_iteration (level, it.x, it.x, it.b, MGD,
             chequerboard_Gauss_Seidel,
             chequerboard_Gauss_Seidel, solve_system_without_
              pivoting);
        it.ip.nr:=it.ip.nr+1 {output of the results possible}
  end; writeln('Should the run be repeated?')
 until not yes_no
end.
```

The last parameter of the procedure `MG_iteration` is the direct solver for $A_{\ell-1}e_{\ell-1} = d_{\ell-1}$. The procedure `solve_system_without_pivoting` used here performs a triangular decomposition without the pivoting of a band matrix (band width $\leqslant$ `Wmax`).

10.2.5 Numerical Examples

As an example, we choose the Poisson-model problem with the two-grid parameters $\nu_1 = 2$, $\nu_2 = 0$, `Galerkin:=false` for the step sizes h_ℓ from (1.7b, c). The error norms $\|x_\ell^m - x_\ell\|_2$ at level $\ell = 5$ with $h_5 = 1/64$ are shown in Table 1. Table 2 yields a summary of the reduction factors $\|x_\ell^m - x_\ell\|_2 / \|x_\ell^{m-1} - x_\ell\|_2$. The last row in Table 2 contains the averaged convergence factors $\varrho := (\|x_\ell^8 - x_\ell\|_2 / \|x_\ell^0 - x_\ell\|_2)^{1/8}$. In contrast to the foregoing iterative methods, the convergence factors hardly depend on step size. Furthermore, the convergence rate of the value ≈ 0.06 is very favourable.

Since the two-grid method depends on the parameters ν_1 and ν_2, their influence on convergence should be discussed. As mentioned in §10.2.2, con-

Table 10.2.1 Errors for $h_\ell = 1/64$

m	$\|x_\ell^m - x_\ell\|_2$
0	$2.935_{10} - 02$
1	$1.210_{10} - 03$
2	$6.206_{10} - 05$
3	$3.378_{10} - 06$
4	$1.939_{10} - 07$
5	$1.152_{10} - 08$
6	$7.058_{10} - 10$
7	$4.432_{10} - 11$
8	$7.188_{10} - 12$

Table 10.2.2 Error quotients $\|x_\ell^m - x_\ell\|_2/\|x_\ell^{m-1} - x_\ell\|_2$ and averaged convergence factors ϱ_ℓ for the two-grid method with $\nu_1 = 2$ and $\nu_2 = 0$

m	$h_\ell =$ 1/4	1/8	1/16	1/32	1/64
1	$6.210_{10}-2$	$5.549_{10}-2$	$4.730_{10}-2$	$4.338_{10}-2$	$4.124_{10}-2$
2	$6.247_{10}-2$	$5.737_{10}-2$	$5.409_{10}-2$	$5.234_{10}-2$	$5.126_{10}-2$
3	$6.249_{10}-2$	$5.850_{10}-2$	$5.804_{10}-2$	$5.565_{10}-2$	$5.443_{10}-2$
4	$6.249_{10}-2$	$5.963_{10}-2$	$6.191_{10}-2$	$5.867_{10}-2$	$5.741_{10}-2$
5	$6.250_{10}-2$	$6.061_{10}-2$	$6.512_{10}-2$	$6.091_{10}-2$	$5.939_{10}-2$
6	$6.250_{10}-2$	$6.143_{10}-2$	$6.767_{10}-2$	$6.293_{10}-2$	$6.125_{10}-2$
7	$6.250_{10}-2$	$6.194_{10}-2$	$6.943_{10}-2$	$6.453_{10}-2$	$6.280_{10}-2$
8	$6.171_{10}-2$	$6.073_{10}-2$	$7.024_{10}-2$	$7.078_{10}-2$	$(1.621_{10}-1)$
ϱ_ℓ	$6.234_{10}-2$	$5.942_{10}-2$	$6.123_{10}-2$	$5.810_{10}-2$	$5.493_{10}-2$

Table 10.2.3 Convergence factors for different smoothing numbers ν

ν	1	2	3	4	5	6	10
$\varrho_3(\nu)$	0.222	0.062	0.04	0.03	0.023	0.0196	0.0133
$\dfrac{0.135}{\nu + 0.135}$	0.119	0.063	0.043	0.033	0.026	0.022	0.0133

vergence depends only on $\nu = \nu_1 + \nu_2$. Therefore, one may choose $\nu_1 = \nu$ and $\nu_2 = 0$ without loss of generality. The convergence factors ϱ_3 for $h_3 = 1/16$ determined as above are shown in Table 3. As expected, convergence improves with increasing ν. In the last row, a comparison with the function $C/(C + \nu)$ for $C = 0.135$ is given. It suggests the asymptotical behaviour $\varrho_\ell(\nu) \approx O(1/\nu)$.

10.3 Analysis for a One-Dimensional Example

In principle, the analysis of two-grid convergence for the Poisson-model problem is possible (cf. Hackbusch [14, §8.1.1]); however, it is not sufficiently transparent for an introductory consideration. Therefore, we consider the tridiagonal equation (1.5a):

$$Ax = b \quad \text{with } A = h^{-2}\,\text{tridiag}\{-1, 2, -1\} = h^{-2}\begin{bmatrix} 2 & -1 & & \\ -1 & 2 & \ddots & \\ & \ddots & \ddots & -1 \\ & & -1 & 2 \end{bmatrix} \tag{10.3.1}$$

discretising the one-dimensional Poisson equation (5b). It should be emphasized that tridiagonal matrices are easy to solve directly. The analysis of iterative methods for these tridiagonal equations is of interest only because of the fact that the convergence properties also carry over to the general case of two or more space dimensions. Furthermore, this chapter serves as a demonstration of how model problems can be investigated by means of Fourier analysis.

10.3.1 Fourier Analysis

We abbreviate the quantities at the levels ℓ and $\ell - 1$ by

$$N = N_\ell, \quad N' = N_{\ell-1}, \quad h = h_\ell = 1/N, \quad h' = h_{\ell-1} = 2h. \tag{10.3.2}$$

The vector $x = (x_k)_{1 \leqslant k \leqslant N-1}$ is formally extended by the components

$$x_0 = x_N = 0. \tag{10.3.3}$$

The vectors (grid functions) e^α with the components

$$e_k^\alpha = \sqrt{2h}\sin(\alpha k \pi h) \qquad (0 \leqslant k \leqslant N) \tag{10.3.4}$$

satisfy the condition (3) for all $\alpha \in \mathbb{Z}$. According to Exercise 4.1.3, the vectors $\{e^\alpha\colon 1 \leqslant \alpha \leqslant N-1\}$ form an orthonormal basis. Therefore, the matrix Q built by e^α as columns is unitary: $Q^H Q = I$ (cf. Lemma 2.7.4):

$$Q := [e^1, e^{N-1}, e^2, e^{N-2}, \ldots, e^\alpha, e^{N-\alpha}, \ldots, e^{(N/2)-1}, e^{(N/2)+1}, e^{N/2}] \tag{10.3.5}$$

Since multiplication by Q or $Q^H = Q^{-1}$ does not change the spectral norm and spectral radius (cf. Lemma 2.9.2), we have

$$\|\hat{M}\|_2 = \|M\|_2 \quad \text{and} \quad \varrho(\hat{M}) = \varrho(M) \quad \text{for } \hat{M} := Q^{-1}MQ. \tag{10.3.6}$$

We will show that the Fourier-transformed iteration matrix $\hat{M}$ of the two-grid method is a block-diagonal matrix of the form

$$\hat{M} = \operatorname{blockdiag}\{M_1, M_2, \ldots, M_{N'-1}, M_{N'}\} \quad \text{with} \tag{10.3.7a}$$

$$M_\alpha \ 2 \times 2\text{-matrices for } 1 \leqslant \alpha \leqslant N' - 1, \quad M_{N'} \ 1 \times 1\text{-matrix} \tag{10.3.7b}$$

Applying (2.5.5b) to $\hat{M}$ and $\hat{M}^H \hat{M}$, one obtains

Lemma 10.3.1. *Matrices of the form* (7a, b) *satisfy*

$$\|\hat{M}\|_2 = \max\{\|M_\alpha\|_2\colon 1 \leqslant \alpha \leqslant N'\}, \quad \varrho(\hat{M}) = \max\{\varrho(M_\alpha)\colon 1 \leqslant \alpha \leqslant N'\}. \tag{10.3.8}$$

Let the Richardson iteration with $\Theta = h^2/4 \approx 1/\varrho(A_\ell)$ be chosen as the smoothing iteration. For the proof of the block structure (7a, b), we transform the iteration matrix

$$M = (I - pA_{\ell-1}^{-1} r A_\ell) S_\ell^\nu \quad \text{with} \quad S_\ell = I - \Theta A_\ell, \Theta = h^{E2}/4 \tag{10.3.9}$$

into

$$\hat{M} = (I - \hat{p}\hat{A}_{\ell-1}^{-1}\hat{r}\hat{A}_\ell)\hat{S}_\ell^\nu \quad \text{with} \quad \hat{A}_\ell := Q^{-1}A_\ell Q, \hat{S}_\ell := Q^{-1}S_\ell Q. \tag{10.3.10a}$$

To define the quantities

$$\hat{p} := Q^{-1}pQ', \quad \hat{A}_{\ell-1} := Q'^{-1}A_{\ell-1}Q', \quad \hat{r} := Q'^{-1}rQ, \tag{10.3.10b}$$

we have to introduce the Fourier transformation Q' at level $\ell - 1$. Replacing $h = h_\ell$ in (4) by $h' = h_{\ell-1}$, one obtains the vectors e'^α with

$$e_k'^\alpha = \sqrt{4h}\sin(2\alpha k\pi h) \qquad (0 \leqslant k \leqslant N'). \tag{10.3.11}$$

Analogously to (5), the vectors e'^α form the matrix

$$Q' = [e'^1, e'^2, \ldots, e'^{N'-1}]. \tag{10.3.12}$$

10.3.2 Transformed Quantities

According to §4.1, $A_\ell e^\alpha = \lambda_\alpha e^\alpha$ holds with $\lambda_\alpha := 4h^{-2}\sin^2(\alpha\pi h/2)$. We introduce

$$s_\alpha^2 = \sin^2(\alpha\pi h/2), \qquad c_\alpha^2 = \cos^2(\alpha\pi h/2). \tag{10.3.13}$$

Noting that $\lambda_{N-\alpha} = s_{N-\alpha}^2 = c_\alpha^2$, we obtain

$$\hat{A}_\ell := Q^{-1}A_\ell Q = \text{blockdiag}\{A_1, \ldots, A_{N'}\} \text{ with the blocks} \tag{10.3.14a}$$

$$A_\alpha = 4h^{-2}\begin{bmatrix} s_\alpha^2 & 0 \\ 0 & c_\alpha^2 \end{bmatrix} \quad \text{for } 1 \leqslant \alpha \leqslant N' - 1, \quad A_{N'} = 2h^{-2}. \tag{10.3.14b}$$

Since $S_\ell = I - \frac{1}{4}h^2 A_\ell$ and $1 - s_\alpha^2 = c_\alpha^2$, $1 - c_\alpha^2 = s_\alpha^2$, (14a, b) yield the result

$$\hat{S}_\ell := Q^{-1}S_\ell Q = \text{blockdiag}\{S_1, \ldots, S_{N'}\} \quad \text{with the blocks} \tag{10.3.15a}$$

$$S_\alpha = \begin{bmatrix} c_\alpha^2 & 0 \\ 0 & s_\alpha^2 \end{bmatrix} \quad \text{for } 1 \leqslant \alpha \leqslant N' - 1, \quad S_{N'} = \frac{1}{2}. \tag{10.3.15b}$$

Because of $A_{\ell-1}e'^\alpha = \lambda'_\alpha e'^\alpha$ with $\lambda'_\alpha = 4h'^{-2}\sin^2(\alpha\pi h'/2) = h^{-2}\sin^2(\alpha\pi h)$ and using $\sin^2(\alpha\pi h) = 4s_\alpha^2 c_\alpha^2$, one obtains the diagonal matrix

$$\hat{A}_{\ell-1} := Q'^{-1}A_{\ell-1}Q' = \text{diag}\{A'_1, \ldots, A'_{N'}\} \quad \text{with } A'_\alpha = 4h^{-2}s_\alpha^2 c_\alpha^2. \tag{10.3.16}$$

Next, we transform p and r. Let p be defined by (1.12a–c). For r we choose the adjoint mapping $r = p^*$:

$$r = \tfrac{1}{2}[\tfrac{1}{2} \quad 1 \quad \tfrac{1}{2}], \quad \text{i.e., } (ru_\ell)(\xi) = \tfrac{1}{4}u_\ell(\xi - h) + \tfrac{1}{2}u_\ell(\xi) + \tfrac{1}{4}u_\ell(\xi + h). \tag{10.3.17a}$$

r and $\hat{r}$ are matrices of the format $(N' - 1) \times (N - 1) = (N' - 1) \times (2N' - 1)$. The representation

$$\hat{r} := Q'^{-1}rQ = [\text{blockdiag}\{r_1, \ldots, r_{N'-1}\}, 0] \quad \text{with } r_\alpha = \sqrt{\tfrac{1}{2}}[c_\alpha^2 \; - s_\alpha^2] \tag{10.3.17b}$$

means that the last column of $\hat{r}$ vanishes (this follows from $re^{N'} = 0$) and that the remaining part of the format $(N'-1) \times 2(N'-1)$ consists of $N'-1$ 1×2 blocks r_α. For the proof of (17b), it must be shown that

$$re^\alpha = c_\alpha^2 e'^\alpha/\sqrt{2}, \quad re^{N-\alpha} = -s_\alpha^2 e'^\alpha/\sqrt{2} \quad \text{for } 1 \leqslant \alpha \leqslant N'-1.$$

(17a) yields $r\sin(\alpha\xi\pi) = [\sin(\alpha(\xi-h)\pi) + 2\sin(\alpha\xi\pi) + \sin(\alpha(\xi+h)\pi)]/4 = [1+\cos(\alpha h\pi)]\sin(\alpha\xi\pi)/2 = \cos(\alpha h\pi/2)^2\sin(\alpha\xi\pi) = c_\alpha^2\sin(\alpha\xi\pi)$ for all α. The different scaling of the vectors e^α, e'^α explains the additional factor in $re^\alpha = c_\alpha^2 e'^\alpha/\sqrt{2}$. Since this identity holds for all α, one may replace α by $N-\alpha$: $re^{N-\alpha} = c_{N-\alpha}^2 e'^{N-\alpha}/\sqrt{2}$. For $0 \leqslant k \leqslant N'$, the equality $\sin(2\alpha k\pi h) = -\sin(2(N-\alpha)k\pi h)$ leads to $e'^{N-\alpha} = -e'^\alpha$ (cf. definition (11)). Further, $c_{N-\alpha}^2 = s_\alpha^2$ proves $re^{N-\alpha} = -s_\alpha^2 e'^\alpha/\sqrt{2}$.

p from (1.12a–c) and r from (17a) are connected by $r = p^*$. From $p^* = \frac{1}{2}p^H$, one derives the representation

$$\hat{p} := Q^{-1}pQ' = Q^{-1}(2r)^H Q' = Q^H(2r)^H Q' = 2[Q'^H rQ]^H = 2\hat{r}^H.$$

Therefore, the result (17b) for $\hat{r}$ proves

$$\hat{p} := Q^{-1}pQ' = \begin{bmatrix} \operatorname{diag}\{p_1, \ldots, p_{N'-1}\} \\ 0 \end{bmatrix} \quad \text{with } p_\alpha = \sqrt{2}\begin{bmatrix} c_\alpha^2 \\ -s_\alpha^2 \end{bmatrix}. \tag{10.3.18}$$

10.3.3 Convergence Results

Since all factors in (10a) have a block-diagonal structure, this carries over to $\hat{M}$ and proves the structure (7a, b). For the 2×2 blocks M_α $(1 \leqslant \alpha \leqslant N'-1)$ and the 1×1 block $M_{N'}$, (14b), (15b), (16), (17b), (18) yield

$$M_\alpha = (I - p_\alpha A_\alpha'^{-1} r_\alpha A_\alpha) S_\alpha^\nu \quad (1 \leqslant \alpha \leqslant N'-1), \quad M_{N'} = 2^{-\nu} \tag{10.3.19a}$$

Inserting the representations of p_α, A'_α, r_α, A_α, S_α^ν, one obtains

$$\begin{aligned} M_\alpha &= \left(\begin{bmatrix} 1 & 0 \\ 0 & 1 \end{bmatrix} - \begin{bmatrix} c_\alpha^2 \\ -s_\alpha^2 \end{bmatrix} \frac{h^2}{4s_\alpha^2 c_\alpha^2}[c_\alpha^2, -s_\alpha^2] 4h^{-2} \begin{bmatrix} s_\alpha^2 & 0 \\ 0 & c_\alpha^2 \end{bmatrix}\right) \begin{bmatrix} c_\alpha^2 & 0 \\ 0 & s_\alpha^2 \end{bmatrix}^\nu \\ &= \left(\begin{bmatrix} 1 & 0 \\ 0 & 1 \end{bmatrix} - \begin{bmatrix} c_\alpha^2 & -c_\alpha^2 \\ -s_\alpha^2 & s_\alpha^2 \end{bmatrix}\right) \begin{bmatrix} c_\alpha^2 & 0 \\ 0 & s_\alpha^2 \end{bmatrix}^\nu = \begin{bmatrix} s_\alpha^2 & c_\alpha^2 \\ s_\alpha^2 & c_\alpha^2 \end{bmatrix} \begin{bmatrix} c_\alpha^2 & 0 \\ 0 & s_\alpha^2 \end{bmatrix}^\nu. \end{aligned} \tag{10.3.19b}$$

The block M_α describes the application of M to the two functions e^α, $e^{N-\alpha}$ (respective columns of the matrix Q, cf. (5)). Since $\alpha < N' < N - \alpha$, e^α corresponds to a *smooth* grid function and $e^{N-\alpha}$ to an *oscillatory* one. Obviously, the inequalities $0 < \alpha < N' < N - \alpha < N$ lead to

$$0 < s_\alpha^2 < \tfrac{1}{2} < c_\alpha^2 < 1. \tag{10.3.20}$$

The two 2×2 matrices in (19b) characterise the coarse-grid correction and the smoothing iteration, respectively. The entries $c_\alpha^2 > s_\alpha^2$ express the fact that the smooth e^α components converge more slowly than the nonsmooth $e^{N-\alpha}$ components. The first matrix reflects complementary behaviour for the coarse-grid correction: The smooth components (s_α^2 in the first column) are better reduced than the oscillatory ones (c_α^2 in the second column).

Exercise 10.3.2. Prove that $\varrho(M_\alpha) = \varrho_\nu(s_\alpha^2)$ and $\|M_\alpha\|_2 = \zeta_\nu(s_\alpha^2)$ with

$$\varrho_\nu(\xi) := \xi(1-\xi)^\nu + (1-\xi)\xi^\nu, \tag{10.3.21a}$$

$$\zeta_\nu(\xi) := \sqrt{2[\xi^2(1-\xi)^{2\nu} + (1-\xi)^2\xi^{2\nu}]}. \tag{10.3.21b}$$

The combination of (6), Lemma 1, and Exercise 2 yields

$$\varrho(M) = \max\{\varrho_\nu(s_\alpha^2)\colon 1 \leqslant \alpha \leqslant N'\}, \qquad \|M\|_2 = \max\{\zeta_\nu(s_\alpha^2)\colon 1 \leqslant \alpha \leqslant N'\}.$$

Since the values of s_α^2 for $1 \leqslant \alpha \leqslant N'$ lie between 0 and $\frac{1}{2}$ (cf. (20)), the following estimates are valid:

$$\varrho(M) \leqslant \varrho_\nu := \max\{\varrho_\nu(\xi)\colon 0 \leqslant \xi \leqslant \tfrac{1}{2}\}, \tag{10.3.22a}$$

$$\|M\|_2 = \zeta_\nu := \max\{\zeta_\nu(\xi)\colon 0 \leqslant \xi \leqslant \tfrac{1}{2}\}. \tag{10.3.22b}$$

The bounds ϱ_ν and ζ_ν of the convergence rate and contraction numbers depend on the smoothing number ν; however, they do not depend on the step size h. Since ϱ_ν and ζ_ν decrease monotonically with increasing ν and $\varrho_1 = \zeta_1 = \frac{1}{2} < 1$, the convergence of the two-grid method for the one-dimensional model problem (1) is proved. A more detailed discussion of the functions $\varrho_\nu(\xi)$, $\zeta_\nu(\xi)$ and their maxima in $[0, \frac{1}{2}]$ yields the following.

Theorem 10.3.3. *Let the two-grid method for solving system* (1) *be characterised by Richardson's iteration with* $\Theta = h^2/4$ *(identical to the Jacobi iteration damped by* $\frac{1}{2}$*) as smoother, by the piecewise linear prolongation p, and the adjoint restriction* (17a). *Then the two-grid method with* $\nu \geqslant 1$ *smoothing steps converges with the rate* ϱ_ν *from* (22a), *which is h-independent. The contraction number (with respect to the Euclidean norm) is bounded by* ζ_ν *from* (22b). *For increasing* ν, *these bounds have the asymptotical behaviour*

$$\varrho_\nu = \frac{1}{e\nu} + O(\nu^{-2}), \qquad \zeta_\nu = \frac{\sqrt{2}}{e\nu} + O(\nu^{-2}). \tag{10.3.23}$$

Some values of ϱ_ν, ζ_ν are listed in Table 1. Obviously, the quantities $\varrho(M)$, $\|M\|_2$ converge with a decreasing step size parameter to their bounds ϱ_ν and

Table 10.3.1 ϱ_ν, ζ_ν

ν	ϱ_ν	ζ_ν
1	1/2	1/2
2	1/4	1/4
3	1/8	0.150
4	0.0832	0.1159
5	0.0671	0.0947
10	0.0350	0.0496

ζ_ν; hence, the given estimates are strict. In §10.6 we will derive a convergence rate for general problems that also behaves like $O(1/\nu)$. Theorem 3 demonstrates that such results about the asymptotical behaviour for large ν are not too pessimistic.

10.4 Multi-Grid Iteration

10.4.1 Algorithm

The two-grid method is not yet suited for practical applications, because one still has to solve one system of equations per iteration at level $\ell - 1$. The problem to be solved in (2.2c) has the form

$$A_{\ell-1} e_{\ell-1} = d_{\ell-1}; \tag{10.4.1}$$

hence, it is of the same structure as the original problem $A_\ell x_\ell = b_\ell$. Instead of solving Eq. (1) exactly, one may approximate the solution iteratively. The iteration of choice is again the two-grid method, now applied to the levels $\ell - 1$ and $\ell - 2$ instead of ℓ and $\ell - 1$. Then new auxiliary problems $A_{\ell-2} e_{\ell-2} = d_{\ell-2}$ arise, for which again the two-grid method (at the level $\ell - 2$) is to be applied until equations $A_0 e_0 = d_0$ arise at the coarsest grid. The arising recursive method is the multi-grid iteration $\Phi_\ell^{\mathrm{MGM}(\nu_1,\nu_2)}$, which has the following algorithmic form:

procedure $\Phi_\ell^{\mathrm{MGM}(\nu_1,\nu_2)}(x_\ell, b_\ell)$;	(10.4.2)
if $\ell = 0$ *then* $x_0 := A_0^{-1} b_0$ *else*	(10.4.2a)
begin for $i := 1$ *to* ν_1 *do* $x_\ell := \mathscr{S}_\ell(x_\ell, b_\ell)$;	(10.4.2b)
$\quad d_{\ell-1} := r(A_\ell x_\ell - b_\ell)$;	(10.4.2c)
$\quad e_{\ell-1}^{(0)} := 0$;	(10.4.2d_1)
$\quad$ *for* $i := 1$ *to* γ *do* $e_{\ell-1}^{(i)} := \Phi_{\ell-1}^{\mathrm{MGM}(\nu_1,\nu_2)}(e_{\ell-1}^{(i-1)}, d_{\ell-1})$;	(10.4.2d_2)
$\quad x_\ell := x_\ell - p e_{\ell-1}^{(\gamma)}$;	(10.4.2e)
$\quad$ *for* $i := 1$ *to* ν_2 *do* $x_\ell := \mathscr{S}_\ell(x_\ell, b_\ell)$;	(10.4.2f)
$\quad \Phi_\ell^{\mathrm{MGM}(\nu_1,\nu_2)} := x_\ell$	(10.4.2g)
end;	

Obviously, the recursive procedure calls terminate after ℓ steps, when level $\ell = 0$ is reached. Hence, the algorithm is well-defined.

ν_1 and ν_2 denote again the number of pre- and post-smoothing steps. In general, $\nu := \nu_1 + \nu_2 > 0$ is assumed. For the iterative solution of the coarse-grid equation (1), γ steps of the iteration $\Phi_{\ell-1}^{\mathrm{MGM}(\nu_1,\nu_2)}$ are applied to the starting value ($2d_1$). We will see that $\gamma = 2$ is sufficient. Therefore, only the cases $\gamma = 1$ and $\gamma = 2$ are of practical interest. The multi-grid iteration with $\gamma = 1$ has the name «V-cycle», whereas the iteration with $\gamma = 2$ is called the «W-cycle» (concerning the reason for these names, compare with Hackbusch [14, §2.5]).

The exact solution of linear equations is not completely avoided in the multi-grid algorithm (2). In (2a) the system $A_0 x_0 = b_0$ corresponding to the coarsest grid is to be solved. Since the coarsest grid has the smallest number of grid points, the solution should not lead to practical difficulties. According to (1.7c), $h_0 = \frac{1}{2}$ would be a possible choice of the coarsest grid size. In this case, $A_0 x_0 = b_0$ represents a single scalar equation.

Formally, the multi-grid method is a product iteration with the factors «smoothing iteration» and «coarse-grid correction», where the latter almost corresponds to a composed method with a secondary iteration as described in §8.4. But different from §8.4, the auxiliary problem, which is to be approximated by the secondary iteration, does not belong to the same space X_ℓ but to the lower-dimensional space $X_{\ell-1}$.

10.4.2 PASCAL Procedures

The multi-grid iteration requires additional parameters: the level number $\ell =$ `level`, the `MG_data MGD`, the possibly different pre- and post-smoothings as well as a direct solver for (2a). According to (1.7a), $\ell = 0$ is the level number of the coarsest grid. Possibly, one would like to choose h_1 instead of h_0 as the coarsest step size. To avoid the renaming of $\{h_1, \ldots, h_\ell\}$ with $\{h_0, \ldots, h_{\ell-1}\}$, the parameter `lmin` (in record `MDG.MI`) indicates the coarsest step size (default value $\ell = 0$). `gamma` represents the number γ from ($2d_2$). These and the parameters already mentioned in §10.2.4 (`Galerkin`, `ny1` and `ny2` for ν_1 and ν_2 from (2b, f)) can be defined interactively by means of the following procedure:

```
procedure define_MG_parameter(var MGD: MG_data); var l:
 integer;
begin with MGD.MI do
  begin writeln; writeln('*** Input of the MG_
   iterationparameter');
   writeln('Direct solution at the coarsest grid level?');
    direct:=yes_no;
   writeln; writeln('Galerkin product for coarse_grid
    matrices?');
   Galerkin:=yes_no; writeln; writeln(
   'What kind of the cycle: gamma=1: V-, 2: W-cycle, >2: not
    recommended');
   repeat write('--> gamma ='); readln(gamma) until gamma>0;
```

```
  writeln('number of pre- (ny1) and postsmoothings (ny2).');
  write('--> ny1 ='); readln(ny1); write('--> ny2 =');
   readln(ny2);
  determine_prolongation_and_restriction(P,R);
  writeln(P.kind,'and',R.kind,'have been chosen'); writeln;
  write('standard value 0 for lowest level lmin?');
  if yes_no then lmin:=0 else
  begin write('--> lmin ='); readln(lmin);
      lmin:=maximum(0,minimum(lmin,Lmax))
  end
 end
end;
```

The following multi-grid procedure does not completely coincide with (2). If `MDG.MI.direct=true`, the equation $A_0 x_0 = b_0$ (more precisely; at level `lmin` instead of 0) is solved exactly as described in (2a). In this case, γ may be replaced by 1 without changing the results in the loop ($2d_2$). However, it is not imperative to solve the system at the lowest level *exactly*. Since even simple iterations are acceptably convergent for lower-dimensional problems, the pre- and post-smoothing at the lowest level (without a coarse-grid correction) can be applied to solve $A_0 x_0 = b_0$ approximately. This choice is indicated by `direct=false`. The procedure `check_levelnumber` tests whether the level number is in the admissible range `lmin` $\leqslant \ell \leqslant$ `Lmax`.

```
procedure MG_iteration(level: integer;
      var xnew,x,b: gridfunction;
      var MGD: MG_data;
      procedure presmoothing(var new: gridfunction;
            var A: data_of_discretisation;
            var x,b: gridfuncdtion;
            var IP: iterationparameter);
      procedure postsmoothing(var new: gridfunction;
            var A: data_of_discretisation;
            var x,b: gridfunction;
            var IP: iterationparameter);
      procedure direct_solver(var x,b: gridfunction;
            var A: data_of_discretisation));
var v,d: gridfunction; i: integer;
begin if check_levelnumber(level,MGD) then with MGD do with
 MI do
 if (level=lmin) and direct then direct_solver(xnew,b,
  AL[lmin]) else
 begin xnew:=x;
```

```
  for i:=1 to ny1 do presmoothing(xnew,AL[level],xnew,b,
   IP[level]);
  if level>0 then with AL[level-1] do
  begin residual(d,AL[level],xnew,b);
   MG_restriction(level,d,d,R,AL);
   if kind=Poisson_model_problem then factor_x_vector
    (nx,ny,d,4,d);
   zero_gridfunction(nx,ny,v);
   i:=gamma;
   if (level-1=lmin) and direct then i:=1 else if level=
    lmin then i:=0;
   for i:=i downto 1 do
   MG_iteration(level-1,v,v,d,MGD,
          presmoothing,postsmoothing,direct_solver);
   MG_prolongation(level,v,v,P,AL);
   vector_plus_vector(AL[level].nx,AL[level].ny,xnew,xnew,v)
  end;
  for i:=1 to ny2 do postsmoothing(xnew,AL[level],xnew,b,
   IP[level])
end end;
```

We remark that the language PASCAL is not optimally suited for the formulation of the multi-grid algorithm, because PASCAL requires the same parameters at all levels and therefore also equally sized arrays of type `gridfunction`. In the actual realisation, this leads to the fact that the $\ell_{\max}+1$ grid functions $\{x_\ell, b_\ell : 0 \leqslant \ell \leqslant \ell_{\max}\}$ require $(\ell_{\max}+1)$ times the maximum storage, whereas the total dimension is much less: $2\sum n_\ell \approx 8n_{\ell_{\max}}/3 \ll 2\ell_{\max} n_{\ell_{\max}}$.

A possible frame program can read as follows:

```
program multigrid_method;
var MGD: MG_data; level,i,itnr: integer; it: data_of_
 iteration;
    v: data_for_comparison; l2,mid: history_of_iteration;
{right_hand_side, zerofunction, exact_solution, boundary_
 value etc. to be inserted}
begin initialise_MG_data(MGD); initialise_it(it); initialise_
 comparison(v);
 with MGD do repeat
 release_MG_data(MGD); release_it(it); release_comparison(v);
 define_MG_discretisation(MGD); define_MG_parameter(MGD);
 write('--> solution at level l='); readln(level);
 set_actual_level(level,MGD); it.A:=AL[level];
```

```
  define_data_of_system(it,boundary_value,right_hand_side);
  define_starting_iterate(it,zerofunction); writeln('Starting
   value defined.');
  define_comparison_solution(v,It.A,exact_solution);
  write('--> Number of iterations ='); readln(itnr);
   it.IP.Nr:=0;
  comparison_with_exact_solution(l2,v,it,Euclidean_norm);
  for i:=1 to itnr do
  begin MG_iteration(level,it.x,it.x,it.b,MGD, chequerboard_
   Gauss_Seidel,
             chequerboard_Gauss_Seidel, solve_system_without_
              pivoting);
   it.ip.nr:=it.ip.nr+1; {call of one multi-grid step}
   comparison_with_exact_solution(l2,v,it,Euclidean_norm);
    {→output}
  end; write('Should the run be repeated?') until not
   yes_no
end.
```

The parameter `MDG.MI.actual_level` is important if `Galerkin=true` is chosen. In this case, according to (2.7b), the matrices A_ℓ for $\ell <$ `actual_level` are determined by means of the Galerkin product (1.26). The procedure `set_actual_level` can be used for the definition of the actual level and computation of the Galerkin products. The necessary data are the matrices $A^{(\mu)}$ contained in `MGD.SH[`μ`]`, $0 \leqslant \mu \leqslant 2$, from which the complete matrix can be defined according to

$$A_\ell := \sum_{\mu=0}^{2} h_\ell^{-\mu} A^{(\mu)} \tag{10.4.3}$$

(the corresponding procedure is `transfer_discretisation`). If `Galerkin=false`, (3) is valid for all ℓ; otherwise, only for $\ell \geqslant$ `actual_level`. The involved auxiliary procedures and additional ones are given below.

```
procedure initialise_MG_data(var MGD: MG_data);
var l: integer;
begin with MGD do begin for l:=0 to Lmax do
 begin initialise_discretisation(AL[l]); initialise_
  iterationparameter(IP[l])
 end;
 MI.lmin:=0; MI.actual_level:=0
end end;

procedure release_MG_data(var MGD: MG_data);
var l: integer;
```

```
begin with MGD do for l:=0 to Lmax do
  begin release_discretisation(AL[l]); release_
   iterationparameter(IP[l]) end;
    initialise_MG_data(MGD)
end;

procedure complete_discretisation_parameter(var A: data_of_
 discretisation);
begin with A do
  begin if kind<fivepoint_formula then
    begin S[-1,0]:=-1; S[1,0]:=-1; S[0,-1]:=-1; S[0,1]:=-1;
     S[0,0]:=4
    end;
    if kind<ninepoint_formula then
    begin S[-1,1]:=0; S[1,1]:=0; S[1,-1]:=0; S[-1,-1]:=0
    end
end end;

procedure compute_star(var A: data_of_discretisation;
 var sh: H_star);
var i,j: integer;
begin with A do if kind=Poisson_model_problem
    then complete_discretisation_parameter(A) else
    for i:=-1 to 1 do for j:=-1 to 1 do S[i,j]:=SH[2,i,j]/h2+
     SH[1,i,j]/h+SH[0,i,j]
end;

procedure transfer_discretisation(var MGD: MG_data);
 var l,i,j: integer;
begin with MGD do for l:=Lmax-1 downto 0 do
    begin release_discretisation(AL[l]); AL[l].kind:=
     AL[Lmax].kind;
          compute_star(AL[l],sh)
end end;

procedure set_actual_level(actual: integer; var MGD:
 MG_data);
var l: integer;
begin with MGD do with MI do
    begin actual_level:=actual; transfer_discretisation(MGD);
        if Galerkin then for l:=actual-1 downto 0 do with
         AL[l] do
        begin Galerkin_star(S,AL[l+1],P,R); kind:=ninepoint_
         formula
end end end;
```

10.4.3 Numerical Examples

The model problem with step size $h = h_5 = 1/64$ is taken as an example. Table 1 shows the Euclidean norm $\|e^m\|_2$ of the errors and the reduction factors $\varrho_{m+1,m} = \|e^m\|_2/\|e^{m-1}\|_2$. The parameters are $\nu_1 = 2$, $\nu_2 = 0$, `lmin=0` (i.e., $h_0 = \frac{1}{2}$), `direct=true`, `Galerkin=false`, nine-point prolongation, and nine-point restriction. As the smoothing iteration, the chequer-board Gauß-Seidel method is chosen. The comparison of the results for $\gamma = 1$ (V-cycle) and $\gamma = 2$ (W-cycle) from Table 1 with the two-grid results (formally $\gamma = \infty$) repeated from Table 2.2 show that $\gamma = 2$ yields almost the same fast convergence as the two-grid method, whereas the V-cycle results are less favourable.

In order to demonstrate that the multi-grid method works well not only for positive definite problems, the next example is the nonsymmetric differential equation

$$-\Delta u + cu_x = f \tag{10.4.4a}$$

in $\Omega = (0,1) \times (0,1)$ with Dirichlet boundary values (1.2.1b), discretised by

$$A_\ell = h_\ell^{-2}\begin{bmatrix} & -1 & \\ -1 & 4 & -1 \\ & -1 & \end{bmatrix} + \frac{1}{2}ch_\ell^{-1}\begin{bmatrix} & 0 & \\ -1 & 0 & 1 \\ & 0 & \end{bmatrix}. \tag{10.4.4b}$$

First, we choose $c = 4$ (for this value, all A_ℓ are M-matrices). $f = u = 0$ are taken as the right-hand side and exact solution, while $x(1 - x + y)$ serves as the starting value. The other parameters are the same as in Table 1. The W-cycle ($\gamma = 2$) shows a convergence rate of ≈ 0.06 (cf. Table 2) and hardly differs from the corresponding rate of the Poisson-model case.

As soon as the coefficient c becomes substantially larger, e.g., $c = 100$, a stability problem arises: The discretisation (4b) yields an M-matrix for $h_5 =$

Table 10.4.1 Multi-grid results for the Poisson-model problem, $h = 1/64$

	$\gamma = 1$ (V-cycle)		$\gamma = 2$ (W-cycle)		$\gamma = \infty$ (two-grid algorithm)
m	$\|e^m\|_2$	$\varrho_{m+1,m}$	$\|e^m\|_2$	$\varrho_{m+1,m}$	$\varrho_{m+1,m}$
1	$1.3274_{10}-1$	$1.727_{10}-1$	$2.9984_{10}-02$	$3.902_{10}-2$	$4.124_{10}-2$
2	$2.2223_{10}-2$	$1.674_{10}-1$	$1.3219_{10}-03$	$4.408_{10}-2$	$5.126_{10}-2$
3	$3.7656_{10}-3$	$1.694_{10}-1$	$6.9050_{10}-05$	$5.223_{10}-2$	$5.443_{10}-2$
4	$6.4110_{10}-4$	$1.702_{10}-1$	$3.7824_{10}-06$	$5.477_{10}-2$	$5.741_{10}-2$
5	$1.0941_{10}-4$	$1.706_{10}-1$	$2.1584_{10}-07$	$5.706_{10}-2$	$5.939_{10}-2$
6	$1.8701_{10}-5$	$1.709_{10}-1$	$1.2689_{10}-08$	$5.879_{10}-2$	$6.125_{10}-2$
7	$3.1996_{10}-6$	$1.710_{10}-1$	$7.6788_{10}-10$	$6.051_{10}-2$	$6.280_{10}-2$

1/64, but not for larger h. A remedy is the matrix-dependent prolongation (1.16a, b) and the corresponding restriction together with the Galerkin product (1.26) for $\ell < 5$ (i.e., `Galerkin=true` is chosen; cf. Hackbusch [14, §10.4]). Table 3 shows the convergence rates $\varrho_{m+1,m}$ for $\gamma = 1$ and $\gamma = 2$. Different from the model case, the results for $\gamma = 2$ are hardly better than those for $\gamma = 1$. Furthermore, the rate ≈ 0.3 is not so favourable. The rate can be improved to ≈ 0.14 by smoothing with the row-wise Gauß-Seidel iteration instead of the (pointwise) chequer-board Gauß-Seidel iteration (Table 3, right column).

In §§9.5.7–8 (cf. Table 9.5.2+4) the indefinite problem with the matrix

$$A_\ell := h_\ell^{-2} \begin{bmatrix} & -1 & \\ -1 & 4 & -1 \\ & -1 & \end{bmatrix} - \begin{bmatrix} & 0 & \\ 0 & 50 & 0 \\ & 0 & \end{bmatrix} \tag{10.4.5}$$

Table 10.4.2 (4b) for $c = 4$, $h = 1/64$

m	$\varrho_{m+1,m}$
1	$3.02452_{10}-2$
2	$4.72247_{10}-2$
3	$5.30817_{10}-2$
4	$5.51039_{10}-2$
5	$5.69383_{10}-2$
6	$5.83535_{10}-2$
7	$5.97048_{10}-2$
8	$6.09182_{10}-2$
9	$6.20609_{10}-2$
10	$6.31206_{10}-2$

Table 10.4.3 Equation (4b) for $c = 100$, $h = 1/64$ smoothing by pointwise and line-Gauß-Seidel iteration

	pointwise Gauß-Seidel iteration		line-G.-S.
m	$\gamma = 1$	$\gamma = 2$	$\gamma = 1$
1	$1.58433_{10}-1$	$2.75119_{10}-2$	$4.65472_{10}-2$
2	$2.60220_{10}-1$	$9.54499_{10}-2$	$9.99392_{10}-2$
3	$3.35135_{10}-1$	$2.73365_{10}-1$	$9.51973_{10}-2$
4	$3.47875_{10}-1$	$3.00345_{10}-1$	$1.31862_{10}-1$
5	$3.35962_{10}-1$	$2.94467_{10}-1$	$1.26717_{10}-1$
6	$3.14166_{10}-1$	$3.06239_{10}-1$	$1.47075_{10}-1$
7	$2.92044_{10}-1$	$3.19984_{10}-1$	$1.30377_{10}-1$
8	$2.71982_{10}-1$	$3.34751_{10}-1$	$1.48727_{10}-1$
9	$2.55278_{10}-1$	$3.25717_{10}-1$	$1.32756_{10}-1$

has been solved. As we will see in §10.6.2, the choice of the coarsest step size is restricted in the case of indefinite problems. Here $h_0 = \frac{1}{2}$ is too coarse, but $h = 1/4$ is possible. However, better results can be obtained with $h = 1/8$ as the coarsest step size. Table 4 shows the results for $h_3 = 1/64$, $h_0 = 1/8$, $\gamma = 2$, `direct=true`, `Galerkin=false`, nine-point prolongation, and nine-point restriction. $\nu_1 = 2$ chequer-board Gauß-Seidel steps are applied ($\nu_2 = 0$) as smoothing. The convergence rate (here 0.442) improves with decreasing grid size. Vice versa, a worse rate 0.613 results for $h = 1/16$.

It is not necessary to choose h_0 sufficiently small, if one uses the Kaczmarz iteration for smoothing, which is also convergent for the indefinite matrix (5). However, for the parameters $h_0 = \frac{1}{2}$, $\nu_1 = 2$, $\nu_2 = 0$, $\gamma = 2$, and $h = 1/64$, one obtains the rather unfavourable convergence rate 0.833 (for $h = 1/16$ even 0.917).

10.4.4 Computational Work

To judge the convergence rates from §10.4.3, one has to take into account the amount of work per iteration, as we know from §3.3. Because of the recursive structure, the amount of work is not obvious. The operations appearing in (2) are $\mathscr{S}_\ell$ in (2b, f), $r(A_\ell x_\ell - b_\ell)$ in (2c), and $x_\ell - pe_{\ell-1}$ in (2e). We denote the corresponding work by

$$C_S n_\ell \quad \text{operations for} \quad x_\ell \mapsto \mathscr{S}_\ell(x_\ell, b_\ell), \tag{10.4.6a}$$

$$C_D n_\ell \quad \text{operations for} \quad x_\ell \mapsto r(A_\ell x_\ell - b_\ell), \tag{10.4.6b}$$

$$C_C n_\ell \quad \text{operations for} \quad x_\ell \mapsto x_\ell - pe_{\ell-1}. \tag{10.4.6c}$$

Proportionality to the dimension n_ℓ is a consequence of the sparsity of the matrix A_ℓ (cf. (3.3.1)). For full matrices, n_ℓ would have to be replaced by n_ℓ^2 in (6a, b) (but compare with Hackbusch [13]).

The dimensions n_ℓ should increase with an increasing level number ℓ at least by a factor C_h:

$$n_{\ell-1} \leqslant n_\ell / C_h \quad \text{for } \ell \geqslant 1. \tag{10.4.7}$$

Otherwise, the difficulty would arise that the auxiliary problems $A_{\ell-1} e_{\ell-1} = d_{\ell-1}$ are of a similar dimension as $A_\ell x_\ell = b_\ell$.

Table 10.4.4 Results for (5)

m	$\|e^m\|_2$	$\varrho_{m+1,m}$
1	$1.301_{10}-1$	0.169309
2	$5.607_{10}-2$	0.430985
3	$2.480_{10}-2$	0.442381
4	$1.097_{10}-2$	0.442503
5	$4.857_{10}-3$	0.442505

Remark 10.4.1. For the standard choice $h_\ell = h_{\ell-1}/2$ and the spatial dimension d: $\Omega_\ell \subset \mathbb{R}^d$, (7) holds with $C_h = 2^d$. In the model case, $d = 2$ is valid.

Theorem 10.4.2. *Assume* (6a–c) *and* (7). *Let* γ *from* (2d$_2$) *satisfy*

$$\gamma < C_h. \tag{10.4.8}$$

Then the work of the multi-grid iteration is proportional to n_ℓ:

$$work(\Phi_\ell^{\mathrm{MGM}(\nu_1,\nu_2)}) \leqslant C(\nu_1 + \nu_2)n_\ell \quad \text{with} \tag{10.4.9a}$$

$$C(\nu) = \frac{\nu C_S + C_D + C_C}{1 - \gamma/C_h} + O((\gamma/C_h)^\ell). \tag{10.4.9b}$$

Proof. Let $C_\ell n_\ell$ be the work for one $\Phi_\ell^{\mathrm{MGM}(\nu_1,\nu_2)}$ step. From the representation (2), one concludes that $C_\ell n_\ell \leqslant (\nu C_S + C_D + C_C)n_\ell + \gamma C_{\ell-1}n_{\ell-1}$. Inequality (7) yields $C_\ell \leqslant (\nu C_S + C_D + C_C) + \vartheta C_{\ell-1}$ with $\vartheta := \gamma/C_h$ and results in the geometrical sum

$$C_\ell \leqslant (\nu C_S + C_D + C_C)(1 + \vartheta + \cdots + \vartheta^{\ell-1}) + \gamma^\ell C_0/n_\ell,$$

where C_0 denotes the work for (2a) (independent with respect to h_ℓ). Since $\gamma^\ell/n_\ell \leqslant \vartheta^\ell/n_1$, (9b) follows. □

Remark 10.4.3. In the two-dimensional case $d = 2$, (8) is satisfied for the interesting values $\gamma = 1, 2$ because of $C_h = 4$ (cf. Remark 1). The following constants are obtained for (9b):

$$C_V(\nu) = \tfrac{4}{3}(\nu C_S + C_D + C_C) + O((1/C_h)^\ell) \quad \text{for } \gamma = 1, \tag{10.4.10a}$$

$$C_W(\nu) = 2(\nu C_S + C_D + C_C) + O((2/C_h)^\ell) \quad \text{for } \gamma = 2. \tag{10.4.10b}$$

Since $\gamma/C_h < 1$ (cf. (8)), formulae (9b) and (10a, b) show that for increasing ℓ, the work for solving $A_0 x_0 = b_0$ in (2a) requires a vanishing portion of the total work.

Exercise 10.4.4. Although the one-dimensional case (3.1) is not of practical interest, one may apply the multi-grid algorithm. Then (8) is not satisfied, because $C_h = 2$ holds for the W-cycle ($\gamma = 2$). Prove that the work equals $O(\ell n_\ell) = O(n_\ell \log n_\ell)$.

For the standard multi-grid parameters as used before, the work amounts to

$$C_S = 2(C_A - 1) \quad \text{for Gauß-Seidel iteration (cf. (4.6.1b))}, \tag{10.4.11a}$$

$$C_D = 2C_A + 11/4 \quad \text{for } r = \text{nine-point restriction (1.20)}, \tag{10.4.11b}$$

$$C_C = 3/2 \quad \text{for } p = \text{nine-point prolongation (1.14e)}. \tag{10.4.11c}$$

C_S and C_D improve for the Poisson-model case ($C_A = 5$):

$$C_S = 5 \qquad \text{for Gauß-Seidel iteration (cf. (4.6.5a))}, \tag{10.4.11a'}$$

$$C_D = 5 + 10/4 \quad \text{for } r = \text{nine-point restriction (1.20).} \tag{10.4.11b'}$$

If the chequer-board Gauß-Seidel method follows the application of r and p, additional operations can be saved (cf. Hackbusch [14, Note 4.3.4]). If we insert (11), the numbers (10a, b) become

$$C_V(\nu) = \tfrac{8}{3}(\nu + 1)C_A + (17 - 8\nu)/3 + O(1/4^\ell) \quad \text{for } \gamma = 1 \text{ or} \tag{10.4.11d}$$

$$C_V(\nu) = 12 + \tfrac{20}{3}\nu + O(1/4^\ell), \tag{10.4.11d'}$$

$$C_W(\nu) = 4(\nu + 1)C_A + (17 - 8\nu)/2 + O(1/2^\ell) \quad \text{for } \gamma = 2 \text{ or} \tag{10.4.11e}$$

$$C_W(\nu) = 18 + 10\nu + O(1/2^\ell). \tag{10.4.11e'}$$

The corresponding *effective work* of the V-[W]-cycle for the Poisson-model problem with $\nu = 2$ is $C_{V[W]}(2)/|C_A \log(\varrho)|$. Using the convergence rates ϱ from Table 1, we obtain

$$Eff(\Phi_\ell^{\mathrm{MGM}(2,0)}) = -C_V(2)/[5\log(0.171)] \approx 2.89 \quad \text{for } \gamma = 1, \tag{10.4.12a}$$

$$Eff(\Phi_\ell^{\mathrm{MGM}(2,0)}) = -C_W(2)/[5\log(0.06)] \approx 2.7 \quad \text{for } \gamma = 2. \tag{10.4.12b}$$

Together with the convergence rate, the effective work is h-independent also. One should compare the numbers from (12a, b) with the competing values from Remark 7.5.11 (for $h = 1/32$). The numbers (11d', e') can also be interpreted as follows: One V-cycle step costs as much as ≈ 5 Gauß-Seidel steps, one W-cycle step corresponds to 7.6 Gauß-Seidel steps.

In the following, the question to be addressed is how many smoothing steps should be performed. The numerical results from §10.4.3 have shown good agreement with the two-grid results. According to Table 2.3, these rates behave asymptotically as $\approx C_\varrho/(1 + \nu)$, where $\nu = \nu_1 + \nu_2$. For simplification, we assume that $C_C + C_D = C_S$. Then the multi-grid work behaves like $work(\Phi_\ell^{\mathrm{MGM}(\nu_1,\nu_2)}) \approx (1 + \nu)C$. With an increasing number ν of smoothing steps, the convergence improves, however, the work increases also. We have to minimise the effective work

$$-\frac{(1 + \nu)C/C_A}{\log(C_\varrho/(1 + \nu))} = C' \frac{1 + \nu}{\log(1 + \nu) - \log C_\varrho}.$$

The minimum is taken for $\nu^* = C_\varrho e - 1$ and has the value $eC_\varrho C/C_A$. This shows at least asymptotically that the faster the multi-grid method (the smaller the C_ϱ), the smaller the number of smoothing steps should be. If, vice versa, it turns out that many smoothing steps are necessary, the multi-grid method is not favourable and one should look for a better suited smoothing iteration.

10.4.5 Iteration Matrix

Since the iteration is defined recursively, the iteration matrix of the multi-grid method is also determined recursively.

Theorem 10.4.5. *Let S_ℓ be the iteration matrix of the (consistent) pre- and post-smoothing iteration $\mathscr{S}_\ell$. Then the multi-grid iteration $\Phi_\ell^{\mathrm{MGM}(\nu_1,\nu_2)}$ is also consistent. Its iteration matrix $M_\ell^{\mathrm{MGM}} = M_\ell^{\mathrm{MGM}}(\nu_1,\nu_2)$ is defined by*

$$M_0^{\mathrm{MGM}} = 0, \qquad M_1^{\mathrm{MGM}} = M_1^{\mathrm{TGM}}(\nu_1,\nu_2), \tag{10.4.13a}$$

$$M_\ell^{\mathrm{MGM}} = M_\ell^{\mathrm{TGM}}(\nu_1,\nu_2) + S_\ell^{\nu_2} p (M_{\ell-1}^{\mathrm{MGM}})^\gamma A_{\ell-1}^{-1} r A_\ell S_\ell^{\nu_1} \quad \text{for } \ell \geqslant 1. \tag{10.4.13b}$$

Proof. Obviously, the coarse-grid correction (2c–e) is consistent. Hence, Exercise 3.2.16a shows the consistency of Φ_ℓ^{MGM}. For $\ell = 0$, Φ_0 describes the exact solution: $M_0 = O$. For $\ell = 1$, the multi- and two-grid methods coincide. This proves (13a). The iteration matrix of the coarse-grid correction (2c–e) is

$$M_\ell^{\mathrm{CGC}} = I - p[I - (M_{\ell-1}^{\mathrm{MGM}})^\gamma] A_{\ell-1}^{-1} r A_\ell, \tag{10.4.13c}$$

because we have $e_{\ell-1}^{(\gamma)} = [I - (M_{\ell-1}^{\mathrm{MGM}}(\nu_1,\nu_2))^\gamma] A_{\ell-1}^{-1} r A_\ell$ in (2e), as can be shown similarly to the proof of (8.4.7b). (3.2.20a) and (2.4) prove (13b). □

10.5 Nested Iteration

10.5.1 Algorithm

The presence of a hierarchy of equations (1.8a): $A_\ell x_\ell = b_\ell$ for all levels $0 \leqslant \ell \leqslant \ell_{\max}$ is necessary for the applicability of the multi-grid method. This situation can even be exploited for another purpose. Obviously, the result x_ℓ^m of an iteration is more desirable, the better the starting iterate x_ℓ^0 is. So far, the choice of a starting iterate has not been studied, because a more or less favourable choice for the starting value depends on information that is not available in iterative analysis.

Assume a hierarchy $A_\ell x_\ell = b_\ell$ of problems. Usually, the solutions x_ℓ approximate a continuous solution x by a certain positive consistency order $O(h_\ell^\varkappa)$. Via the triangle inequality, x_ℓ and $x_{\ell-1}$ should also differ only by $O(h_\ell^\varkappa + h_{\ell-1}^\varkappa)$. This assumption can be expressed as follows:

$$\|x_\ell - \tilde{p} x_{\ell-1}\| \leqslant C_1 h_\ell^\varkappa \qquad (\varkappa > 0,\ x_\ell, x_{\ell-1} \text{ solutions to (1.8a)}). \tag{10.5.1}$$

Here, $\tilde{p}\colon X_{\ell-1} \to X_\ell$ is a suitable prolongation, which may not necessarily coincide with p from §10.1.3 or p from (4.2e).

Inequality (1) suggests using approximations to $x_{\ell-1}$ as the starting iterate of the iteration at level ℓ. The algorithm as proposed by Kronsjö–Dahlquist [1] reads

$\tilde{x}_0 :=$ suitable approximation of the solution of $A_0 x_0 = b_0$;
for $\ell := 1$ *to* $\ell_{\max}$ *do*
begin $\tilde{x}_\ell := \tilde{p}\tilde{x}_{\ell-1}$; (10.5.2a)
for $i := 1$ *to* m_ℓ *do* $\tilde{x}_\ell := \Phi_\ell(\tilde{x}_\ell, b_\ell)$
end;

Note that (2a) is not an iteration in the proper sense, but a *finite* process. The optimal choice of the step number m_ℓ will be the subject of Exercise 3. Here we are only interested in the multi-grid iteration $\Phi_\ell = \Phi_\ell^{\mathrm{MGM}}$, for which the numbers m_ℓ can be chosen indepenently of ℓ:

$$m_\ell = m \qquad (\ell \geqslant 1). \tag{10.5.2b}$$

In Remark 2 we will see that even the smallest possible number $m = 1$ is of practical interest.

10.5.2 Error Analysis

Multi-grid methods (and other methods with an h-independent convergence rate) satisfy the condition

$$\|M_\ell^\Phi\| \leqslant \zeta < 1 \quad \text{for } \ell \geqslant 1,\ M_\ell^\Phi \text{ iteration matrix of } \Phi_\ell. \tag{10.5.3}$$

The inequality opposite to condition (4.7): $n_{\ell-1} \leqslant n_\ell / C_h$ is

$$n_\ell \leqslant \bar{C}_h n_{\ell-1}. \tag{10.5.4}$$

By $n_\ell / n_{\ell-1} \approx (h_{\ell-1}/h_\ell)^d$, inequality (4) also gives an estimate of $h_{\ell-1}/h_\ell$ from above. Together with the norm of $\tilde{p}$, one obtains an estimate of the form

$$\|\tilde{p}\|\,(h_{\ell-1}/h_\ell)^\varkappa \leqslant C_2 \qquad (\tilde{p}\colon X_{\ell-1} \to X_\ell) \tag{10.5.5}$$

with $\varkappa$ from (1).

Theorem 10.5.1. *Assume* (1), (3), (5). *The iteration number* $m_\ell = m$ (*cf.* (2b)) *should be sufficiently large so that*

$$C_2 \zeta^m < 1. \tag{10.5.6}$$

Then the nested iteration (2a, b) *produces results* $\tilde{x}_\ell$ *for all levels* $0 \leqslant \ell \leqslant \ell_{\max}$ *satisfying the error estimates*

$$\|\tilde{x}_\ell - x_\ell\| \leqslant C_3(\zeta^m) C_1 h_\ell^\varkappa \quad \textit{with} \tag{10.5.7a}$$

$$C_3(\zeta^m) := \zeta^m / (1 - C_2 \zeta^m), \tag{10.5.7b}$$

provind that the starting iterate $\tilde{x}_0$ *satisfies inequality* (7a) *for* $\ell = 0$.

Proof. By assumption, (7a) holds for $\ell = 0$. Assume (7a) for levels $\leqslant \ell - 1$. The starting iterate $x_\ell^0 := \tilde{p}\tilde{x}_{\ell-1}$ has an error that can be bounded by

$$\begin{aligned}\|x_\ell^0 - x_\ell\| &\leqslant \|\tilde{p}x_{\ell-1} - x_\ell\| + \|\tilde{p}(\tilde{x}_{\ell-1} - x_{\ell-1})\| \\ &\leqslant \|\tilde{p}x_{\ell-1} - x_\ell\| + \|\tilde{p}\|\,\|\tilde{x}_{\ell-1} - x_{\ell-1}\| \underset{(1),(7a)}{\leqslant} \\ &\leqslant C_1 h_\ell^\varkappa + \|\tilde{p}\|\, C_3(\zeta^m) C_1 h_{\ell-1}^\varkappa \\ &\leqslant C_1 h_\ell^\varkappa [1 + \|\tilde{p}\| (h_{\ell-1}/h_\ell)^\varkappa C_3(\zeta^m)] \underset{(5)}{\leqslant} C_1 h_\ell^\varkappa [1 + C_2 C_3(\zeta^m)].\end{aligned}$$

After m iteration steps, the error is reduced to $\|x_\ell^m - x_\ell\| \leqslant \zeta^m \|x_\ell^0 - x_\ell\| \leqslant C_1 h_\ell^\varkappa \{\zeta^m [1 + C_2 C_3(\zeta^m)]\}$ because of (3). Definition (7b) shows $\{\ldots\} = C_3(\zeta^m)$ and proves (7a) for ℓ. □

Estimate (7a) has an important practical interpretation. It describes that the *iteration* error $\tilde{x}_\ell - x_\ell$ coincides up to a factor $C_3(\zeta^m)$ with the bound $C_1 h_\ell^\varkappa$ from (1), which is related to the *discretisation* error. We recall Remark 3.3.4: As long as x_ℓ is only considered an approximation to a continuous solution of a differential equation, it makes no sense to compute x_ℓ more precisely than by the discretisation error. The nested iteration provides a convenient way to compute $\tilde{x}_\ell$ with

$$\|\tilde{x}_\ell - x_\ell\| \leqslant C_3(\zeta^m) * \text{discretisation error}, \tag{10.5.8}$$

where the constant $C_3(\zeta^m)$ can be controlled without knowing the discretisation error bound $C_1 h_\ell^\varkappa$ quantitatively.

Remark 10.5.2. The standard choice $h_\ell = h_{\ell-1}/2$ and inequality $\|\tilde{p}\| \leqslant 1$ which is valid for the standard interpolations, yield the constant $C_2 = 2^\varkappa$ in (5). The consistency order of the model case is $\varkappa = 2$, from which $C_2 = 4$. Therefore, the factor $C_3(\zeta^m)$ equals

$$C_3(\zeta^m) = \zeta^m/(1 - 4\zeta^m). \tag{10.5.9}$$

For multi-grid methods with convergence rates $\leqslant \zeta = 0.2$ (see the results in §10.4.3), condition (6) is satisfied for only *one* iteration step ($m = 1$) and produces the value $C_3(0.2) = 1$.

Exercise 10.5.3. How should one choose m_ℓ in (7a), in case the contraction number $\zeta = \zeta_\ell < 1$ depends on h_ℓ like $\zeta_\ell = 1 - C(h_\ell)^\varrho$ with $\varrho > 0$?

10.5.3 Amount of Computational Work

Let Cn_ℓ be the work required by the iteration Φ_ℓ at level ℓ and assume (7): $n_{\ell-1} \leqslant n_\ell / C_h$. The work for $\tilde{x}_{\ell-1} \mapsto \tilde{p}\tilde{x}_{\ell-1}$ is considered negligible. The total work amounting to $C_{\text{nested it}} n_\ell \leqslant mC(n_1 + n_2 + \cdots + n_\ell)$ can be estimated, using the geometrical sum $n_1 + n_2 + \cdots + n_\ell \leqslant n_\ell \sum C_h^{-k} \leqslant C_h n_\ell/(C_h - 1)$, by

$$C_{\text{nested it}} \leqslant mCC_h/(C_h - 1).$$

For the standard case $C_h = 2^d = 4$ (cf. Remark 4.1), we obtain the result

$$work_{\text{nested it (2a, b)}} \leqslant \frac{4m}{3} work(\Phi_{\ell_{\max}}). \tag{10.5.10}$$

If we tried to achieve the accuracy $\varepsilon = Ch^{\varkappa}$ at the level $\ell = \ell_{\max}$ with the starting iterate $\tilde{x}_\ell := 0$ by iterating with Φ_ℓ, the work would be proportional to $O(\log \varepsilon) = O(|\log h_\ell|)$ (cf. (3.3.4b)). According to Remark 2, $m = 1$ is a realistic choice. Inequality (10) shows that sufficient accuracy for *all* levels $0 \leqslant \ell \leqslant \ell_{\max}$ can be attained with the 4/3-fold work of a single Φ_ℓ step.

Together with the numbers from (4.11d′, e′) and Table 4.1 (with $\nu_1 = 2$, $\nu_2 = 0$, $m = 1$), we obtain the following results:

V-cycle ($\gamma = 1$) requires $34 n_{\ell_{\max}}$ operations to produce

$$\|\tilde{x}_\ell - x_\ell\| \leqslant 0.53 C_1 h_\ell^{\varkappa} \quad \text{for } 0 \leqslant \ell \leqslant \ell_{\max}. \tag{10.5.11a}$$

W-cycle ($\gamma = 2$) requires $51 n_{\ell_{\max}}$ operations to produce

$$\|\tilde{x}_\ell - x_\ell\| \leqslant 0.08 C_1 h_\ell^{\varkappa} \quad \text{for } 0 \leqslant \ell \leqslant \ell_{\max}. \tag{10.5.11b}$$

The following comparison may give the reader an idea of these numbers: The work given in (11b) corresponds to ≈ 10 Gauß-Seidel iteration steps.

Since the nested iteration (2a) is a finite process and not an iteration, the considerations of §3.3 are not applicable. How many operations are necessary, depends on the accuracy requirements of the user.

10.5.4 Pascal Procedures

The following procedure performs the step $\tilde{x}_{\ell-1} \mapsto \tilde{x}_\ell$ if $\ell > 0$. For $\ell = 0$, $\tilde{x}_0$ is determined as the exact solution.

```
procedure nested_iteration(l: integer;
 var xl,xlminus1: gridfunction;
        m: integer; var MGD: MG_data;
        function right_hand_side(x,y: real): real;
        function boundary_values(x,y: real): real;
        procedure interpolation(l: integer;var px,x:
         gridfunction;
         var AL: hierarchy_of_discretisation;
         function boundary_values(x,y: real): real);
        procedure presmoothing(var new: gridfunction;
         var A: data_of_discretisation;
         var x,b: gridfunction; var IP:
         iterationparameter);
```

```
          procedure postsmoothing(var new: gridfunction;
            var A: data_of_discretisation;
            var x,b: gridfunction; var IP:
            iterationparameter);
          procedure direct_solver(var x,b:gridfunction;
            var A: data_of_discretisation));
var b: gridfunction; i: integer;
begin if l<0 then writeln('l<0 in nested iteration. No
 action') else
  if l>Lmax then writeln('l>Lmax in nested iteration. No
   action')
  else
  with MGD do with AL[l] do
  begin set_actual_level(l,MGD);
    define_inner_points(nx,ny,b,right_hand_side);
    if kind=Poisson_model_problem then factor_x_vector
     (nx,ny,b,h2,b);
    {problem defined at level l}
    if l=0 then
    begin direct_solver(xl,b,AL[l]);
          define_boundary_values(nx,ny,xl,boundary_values)
    end else {case 0<l≤Lmax}
    begin interpolation(l,xl,xlminus1,AL,boundary_values);
        for i:=1 to m do
        MG_iteration(l,xl,xl,b,MGD,presmoothing,
               postsmoothing,direct_solver)
end end end;
```

The meaning of the parameter is `l` $= \ell$, `xl` $= \tilde{x}_\ell$ (output), `xlminus1` $= \tilde{x}_{\ell-1}$ (input), `m` $= m$ from (2b), `MGD`: the `MG_data` defined above. `MGD` contains the information about A_ℓ. The functions `right_hand_side` and `boundary_values` are used to define b_ℓ. The parameter procedures `presmoothing`, `postsmoothing`, and `direct_solver` are the same as for the multi-grid iteration from §10.4.2. $\tilde{p}$ is performed by `interpolation`. For example, one may choose the linear or cubic prolongation:

```
procedure linear_interpolation(l: integer; var px,x:
 gridfunction;
                   var AL: hierarchy_of_discretisation;
                   function boundary_values(x,y: real): real);
begin transfer_for_interpolation(l,px,x,AL,boundary_values);
      interpolate(px,AL[l],linear_interpolation_F)
end;
```

```
procedure cubic_interpolation(l: integer;
 var px,x: gridfunction;
                    var AL: hierarchy_of_discretisation;
                    function boundary_values(x,y: real): real);
begin transfer_for_interpolation(l,px,x,AL,boundary_values);
      interpolate(px,AL[l],cubic_interpolation_F)
end;
```

The following procedures and functions are needed:

```
function linear_interpolation_F(LL,L,R,RR: real; il,ir:
 integer): real;
begin linear_interpolation_F:=(R+L)/2
end;

function quadratic_interpolation_F(LL,L,R,RR: real; il,ir:
 integer): real;
begin if il>=3 then quadratic_interpolation_F:=(3*R+6*L-LL)/8
       else
      if ir>=3 then quadratic_interpolation_F:=(3*L+6*R-RR)/8
       else
      quadratic_interpolation_F:=linear_interpolation_F
       (LL,L,R,RR,il,ir)
end;

function cubic_interpolation_F(LL,L,R,RR: real; il,ir:
 integer): real;
begin if (il>=3)and(ir>=3) then cubic_interpolation_F:=
 (9*(L+R)-LL-RR)/16
      else cubic_interpolation_F:=
       quadratic_interpolation_F(LL,L,R,RR,il,ir)
end;

procedure interpolate(var x: gridfunction; var A: data_of_
 discretisation;
                     function interpolation(LL,L,R,RR: real;
                      il,ir: integer): real);
var i,j: integer;
begin with A do begin i:=2; while i<nx do
  begin j:=1; while j<ny do
     begin x[i,j]:=interpolation(x[i,maximum(0,j-3)],
                                 x[i,j-1],x[i,j+1],
                                 x[i,minimum(ny,j+3)],j,
                                 ny-j);
           j:=j+2
     end;
     i:=i+2
```

```
  end; {interpolation in y direction}
  i:=1; while i<nx do
  begin for j:=1 to ny-1 do
     x[i,j]:=interpolation(x[maximum(0,i-3),j],x[i-1,j],
                           x[i+1,j],
                          x[minimum(nx,i+3),j],i,nx-i);
     i:=i+2 {interpolation in x direction}
end end end;
```

```
procedure transfer_for_interpolation(l: integer; var px,x:
 gridfunction;
                  var AL: hierarchy_of_discretisation;
                  function boundary_values(x,y: real): real);
var i,j,ii: integer;
begin with AL[l-1] do for i:=nx downto 0 do
      begin ii:=2*i; for j:=ny downto 0 do px[ii,2*j]:=x[i,j]
      end;
      with AL[l] do define_boundary_values(nx,ny,px,boundary_
       values)
end;
```

The interpolation functions (e.g., `cubic_interpolation_F`) compute the funtion value $u(x_i)$ from the respective values LL,L,R,RR at x_{i-3}, x_{i-1}, x_{i+1}, x_{i+3}. In the case of cubic interpolation, il $= i$ and ir $= N - i$ must be $\geqslant 3$. Otherwise (close to the boundary), the quadratic interpolation is used (cf. Hackbusch [14, §3.4.3]).

A possible frame program may be as follows:

```
program nested_iteration;
var MGD: MG_data; l,level,m: integer; it: data_of_iteration;
    v: data_for_comparison; max: history_of_iteration;
{functions exact_solution, right_hand_side, etc. are to be
 inserted}
begin initialise_MG_data(MGD); initialise_it(it); initialise_
 comparison(v);
 with MGD do repeat
   release_MG_data(MGD); release_it(it); release_
    comparison(v);
   define_MG_discretisation(MGD); define_MG_parameter(MGD);
   write('--> maximal level ='); readln(level);
   write('--> Number m of iteration steps per level =');
    readln(m);
   for l:=0 to level do with AL[l] do with it do
   begin writeln; writeln('*** nested iteration at the
    level',l,'***');
```

```
      nested_iteration(l,x,x,m,MGD,right_hand_side,boundary_
       value,
               linear_interpolation, chequerboard_Gauss_
                Seidel,
               chequerboard_Gauss_Seidel,solve_system_
                without_pivoting);
      A:=AL[l]; IP.Nr:=0; define_comparison_solution(v,AL[l],
       exact_solution);
      comparison_with_exact_solution(max,v,it,maximum_norm);
      {here possibility for storing the results}
   end; write('Should the run be repeated?')
  until not yes_no
end.
```

10.5.5 Numerical Examples

First, the nested iteration is applied to the differential equation

$$-\Delta u = f := -\Delta(e^{x+y^2}) \tag{10.5.12a}$$

with boundary values $\varphi = e^{x+y^2}$. The negative Laplacean $-\Delta$ is discretised at all levels by the standard five-point star. The interpolation $\tilde{p}$ is the cubic one. Let x_ℓ^* be the restriction of the exact solution e^{x+y^2} of (12a) to the grid Ω_ℓ. Note that x_ℓ^* does not coincide with the discrete solution x_ℓ of the system $A_\ell x_\ell = b_\ell$ corresponding to (12a). In Table 1, the results $\tilde{x}_\ell$ of the nested iteration (2a) are compared with x_ℓ^*, because this is the error most interesting in practice. The maximum norm $\|\tilde{x}_\ell - x_\ell^*\|_\infty$ of these errors is given for the cases $m = 1$ and $m = 2$. For comparison, the last column shows the discretisation error $\|x_\ell - x_\ell^*\|_\infty$, which formally corresponds to $m = \infty$. The multi-grid method used in (2a) has the same parameters as the W-cycle ($\gamma = 2$) in Table 4.1. One understands from Table 1 that the choice $m = 1$ is sufficient. Doubling the work by using $m = 2$, we cannot improve the total error $\|\tilde{x}_\ell - x_\ell^*\|_\infty$ substantially.

Table 10.5.1 Errors $\|\tilde{x}_\ell - x_\ell\|_\infty$ of the nested iteration for (12a)

ℓ	h_ℓ	$m = 1$	$m = 2$	$m = \infty$
0	1/2	$7.9944658_{10}-2$	$7.9944658_{10}-2$	$7.9944658_{10}-2$
1	1/4	$3.9908756_{10}-2$	$2.9215605_{10}-2$	$2.8969488_{10}-2$
2	1/8	$1.5788721_{10}-2$	$8.1023136_{10}-3$	$8.0307789_{10}-3$
3	1/16	$3.2919346_{10}-3$	$2.0768391_{10}-3$	$2.0729855_{10}-3$
4	1/32	$5.7591549_{10}-4$	$5.2253758_{10}-4$	$5.2247399_{10}-4$
5	1/64	$1.3291689_{10}-4$	$1.3093946_{10}-4$	$1.3093956_{10}-4$

Analogous data are given in Table 2 for the differential equation

$$-\Delta u = f := -\Delta(y \sin(10x)) \qquad (10.5.12b)$$

with the solution $y\sin(10x)$ being oscillatory in the x direction. By the non-smooth behaviour of the solution, the discretisation error (last column) for problem (12b) is nearly one digit worse than for (12a). Therefore, the additional error $O(h_\ell^2)$ of the *linear* interpolation $\tilde{p}$, which is used instead of the cubic one, is of minor consequence. Also for this example, it does not pay to perform $m = 2$ iterations per level.

10.5.6 Comments

Additional variants for the nested iteration (e.g., combinations with extrapolation techniques) are discussed in Hackbusch [14, §5.4, §9.3.4, §16.4] and [16, §5.6.5].

Although nonlinear systems are not the subject of this book, we note that the nested iteration is of even greater importance for nonlinear systems of equations. In the linear case, it helps to save computer time. However, for nonlinear iterations the availability of sufficiently good starting iterates often determines convergence (against the desired solution) or divergence. The nested iteration with its starting value $\tilde{x}_\ell := \tilde{p}\tilde{x}_{\ell-1}$ is a suitable technique for generating such starting iterates.

A description and analysis of the nonlinear multi-grid method and the corresponding nested iteration can be found in Hackbusch [14, §9], [20].

10.6 Convergence Analysis

10.6.1 Summary

The convergence proof of multi-grid methods differs from the convergence considerations used before, because here the relationship between the equations $A_\ell x_\ell = b_\ell$ and $A_{\ell-1}x_{\ell-1} = b_{\ell-1}$ plays an important rôle.

Table 10.5.2 Errors $\|\tilde{x}_\ell - x_\ell\|_\infty$ of the nested iteration for (12b)

ℓ	h_ℓ	$m = 1$	$m = 2$	$m = \infty$
0	1/2	$2.8249099_{10}+0$	$2.8249099_{10}+0$	$2.8249099_{10}+0$
1	1/4	$5.0876212_{10}-1$	$4.6124302_{10}-1$	$4.7880033_{10}-1$
2	1/8	$9.5881341_{10}-2$	$1.0330948_{10}-1$	$1.0308770_{10}-1$
3	1/16	$2.7648979_{10}-2$	$2.6636710_{10}-2$	$2.6689213_{10}-2$
4	1/32	$6.8798570_{10}-3$	$6.6486368_{10}-3$	$6.6506993_{10}-3$
5	1/64	$1.6998365_{10}-3$	$1.6716069_{10}-3$	$1.6714014_{10}-3$

As sufficient criteria, we introduce and discuss two conditions in §10.6.2–3: the smoothing and approximation property. The smoothing property is of an algebraic nature, whereas the proof of the approximation property involves the continuous problem, whose discretisation is described by $A_\ell x_\ell = b_\ell$. The smoothing and approximation property together yield the convergence statement for the two-grid methods (§10.6.4). For $\gamma \geqslant 2$, multi-grid convergence can be concluded directly from the two-grid convergence (§10.6.5).

For positive definite A_ℓ, the multi-grid method can be designed as a symmetric iteration. In §10.7 we will achieve for this case even better convergence results, which include the V-cycle ($\gamma = 1$). These results are generalised in Theorem 7.17 to the nonsymmetric case.

The analysis represented below is as strongly simplified compared with that of Hackbusch [14], as here we base our considerations mostly on the Euclidean and spectral norm. Other norms are mentioned in §10.6.6 and §10.7.2.

In contrast to what has been said above, there are multi-grid methods for which convergence proofs can be performed by purely algebraic considerations. These variants will be discussed in §11.6.

10.6.2 Smoothing Property

In §10.1.1 we called a grid function $x_\ell = \sum \xi_{\alpha\beta} e^{\alpha\beta}$ (cf. (1.2b)) «smooth» if the coefficients $\xi_{\alpha\beta}$ of high frequencies α, β (corresponding to the large eigenvalues $\lambda_{\alpha\beta}$) are small. Quantitatively, one may measure the smoothness by $\|A_\ell x_\ell\|_2 = (\sum |\lambda_{\alpha\beta}\xi_{\alpha\beta}|^2)^{1/2}$. If the smoothing step (2.2a) really leads to a smoothing of the errors $e_\ell = x_\ell^0 - x_\ell$, the error $S_\ell^\nu e_\ell$ obtained from the smoothing step must have a better smoothing measure $\|A_\ell S_\ell^\nu e_\ell\|_2$ then e_ℓ. Therefore, smoothing ability is characterised by the spectral norm $\|A_\ell S_\ell^\nu\|_2$. Before defining the smoothing property, we analyse $\|A_\ell S_\ell^\nu\|_2$ for Richardson's iteration with positive definite A_ℓ:

$$\mathscr{S}_\ell(x_\ell, b_\ell) := x_\ell - \Theta(A_\ell x_\ell - b_\ell) \quad \text{with} \tag{10.6.1a}$$

$$\Theta = \Theta_\ell = 1/\varrho(A_\ell) = 1/\|A_\ell\|_2. \tag{10.6.1b}$$

We have $\|A_\ell S_\ell^\nu\|_2 = \|A_\ell(I - \Theta A_\ell)^\nu\|_2 = \|X(I - X)^\nu\|_2/\Theta$ with $X := \Theta A_\ell$. The following lemma can be applied to $X(I - X)^\nu$.

Lemma 10.6.1. (a) *For all matrices X with $0 \leqslant X \leqslant I$, the inequality*

$$\|X(I - X)^\nu\|_2 \leqslant \eta_0(\nu) \qquad (\nu \geqslant 0) \tag{10.6.2a}$$

holds, where the function $\eta_0(\nu)$ is defined by

$$\eta_0(\nu) := \nu^\nu/(\nu + 1)^{\nu+1}. \tag{10.6.2b}$$

(b) *The asymptotical behaviour of $\eta_0(\nu)$ for $\nu \to \infty$ is*

$$\eta_0(\nu) = \frac{1}{e\nu} + O(\nu^{-2}). \tag{10.6.2c}$$

Proof. Let $f(\xi) := \xi(1-\xi)^\nu$. According to Lemma 2.4.7a, we have

$$\|X(I-X)^\nu\|_2 = \varrho(X(I-X)^\nu) = \max\{|f(\xi)|: \xi \in \sigma(X)\}.$$

By $f(\xi) \leqslant f(1/(\nu+1)) = \eta_0(\nu)$ for all $\xi \in [0,1] \supset \sigma(X)$, part (a) is proved. The discussion of $\eta_0(\nu)$ yields statement (b). □

Remark 10.6.2. For $A_\ell > 0$, Richardson's method (1a, b) leads to

$$\|A_\ell S_\ell^\nu\|_2 \leqslant \eta_0(\nu)\|A_\ell\|_2 \quad \text{for all } \nu \geqslant 0, \ell \geqslant 0. \tag{10.6.3}$$

Note that the factor $\eta_0(\nu)$ is independent of h_ℓ and ℓ. The smoothing property, which we are going to define, is an estimate with a form similar to (3). Instead of $\eta_0(\nu)$, we may take an arbitrary sequence $\eta(\nu) \to 0$. Furthermore, it is neither necessary nor desirable to require an inequality like (3) for *all* $\nu \geqslant 0$.

Definition 10.6.3. An iteration $\mathscr{S}_\ell$ ($\ell \geqslant 0$) satisfies the *smoothing property*, if there are functions $\eta(\nu)$ and $\bar{\nu}(h)$ indepenent of ℓ with

$$\|A_\ell S_\ell^\nu\|_2 \leqslant \eta(\nu)\|A_\ell\|_2 \quad \text{for all } 0 \leqslant \nu < \bar{\nu}(h_\ell), \ell \geqslant 1, \tag{10.6.4a}$$

$$\lim_{\nu\to\infty} \eta(\nu) = 0, \tag{10.6.4b}$$

$$\lim_{h\to 0} \bar{\nu}(h) = \infty \quad \text{or} \quad \bar{\nu}(h) = \infty. \tag{10.6.4c}$$

$\bar{\nu}(h) = \infty$ in (4c) expresses the fact that (4a) holds for *all* ν. This happens only for *convergent* iterations $\mathscr{S}_\ell$, as shown in

Remark 10.6.4. The conditions (4a, b) with $\bar{\nu}(h) = \infty$ imply the convergence of $\mathscr{S}_\ell$.

Proof. By $\eta(\nu) \to 0$, we have $\varrho(S_\ell^\nu) \leqslant \|S_\ell^\nu\|_2 \leqslant \|A_\ell^{-1}\|_2 \|A_\ell S_\ell^\nu\|_2 \leqslant \eta(\nu)\,\mathrm{cond}_2(A_\ell) < 1$ for sufficiently large ν. Hence, also $\varrho(S_\ell) < 1$. □

From Remark 2 one concludes

Theorem 10.6.5. *For* $A_\ell > 0$, *Richardson's method* (1a, b) *satisfies the smoothing property* (4a–c) *with* $\eta(\nu) := \eta_0(\nu)$ *and* $\bar{\nu}(h) = \infty$.

The reason for the more general condition (4c) instead of $\bar{\nu}(h) = \infty$ is that the smoothing property can also be formulated for nonconvergent iterations. Examples of nonconvergent iterations are the Gauß-Seidel method for the indefinite problem (4.5), as well as Richardson's iteration in

Remark 10.6.6. Assume that the indefinite matrix $A_\ell = A_\ell^H$ has the spectrum $\sigma(A_\ell) \subset [-\alpha_\ell, \beta_\ell]$ with $0 < \alpha_\ell \leqslant \beta_\ell$, $\lim_{\ell\to\infty} \alpha_\ell/\beta_\ell = 0$. Although the Richardson method is divergent, it satisfies the smoothing property.

Proof. The damping factor is $\Theta = 1/\beta_\ell$. As in the proof for Lemma 1, we have $\|A_\ell(I - \Theta A_\ell)^\nu\|_2 \leqslant \max\{\eta_0(\nu), (\alpha_\ell/\beta_\ell)(1 + \alpha_\ell/\beta_\ell)^\nu\} \|A_\ell\|_2$. Choose $\bar{\nu}(h_\ell) := \beta_\ell/\alpha_\ell \to \infty$. For $\nu < \bar{\nu}(h_\ell)$, the inequalities

$$(\alpha_\ell/\beta_\ell)(1 + \alpha_\ell/\beta_\ell)^\nu \leqslant (\alpha_\ell/\beta_\ell)\exp\{\nu\alpha_\ell/\beta_\ell\} = \frac{1}{\nu}\{(\nu/\bar{\nu})\exp(\nu/\bar{\nu})\} \leqslant e/\nu$$

follow. Hence, (4a–c) is satisfied with $\eta(\nu) := \max\{\eta_0(\nu), e/\nu\} = e/\nu$. □

The assumptions of Remark 6 are fulfilled for the discretisation of the Helmholtz equation $-\Delta u - cu = f$ $(c > 0)$, because $O(\alpha_\ell/\beta_\ell) = O(h_\ell^2)$.

The following theorem can be considered a perturbation lemma. It shows that the smoothing property remains valid under the perturbation of the matrix A'_ℓ into $A_\ell = A'_\ell + A''_\ell$, where A_ℓ may be indefinite and nonsymmetric.

Theorem 10.6.7. *Let $A_\ell = A'_\ell + A''_\ell$ and $\mathscr{S}_\ell$ and $\mathscr{S}'_\ell$ be the smoothing iterations corresponding to A_ℓ and A'_ℓ, respectively. Their iteration matrices are denoted by S_ℓ and S'_ℓ with $S''_\ell := S_\ell - S'_\ell$. Assume that*

$$A'_\ell \text{ and } S'_\ell \text{ satisfy the smoothing property with } \eta'(\nu),\ \bar{\nu}'(h), \tag{10.6.5a}$$

$$\|S'_\ell\|_2 \leqslant C'_S \quad \text{for all } \ell \geqslant 1, \tag{10.6.5b}$$

$$\lim_{\ell\to\infty} \|S''_\ell\|_2 = 0, \tag{10.6.5c}$$

$$\lim_{\ell\to\infty} \|A''_\ell\|_2/\|A'_\ell\|_2 = 0. \tag{10.6.5d}$$

Then the iteration $\mathscr{S}_\ell$ for A_ℓ also satisfies the smoothing property. The corresponding bound $\eta(\nu)$ can be chosen, e.g., as $\eta(\nu) := 2\eta'(\nu)$.

Proof. $C_S := C'_S + \max\{\|S''_\ell\|_2 : \ell \geqslant 1\}$ satisfies $\|S_\ell\|_2 \leqslant C_S$ for all $\ell \geqslant 1$. Without loss of generality, we may suppose that $C_S \geqslant 1$. S_ℓ^ν can be split into $S_\ell'^\nu + S_\ell''^{(\nu)}$ with

$$\|S_\ell''^{(\nu)}\|_2 = \|S_\ell^\nu - S_\ell'^\nu\|_2 = \left\| \sum_{\mu=0}^{\nu-1} S_\ell^\mu (S_\ell - S'_\ell) S_\ell'^{\nu-1-\mu} \right\|_2 = \left\| \sum_{\mu=0}^{\nu-1} S_\ell^\mu S''_\ell S_\ell'^{\nu-1-\mu} \right\|_2$$

$$\leqslant \left(\sum_{\mu=0}^{\nu-1} C_S^\mu C_S'^{\nu-1-\mu} \right) \|S''_\ell\|_2 \leqslant \nu C_S^{\nu-1} \|S''_\ell\|_2 \underset{(5c)}{\to} 0 \quad \text{for } \ell \to \infty. \tag{10.6.5e}$$

For $1 \leqslant \nu \leqslant \bar{\nu}'(h_\ell)$, we have

$$\begin{aligned} \|A_\ell S_\ell^\nu\|_2 &\leqslant \|A'_\ell S_\ell'^\nu\|_2 + \|A''_\ell\|_2 \|S_\ell^\nu\|_2 + \|A'_\ell\|_2 \|S_\ell''^{(\nu)}\|_2 \\ &\leqslant \eta'(\nu)\|A'_\ell\|_2 + C_S^\nu \|A''_\ell\|_2 + \nu C_S^{\nu-1} \|S''_\ell\|_2 \|A'_\ell\|_2 \\ &= \eta'(\nu)\|A_\ell\|_2 \left\{ \frac{\|A'_\ell\|_2}{\|A_\ell\|_2} + C_S^\nu \frac{\|A''_\ell\|_2}{\|A_\ell\|_2} + \nu C_S^{\nu-1} \frac{\|A'_\ell\|_2}{\|A_\ell\|_2} \|S''_\ell\|_2 \right\}. \end{aligned} \tag{10.6.5f}$$

By $\|A''_\ell\|_2/\|A'_\ell\|_2 \to 0$, $\|S''_\ell\|_2 \to 0$, $\|A'_\ell\|_2/\|A_\ell\|_2 \to 1$, $\|A''_\ell\|_2/\|A_\ell\|_2 \to 0$, the expression $\{\dots\}$ converges for $\ell \to \infty$ (i.e., for $h = h_\ell \to 0$) and fixed v to 1. This proves that $\bar{v}''(h) \to \infty$ $(h \to 0)$ for

$$\bar{v}''(h) := \sup\left\{v > 0: \frac{\|A'_\ell\|_2}{\|A_\ell\|_2} + C_S^v \frac{\|A''_\ell\|_2}{\|A_\ell\|_2} + vC_S^{v-1} \frac{\|A'_\ell\|_2}{\|A_\ell\|_2} \|S''_\ell\|_2 \leqslant 2 \text{ for } h_\ell \leqslant h\right\}.$$

We define $\eta(v) := 2\eta'(v)$ and $\bar{v}(h) := \min\{\bar{v}'(h_\ell), \bar{v}''(h)\}$. For $v \leqslant \bar{v}(h)$ (5f) proves the smoothing property $\|A_\ell S_\ell^v\|_2 \leqslant \eta(v)\|A_\ell\|_2$. □

Usually, discretisations of elliptic differential equations satisfy the following conditions:

There is an h-independent constant c_0 such that

$$A'_\ell := \tfrac{1}{2}(A_\ell + A_\ell^H) + c_0 I \text{ is positive definite,} \tag{10.6.6a}$$

$$\underline{C}h_\ell^{-2m} \leqslant \|A'_\ell\|_2 \leqslant \bar{C}h_\ell^{-2m} \quad (2m\text{: order of the differential eq.}), \tag{10.6.6b}$$

$$\|A''_\ell\|_2 \leqslant Ch_\ell^{1-2m} \quad \text{for } A''_\ell := A_\ell - A'_\ell = \tfrac{1}{2}(A_\ell - A_\ell^H) - c_0 I \tag{10.6.6c}$$

(cf. Hackbusch [14, 15]). To apply Theorem 7, one proves the smoothing property for the positive definite matrix A'_ℓ and transfers this property to A_ℓ by means of Theorem 7. Condition (5d) follows from (6b, c) by $\|A''_\ell\|_2/\|A'_\ell\|_2 \leqslant O(h_\ell) \to 0$. Since $S''_\ell = -\Theta A''_\ell = -A''_\ell/\|A'_\ell\|_2$ in the case of Richardson's method, (5d) also implies (5c). (5b) is always satisfied with $C_S = 2$, because $S'_\ell = I - A'_\ell/\|A'_\ell\|_2$ (even $C_S = 1$, if $A'_\ell \geqslant 0$).

The smoothing property can be proved not only for the Richardson method but also for the *damped* (block-)Jacobi iteration, the 2-cyclic Gauß-Seidel iteration (in particular, the chequer-board Gauß-Seidel method for five-point formulae), and the Kaczmarz iteration. Furthermore, symmetric iterations like the symmetric Gauß-Seidel method, SSOR, and the ILU iteration belong to this class. The symmetric case will be considered in §10.7.3. The smoothing property does *not* hold, e.g., for the nondamped Jacobi method or the SOR method with $\omega \geqslant \omega_{\text{opt}}$. For the smoothing analysis of the iterations mentioned above, compare with Hackbusch [14, §6.2].

The proof of Lemma 1 is based on the properties of the spectral norm for normal matrices. Correspondingly, statements for general matrices are proved via perturbation arguments. Nevertheless, it is possible to obtain the smoothing property for general matrices directly. Even other norms than the spectral norm may be used.

Theorem 10.6.8 (Reusken [2]). *Let $\|\cdot\|$ be a matrix norm corresponding to a vector norm. Let $S_\ell = I - W_\ell^{-1}A_\ell$ be the iteration matrix of the smoother and assume*

$$\|I - 2W_\ell^{-1}A_\ell\| \leqslant 1, \tag{10.6.7a}$$

$$\|W_\ell\| \leqslant C\|A_\ell\| \tag{10.6.7b}$$

with a constant C independent of ℓ. *Then the smoothing property* (7c) *holds*:

$$\|A_\ell S_\ell^\nu\| \leqslant C\sqrt{2/(\pi\nu)}\,\|A_\ell\| \quad \text{for all } \nu \geqslant 1. \tag{10.6.7c}$$

The proof is based on the following

Lemma 10.6.9. *Let the matrix B satisfy* $\|B\| \leqslant 1$ *with respect to a matrix norm corresponding to a vector norm. Then*

$$\|(I-B)(I+B)^\nu\| \leqslant 2\binom{\nu}{[\nu/2]} \leqslant 2^{\nu+1}\sqrt{2/(\pi\nu)}. \tag{10.6.8}$$

Proof. Note that

$$\begin{aligned}(I-B)(I+B)^\nu &= (I-B)\sum_{\mu=0}^{\nu}\binom{\nu}{\mu}B^\mu \\ &= I + \sum_{\mu=1}^{\nu}\binom{\nu}{\mu}B^\mu - \sum_{\mu=0}^{\nu-1}\binom{\nu}{\mu}B^{\mu+1} - B^{\nu+1} \\ &= (I-B^{\nu+1}) + \sum_{\mu=1}^{\nu}\left[\binom{\nu}{\mu} - \binom{\nu}{\mu-1}\right]B^\mu.\end{aligned}$$

By $\|B^\mu\| \leqslant 1$, $\binom{\nu}{\mu-\alpha} = \binom{\nu}{\alpha}$ and $\binom{\nu}{\mu} \geqslant \binom{\nu}{\mu-1}$ for $\mu \leqslant [\nu/2]$ ($[\dots]$ is the truncation to the next integer) one obtains

$$\begin{aligned}\|(I-B)(I+B)^\nu\| &\leqslant 2 + 2\sum_{\mu=1}^{[\nu/2]}\left|\binom{\nu}{\mu} - \binom{\nu}{\mu-1}\right| \\ &= 2 + 2\sum_{\mu=1}^{[\nu/2]}\left\{\binom{\nu}{\mu} - \binom{\nu}{\mu-1}\right\} = 2\binom{\nu}{[\nu/2]} - 2\binom{\nu}{0} \\ &= 2\binom{\nu}{[\nu/2]}.\end{aligned}$$

One may check that the sequence $a_k := \binom{2k}{k}\sqrt{k}/2^{2k}$ is monotonically increasing and tending to $\lim a_k := 1\sqrt{\pi}$. $\binom{\nu}{[\nu/2]} = a_{\nu/2}2^\nu/\sqrt{\nu/2}$ for even ν leads to the desired estimate $a_k \leqslant 1/\sqrt{\pi}$. For odd ν note that $\binom{\nu}{[\nu/2]} = \frac{1}{2}\binom{\nu+1}{(\nu+1)/2}$.

Proof of Theorem 8. One may write $(I-B)(I+B)^\nu = 2^{\nu+1}W_\ell^{-1}A_\ell S_\ell^\nu$ with $B := I - 2W_\ell^{-1}A_\ell$; hence,

$$\|A_\ell S_\ell^\nu\| = 2^{-\nu-1}\|W_\ell(I-B)(I+B)^\nu\| \leqslant 2^{-\nu-1}\|W_\ell\|\,\|(I-B)(I+B)^\nu\|.$$

Assumption (7b) and Lemma 9 yield the assertion. □

Example 10.6.10. (a) Let C_i $(1 \leqslant i \leqslant 4)$ be positive constants independent of ℓ with

$$C_1 I \leqslant \tfrac{1}{2}(A_\ell + A_\ell^H) \leqslant C_2 h_\ell^{-2} I, \tag{10.6.9a}$$

$$\|\tfrac{1}{2}(A_\ell - A_\ell^H)\|_2 \leqslant C_3 h_\ell^{-1} I, \tag{10.6.9b}$$

$$\|A_\ell\|_2 \geqslant C_4 h_\ell^{-2} I. \tag{10.6.9c}$$

Let $\Theta = \Theta_\ell := h_\ell^2 C_1/(C_1 C_2 + C_3^2)$ and $C := (C_1 C_2 + C_3^2)/(C_1 C_4)$. Then the Richardson method damped by Θ_ℓ satisfies the smoothing property (7c) with the constant C from above.
(b) Let $\mathscr{S}_\ell$ be the Jacobi or Gauß-Seidel method damped by $\vartheta = \frac{1}{2}$. Furthermore, A_ℓ is assumed to be weakly diagonally dominant. Then the smoothing property (7c) holds with $C = 2$ with respect to the row-sum norm $\|\cdot\|_\infty$.

Proof. (i) Theorem 4.4.8 proves (7a). (7b) follows with $C = 1/\Theta$.
(ii) Since $\vartheta = \frac{1}{2}$, (7a) is the estimation of the nondamped methods. Weak diagonal dominance implies (7a). From $\|D_\ell\|_\infty \leqslant \|D_\ell - E_\ell\|_\infty \leqslant \|A_\ell\|_\infty$ for $A_\ell = D_\ell - E_\ell - F_\ell$ (cf. (4.2.7a–d)) one concludes (7b) with $C = 1/\vartheta$. □

10.6.3 Approximation Property

10.6.3.1 Formulation. For the coarse-grid correction, the fine-grid solution e_ℓ of $A_\ell e_\ell = d_\ell$ is replaced by $pe_{\ell-1}$ from $A_{\ell-1}e_{\ell-1} = d_{\ell-1} := rd_\ell$. Therefore, $pe_{\ell-1} \approx e_\ell$, i.e., $pA_{\ell-1}^{-1}rd_\ell \approx A_\ell^{-1}d_\ell$, should be valid. We quantify this requirement by

$$\|pA_{\ell-1}^{-1}rd_\ell - A_\ell^{-1}d_\ell\|_2 \leqslant C_A\|d_\ell\|_2/\|A_\ell\|_2 \quad \text{for all } \ell \geqslant 1,\ d_\ell \in X_\ell. \tag{10.6.10}$$

(10) can be rewritten by means of the matrix norm (spectral norm) as

approximation property
$\|A_\ell^{-1} - pA_{\ell-1}^{-1}r\|_2 \leqslant C_A/\|A_\ell\|_2 \quad \text{for all } \ell \geqslant 1.$ (10.6.11)

In general, proofs of the approximation property (11) are not of an algebraic nature but use (at least indirectly) properties of the underlying boundary value problem. One possible route to the proof is as follows. Assume that $A_{\ell-1} = rA_\ell p$ according to (1.26). For an arbitrary restriction $r'\colon X_\ell \to X_{\ell-1}$, the following representation holds:

$$A_\ell^{-1} - pA_{\ell-1}^{-1}r = (A_\ell^{-1} - pA_{\ell-1}^{-1}r)(I - pr') = (I - pA_{\ell-1}^{-1}rA_\ell)(I - pr')A_\ell^{-1}.$$

Considerations concerning the «discrete regularity» of A_ℓ (cf. Hackbusch [8], [9], [14, §6.3.2.1], [15, §9.2]) allow us to prove that $v_\ell := A_\ell^{-1}f_\ell$ is sufficiently smooth. Hence, the interpolation error $\delta_\ell = (I - pr')v_\ell = v_\ell - pr'v_\ell$ can be estimated by $\|\delta_\ell\|_2 \leqslant C\|f_\ell\|_2/\|A_\ell\|_2$. Discrete regularity can also be used to show $\|I - pA_{\ell-1}^{-1}rA_\ell\| \leqslant$ const. Together, one obtains the approximation property (11). In case $A_{\ell-1}$ is not the Galerkin product, compare Hackbusch [14, Criteria 6.3.35 and 38].

The easiest proof of the approximation property can be given for Galerkin discretisations. For this purpose, the discretisation is explained in §10.6.3.2. In §§10.6.3.3–4, the connections between the Galerkin subspaces and vector spaces X_ℓ are described and in §10.6.3.5 the standard error estimates discussed. §10.6.3.6 contains the crucial part of the proof of the approximation property.

10.6.3.2 Galerkin Discretisation. Let V be a Hilbert space with a scalar product $(\cdot,\cdot)_V$ and norm $\|\cdot\|_V$ that can be embedded continuously into another Hilbert space U, i.e., $V \subset U$ and $\sup\{\|v\|_V/\|v\|_U: 0 \neq v \in V\} < \infty$. The dual space V' (of the continuous linear mappings from V into $\mathbb{K}$) is endowed with the *dual norm*

$$\|f\|_{V'} := \|f\|_{\mathbb{K}\leftarrow V} = \sup\{|f(v)|/\|v\|_V: 0 \neq v \in V\} \quad \text{for } f \in V'.$$

One may identify the Hilbert space U with its dual space U'. Then one obtains the inclusions

$$V \subset U = U' \subset V' \qquad \text{(continuous embeddings).} \tag{10.6.12}$$

For $f(v)$ $(v \in V, f \in V')$, one may also write $(f, v)_U$, because the scalar product of U can be extended continuously from $U \times U$ onto $V' \times V$ and $V \times V'$ (cf. Hackbusch [15, §6.3]).

Let a: $V \times V \to \mathbb{K}$ be a continuous bilinear form (or sesquilinear form, if $\mathbb{K} = \mathbb{C}$). The problem to be solved reads (in the weak formulation)

$$\text{find } v \in V \text{ with } a(v, w) = (f, w)_U \quad \text{for all } w \in V, \tag{10.6.13}$$

where $f \in V'$. The Poisson-model problem corresponds to (13) with

$$V = H_0^1(\Omega), \quad U = L^2(\Omega), \quad a(v,w) = \int_\Omega \langle \nabla v, \nabla w \rangle \, dx, \quad (f,w)_U = \int_\Omega fw \, dx.$$

For the notation of the Sobolev space $H_0^1(\Omega)$ and further details concerning the formulation (13), we refer to Hackbusch [15, §§6–7], [14], [16, §8.3].

For the discretisation, we introduce a *hierarchy of finite-dimensional subspaces*:

$$V_0 \subset V_1 \subset \cdots \subset V_{\ell-1} \subset V_\ell \subset \cdots \subset V. \tag{10.6.14a}$$

The *Galerkin solution* v_ℓ is defined by problem (14b):

$$\text{find } u_\ell \in V_\ell \text{ with } a(u_\ell, w) = (f, w)_U \quad \text{for all } w \in V_\ell. \tag{10.6.14b}$$

10.6.3.3 Hierarchy of the Systems of Equations. For a concrete description of the Galerkin solution v_ℓ, a *basis* $\{b_{\ell,\alpha}: \alpha \in I_\ell\}$ of V_ℓ must be chosen. $n_\ell = \# I_\ell = \dim V_\ell$ denotes the dimension. Given $v_\ell \in V_\ell$, we denote the vector

$x_\ell = (x_{\ell,\alpha})_{\alpha \in I_\ell} \in X_\ell := \mathbb{R}^{I_\ell}$ with

$$P_\ell x_\ell := \sum_{\alpha \in I_\ell} x_{\ell,\alpha} b_{\ell,\alpha} = v_\ell \in V_\ell. \tag{10.6.15}$$

as the *coefficient vector*. Proofs of the next exercises can be found, e.g., in Hackbusch [14, 15]:

Exercise 10.6.11. Prove that by the connection $u_\ell = P_\ell x_\ell$, the Galerkin discretisation (14b) is equivalent to system (16a):

$$A_\ell x_\ell = b_\ell \quad \text{with} \tag{10.6.16a}$$

$$A_{\ell,\alpha\beta} = a(b_{\ell,\beta}, b_{\ell,\alpha}), \quad f_{\ell,\alpha} = (f_\ell, b_{\ell,\alpha})_U \qquad (\alpha, \beta \in I_\ell). \tag{10.6.16b}$$

Usually, the basis $\{b_{\ell,\alpha}: \alpha \in I_\ell\}$ is chosen such that the coefficient vector x_ℓ and the represented function $v_\ell = P_\ell x_\ell$ are comparable with respect to their X_ℓ and U norm, respectively. More precisely, we require that

$$\underline{C}_P^{-1} \|x_\ell\|_2 \leqslant \|P_\ell x_\ell\|_U \leqslant \overline{C}_P \|x_\ell\|_2 \quad \text{for all } x_\ell \in X_\ell, \quad \ell \geqslant 0. \tag{10.6.17}$$

For this purpose, one may scale the ℓ_2 norm suitably (cf. (1.21)). Since X_ℓ and U are Hilbert spaces (with the scalar products $\langle \cdot, \cdot \rangle = \langle \cdot, \cdot \rangle_\ell$ and $(\cdot, \cdot)_U$), there is adjoint mapping to $P_\ell: X_\ell \to U$:

$$R_\ell = P_\ell^*: U \to X_\ell, \tag{10.6.18}$$

defined by $\langle R_\ell v, x_\ell \rangle = (v, P_\ell x_\ell)_U$.

Any continuous bilinear form $a(\cdot, \cdot)$ uniquely defines an operator

$$A: V \to V' \quad \text{with} \quad a(u, v) = (Au, v)_U \quad \text{for all } u, v \in V. \tag{10.6.19a}$$

Exercise 10.6.12. Prove the representations (19b) for A_ℓ, f_ℓ from (16a, b):

$$A_\ell = R_\ell A P_\ell, \qquad f_\ell = R_\ell f. \tag{10.6.19b}$$

As $R_\ell P_\ell: X_\ell \to X_\ell$ is positive definite, the mappings

$$\hat{P}_\ell := P_\ell (R_\ell P_\ell)^{-1}, \qquad \hat{R}_\ell = \hat{P}_\ell^* := (R_\ell P_\ell)^{-1} R_\ell \tag{10.6.20a}$$

exist. By construction, we have

$$R_\ell \hat{P}_\ell = \hat{R}_\ell P_\ell = I: \quad X_\ell \to X_\ell. \tag{10.6.20b}$$

Lemma 10.6.13. (20c) *holds with* $\underline{C}_P$ *from* (17) *and vice versa*:

$$\|\hat{P}_\ell\|_{U \leftarrow X_\ell} = \|\hat{R}_\ell\|_{X_\ell \leftarrow U} = \|(R_\ell P_\ell)^{-1}\|_2^{1/2} \leqslant \underline{C}_P. \tag{10.6.20c}$$

Proof. The first equality in (20c) follows from $\hat{R}_\ell = \hat{P}_\ell^*$. By $\|\hat{P}_\ell x_\ell\|_U^2 = (\hat{P}_\ell x_\ell, \hat{P}_\ell x_\ell)_U = \langle \hat{R}_\ell \hat{P}_\ell x_\ell, x_\ell \rangle_\ell = \langle (R_\ell P_\ell)^{-1} x_\ell, x_\ell \rangle_\ell$, the second one is proved. The estimate $\|(R_\ell P_\ell)^{-1}\|_2 = 1/\lambda$ holds with $\lambda = \lambda_{\min}(R_\ell P_\ell)$. The corresponding eigenvector x_ℓ satisfies

$$\lambda \|x_\ell\|_2^2 = \lambda\langle x_\ell, x_\ell\rangle_\ell = \langle R_\ell P_\ell x_\ell, x_\ell\rangle_\ell = (P_\ell x_\ell, P_\ell x_\ell)_U$$
$$= \|P_\ell x_\ell\|_U^2 \geqslant \|x_\ell\|_2^2/\underline{C}_P^2;$$

hence, $\|(R_\ell P_\ell)^{-1}\|_2 \leqslant \underline{C}_P^2$ implies the last inequality in (20c). □

Exercise 10.6.14. The matrix $R_\ell P_\ell$ is called the «mass matrix». Prove that condition (17) with suitable scaling of $\langle\cdot,\cdot\rangle_\ell$ is equivalent to the property that the condition $\varkappa(R_\ell P_\ell)$ of the mass matrix is independent of the level number (and therefore, of the step size h_ℓ). *Hint:* Let $C_\ell := \varrho(R_\ell P_\ell)$. If $\sup C_\ell = \infty$ or $\inf C_\ell = 0$, replace $\langle\cdot,\cdot\rangle_\ell$ by $\langle\cdot,\cdot\rangle_\ell/C_\ell$.

10.6.3.4 Canonical Prolongation and Restriction. The vector $x_{\ell-1} \in X_{\ell-1}$ is associated with the function $v_{\ell-1} := P_{\ell-1}x_{\ell-1} \in V_{\ell-1}$, which by $V_{\ell-1} \subset V_\ell$ (cf. (14a)) also belongs to V_ℓ and therefore admits a representation $v_{\ell-1} = P_\ell x_\ell$ with a coefficient vector $x_\ell \in X_\ell$. We set $px_{\ell-1} := x_\ell$. This defines the *canonical prolongation*, which formally can be written as $p := P_\ell^{-1}P_{\ell-1}\colon X_{\ell-1} \to X_\ell$ or

$$P_\ell p = P_{\ell-1}. \tag{10.6.21a}$$

According to (1.22), the *canonical restriction* is defined as an adjoint of p:

$$r = p^*, \qquad rR_\ell = R_{\ell-1}. \tag{10.6.21b}$$

Exercise 10.6.15. Prove the following: (a) $p = \hat{R}_\ell P_{\ell-1}$ and the estimates

$$\|p\|_{X_\ell \leftarrow X_{\ell-1}} = \|r\|_{X_{\ell-1}\leftarrow X_\ell} \leqslant \underline{C}_P \overline{C}_P \quad \text{for all } \ell \geqslant 1. \tag{10.6.21c}$$

(b) Galerkin's discretisation leads to the product representation (1.26):

$$A_{\ell-1} = rA_\ell p \quad \text{for all } \ell \geqslant 1. \tag{10.6.21d}$$

10.6.3.5 Error Estimate of the Galerkin Solution. If A from (19a) corresponds to the (differentiation) order $2m$, V is a (subspace of the) Sobolev space $H^m(\Omega)$, whose norm is now denoted by $\|\cdot\|_V = \|\cdot\|_m$. By arguments from approximation theory, one finds estimates of the form

$$\|u - u_\ell\|_m \leqslant C_m h_\ell^m \|u\|_{2m} \quad (u,\, u_\ell\text{: solutions of (13), (14b)}). \tag{10.6.22a}$$

Here, h_ℓ is the discretisation parameter (e.g., the grid size) associated with V_ℓ. $\|\cdot\|_{2m}$ is the norm of the Sobolev space $H^{2m}(\Omega) \cap V$.

The problem (13) is called *2m-regular* if

$$\|u\|_{2m} \leqslant C_{\mathrm{reg}}\|f\|_U \quad \text{for } u = A^{-1}f \text{ and all } f \in U. \tag{10.6.22b}$$

Together, (22a) and (22b) yield the inequality

$$\|u - u_\ell\|_V = \|u - u_\ell\|_m \leqslant Ch_\ell^m \|f\|_U. \tag{10.6.22c}$$

Exercise 11 and (19b) show that $u_\ell = P_\ell A_\ell^{-1} R_\ell f$; hence, $u - u_\ell = E_\ell f$ with

$$E_\ell := A^{-1} - P_\ell A_\ell^{-1} R_\ell. \tag{10.6.22d}$$

(22c) can be expressed as

$$\|E_\ell\|_{V \leftarrow U} \leqslant C h_\ell^m. \tag{10.6.22e}$$

If the bilinear form (sesquilinear form) a is not symmetric, we require that the *adjoint problem*

$$\text{find } v^* \in V \text{ with } a(w, v^*) = (w, f)_U \quad \text{for all } w \in V$$

has the same properties (22a, b) as the original problem (13). Analogously to (22e), one obtains

$$\|E_\ell^*\|_{V \leftarrow U} \leqslant C^* h_\ell^m. \tag{10.6.22e*}$$

As $\|E_\ell^*\|_{V \leftarrow U} = \|E_\ell\|_{U \leftarrow V'}$ (by $U = U'$), it follows that

$$\|E_\ell\|_{U \leftarrow V'} \leqslant C^* h_\ell^m. \tag{10.6.22f}$$

The *inverse estimate* is the inequality (23a):

$$\|v_\ell\|_V \leqslant C_{\text{inv}} h_\ell^{-m} \|v_\ell\|_U \quad \text{for all } v_\ell \in V_\ell, \ell \geqslant 0. \tag{10.6.23a}$$

Exercise 10.6.16. Assume (20c), (23a), and the boundedness

$$|a(u, v)| \leqslant C_a \|u\|_V \|v\|_V \quad \text{for all } u, v \in V \tag{10.6.23b}$$

of a. Prove (23c) with $C_K := C_a (C_{\text{inv}} \bar{C}_P)^2$:

$$\|A_\ell\|_2 \leqslant C_K h_\ell^{-2m}. \tag{10.6.23c}$$

A last condition for the approximation property is almost identical to inequality (5.4):

$$h_{\ell-1} \leqslant C_h h_\ell \quad \text{for all } \ell \geqslant 1. \tag{10.6.23d}$$

10.6.3.6 Proof of the Approximation Property. Mainly, the assumptions in the next theorem are the h_ℓ independence of the condition of the mass matrix $R_\ell P_\ell$, the error estimate (22a), the $2m$-regularity (22b), the inverse estimate (23a), and condition (23d), which usually holds with $C_h = 2$.

Theorem 10.6.17. *Let A_ℓ be the matrices* (16b) *of the Galerkin discretisation. Choose the canonical p and r. Assume* (20c), (22e, e*), *and* (23c, d). *Then the approximation property* (11) *holds.*

Proof. $A_{\ell-1} = r A_\ell p$ implies $E_\ell A E_\ell = E_\ell$ and

$$\|E_\ell\|_{U \leftarrow U} \leqslant \|E_\ell\|_{U \leftarrow V'} \|A\|_{V' \leftarrow V} \|E_\ell\|_{V \leftarrow U} \leqslant C_a C C^* h_\ell^{2m}, \tag{10.6.24a}$$

because $\|A\|_{V'\leftarrow V} \leqslant C_a$ follows from (23b) and (22f) from (22e*). The triangle inequality applied to $E_{\ell-1} - E_\ell = P_\ell A_\ell^{-1} R_\ell - P_{\ell-1} A_{\ell-1}^{-1} R_{\ell-1}$ yields the estimate

$$\|P_\ell A_\ell^{-1} R_\ell - P_{\ell-1} A_{\ell-1}^{-1} R_{\ell-1}\|_{U\leftarrow U} \leqslant C_a C C^* (h_\ell^{2m} + h_{\ell-1}^{2m}). \quad (10.6.24b)$$

(23d) gives $h_{\ell-1}^{2m} \leqslant C_h^{2m} h_\ell^{2m}$. From $h_\ell^{2m} \leqslant C_K / \|A_\ell\|_2$ (cf. (23c)) and $P_\ell = P_{\ell-1} p$, $R_\ell = r R_{\ell-1}$ (cf. (21a, b)) one concludes that

$$\|P_\ell (A_\ell^{-1} - p A_{\ell-1}^{-1} r) R_\ell\|_{U\leftarrow U} \leqslant C' / \|A_\ell\|_2 \quad (10.6.24c)$$

with $C' := C_a C C^* (1 + C_h^{2m}) C_K$. Multiplying $P_\ell (A_\ell^{-1} - p A_{\ell-1}^{-1} r) R_\ell$ by $\hat{R}_\ell$ from the left and by $\hat{P}_\ell$ from the right and using (20b, c), we obtain $\|A_\ell^{-1} - p A_{\ell-1}^{-1} r\|_2 \leqslant \|\hat{R}_\ell\|_{X_\ell \leftarrow U} \|P_\ell (A_\ell^{-1} - p A_{\ell-1}^{-1} r) R_\ell\|_{U\leftarrow U} \|\hat{P}_\ell\|_{U\leftarrow X_\ell} \leqslant C' \underline{C}_P^2 / \|A_\ell\|_2$. Hence, the approximation property with $C_A := C' \underline{C}_P^2$. □

10.6.4 Convergence of the Two-Grid Iteration

As mentioned in §10.2.2, $\varrho(M_\ell^{\mathrm{TGM}}(\nu_1, \nu_2)) = \varrho(M_\ell^{\mathrm{TGM}}(\nu, 0))$ holds for $\nu = \nu_1 + \nu_2$, so that we may restrict our considerations to $\nu_1 > 0$, $\nu_2 = 0$. This choice is optimal for statements concerning the contraction number $\|M_\ell^{\mathrm{TGM}}(\nu, 0)\|_2$ with respect to the spectral norm. The following Theorems 18 and 19 correspond to the cases $\bar{\nu}(h) = \infty$ and $\bar{\nu}(h) < \infty$, respectively.

Theorem 10.6.18. *Assume the smoothing and approximation properties* (4a–c), (11) *with* $\bar{\nu}(h) = \infty$. *For given* $0 < \zeta < 1$, *there exists a lower bound* $\underline{\nu}$ *such that*

$$\|M_\ell^{\mathrm{TGM}}(\nu, 0)\|_2 \leqslant C_A \eta(\nu) \leqslant \zeta \quad \text{for all } \nu \geqslant \underline{\nu},\ \ell \geqslant 1. \quad (10.6.25)$$

Here, C_A *and* $\eta(\nu)$ *are the quantities from* (11) *and* (4a, b). *By* $\zeta < 1$, (25) *implies the convergence of the two-grid iteration. Note that the contraction bound* $C_A \eta(\nu)$ *is* h_ℓ*-independent.*

Proof. The two-grid iteration matrix reads

$$M_\ell^{\mathrm{TGM}}(\nu, 0) = (I - p A_{\ell-1}^{-1} r A_\ell) S_\ell^\nu = [A_\ell^{-1} - p A_{\ell-1}^{-1} r][A_\ell S_\ell^\nu]$$

(cf. Lemma 2.1). Estimating both factors by (4a) and (11), we obtain (25). □

Theorem 10.6.19. *Assume the smoothing and approximation properties* (4a–c), (11), *including the case* $\bar{\nu}(h) < \infty$. *For all* $0 < \zeta < 1$, *there exist bounds* $\bar{h} > 0$ *and* $\underline{\nu}$ *such that* (25) *holds for all* $\nu \in [\underline{\nu}, \bar{\nu}(h))$ *and all* ℓ *with* $h_\ell \leqslant \bar{h}$, *where the interval* $[\underline{\nu}, \bar{\nu}(h)]$ *is not empty (i.e.,* $\underline{\nu} < \bar{\nu}(h)$).

Proof. Choose $\underline{\nu}$ as in Theorem 18. Because of $\bar{\nu}(h) \to \infty$ as $h \to 0$, $\bar{h}$ can be chosen such that $\bar{\nu}(h_\ell) > \underline{\nu}$ for all $h_\ell \leqslant \bar{h}$. □

10.6.5 Convergence of the Multi-Grid Iteration

In Theorem 4.5 the representation $M_\ell^{\mathrm{MGM}}(\nu,0) = M_\ell^{\mathrm{TGM}}(\nu,0) - \cdots$ of the multi-grid iteration matrix has been shown. We are exploiting the fact that the perturbation «...» is sufficiently small; hence, two-grid convergence implies multi-grid convergence. Besides the smoothing and approximation properties, we require additional conditions, which however are easy to satisfy. The first one is

$$\|S_\ell^\nu\|_2 \leqslant C_S \quad \text{for all } \ell \geqslant 1, 0 < \nu < \bar{\nu} := \min_{\ell \geqslant 1} \bar{\nu}(h_\ell) \tag{10.6.26}$$

with $\bar{\nu}(h_\ell)$ from (4c).

Exercise 10.6.20. Assume $S_\ell := S'_\ell + S''_\ell$. Let (26) hold for $S_\ell'^\nu$ and assume (5c): $\lim_{\ell\to\infty} \|S''_\ell\|_2 = 0$. Analogously to Theorem 7, show (26) for S_ℓ.

Exercise 10.6.21. Prove the inequalities

$$\underline{C}_p^{-1} \|x_{\ell-1}\|_2 \leqslant \|px_{\ell-1}\|_2 \leqslant \bar{C}_p \|x_{\ell-1}\|_2 \quad \text{for all } x_{\ell-1} \in X_{\ell-1}, \ell \geqslant 1 \tag{10.6.27}$$

for the canonical choice (21a, b) by means of (17) with $\underline{C}_p = \bar{C}_p := \underline{C}_P \bar{C}_P$.

The identity $pA_{\ell-1}^{-1} r A_\ell S_\ell^\nu = S_\ell^\nu - [A_\ell^{-1} - pA_{\ell-1}^{-1} r] A_\ell S_\ell^\nu = S_\ell^\nu - M_\ell^{\mathrm{TGM}}(\nu,0)$ implies

Lemma 10.6.22. *Let* (26) *and* (27) *be valid. Then*

$$\|A_{\ell-1}^{-1} r A_\ell S_\ell^\nu\|_2 \leqslant \underline{C}_p (C_S + \|M_\ell^{\mathrm{TGM}}(\nu,0)\|_2). \tag{10.6.28}$$

Suppose $\nu = \nu_1 > 0$ *and* $\nu_2 = 0$ *as in* §10.6.4. *Using* (27) *and* (28), *one can estimate the multi-grid iteration matrix from* (4.13a, b) *by*

$$\|M_\ell^{\mathrm{MGM}}(\nu,0)\|_2 \leqslant \|M_\ell^{\mathrm{TGM}}(\nu,0)\|_2 + C^* \|M_{\ell-1}^{\mathrm{MGM}}(\nu,0)\|_2^\gamma \quad \text{for } \ell \geqslant 1, \tag{10.6.29a}$$

$$C^* := \underline{C}_p \bar{C}_p (C_S + 1). \tag{10.6.29b}$$

Here, ν is assumed to be chosen large enough so that $\|M_\ell^{\mathrm{TGM}}(\nu,0)\|_2 \leqslant 1$ according to Theorem 18 or 19. Together with $M_0^{\mathrm{MGM}} = O$ (cf. (4.13a)), inequality (29a) leads to the recursive inequalities (30a) for the quantities ζ_ℓ:

$$\zeta_\ell := \|M_\ell^{\mathrm{MGM}}(\nu,0)\|_2, \tag{10.6.29c}$$

$$\zeta_0 := 0, \qquad \zeta_\ell \leqslant \zeta + C^*(\zeta_{\ell-1})^\gamma \quad \text{for } \ell \geqslant 1. \tag{10.6.30a}$$

ζ is the ℓ-independent bound for the two-grid convergence, whose existence is stated by Theorem 18 or 19:

$$\|M_\ell^{\mathrm{TGM}}(\nu,0)\|_2 \leqslant \zeta. \tag{10.6.30b}$$

An analysis of the fixed-point equation $x = \zeta + C^* x^\gamma$ yields

Lemma 10.6.23. *Assume* $\gamma \geqslant 2$, $C^*\gamma > 1$, *and* $\zeta \leqslant \frac{\gamma - 1}{\gamma} \Big/ (C^*\gamma)^{1/(\gamma-1)}$. *Then all solutions of the inequalities* (30a) *are bounded by*

$$\zeta_\ell \leqslant \zeta^* \leqslant \frac{\gamma}{\gamma - 1}\zeta < 1 \quad \textit{for all } \ell \geqslant 0. \tag{10.6.31a}$$

Exercise 10.6.24. Prove that for the most interesting case $\gamma = 2$, one has

$$\zeta^* = 2\zeta/(1 + \sqrt{1 - 4C^*\zeta}) \quad \text{for } \zeta^* \text{ from (31a).} \tag{10.6.31b}$$

Since, by (29c), ζ_ℓ are the contraction number bounds, we obtain

Theorem 10.6.25. *Assume the smoothing and the approximation properties* (4a–c) *and* (11), *the conditions* (26), (27), *and in addition* $\gamma \geqslant 2$. *As in Theorems* 18 *and* 19 *for every* $0 < \zeta' < 1$, *there are* $\underline{v}$ *and* $\bar{h} > 0$, such that

$$\|M_p^{\mathrm{MGM}}(v, 0)\|_2 \leqslant \zeta' < 1 \qquad \text{for} \quad \underline{v} \leqslant v < \bar{v} := \min_{l \geqslant 1} \bar{v}(h_l), \tag{10.6.32}$$

provided $h_1 \leqslant \bar{h}$. Here, $\underline{v} < \bar{v}$ holds. In the case of $\bar{v}(h) = \infty$, one may set $\bar{h} := \infty$ *(i.e., the choice of the grid size is not restricted).*

Proof. Choose $\zeta := \frac{\gamma - 1}{\gamma}\zeta'$ small enough, so that ζ fulfils the assumptions of Lemma 23. According to Theorem 19, $\underline{v}$ and $\bar{h}$ are to be chosen in such a way that (30b): $\|M_\ell^{\mathrm{TGM}}(v, 0)\|_2 \leqslant \zeta$ for $\underline{v} \leqslant v < \bar{v}$ holds. Lemmata 22 and 23 give

$$\zeta_\ell = \|M_\ell^{\mathrm{MGM}}(v, 0)\|_2 \leqslant \frac{\gamma}{\gamma - 1}\zeta \leqslant \zeta'. \qquad \square$$

10.6.6 Case of Weaker Regularity

The proof of the approximation property has made use of $2m$-regularity (cf. (22b)), which in the case of the Poisson equation reads $A^{-1} = (-\Delta)^{-1}\colon U = L^2(\Omega) = H^0(\Omega) \to H^2(\Omega) \cap H_0^1(\Omega)$. This assumption is true for the unit square $\Omega = (0, 1) \times (0, 1)$, but it does not hold, e.g., for domains with re-entrant corners. In the general case, one obtains only statements of the form

$$A^{-1}\colon H^{-\sigma m}(\Omega) \to H^{(2-\sigma)m}(\Omega) \cap H_0^m(\Omega) \quad \text{for some } \sigma \in (0, 1) \tag{10.6.33}$$

(cf. Hackbusch [15, §9.1]). A similar statement may be assumed for A^*. If $\sigma < 1$, the approximation property (11) cannot be proved but has to be formulated by means of other norms.

Let $|\cdot|_t$ for $-1 \leqslant t \leqslant 1$ be a discrete analogue of the Sobolev norm of $H^{tm}(\Omega)$. We define $U_\ell := (X_\ell, |\cdot|_\sigma)$ and $F_\ell := (X_\ell, |\cdot|_{-\sigma})$. Then

$$\|A_\ell^{-1} - pA_{\ell-1}^{-1}r\|_{U_\ell \leftarrow F_\ell} \leqslant (C_A/\|A_\ell\|_2)^{1-\sigma} \tag{10.6.34}$$

can be shown (cf. Hackbusch [14, §6.3.1.3]). For the notation of the norm compare (2.6.11). For $A_\ell > 0$, the norms can be defined by

$$\|x_\ell\|_{U_\ell} = |x_\ell|_\sigma := \|A_\ell^{\sigma/2} x_\ell\|_2, \qquad \|f_\ell\|_{F_\ell} = |f_\ell|_{-\sigma} := \|A_\ell^{-\sigma/2} f_\ell\|_2. \quad (10.6.35)$$

In the general case, replace A_ℓ in (35) by the positive definite part $A'_\ell := \frac{1}{2}(A_\ell + A_\ell^H) + c_0 I$ (cf. (6a)).

Part (4a) of the smoothing property (4a–c) has to be adapted to the new norms. (4a) thus becomes

$$\|A_\ell S_\ell^\nu\|_{F_\ell \leftarrow U_\ell} \leqslant \eta(\nu) \|A_\ell\|_2^{1-\sigma} \quad \text{for } 0 \leqslant \nu \leqslant \bar{\nu}(h_\ell). \quad (10.6.36)$$

Exercise 10.6.26. Let $\mathscr{S}_\ell$ be the Richardson iteration (1a, b) and assume $A_\ell > 0$. Using the norms from (35), prove for all $\nu \geqslant 0$ that

$$\|A_\ell S_\ell^\nu\|_{F_\ell \leftarrow U_\ell} = \|A_\ell^{1-\sigma}(I - \Theta A_\ell)^\nu\|_2 \leqslant \left(\eta_0\left(\frac{\nu}{1-\sigma}\right)\|A_\ell\|_D{}^2\right)^{1-\sigma}. \quad (10.6.37)$$

The two-grid contraction number with respect to $\|\cdot\|_{U_\ell}$ can be concluded from the product of (34) and (36):

$$\|M_\ell^{\mathrm{TGM}}(\nu, 0)\|_{U_\ell \leftarrow U_\ell} \leqslant \eta(\nu) C_A^{1-\sigma}. \quad (10.6.38)$$

Similarly to §10.6.5, one obtains a corresponding convergence result for the multi-grid method.

The special bounds in (37) yield the right-hand side $\left(\eta_0\left(\frac{\nu}{1-\sigma}\right)C_A\right)^{1-\sigma}$ in (38). For the standard case discussed in §§10.6.2–5, we had $\sigma = 0$ and the bound in (38) had behaved like $O(1/\nu)$. For $0 < \sigma < 1$, the contraction number behaves only as $O(1/\nu^{1-\sigma})$. The value $\sigma = 1$ is not sufficient, because the right-hand side in (37) fails to fulfil (4b).

10.7 Symmetric Multi-Grid Methods

Analysis of the multi-grid iteration in §10.6 is performed for the general (nonsymmetric) case in order to emphasize that multi-grid iterations are not restricted to symmetric or even only positive definite problems. However, the symmetric case admits some stronger statements that are covered in this chapter.

10.7.1 Symmetric Multi-Grid Algorithm

The required symmetry conditions are

$$r = p^* \quad \text{and } A_\ell > 0 \quad \text{for all } \ell \geqslant 0, \quad (10.7.1a)$$

(cf. (1.22)). Further, we use the variant (2.3b) with the pre-smoothing $\mathscr{S}_\ell$ and the post-smoothing $\hat{\mathscr{S}}_\ell$, where the post-smoothing is adjoint to $\mathscr{S}_\ell$:

$$\hat{\mathscr{S}}_\ell = \mathscr{S}_\ell^* \quad \text{for all } \ell \geqslant 0 \tag{10.7.1b}$$

(cf. §4.8.4). The number of pre- and post-smoothing steps should be

$$\nu_1 = \nu_2 = \nu/2 > 0. \tag{10.7.1c}$$

Occasionally, we need the Galerkin product property:

$$A_{\ell-1} = r A_\ell p. \tag{10.7.1d}$$

Lemma 10.7.1. *Assume* (1a–c). (a) *The two- and multi-grid iterations* $\Phi_\ell^{\mathrm{TGM}}\left(\frac{\nu}{2},\frac{\nu}{2}\right)$ *and* $\Phi_\ell^{\mathrm{MGM}}\left(\frac{\nu}{2},\frac{\nu}{2}\right)$ *are symmetric: Their transformed iteration matrices* $A_\ell^{1/2} M_\ell A_\ell^{-1/2}$ *are Hermitian (here, we use the abbreviations* $M_\ell = M_\ell^{\mathrm{TGM}}\left(\frac{\nu}{2},\frac{\nu}{2}\right)$ *and* $M_\ell^{\mathrm{MGM}}\left(\frac{\nu}{2},\frac{\nu}{2}\right)$*)*.

(b) *If the iteration converges, it converges* monotonically with respect to the energy norm $\|\cdot\|_{A_\ell}$, *and the matrices* $W_\ell = W_\ell^{\mathrm{TGM}}\left(\frac{\nu}{2},\frac{\nu}{2}\right)$ *and* $W_\ell^{\mathrm{MGM}}\left(\frac{\nu}{2},\frac{\nu}{2}\right)$, *respectively, of the third normal form are positive definite and fulfil*

$$(1-\varrho_\ell) W_\ell \leqslant A_\ell \leqslant (1+\varrho_\ell) W_\ell \quad \text{with } \varrho_\ell = \varrho(M_\ell) = \|A_\ell^{1/2} M_\ell A_\ell^{-1/2}\|_2. \tag{10.7.2a}$$

If, according to Theorems 6.3 *or* 6.18, $\varrho_\ell \leqslant \varrho < 1$ *is* h_ℓ*-independent, then the condition* (2b) *is* h_ℓ*-independent also*:

$$\varkappa(W_\ell^{-1} A_\ell) \leqslant \frac{1+\varrho_\ell}{1-\varrho_\ell} \leqslant \frac{1+\varrho}{1-\varrho}. \tag{10.7.2b}$$

(c) *In the case of* (1d), *the inequalities* (2a, b) *can be improved*:

$$(1-\varrho_\ell) W_\ell \leqslant A_\ell \leqslant W_\ell, \qquad \varkappa(W_\ell^{-1} A_\ell) \leqslant 1/(1-\varrho_\ell) \leqslant 1/(1-\varrho). \tag{10.7.2c}$$

Proof. The representations (2.4) and (4.13a, b) show that $A_\ell M_\ell A_\ell^{-1} = M_\ell^H$, because $A_\ell \hat{S}_\ell A_\ell^{-1} = S_\ell^H$ according to (4.8.10). This proves part (a). Part (b) is obtained from Remark 4.8.3c. Part (c) is based on the property $A_\ell^{1/2} M_\ell A_\ell^{-1/2} \geqslant 0$, which will be proved in (13b). □

10.7.2 Two-Grid Convergence for $\nu_1 > 0$, $\nu_2 > 0$

The case $\nu_1 = \nu > 0$ and $\nu_2 = 0$ has been treated in §10.6. The technique used there can also be applied to the general case $\nu_1 \geqslant 0$, $\nu_2 \geqslant 0$, $\nu := \nu_1 + \nu_2 > 0$, and especially to $\nu_1 = \nu_2 = \nu/2$.

Exercise 10.7.2. Assume (1a, b). Prove

$$\Phi_\ell^{\mathrm{TGM}}(0,\nu_2) = (\Phi_\ell^{\mathrm{TGM}}(\nu_2,0))^*, \quad \Phi_\ell^{\mathrm{MGM}}(0,\nu_2) = (\Phi_\ell^{\mathrm{MGM}}(\nu_2,0))^*, \tag{10.7.3a}$$

$$\Phi_\ell^{\mathrm{TGM}}(\nu_1,\nu_2) = \Phi_\ell^{\mathrm{TGM}}(0,\nu_2) \circ \Phi_\ell^{\mathrm{TGM}}(\nu_1,0) \quad \text{in the case (1d).} \tag{10.7.3b}$$

Under the assumption (1d), the statements (3c, d) for the two-grid iteration matrices $M_\ell := M_\ell^{\mathrm{TGM}}$ follow from (3a, b):

$$M_\ell(\nu_1,\nu_2) = M_\ell(0,\nu_2)M_\ell(\nu_1,0) = A_\ell^{-1}M_\ell(\nu_2,0)^H A_\ell M_\ell(\nu_1,0), \tag{10.7.3c}$$

$$A_\ell^{1/2}M_\ell(\nu_1,\nu_2)A_\ell^{-1/2} = (A_\ell^{1/2}M_\ell(\nu_2,0)A_\ell^{-1/2})^H(A_\ell^{1/2}M_\ell(\nu_1,0)A_\ell^{-1/2}). \tag{10.7.3d}$$

For estimating $A_\ell^{1/2}M_\ell(\nu,0)A_\ell^{-1/2}$, one may use the approximation property (4a) and the smoothing property (4b):

$$\|A_\ell^{1/2}(A_\ell^{-1} - pA_{\ell-1}^{-1}r)\|_2 \leqslant \sqrt{C_A/\|A_\ell\|_2}, \tag{10.7.4a}$$

$$\|A_\ell S_\ell^\nu A_\ell^{-1/2}\|_2 \leqslant \sqrt{\eta(2\nu)\,\|A_\ell\|_2}, \tag{10.7.4b}$$

which correspond to (6.34) and (6.36) for the energy norm $\|\cdot\|_{U_\ell} = \|\cdot\|_{A_\ell}$ and the Euclidean norm $\|\cdot\|_{F_\ell} = \|\cdot\|_2$. Under the assumption (1d), inequality (4a) is equivalent to the approximation property (6.11). In the case of Richardson's iteration (6.1a, b), inequality (4b) holds because of

$$\|A_\ell S_\ell^\nu A_\ell^{-1/2}\|_2^2 = \|A_\ell^{1/2}S_\ell^\nu\|_2^2 = \|A_\ell S_\ell^{2\nu}\|_2 \leqslant \eta_0(2\nu)\,\|A_\ell\|_2$$

with $\eta(2\nu) = \eta_0(2\nu)$ (η_0 from (6.2b)). (4a, b) yield the estimate

$$\|M_\ell^{\mathrm{TGM}}(\nu,0)\|_{A_\ell} = \|A_\ell^{1/2}M_\ell^{\mathrm{TGM}}(\nu,0)A_\ell^{-1/2}\|_2 \leqslant \sqrt{\eta(2\nu)C_A}. \tag{10.7.4c}$$

Using (3d), one finally proves the following convergence theorem.

Theorem 10.7.3. *Assume* (1a, b, d). *The smoothing and approximation properties* (4a, b) *imply*

$$\|M_\ell^{\mathrm{TGM}}(\nu_1,\nu_2)\|_{A_\ell} \leqslant C_A\sqrt{\eta(2\nu_1)\eta(2\nu_2)}, \qquad \left\|M_\ell^{\mathrm{TGM}}\left(\frac{\nu}{2},\frac{\nu}{2}\right)\right\|_{A_\ell} \leqslant C_A\eta(\nu). \tag{10.7.5}$$

Two-grid convergence follows as in Theorems 6.18 *or* 6.19.

As in §10.6.5, multi-grid convergence can be concluded from the two-grid convergence. However, it has to be emphasized that the proof technique of §10.6.5 requires $\gamma \geqslant 2$ and therefore excludes the V-cycle ($\gamma = 1$).

10.7.3 Smoothing Property in the Symmetric Case

In particular, the condition (1b): $\hat{\mathscr{S}}_\ell = \mathscr{S}_\ell^*$ is satisfied if $\hat{\mathscr{S}}_\ell = \mathscr{S}_\ell$ is a *symmetric* smoothing iteration. For symmetric iterations, the proof of the smoothing property is rather easy.

Lemma 10.7.4. *Let $S_\ell = I - W_\ell^{-1}A_\ell$ be the iteration matrix of $\mathcal{S}_\ell$ and assume that $\gamma W_\ell \leqslant A_\ell \leqslant \Gamma W_\ell$ for all $\ell \geqslant 0$ with $0 \leqslant \gamma \leqslant \Gamma < 2$. Then*

$$\|A_\ell S_\ell^\nu\|_2 \leqslant \|W_\ell\|_2 \max\{\eta_0(\nu), \Gamma|1-\Gamma|^\nu\}. \tag{10.7.6a}$$

(6a) *implies the smoothing property* (6.4a–c) *with* $\bar{\nu}(h) = \infty$, *if*

$$\|W_\ell\|_2 \leqslant C_W \|A_\ell\|_2 \quad \text{for all } \ell \geqslant 0. \tag{10.7.6b}$$

Proof. Define $Y := W_\ell^{-1/2} R_\ell W_\ell^{-1/2}$ with R_ℓ from $A_\ell = W_\ell - R_\ell$ and note that

$$\|A_\ell S_\ell^\nu\|_2 = \|W_\ell^{1/2}(I-Y)Y^\nu W_\ell^{1/2}\|_2 \leqslant \|W_\ell^{1/2}\|_2^2 \|(I-Y)Y^\nu\|_2.$$

The first factor is $\|W_\ell\|_2$, the second can be estimated as in Lemma 6.1 by $\max\{\eta_0(\nu), \Gamma|1-\Gamma|^\nu\}$ because of $(1-\Gamma)I \leqslant Y \leqslant (1-\gamma)I$. □

The following variant of the estimate is due to Wittum [5]. The estimate is helpful, if good bounds for $\|R_\ell\|_2$ are known.

Lemma 10.7.5. *In addition to the assumptions of Lemma* 4, *assume* $\nu \geqslant 2$. *Define* $R_\ell := W_\ell - A_\ell$. *Then*

$$\|A_\ell S_\ell^\nu\|_2 \leqslant \|S_\ell\|_2 \|R_\ell\|_2 \max\{\eta_0(\nu-2), \Gamma|1-\Gamma|^{\nu-2}\}. \tag{10.7.6c}$$

Proof. Define Y as above, estimate $\|A_\ell S_\ell^\nu\|_2$ by $\|W_\ell^{1/2}Y\|_2^2 \|(I-Y)Y^{\nu-2}\|_2$, and use $\|W_\ell^{1/2}Y\|_2^2 = \|W_\ell^{-1/2}R_\ell^2 W_\ell^{-1/2}\|_2 = \varrho(W_\ell^{-1/2}R_\ell^2 W_\ell^{-1/2}) = \varrho(W_\ell^{-1}R_\ell^2) \leqslant \|W_\ell^{-1}R_\ell\|_2 \|R_\ell\|_2 = \|S_\ell\|_2 \|R_\ell\|_2$. □

Exercise 10.7.6. Under the same assumption as in Lemma 4, prove (6d) for the modified condition (4b):

$$\|A_\ell S_\ell^\nu A_\ell^{-1/2}\|_2 \leqslant \sqrt{\|W_\ell\|_2 \max\{\eta_0(2\nu), \Gamma(1-\Gamma)^{2\nu}\}}. \tag{10.7.6d}$$

The condition $\Gamma < 2$ in $\gamma W_\ell \leqslant A_\ell \leqslant \Gamma W_\ell$ coincides with the convergence condition (4.8.3a). However, $\gamma = 0$ is sufficient for the smoothing property, although the convergence rate $\varrho(S_\ell)$ becomes worse the smaller γ is. Since damping corresponds to the replacement of W_ℓ by $\vartheta^{-1}W_\ell$, one obtains

Remark 10.7.7. After a possibly necessary damping, all symmetric iterations satisfy the assumption $\gamma W_\ell \leqslant A_\ell \leqslant \Gamma W_\ell$ with $0 \leqslant \gamma \leqslant \Gamma < 2$.

10.7.4 Strengthened Two-Grid Convergence Estimates

To simplify the following considerations, the smoothing iteration $\mathcal{S}_\ell$ is assumed to satisfy the inequality $\gamma W_\ell \leqslant A_\ell \leqslant \Gamma W_\ell$ with $0 \leqslant \gamma \leqslant \Gamma \leqslant 1$. As mentioned in Remark 7, this assumption can always be achieved by damping. However, the following statements also hold in a somewhat modified form for $0 \leqslant \gamma \leqslant \Gamma < 2$. Hence, the assumptions are

$$\hat{\mathscr{S}}_\ell = \mathscr{S}_\ell, \qquad \hat{S}_\ell = S_\ell = I - W_\ell^{-1} A_\ell, \qquad 0 < A_\ell \leqslant W_\ell. \tag{10.7.7}$$

The approximation property is required in the form (6.34) with $\|\cdot\|_{U_\ell} := \|\cdot\|_{W_\ell}$, $\|\cdot\|_{F_\ell} := \|\cdot\|_{W_\ell^{-1}}$ (compare the following Remark 13):

$$\| W_\ell^{1/2} (A_\ell^{-1} - p A_{\ell-1}^{-1} r) W_\ell^{1/2} \|_2 \leqslant C_A \quad \text{for all } \ell \geqslant 1. \tag{10.7.8a}$$

Lemma 10.7.8 *Assume* (1a, d). *The approximation property* (8a) *is equivalent to inequality* (8b):

$$0 \leqslant A_\ell^{-1} - p A_{\ell-1}^{-1} r \leqslant C_A W_\ell^{-1} \quad \textit{for all } \ell \geqslant 1. \tag{10.7.8b}$$

Proof. (2.10.3f) yields $-C_A I \leqslant W_\ell^{1/2} (A_\ell^{-1} - p A_{\ell-1}^{-1} r) W_\ell^{1/2} \leqslant C_A I$. Multiplication by $W_\ell^{-1/2}$ from both sides results in the bounds $\pm C_A W_\ell^{-1}$ for $A_\ell^{-1} - p A_{\ell-1}^{-1} r$. The lower bound $-C_A I$ can be replaced by 0, as can be concluded from Lemma 9 that follows. □

The proof of the modified approximation property (8a) is postponed until Remark 13. First, we transform all quantities into a form better suited to symmetry:

$$\check{p} := A_\ell^{1/2} p A_{\ell-1}^{-1/2}, \quad \check{r} := \check{p}^* = A_{\ell-1}^{-1/2} r A_\ell^{1/2}, \quad Q_\ell := I - \check{p}\check{r}, \tag{10.7.9a}$$

$$X_\ell := A_\ell^{1/2} W_\ell^{-1} A_\ell^{1/2}, \quad \check{S}_\ell := A_\ell^{1/2} S_\ell A_\ell^{-1/2} = I - X_\ell. \tag{10.7.9b}$$

Since (1d) can be rewritten as $\check{r}\check{p} = I$, the following lemma may be concluded.

Lemma 10.7.9. *Under the assumption* (1a, d), $Q_\ell = I - \check{p}\check{r}$ *is an* orthogonal projection: $Q_\ell^2 = Q_\ell = Q_\ell^H$. *Like any orthogonal projection, it fulfils*

$$0 \leqslant Q_\ell \leqslant I \quad \textit{for all } \ell \geqslant 1. \tag{10.7.10a}$$

$Q_\ell \geqslant 0$ also implies $0 \leqslant A_\ell^{-1/2} Q_\ell A_\ell^{-1/2} = A_\ell^{-1} - p A_{\ell-1}^{-1} r$, so that the proof of the first inequality in (8b) *is completed. Multiplication of* (8b) *by $A_\ell^{1/2}$ from both sides yields*

Lemma 10.7.10. *Assume* (1a, d). *The statements* (8a) *or* (8b) *are equivalent to*

$$0 \leqslant Q_\ell \leqslant C_A X_\ell \quad \textit{for all } \ell \geqslant 1. \tag{10.7.10b}$$

According to (2.4), the *transformed two-grid iteration matrix* is

$$\check{M}_\ell(\nu_1, \nu_2) := A_\ell^{1/2} M_\ell^{\mathrm{TGM}}(\nu_1, \nu_2) A_\ell^{-1/2} = \check{S}_\ell^{\nu_2} Q_\ell \check{S}_\ell^{\nu_1}. \tag{10.7.11}$$

In contrast to Theorems 6.18–19, it is now possible to prove convergence for *all* $\nu > 0$.

Theorem 10.7.11. *Assume* (1a–d), (7), *and the approximation property* (8a). *Then the two-grid iteration converges monotonically with respect to the energy norm* $\|\cdot\|_{A_\ell}$:

$$\varrho\left(M_\ell^{\mathrm{TGM}}\left(\frac{\nu}{2},\frac{\nu}{2}\right)\right) = \left\|M_\ell^{\mathrm{TGM}}\left(\frac{\nu}{2},\frac{\nu}{2}\right)\right\|_{A_\ell} = \left\|\check{M}_\ell\left(\frac{\nu}{2},\frac{\nu}{2}\right)\right\|_2$$
$$\leqslant \begin{Bmatrix} C_A\eta_0(\nu) & \text{if } C_A \leqslant 1+\nu \\ (1-1/C_A)^\nu & \text{if } C_A > 1+\nu \end{Bmatrix} < 1. \qquad (10.7.12)$$

Proof. It remains to show the inequality «$\leqslant$» in (12). The inequality (13a) following from (10a, b) can be inserted into (11) and yields (13b):

$$0 \leqslant Q_\ell \leqslant \alpha C_A X_\ell + (1-\alpha)I \qquad \text{for all } 0 \leqslant \alpha \leqslant 1, \quad (10.7.13\text{a})$$
$$0 \leqslant \check{M}_\ell \leqslant \check{S}_\ell^{\nu/2}[\alpha C_A X_\ell + (1-\alpha)I]\check{S}_\ell^{\nu/2} \quad \text{for all } 0 \leqslant \alpha \leqslant 1. \quad (10.7.13\text{b})$$

Since $\check{S}_\ell = I - X_\ell$, the right-hand side of (13b) is a polynomial $f(X_\ell;\alpha)$, where

$$f(\xi;\alpha) := (1-\xi)^\nu(1-\alpha+\alpha C_A\xi). \qquad (10.7.13\text{c})$$

For *all* $0 \leqslant \alpha \leqslant 1$, inequality $0 \leqslant X_\ell \leqslant I$ (cf. (7)) implies the estimate

$$\|\check{M}_\ell\|_2 \leqslant \|f(X_\ell;\alpha)\|_2 \leqslant m(\alpha) := \max\{f(\xi;\alpha): 0 \leqslant \xi \leqslant 1\}. \quad (10.7.13\text{d})$$

In particular, for $\alpha = 1$, the bound $C_A\eta_0(\nu)$ results. If $1+\nu < C_A$, the value $\alpha^* := \nu/(C_A - 1)$ belongs to $[0,1]$ and yields the better bound $m(\alpha^*) = (1-1/C_A)^\nu$. □

Exercise 10.7.12. Prove the statements of Lemmata 8, 10 and Theorem 11 under the assumption $rA_\ell p \leqslant A_{\ell-1}$ instead of (1d).

It remains to discuss the approximation property (8a).

Remark 10.7.13. Assume the approximation property in the original form (6.11): $\|A_\ell^{-1} - pA_{\ell-1}^{-1}r\|_2 \leqslant C'_A/\|A_\ell\|_2$. Furthermore, let (6b): $\|W_\ell\|_2 \leqslant C_W\|A_\ell\|_2$ be valid. Then (8a) is satisfied with $C_A := C'_A C_W$.

Exercise 10.7.14. Prove that under the assumptions (1a, b, d), (7), and $\nu = \nu_1 + \nu_2 > 0$, the two-grid iteration $\Phi_\ell^{\mathrm{TGM}}(\nu_1,\nu_2)$ converges monotonically with respect to the energy norm $\|\cdot\|_{A_\ell}$. What is the h_ℓ-independent contraction number? *Hint:* First, use (3d) to estimate $\left\|\check{M}_\ell\left(\frac{\nu}{2},0\right)\right\|_2$ and thereafter apply (3d) to $\|\check{M}_\ell(\nu_1,\nu_2)\|_2$.

Also in the proof of Theorem 11, the smoothing property is used indirectly; however, now it is formulated by means of the polynomial (13c) for arbitrary $0 \leqslant \alpha \leqslant 1$ instead of $\alpha = 1$.

10.7.5 V-Cycle Convergence

We apply the technique of §10.7.4 to the multi-grid method. Since Theorem 6.25 excludes the V-cycle ($\gamma = 1$), we concentrate on this case.

Theorem 10.7.15. *Under the same assumptions* (1a–d), (7), (8a) *as in Theorem* 11, *the* V-*cycle* ($\gamma = 1$) *converges monotonically with respect to the energy norm* $\|\cdot\|_{A_\ell}$ *with the rate*

$$\varrho\left(M_\ell^{\mathrm{V}}\left(\frac{\nu}{2},\frac{\nu}{2}\right)\right) = \left\|M_\ell^{\mathrm{V}}\left(\frac{\nu}{2},\frac{\nu}{2}\right)\right\|_{A_\ell} \leqslant \frac{C_A}{C_A+\nu} < 1. \tag{10.7.14}$$

Proof. For $\gamma = 1$, abbreviate M_ℓ^{MGM} by M_ℓ^{V}. For $\gamma = 1$, the recursive equations (4.13a, b) become

$$M_0^{\mathrm{V}}(\nu_1,\nu_2) = 0, \quad M_\ell^{\mathrm{V}}(\nu_1,\nu_2) = M_\ell^{\mathrm{TGM}}(\nu_1,\nu_2) + S_\ell^{\nu_2} p M_{\ell-1}^{\mathrm{V}}(\nu_1,\nu_2) A_{\ell-1}^{-1} r A_\ell S_\ell^{\nu_1}.$$

Transformation to the symmetric form yields

$$\begin{aligned} \check{M}_\ell^{\mathrm{V}} &:= A_\ell^{1/2} M_\ell^{\mathrm{V}}(\nu_1,\nu_2) A_\ell^{-1/2} = \check{M}_\ell^{\mathrm{TGM}}(\nu_1,\nu_2) + \check{S}_\ell^{\nu_2}\check{p}\check{M}_{\ell-1}^{\mathrm{V}}\check{r}\check{S}_\ell^{\nu_1} \underset{(11)}{=} \\ &= \check{S}_\ell^{\nu_2}\{I - \check{p}[I - \check{M}_{\ell-1}^{\mathrm{V}}]\check{r}\}\check{S}_\ell^{\nu_1}, \\ \check{M}_0^{\mathrm{M}} &= 0. \end{aligned} \tag{10.7.15}$$

In the following, choose $\check{M}_\ell^{\mathrm{V}} = \check{M}_\ell^{\mathrm{V}}\left(\frac{\nu}{2},\frac{\nu}{2}\right)$, i.e., $\nu_1 = \nu_2 = \frac{\nu}{2}$. Using (15), we conclude that

$$\check{M}_\ell^{\mathrm{V}} \geqslant 0 \tag{10.7.16a}$$

inductively from $\check{M}_0^{\mathrm{V}} = 0$ and $I - \check{p}[I - \check{M}_{\ell-1}^{\mathrm{V}}]\check{r} \geqslant I - \check{p}\check{r} = Q_\ell \geqslant 0$. Hence, the statements (16b) and (16c) are equivalent:

$$\left\|M_\ell^{\mathrm{V}}\left(\frac{\nu}{2},\frac{\nu}{2}\right)\right\|_{A_\ell} = \left\|\check{M}_\ell^{\mathrm{V}}\left(\frac{\nu}{2},\frac{\nu}{2}\right)\right\|_2 \leqslant \zeta_\ell, \tag{10.7.16b}$$

$$0 \leqslant \check{M}_\ell^{\mathrm{V}} \leqslant \zeta_\ell I. \tag{10.7.16c}$$

The induction hypothesis is $0 \leqslant \check{M}_{\ell-1}^{\mathrm{V}} \leqslant \zeta_{\ell-1} I$ with $\zeta_{\ell-1} := \dfrac{C_A}{C_A+\nu}$. Inserting this inequality into (15), we arrive at

$$\begin{aligned} 0 &\leqslant \check{M}_\ell^{\mathrm{V}} \leqslant \check{S}_\ell^{\nu/2}\{I - (1-\zeta_{\ell-1})\check{p}\check{r}\}\check{S}_\ell^{\nu/2} = \check{S}_\ell^{\nu/2}\{(1-\zeta_{\ell-1})Q_\ell + \zeta_{\ell-1}I\}\check{S}_\ell^{\nu/2} \underset{(13a)}{\leqslant} \\ &\leqslant \check{S}_\ell^{\nu/2}\{(1-\zeta_{\ell-1})[\alpha C_A X_\ell + (1-\alpha)I] + \zeta_{\ell-1}I\}\check{S}_\ell^{\nu/2} \quad \text{for all } 0 \leqslant \alpha \leqslant 1. \end{aligned}$$

$\beta := (1-\zeta_{\ell-1})(1-\alpha) + \zeta_{\ell-1}$ varies for $\alpha \in [0,1]$ in $[\zeta_{\ell-1}, 1]$. Substitution of α by β yields

$$0 \leqslant \check{M}_\ell^{\mathrm{V}} \leqslant \check{S}_\ell^{\nu/2}\{(1-\beta)C_A X_\ell + \beta I\}\check{S}_\ell^{\nu/2} \quad \text{for all } \zeta_{\ell-1} \leqslant \beta \leqslant 1.$$

The right-hand side is the polynomial $f(\xi;\beta) := (1-\xi)^\nu[\beta + (1-\beta)C_A\xi]$ for $\xi = X_\ell$ and can be estimated by $\|f(X_\ell;\beta)\|_2 \leqslant m(\beta) := \max\{|f(\xi;\beta)|: 0 \leqslant \xi \leqslant 1\}$ (cf. (13c, d)). For $\beta = \zeta_{\ell-1} = C_A/(C_A+\nu)$, one finds $m(\beta) = f(0;\beta) = \beta = C_A/(C_A+\nu)$. Hence, (16c) holds with $\zeta_\ell = C_A/(C_A+\nu)$. □

Exercise 10.7.16. (a) Under the same assumptions, prove

$$\check{M}_\ell^{\mathrm{V}}(0,\nu_2)\check{M}_\ell^{\mathrm{V}}(\nu_1,0) = \check{M}_\ell^{\mathrm{V}}(\nu_1,\nu_2) \tag{10.7.17}$$

and discuss convergence for $\nu = \nu_1 + \nu_2 > 0$.
(b) Prove the statement of Theorem 15 under the condition $rA_\ell p \leqslant A_{\ell-1}$ instead of (1d).

The condition $A_\ell \leqslant W_\ell$ in (7) can be generalised to $A_\ell \leqslant \sqrt{2}W_\ell$ (cf. Wittum [4, Proposition 4.2.4]).

Obviously, monotone and h_ℓ-independent convergence can also be shown for the W-cycle (more generally, $\gamma \geqslant 2$). For this case (assuming $C_A \geqslant 1$), one finds, e.g., the estimate

$$\left\| M_\ell^{\mathrm{W}}\left(\frac{\nu}{2},\frac{\nu}{2}\right)\right\|_{A_\ell} \leqslant \sqrt{C_A}/(\sqrt{C_A}+\nu). \tag{10.7.18}$$

In the case of weaker regularity (cf. §10.6.6) and for $\gamma = 2$, one can still prove $\left\| M_\ell^{\mathrm{W}}\left(\frac{\nu}{2},\frac{\nu}{2}\right)\right\|_{A_\ell} = O(\nu^{\sigma-1}) < 1$ for all $\nu > 0$.

10.7.6 Multi-Grid Convergence for All $\nu > 0$ in the Nonsymmetric Case

The analysis of §10.6 showed multi-grid convergence for sufficiently large $\nu \geqslant \underline{\nu}$. In the symmetric case, §10.7.5 ensured convergence for all $\nu > 0$ and arbitrarily coarse h_0. In the general case, one can still obtain convergence for all $\nu = \nu_1 + \nu_2 > 0$; however, h_0 must be sufficiently small: $h_0 \leqslant \bar{h}$. The proof technique is the same as for Theorem 6.7.

Theorem 10.7.17. *Let the matrices A_ℓ ($\ell \geqslant 0$) be split into $A_\ell = A'_\ell + A''_\ell$ such that $A'_\ell > 0$. Let S_ℓ and S'_ℓ be the iteration matrices of the corresponding smoothing iterations $\mathscr{S}_\ell$ and $\mathscr{S}'_\ell$. For A''_ℓ and $S''_\ell := S_\ell - S'_\ell$ assume*

$$\|A_\ell'^{-1/2}A''_\ell A_\ell'^{-1/2}\|_2 \leqslant C_1 h_\ell^\varkappa, \quad \|A_\ell'^{1/2}S''_\ell A_\ell'^{-1/2}\|_2 \leqslant C_2 h_\ell^\varkappa \tag{10.7.19a}$$

with $\varkappa > 0$. Assume that the following norms are bounded by 1:

$$\|A_\ell'^{1/2}S'_\ell A_\ell'^{-1/2}\|_2, \quad \|A_\ell'^{1/2}pA_{\ell-1}'^{-1/2}\|_2, \quad \|A_{\ell-1}'^{-1/2}rA_\ell'^{1/2}\|_2 \leqslant 1 \tag{10.7.19b}$$

for all $\ell \geqslant 1$ *and that the two- or multi-grid method for* A'_ℓ *(with fixed parameters* γ, ν_1, ν_2*) converges monotonically with respect to the energy norm* $\|\cdot\|_{A'_\ell}$ *with the contraction number* ζ'. *Further, let* $\sup\{h_\ell/h_{\ell-1}: \ell \geqslant 1\} < 1$ *(cf.* (4.7)) *and* $\varepsilon \in (0, 1-\zeta')$ *be valid. Then the two- and multi-grid method for* A_ℓ *also converge monotonically with respect to the energy norm* $\|\cdot\|_{A'_\ell}$ *with the contraction number* $\zeta = \zeta' + \varepsilon$, *provided that* $h_0 \leqslant \bar{h}$ *holds with sufficiently small* $\bar{h}$.

Proof. First, the two-grid case is considered. The transformed iteration matrix $A_\ell'^{1/2} M_\ell' A_\ell'^{-1/2}$ (of the iteration for $A'_\ell x'_\ell = b'_\ell$) is the product

$$[A_\ell'^{1/2} S'_\ell A_\ell'^{-1/2}]^{\nu_2} \times [A_\ell'^{1/2}(A_\ell'^{-1} - pA_{\ell-1}'^{-1} r)A_\ell'^{1/2}] \times [A_\ell'^{-1/2} A'_\ell A_\ell'^{-1/2}]$$
$$\times [A_\ell'^{1/2} S'_\ell A_\ell'^{-1/2}]^{\nu_1}.$$

Because of (19a), the perturbations of S'_ℓ and A'_ℓ in the 1st, 3rd, and 4th factor by S''_ℓ and A''_ℓ, respectively, enlarge the spectral norm only by $O(h_\ell^\varkappa)$. A similar statement holds for the second factor, because

$$[A_\ell'^{1/2}(A_\ell^{-1} - pA_{\ell-1}^{-1} r)A_\ell'^{1/2}] - [A_\ell'^{1/2}(A_\ell'^{-1} - pA_{\ell-1}'^{-1} r)A_\ell'^{1/2}]$$
$$= A_\ell'^{-1/2} A''_\ell A_\ell^{-1} A_\ell'^{1/2} + A_\ell'^{1/2} p A_{\ell-1}^{-1} A''_{\ell-1} A_{\ell-1}'^{-1} r A_\ell'^{1/2}.$$

Let M_ℓ be the two-grid iteration matrix for A_ℓ. The assertion follows from $|\|M_\ell\|_{A'_\ell} - \|M'_\ell\|_{A'_\ell}| \leqslant Ch_\ell^\varkappa \leqslant C\bar{h}^\varkappa$ for the choice $\bar{h} := (\varepsilon/C)^{1/\varkappa}$. In the multi-grid case, the following recursive estimate holds:

$$|\|M_\ell\|_{A'_\ell} - \|M'_\ell\|_{A'_\ell}| \leqslant C_0 h_\ell^\varkappa + |\|M_\ell\|_{A'_{\ell-1}} - \|M'_\ell\|_{A'_{\ell-1}}|,$$

which by $h_\ell/h_{\ell-1} \leqslant C_h < 1$ leads to $|\|M_\ell\|_{A'_\ell} - \|M'_\ell\|_{A'_\ell}| \leqslant Ch_0^\varkappa \leqslant C\bar{h}^\varkappa \leqslant \varepsilon$. □

Remark 10.7.18. The conditions (19b) are satisfied, if $S'_\ell = I - W_\ell'^{-1} A'_\ell$ with $2W'_\ell \geqslant A'_\ell$ (cf. (7)) and $r = p^*$ and $rA'_\ell p \leqslant A'_{\ell-1}$ (cf. Exercise 12 and 16b). (1d) is sufficient for $rA'_\ell p \leqslant A'_{\ell-1}$.

The statement of Theorem 17 is not yet uniform with respect to $\nu = \nu_1 + \nu_2$. In particular, $\bar{h}$ might depend on ν. A ν-independent $\bar{h}$ can be obtained as follows: Theorem 6.25 (modified according to §10.7.2 to the energy norm $\|\cdot\|_{A'_\ell}$) shows convergence for $\nu \geqslant \underline{\nu}$ as long as $h_0 \leqslant \bar{h}_0$. For the finitely many $\nu = 1, \ldots, \underline{\nu} - 1$, one concludes convergence for $h_0 \leqslant \bar{h}_\nu$ from Theorem 17 with suitable $\bar{h}_\nu$. For $h_0 \leqslant \bar{h} := \min\{\bar{h}_\nu: 0 \leqslant \nu \leqslant \underline{\nu} - 1\}$, one obtains convergence for all $\nu > 0$.

For related results, compare Mandel [1] and Bramble–Pasciak–Xu [1].

10.8 Combination of Multi-Grid Methods with Semi-Iterations

10.8.1 Semi-Iterative Smoothers

Thus far, only the ν-fold application of a smoothing iteration $\mathscr{S}_\ell$ has been considered as a smoothing step (2.4b, f). An alternative is semi-iterative smoothing, where S_ℓ^ν is replaced by a polynomial $P_\nu(S_\ell)$ of degree ν with $P_\nu(1) = 1$. However, one should *not* choose the polynomials that were found to be optimal in §7, because those minimise $\|P_\nu(S_\ell)\|_2$ (more precisely, $\max\{|P_\nu(\xi)|: a \leqslant \xi \leqslant b < 1\}$). Using $\mathscr{S}_\ell$ for smoothing, we do not primarily want to make the error small, but smooth. The smoothing property (6.4a) leads us to (1a, b):

$$\text{minimise } \|A_\ell P_\nu(S_\ell)\|_2 \text{ over all polynomials with} \tag{10.8.1a}$$

$$\text{degree } P_\nu \leqslant \nu,\ P_\nu(1) = 1. \tag{10.8.1b}$$

The semi-iterative Richardson method ($S_\ell = I - A_\ell/\|A_\ell\|_2$, $A_\ell > 0$) with $\sigma(A_\ell) \subset \sigma_M := [0, \|A_\ell\|_2]$ leads to the optimisation problem

$$\text{minimise } \max\{|\xi P_\nu(1 - \xi/\|A_\ell\|_2)|: 0 \leqslant \xi \leqslant \|A_\ell\|_2\} \text{ with (1b)} \tag{10.8.2}$$

(analogously to (7.3.9)). The solution reads as follows.

Theorem 10.8.1. *Let $A_\ell > 0$. The minimisation problem (2) is solved by the polynomial P_ν that is obtained from the Chebyshev polynomial $T_{\nu+1}$ (cf. Lemma 7.3.3) by*

$$\tau P_\nu(1 - \tau) = \eta(\nu) T_{\nu+1}\left(\tau - (1 - \tau)\cos\frac{\pi}{2\nu + 2}\right) \tag{10.8.3a}$$

with $\eta(\nu)$ from (3d). P_ν is the product $P_\nu(1 - \tau) = \prod_{\mu=1}^{\nu} (1 - \omega_\mu \tau)$ with

$$\omega_\mu = \left(1 + \cos\frac{\pi}{2\nu + 2}\right)\Big/\left(\cos\frac{\pi}{2\nu + 2} - \cos\frac{(2\mu + 1)\pi}{2\nu + 2}\right). \tag{10.8.3b}$$

The expression (2) to be minimised takes the value

$$\|A_\ell P_\nu(1 - A_\ell/\|A_\ell\|_2)\|_2 \leqslant \eta(\nu)\, \|A_\ell\|_2 \quad \text{with} \tag{10.8.3c}$$

$$\eta(\nu) = \frac{1}{\nu + 1}\frac{\sin(\pi/(2\nu + 2))}{1 + \cos(\pi/(2\nu + 2))} \leqslant \frac{2(\sqrt{2} - 1)}{(\nu + 1)^2} \quad \text{for all } \nu \geqslant 1. \tag{10.8.3d}$$

Proof. One verifies that $P_\nu(1) = 1$ and that the right-hand side in (3a) takes the values $\pm\eta(\nu)$ «equi-oscillating» in $[0, 1]$. □

Concerning this result the following should be mentioned:

(i) The semi-iterative smoothing achieves an *order reduction*. While the smoothing factor $\eta(v)$ of the stationary Richardson method behaves like $O(1/(v+1))$, the order becomes $O(1/(v+1)^2)$ in the semi-iterative case.

(ii) The application of the Chebyshev method requires knowledge of the interval $\sigma_M = [a, b]$ containing the spectrum of S_ℓ. Especially, the estimation of $b = 1 - \lambda_\ell/\|A_\ell\|_2$ with $\lambda_\ell = \lambda_{\min}(A_\ell)$ is of decisive importance. An over-estimation of the uper bound $\Lambda_\ell = \|A_\ell\|_2$ in $\lambda_\ell I \leqslant A_\ell \leqslant \Lambda_\ell I$ is less sensitive (proof is that $\Lambda_\ell/\lambda_\ell$ is the essential quantity). A different situation arises in Theorem 1, where we estimate the spectrum of A_ℓ simply by $0 \leqslant A_\ell \leqslant \Lambda_\ell I$, $\Lambda_\ell = \|A_\ell\|_2$, i.e., the lower bound λ_ℓ is trivially chosen as $\lambda_\ell := 0$. The replacement of $0 \leqslant A_\ell$ by $\lambda_\ell I \leqslant A_\ell$ with $\lambda_\ell = \lambda_{\min}(A_\ell)$ would yield only an imperceptible improvement.

(iii) The statements from (ii) clarify the fact that the condition $\varkappa(W_\ell^{-1}A_\ell)$ is not the essential quantity for ability as a smoothing iteration.

(iv) The product representation $\Pi(1 - \omega_\mu \tau)$ seems to disregard the warnings of §7.3.4 concerning instabilities. The contradiction is solved by the fact that the number v of the smoothing steps should be relatively small according to the discussion in §10.4.4. However, if one restricts onself, e.g., to $v \leqslant 4$, stability problems cannot arise.

For the general symmetric smoothing iteration with $S_\ell = I - W_\ell^{-1}A_\ell$, one obtains analogous results for the minimisation of the expression $\|W_\ell^{-1/2}A_\ell P_v(S_\ell)W_\ell^{-1/2}\|_2$, which corresponds to the approximation property (7.8a) concerning the choice of norms. Corresponding to the smoothing property (7.4b), the minimisation of

$$\|W_\ell^{-1/2}A_\ell P_v(S_\ell)A_\ell^{1/2}\|_2 = \|Y^{1/2}P_v(I - Y)\|_2 \quad \text{with } Y := A_\ell^{1/2}W_\ell^{-1}A_\ell^{1/2}$$

is also of interest. The corresponding optimal polynomial can be found in Hackbusch [14, Proposition 6.2.35]. The bound $O(1/\sqrt{v})$ from (7.4b) improves to $1/(2v+1)$. The ADI parameters (cf. §7.5.3 and Hackbusch [14, §3.3.4 and Lemma 6.2.36]) have also to be chosen differently for optimising the smoothing effect.

The *conjugate gradient* method is only conditionally applicable. The standard cg method minimises $\|P_v(S_\ell)e_\ell\|_{A_\ell} = \|A_\ell^{1/2}P_v(S_\ell)e_\ell\|_2$, where e_ℓ is the error before smoothing and $P_v(\xi)$ the corresponding optimal polynomial (cf. Remark 9.4.8). However, since not the energy norm but the residual $\|A_\ell P_v(S_\ell)e_\ell\|_2$ is to be minimised, the method of the conjugate residuals (cf. §9.5) or the conjugate gradient method for the «squared equation» $A_\ell^H A_\ell x_\ell = A_\ell^H b_\ell$ is better suited. These remarks apply to the pre-smoothing part only. The conjugate gradient methods do not seem to make much sense for the post-smoother. In any case, an unsymmetric multi-grid iteration results. Cf. Bank–Douglas [1].

The smoothing property of conjugate gradient methods has also been mentioned by Il'in [1].

10.8.2 Damped Coarse-Grid Corrections

The treatment of nonlinear equations suggests damping the coarse-grid correction as known from gradient methods, in order to obtain a descent method (cf. Hackbusch–Reusken [1]). It turns out that in the linear case, it is also possible to improve convergence. In particular, V-cycle convergence can be accelerated (cf. Reusken [1], Braess [3]). The optimally damped coarse-grid correction step reads

$$x_\ell^{\text{new}} := x_\ell - \lambda p_\ell \quad \text{with } \lambda := \frac{\langle d_\ell, p_\ell \rangle_\ell}{\langle p_\ell, A_\ell p_\ell \rangle_\ell},\ d_\ell := A_\ell x_\ell - b_\ell,\ p_\ell := p\tilde{e}_{\ell-1}, \tag{10.8.4}$$

where $\tilde{e}_{\ell-1}$ is the approximation of the solution of the coarse-grid equation $A_{\ell-1} e_{\ell-1} = d_{\ell-1} := r d_\ell$.

Exercise 10.8.2. Prove that (a) $\lambda = 1$ is optimal for the two-grid method. (b) If $r = p^*$ and $A_{\ell-1} = r A_\ell p$, λ from (4) can be written in the form

$$\lambda = \langle d_{\ell-1}, \tilde{e}_{\ell-1} \rangle_{\ell-1} / \langle A_{\ell-1} \tilde{e}_{\ell-1}, \tilde{e}_{\ell-1} \rangle_{\ell-1}. \tag{10.8.5}$$

Another possibility is the damping of the complete multi-grid method. The inequality $M_\ell^{\text{MGM}} \geqslant 0$ (cf. (7.16a)) holding in the symmetric case and implying $\sigma(M_\ell^{\text{MGM}}) \subset [0, \rho(M_\ell^{\text{MGM}})]$ indicates that damping (extrapolation) with $\Theta := 2/(2 - \rho(M_\ell^{\text{MGM}})) \approx 1 + \frac{1}{2}\rho(M_\ell^{\text{MGM}})$ leads to the nearly halved convergence rate $\rho(M_\ell^{\text{MGM}})/(2 - \rho(M_\ell^{\text{MGM}}))$ (cf. Exercise 8.3.1).

10.8.3 Multi-Grid Iteration as Basis of the Conjugate Gradient Method

As shown in §10.7.1, the multi-grid method for $A_\ell > 0$ can be designed as a symmetric iteration. The convergence statement $\sigma(M_\ell^{\text{MGM}}) \subset [0, \rho_\ell]$ with $\rho_\ell := \rho(M_\ell^{\text{MGM}})$ with $\rho_\ell := \rho(M_\ell^{\text{MGM}})$ corresponds to the inclusion

$$\gamma W_\ell^{\text{MGM}} \leqslant A_\ell \leqslant W_\ell^{\text{MGM}} \quad \text{with } \gamma := 1 - \rho_\ell \tag{10.8.6}$$

for the matrix W_ℓ^{MGM} of the third normal form of the multi-grid iteration (cf. Remark 4.8.3c). Applying the cg method to Φ_ℓ^{MGM}, after m steps, one obtains an improvement by $2[(\sqrt{\varkappa} - 1)/(\sqrt{\varkappa} + 1)]^m$, where $\varkappa$ is the condition $\varkappa = \Gamma/\gamma = 1/(1 - \rho_\ell)$. A simple rewriting yields

$$2\left(\frac{\sqrt{\varkappa} - 1}{\sqrt{\varkappa} + 1}\right)^m = 2\rho_\ell^m/(1 + \sqrt{1 - \rho_\ell})^{2m} \approx 2\left(\frac{\rho_\ell}{4} + O(\rho_\ell^2)\right)^m.$$

Since ρ_ℓ may be assumed to be small (cf. §10.4.3), the convergence rate ρ_ℓ of the multi-grid method can be accelerated by means of the conjugate gradient method to $\rho_\ell/4$ (cf. Braess [4], Kettler [1]).

However, the use of the conjugate gradient method is of practical interest only if the multi-grid convergence rate is relatively unfavourable (e.g., $\geqslant 0.4$). The reason for this are the considerations from §10.5.2: In the case of a good convergence rate, the nested iteration (5.2a) requires only very few multi-grid steps per level. For fast multi-grid methods, the iteration number $m = 1$ has proved in §10.5.5 to be sufficient. For this value, the gradient and conjugate gradient methods still coincide. Only for $m \geqslant 2$, does the conjugate gradient method deserve attention.

10.9 Further Comments

10.9.1 Multi-Grid Method of the Second Kind

The discretisation of Fredholm's integral equations of the second kind leads to fixed-point equations of the form $x_\ell = K_\ell x_\ell + b_\ell$, i.e.,

$$A_\ell x_\ell = b_\ell \quad \text{with } A_\ell = I - K_\ell. \tag{10.9.1}$$

Here, K_ℓ does not characterise a difference operator as in the case of discretised differential equations, but an integration. Therefore, the *Picard iteration* $x_\ell^{m+1} := K_\ell x_\ell^m + b_\ell$ (corresponding to Richardson's iteration) has a substantially better smoothing property. This implies that the multi-grid method ($\gamma = 2$) has a convergence rate $\varrho_\ell = O(h_\ell^\varkappa)$ with a positive exponent $\varkappa$. Hence, different from the situation considered before, the convergence is better the larger the dimension of the problems is. Since the absolute value of ϱ_ℓ may be of the order of 10^{-3} to 10^{-6}, the multi-grid method is close to a direct solver. The work of the method is still proportional to the work of the Picard iteration.

Its application is not restricted to discrete integral equations. In the example of §8.4.1, the equation $Ax = b$ was preconditioned by means of B, where B as well as A were discretisations of the differential equation. If both differential equations share the same principal part (i.e., if the terms of the highest order of differentiation coincide), the equation $A'x = b' := B^{-1}b$ with $A' := B^{-1}A$ leads to the fast multi-grid convergence mentioned above. The application of the multi-grid method of the second kind to $A'x = b'$ requires that the Picard iteration $x^{m+1} = Kx^m + b' = x^m - B^{-1}(Ax - b)$ be performed. For example, B could be the five-point formula of the Poisson-model problem, whereas A discretises the equation $-\Delta u + c_1 u_x + c_2 u_y + cu = f$.

An exact description and analysis of the multi-grid method of the second kind as well as many examples of application can be found in Hackbusch [14, §16], [16], [13].

10.9.2 History of the Multi-Grid Method

The first two-grid method was described by Brakhage [1] in 1960. More precisely, it was a two-grid method of the second kind, because it applied to the problems mentioned in §10.9.1. In 1961 Fedorenko [1] published a two- and in 1964 a multi-grid method (cf. Fedorenko [2]) for the Poisson-model problem. In 1966 Bachvalov [1] proved the typical convergence properties for a more complicated situation. Additional early publications were due to Astrachancev [1] (1971), Hackbusch [1] (1976), Bank–Dupont (in a 1977 report that later was split into [1,2]), Brandt [1] (1977), and Nicolaides [1] (1977). Further details concerning these and other papers by Frederickson, Wesseling, Hemker, and Braess are mentioned in Hackbusch [14, §2.6.5].

An extensive list of multi-grid literature published before 1985 is contained in Hackbusch [14] and the volume edited by McCormick [1]. We also refer the interested reader to the proceedings volumes of Hackbusch–Trottenberg [1, 2, 3], Braess–Hackbusch–Trottenberg [1], and Hackbusch [17], [18].

10.9.3 Robust Methods

In order to make our presentation brief, smoothers other than the Richardson and the chequer-board Gauß-Seidel iterations have been mentioned only marginally. In applications to more complicated systems, the problem of robustness arises: Are those convergence rates known from the Poisson-model problem uniformly valid in a larger class of problems? In the simplest case, the equation $A(\varepsilon)x = b$ depends on one parameter $\varepsilon \in (0, \infty)$. If an iteration method achieves convergence rates not only of the form $\varrho(\varepsilon) \leqslant 1 - C(\varepsilon)h^{\varkappa}$, but with $C = C(\varepsilon)$ uniformly in $\varepsilon \in (0, \infty)$: $\varrho(\varepsilon) \leqslant 1 - Ch^{\varkappa}$, then this iteration is *robust* with respect to the class of problems. Robust multi-grid methods have to satisfy $\varrho(\varepsilon) \leqslant \zeta < 1$ (ζ h_ℓ- and ε-independent; cf. Hackbusch [14, §10]).

Good experiences concerning robustness have been made with incomplete LU decompositions, which were introduced by Wesseling [1,2] as multi-grid smoothers (cf. Kettler [1]). Robustness holds for cg methods applied to the modified ILU iteration ($\omega = -1$), as well as for multi-grid methods using point- or blockwise ILU iterations as the smoother (then with $\omega = 0$ or even $\omega = 1$; cf. Wittum [5], Kettler [1]).

Another approach is the *frequency decomposition multi-grid method* of Hackbusch [11], which uses not only one but several coarse-grid corrections with different coarse-grid equations. The prolongations from the different coarse grids into the fine grid are constructed in such a way that the corrections cover different frequency intervals.

The construction of the coarse-grid equation at level $\ell - 1$ requires more information than given only by the system $A_\ell x_\ell = b_\ell$ for $\ell = \ell_{\max}$. This fact may lead to difficulties when the multi-grid method is wanted as a blackbox solver. Therefore, it is remarkable that there are variants, the so-called *algebraic multi-grid methods*, in which the coarse-grid matrix $A_{\ell-1}$ is constructed only by the information contained in the entries of the matrix A_ℓ (cf. Stüben [1]).

10.9.4 Frequency Filtering Decompositions

An essential characteristic of the multi-grid method, besides the use of a coarser grid, is the product form $\Phi_\ell^{CGC} \circ \mathscr{S}_\ell^\nu$ with the different frequency intervals, in which the coarse-grid correction and the smoothing step are active. Although many methods can be used as the smoother, the question remains as to whether there exists an alternative to Φ_ℓ^{CGC}. It would be desirable to have a method filtering out the coarse frequencies and needing no hierarchy of grids. Such a method was proposed by Wittum [6, 7] and is based on a sequence of partial steps Φ_ν reducing certain frequency intervals.

First, we describe the standard *blockwise ILU decomposition*. Suppose that A has the block-tridiagonal form

$$A = A^H = \text{blocktridiag}\{L_i, D_i, L_i^H: i = 1, \dots, N-1\} \qquad (10.9.2a)$$

(cf. (1.2.8), (2.5.4)), which, e.g., holds for five- or nine-point formulae. As in (1.2.2), $N - 1 = h^{-1} - 1$ is the number of inner grid points per row. The *exact* LU decomposition is $A = LD^{-1}L^H$ with

$$L = \text{blocktridiag}\{L_i, T_i, 0\}, \qquad D = \text{blockdiag}\{T_i\}, \qquad (10.9.2b)$$

$$T_1 := D_1, \quad T_i := D_i - L_i T_{i-1}^{-1} L_i^H \qquad (2 \leqslant i \leqslant N-1). \qquad (10.9.2c)$$

Even if the blocks D_i are tridiagonal (cf. (1.2.8)), the arising matrices T_i are full. The usual block-ILU decomposition is obtained from (2c) by replacing the full inverse of T_{i-1} with the tridiagonal part tridiag$\{T_{i-1}^{-1}\}$ of the exact inverse.

Another approach goes back to Axelsson–Polman [1]. Let $t^{(1)}$ and $t^{(2)}$ be two «test vectors». The matrices T_i are defined by

Lemma 10.9.1. *Assume that* $t^{(1)}, t^{(2)} \in \mathbb{R}^{N-1}$ *satisfy*

$$\det((t_{i+j}^{(k)})_{j=0,1}^{k=1,2}) \neq 0 \quad \textit{for all } 1 \leqslant i \leqslant N-2.$$

The vectors $c^{(1)}, c^{(2)} \in \mathbb{R}^{N-1}$ *may be arbitrary. Then there is a unique symmetric tridiagonal matrix* T *satisfying the equations* $Tt^{(k)} = c^{(k)}$ *for* $k = 1, 2$.

Hence, we can uniquely define symmetric tridiagonal matrices T_i by

$$T_1 := D_1, \quad T_i t^{(k)} = (D_i - L_i T_{i-1}^{-1} L_i^H) t^{(k)} \quad \text{for } k = 1, 2, \quad 2 \leqslant i \leqslant N-1. \qquad (10.9.3)$$

The matrices T_i inserted into (2b) yield a new incomplete blockwise triangular decomposition $A = LD^{-1}L^H - C$. Definition (3) means that T_i is exact with respect to the «test subspace» span$\{t^{(1)}, t^{(2)}\}$. The corresponding iteration is

$$\Phi(x, b; t^{(1)}, t^{(2)}) := x - L^{-H} D L^{-1}(Ax - b) \quad \text{with } L, D \text{ from (2b), (3).} \qquad (10.9.4)$$

Wittum [6, 7] has proposed taking the sine functions e^ν from (3.4) with different frequencies ν as test vectors:

$$\Phi_\nu := \Phi(\cdot, \cdot; e^\nu, e^{\nu+1}) \quad \text{with } \nu \in [1, N-2]. \qquad (10.9.5)$$

Table 10.9.1 Frequency filtering iteration for the Poisson-model problem

m	$h = 1/8, k = 3$		$h = 1/16, k = 4$		$h = 1/32, k = 5$		$h = 1/64, k = 6$	
	$\|e^m\|_2$	$\varrho_{m+1,m}$	$\|e^m\|_2$	$\varrho_{m+1,m}$	$\|e^m\|_2$	$\varrho_{m+1,m}$	$\|e^m\|_2$	$\varrho_{m+1,m}$
0	$6.3_{10}-01$	$4.1_{10}-7$	$7.0_{10}-01$	$2.6_{10}-6$	$7.4_{10}-01$	$1.9_{10}-5$	$7.6_{10}-01$	$7.3_{10}-5$
1	$2.6_{10}-07$	$3.9_{10}-6$	$1.8_{10}-06$	$1.5_{10}-5$	$1.4_{10}-05$	$1.0_{10}-4$	$5.6_{10}-05$	$3.9_{10}-4$
2	$1.0_{10}-12$	$(4.0_{10}-1)$	$2.8_{10}-11$	$(2.3_{10}-2)$	$1.5_{10}-09$	$8.2_{10}-4$	$2.2_{10}-08$	$4.1_{10}-4$
3	$4.1_{10}-13$		$6.7_{10}-13$		$1.2_{10}-12$		$9.3_{10}-12$	

By means of a factor $\alpha > 1$, which, e.g., may be chosen as $\alpha = 2$, a geometrical sequence of frequencies is selected:

$$\nu_1 := 1, \quad \nu_{i+1} := \max\{\nu_i + 2, [\alpha\nu_i]\} \quad \text{as long as } \nu_{i+1} \leqslant N - 2. \tag{10.9.6a}$$

Here, $[\ldots]$ denotes rounding to an integer. Let k be the number of frequencies determined in (6a). Obviously, this number equals $k = O(\log N) = O(|\log h|) = O(\log n)$. The iteration of the frequency filtering decomposition is defined by the product (6b):

$$\Phi_\alpha^{\mathrm{ffd}} := \Phi_{\nu_k} \circ \cdots \circ \Phi_{\nu_3} \circ \Phi_{\nu_2} \circ \Phi_{\nu_1} \quad (\alpha > 1 \text{ with } \nu_i \text{ from (6a)}). \tag{10.9.6b}$$

The work of one iteration $\Phi_\alpha^{\mathrm{ffd}}$ amounts to $O(n \log n)$. The numerical results (cf. Wittum [6, 7]) demonstrate the very fast convergence of this iteration. Its effectiveness even can exceed that of the standard multi-grid methods.

Convergence analysis is known for the case of a nine-point formula $A > 0$ with constant coefficients $D_i = D_{i+1}$, $L_i = L_{i+1} = L_i^H$ (cf. Wittum [7]). One step of the proof concerns the monotone convergence of Φ_ν with respect to the energy norm for all ν. However, the more characteristic step is a so-called «neighbourhood property». According to its definition, Φ_ν eliminates error components in $\mathrm{span}\{e^\nu, e^{\nu+1}\}$. It is essential that Φ_ν yields a uniform and h-independent contraction number for all frequencies $\nu \leqslant \mu \leqslant \alpha\nu$, i.e., that Φ_ν also acts effectively in a certain neighbourhood of the «gauge frequency» ν.

The idea of frequency filtering decompositions can also be generalised to nonsymmetric or even nonlinear problems (cf. Wittum [7]).

Table 1 shows the iteration error $\|e^m\|_2 = \|x^m - x\|_2$ of the frequency filtering decomposition method $\Phi_\alpha^{\mathrm{ffd}}$ for $\alpha = 2$ applied to the Poisson-model problem (for the Pascal program, cf. [Prog]). The arising number k of partial steps is between 3 and 6. After 2 to 3 steps, machine accuracy is reached. One observes that with decreasing h, the convergence speed is bounded from above and is hence h-independent.

11 Domain Decomposition Methods

11.1 Introduction

Various iterative methods can be classified as domain decomposition methods. Although a prototype of this iteration due to H. A. Schwarz (Viertelsjahresschrift der Naturforschenden Gesellschaft in Zürich, Vol. 15, 1870) is already 120 years old, this class of algorithms has attracted the interest of numerical analysts only recently.

Let the system of equations $Ax = b$ represent a discretisation of a boundary value problem in the domain Ω (cf. §1.2). The naming characteristic of the domain decomposition method is a decomposition of the complete problem into smaller systems of equations corresponding to boundary value problems in subdomains $\Omega' \subset \Omega$.

The choice of subdomains can be influenced by different motivations. In former times, when fast methods like, e.g., the multi-grid iteration had not yet been (sufficiently) known and only fast direct solvers for special problems were available, one tried to decompose complex domains into disjoint rectangles (cf. Fig. 1a) or overlapping rectangles (cf. Fig. 1b), since simple problems like the Poisson-model problem on rectangles were directly solvable.

On the other hand, it might be possible to split the complete problem in a natural way into disjoint subproblems (e.g., when the subdomains model physically different materials). In particular, users of *parallel computers* are looking for a decomposition of entire problems into separate subproblems, which should be equally sized (because of the load-balancing of processors) and for programming reasons structured as simply as possible.

Here it must be emphasized that the solution of subproblems can never solve a complete problem, but only represents a partial step of the algorithm

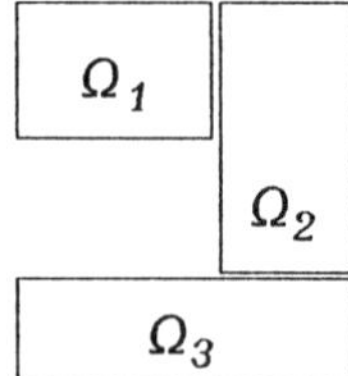

Fig. 11.1.1a Disjoint subdomains

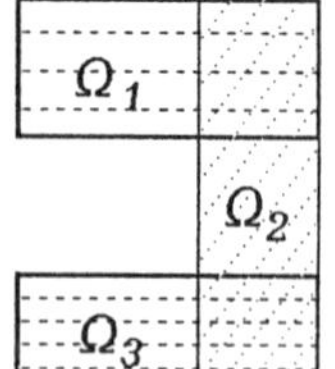

Fig. 11.1.1b Overlapping subdomains

that also has to establish the coupling of subproblems. The major part of computational work and programming effort, however, should focus on the solution of subproblems.

The «capacitance matrix method» as well as the «method of fictitious domains» also belong to the class of domain decomposition methods. These will be mentioned together with the «Schur complement methods» in §11.7.

In the course of the development of the domain decomposition method, the term «subdomain» was generalised to «subspace», in particular, to a subspace of the Galerkin method. If the subspace is spanned by those finite element functions that differ from zero only in one subdomain $\Omega' \subset \Omega$, the terms «subdomain» and «subspace» coincide. Other subspaces, however, can also be constructed and deserve practical interest.

Since in this book we start from the algebraic equation $Ax = b$ and do not discuss the construction of $Ax = b$ by means of discretisation, we have chosen a presentation of the domain decomposition method that is based on a *decomposition of the vectors* $x \in X = \mathbb{K}^I$. The connection to a traditional variational formulation is shown in §11.5.3.

11.2 Formulation of the Domain Decomposition Method

11.2.1 General Construction

Let $X = \mathbb{K}^I$ be the linear space containing the solution x of $Ax = b$. The subproblems, which we index by $\varkappa \in J$, correspond to lower-dimensional problems represented by vectors $x^\varkappa \in X_\varkappa = \mathbb{K}^{I_\varkappa}$. The solution x of the system $Ax = b$ is composed of the partial solutions $x^\varkappa$. For that purpose, we choose linear and injective mappings, which may be called *prolongations*:

$$p_\varkappa \colon X_\varkappa \to X \qquad (\varkappa \in J). \tag{11.2.1}$$

The true solution $x = A^{-1}b$ is sought in the form

$$x = \sum_{\varkappa \in J} p_\varkappa x^\varkappa. \tag{11.2.2}$$

This is possible only if

$$\sum_{\varkappa \in J} \operatorname{range}(p_\varkappa) = X \tag{11.2.3}$$

holds. Here, $\operatorname{range}(p_\varkappa) = \{p_\varkappa x^\varkappa : x^\varkappa \in X_\varkappa\}$ denotes the *image space* of $p_\varkappa$; furthermore, the sum $\sum V_\varkappa$ of subspaces $V_\varkappa \subset X$ denotes the space spanned by these subspaces:

$$\sum_{\varkappa \in J} V_\varkappa = \operatorname{span}\{V_\varkappa : \varkappa \in J\} = \left\{x \in X : x = \sum v^\varkappa : v^\varkappa \in V_\varkappa\right\}.$$

$p_\varkappa$ from (1) is represented by a rectangular matrix. Its Hermitian transposed matrix $p_\varkappa^H$ is a mapping from X onto $X_\varkappa$:

$$r_\varkappa := p_\varkappa^H : X \to X_\varkappa. \qquad (\varkappa \in J). \tag{11.2.4}$$

For each $\varkappa \in J$, we define the square matrices

$$A_\varkappa := r_\varkappa A p_\varkappa \tag{11.2.5}$$

Their size is $n_\varkappa \times n_\varkappa$, where

$$n_\eta := \dim X_\varkappa. \tag{11.2.6}$$

The lower-dimensional subproblems are equations of the form

$$A_\varkappa y^\varkappa = c^\varkappa \qquad (\varkappa \in J, y^\varkappa, c^\varkappa \in X_\varkappa). \tag{11.2.7}$$

For the present, we assume that problems of the form (7) can be solved exactly. The regularity of $A_\varkappa$ is not guaranteed without additional conditions on A. A sufficient condition is given in

Exercise 11.2.1. Assume that $A > 0$ and $p_\varkappa$ is injective for all $\varkappa \in J$. Prove $A_\varkappa > 0$.

We associate the prolongations $p_\varkappa$ with the *restrictions* $r_\varkappa$ from (4) and the *projections*

$$P_\varkappa := p_\varkappa A_\varkappa^{-1} r_\varkappa A = p_\varkappa A_\varkappa^{-1} p_\varkappa^H A : X \to X. \tag{11.2.8a}$$

Exercise 11.2.2. Prove that (a) $P_\varkappa$ ($\varkappa \in J$) are projections onto range ($p_\varkappa$).
(b) Let $A > 0$. $P_\varkappa$ is an A-orthogonal projection, i.e., $P_\varkappa$ is self-adjoint with respect to the A-scalar product (9.1.11a):

$$\langle P_\varkappa x, y\rangle_A = \langle x, P_\varkappa y\rangle_A \quad \text{for all } x, y \in X, \text{ where} \tag{11.2.8b}$$

$$\langle x, y\rangle_A := \langle Ax, y\rangle \qquad (x, y \in X; \text{ cf. } (9.1.11a)) \tag{11.2.8c}$$

(c) Let $A = A^H$. For all symmetric iterations with an iteration matrix M, the identity $\langle AMx, y\rangle = \langle Ax, My\rangle$ holds for all $x, y \in X$.

11.2.2 The Prolongations

Let $n_\varkappa$ from (6) be the partial dimension and $n = \dim X$ the entire one. A first classification is given by the alternatives (9a) and (9b):

$$\sum_{\varkappa \in J} n_\varkappa = n \qquad \text{(«disjoint subdomains»)} \tag{11.2.9a}$$

$$\sum_{\varkappa \in J} n_\varkappa > n \qquad \text{(«overlapping subdomains»).} \tag{11.2.9b}$$

Note that $\sum n_\varkappa < n$ is excluded because of (3).

Remark 11.2.3. In case (9a), each $x \in X$ allows a unique decomposition (2); in case (9b), more than one representation (2) is possible.

A particularly simple situation of the form (9a) arises from the block representation of x. Let $\{I_\varkappa : \varkappa \in J\}$ describe a block structure (2.5.1): $I = \bigcup_{\varkappa \in J} I_\varkappa$. We choose

$$X_\varkappa = \mathbb{K}^{I_\varkappa}, \qquad (p_\varkappa)_{\alpha\beta} = \delta_{\alpha\beta} \quad \text{for } \alpha \in I,\ \beta \in I_\varkappa \tag{11.2.10a}$$

(δ: Kronecker symbol). Therefore, $\sum_{\varkappa \in J} p_\varkappa x^\varkappa$ is only another notation of the vector $x = (x^\varkappa)_{\varkappa \in J}$ composed of the blocks $x^\varkappa$. In the case of ordered indices, the matrix $p_\varkappa$ takes the form

$$p_\varkappa = \begin{bmatrix} 0 \\ \vdots \\ 0 \\ I \\ 0 \\ \vdots \\ 0 \end{bmatrix} \}\text{block of index } \varkappa, \qquad r_\varkappa = p_\varkappa^H = [0, \ldots, 0, I, 0, \ldots, 0]. \tag{11.2.10b}$$

Remark 11.2.4. If $p_\varkappa$ is defined accordng to (10a, b), the matrices $A_\varkappa$ from (5) coincide with the diagonal-block $A^{\varkappa\kappa}$ of the matrix A (cf. §2.5).

In the case (9b), $p_\varkappa$ can still be defined by (10a), but the subsets $I_\varkappa \subset I$ are no longer disjoint and therefore do not describe a block decomposition. A property still remaining is mentioned in

Remark 11.2.5. If the index set I is decomposed into $I = \bigcup_{\varkappa \in J} I_\varkappa$ (possibly nondisjointly) and the prolongations $p_\varkappa$ are defined by (10a), the matrices $A_\varkappa$ from (5) represent the principal submatrices of the matrix A belonging to the index subset $I_\varkappa$.

11.2.3 Multiplicative and Additive Schwarz Iteration

The projections $P_\varkappa$ from (8a) are associated with the iterations

$$\Phi_\varkappa(x,b) := x - p_\varkappa A_\varkappa^{-1} r_\varkappa(Ax - b) \qquad (\varkappa \in J). \tag{11.2.11}$$

Exercise 11.2.6. Prove that the iteration matrix belonging to (11) is $M_\varkappa = I - P_\varkappa$. Like $P_\varkappa$, $M_\varkappa$ is a projection. In the case of $A > 0$, $M_\varkappa$ is A-orthogonal (cf. Exercise 2). $\Phi_\varkappa$ is a symmetric iteration.

Let the index set $J = \{1, \ldots, k\}$ be ordered. The *multiplicative Schwarz iteration* is the k-fold product

$$\Phi^{\mathrm{multSI}} := \Phi_k \circ \Phi_{k-1} \circ \cdots \circ \Phi_2 \circ \Phi_1 \qquad (\Phi_\varkappa \text{ from (11)}). \tag{11.2.12}$$

In contrast, the *additive Schwarz iteration* (with a damping factor Θ) is defined by

$$\Phi_\Theta^{\mathrm{addSI}}(x,b) := x - \Theta \sum_{\varkappa \in J} p_\varkappa A_\varkappa^{-1} r_\varkappa(Ax - b), \tag{11.2.13}$$

where the index set J need not be ordered.

Lemma 11.2.7. (a) *The iteration matrices of Φ^{multSI} and $\Phi_\Theta^{\mathrm{addSI}}$ are*

$$M^{\mathrm{multSI}} = (I - P_k)(I - P_{k-1}) \cdot \ldots \cdot (I - P_1), \tag{11.2.14a}$$

$$M_\Theta^{\mathrm{addSI}} = I - \Theta\left(\sum_{\varkappa \in J} p_\varkappa A_\varkappa^{-1} r_\varkappa\right) A = I - \Theta \sum_{\varkappa \in J} P_\varkappa. \tag{11.2.14b}$$

(b) *Let $A > 0$. The matrix of the second normal form of $\Phi_\Theta^{\mathrm{addSI}}$ is*

$$\Theta N^{\mathrm{addSI}} \quad \textit{with } N^{\mathrm{addSI}} = \sum_{\varkappa \in J} p_\varkappa A_\varkappa^{-1} r_\varkappa. \tag{11.2.15a}$$

Under assumption (3), *N^{addSI} is regular, so that the matrix $W_\Theta^{\mathrm{addSI}} = \Theta^{-1} W^{\mathrm{addSI}}$ of the third normal form exists. It satisfies*

$$A \leqslant k W^{\mathrm{addSI}} \qquad (k := \#J = \textit{number of «subdomains»}). \tag{11.2.15b}$$

Proof of (b). (i) Set $N := N^{\mathrm{addSI}}$. $Nx = 0$ implies $r_\varkappa x = 0$ because of $\langle x, Nx \rangle = \sum \langle A_\varkappa^{-1} r_\varkappa x, r_\varkappa x \rangle$. Since kernel $(r_\varkappa)$ = range $(p_\varkappa)^\perp$, the orthogonality $x \perp$ range $(p_\varkappa)$ follows for all $\varkappa \in J$. From (3) one concludes $x = 0$.
(ii) $A^{1/2} p_\varkappa A_\varkappa^{-1} r_\varkappa A^{1/2}$ is $\leqslant I$, since it is an orthogonal projection. Summation yields $A^{1/2} N A^{1/2} \leqslant kI$. According to (2.10.3g), inequality (15b) follows from $N \leqslant kA^{-1}$. □

Exercise 11.2.8. Let $A > 0$. Prove the following: (a) The iteration adjoint to Φ^{multSI} is $\Phi_1 \circ \cdots \circ \Phi_k$. The corresponding symmetric iteration (cf. (4.8.11)) is $\Phi^{\mathrm{symmultSI}} := \Phi_1 \circ \cdots \circ \Phi_{k-1} \circ \Phi_k \circ \Phi_{k-1} \circ \cdots \circ \Phi_1$.
(b) $\Phi_\Theta^{\mathrm{addSI}}$ is symmetric.

11.2.4 Interpretation as Gauß-Seidel and Jacobi Iteration

Assume the case (9a) («disjoint subdomains»). Moreover, let $p_\varkappa$ be given by (10a). Then $\Phi_\varkappa$ from (11) describes the solution of equation $Ax = b$ with respect to the block of index $\varkappa \in J$. This proves part (a) of

Lemma 11.2.9. (a) *Under the assumptions* (9a) *and* (10a), *the multiplicative Schwarz iteration coincides with the block-Gauß-Seidel iteration, whereas the additive Schwarz iteration coincides with the damped block-Jacobi iteration.*
(b) *Let* $A > 0$. *The multiplicative Schwarz iteration converges. The additive Schwarz iteration converges for sufficiently small* $\Theta > 0$ ($\Theta < 2k$ *is always sufficient*).

Proof of (b). Use (15b) and Theorems 4.4.11, 4.4.18, and 4.5.4. □

Now we assume again (9a), but $p_\varkappa$ need not be of the form (10a). The vectors $x^\varkappa$ (regarded as block-vectors) can be gathered into the block-vector $\hat{x} = (x^\varkappa)_{\varkappa \in J}$. The linear mapping $\hat{x} = (x^\varkappa)_{\varkappa \in J} \mapsto x = \sum p_\varkappa x^\varkappa$ describes a regular transformation $x = T\hat{x}$. We define $\hat{p}_\varkappa$ by $(\hat{p}_\varkappa)_{\alpha\beta} = \delta_{\alpha\beta}$ for $\alpha \in I$, $\beta \in I_\varkappa$ (cf. (10a)). Obviously, we have

$$p_\varkappa = T\hat{p}_\varkappa, \qquad x = \sum p_\varkappa x^\varkappa = T\hat{x}. \tag{11.2.16a}$$

Simultaneously, we transform the system $Ax = b$ into

$$\hat{A}\hat{x} = \hat{b} \quad \text{with } \hat{A} = T^H A T, \quad \hat{b} = T^H b. \tag{11.2.16b}$$

Define $\hat{\Phi}_\varkappa$ by means of $\hat{p}_\varkappa$, $\hat{A}$, and $\hat{r}_\varkappa = \hat{p}_\varkappa^H$. Using the notation from §8.1.4, one verifies that $\Phi_\varkappa = T \circ \hat{\Phi}_\varkappa \circ T^H$. Since $\hat{\Phi}_\varkappa$ satisfies the assumptions (9a) and (10a), Lemma 9a yields $\hat{\Phi}^{\text{multSI}} = \Phi^{\text{blockGS}}$ and $\hat{\Phi}_\Theta^{\text{addSI}} = \Phi_\Theta^{\text{blockJac}}$. Hence, Φ^{multSI} and $\Phi_\Theta^{\text{addSI}}$ are the two-sided transformed blockwise Gauß-Seidel and Jacobi iterations:

$$\Phi^{\text{multSI}} = T^{-H} \circ \Phi^{\text{blockGS}} \circ T^{-1}, \qquad \Phi_\Theta^{\text{addSI}} = T^{-H} \circ \Phi_\Theta^{\text{blockJac}} \circ T^{-1}. \tag{11.2.17}$$

In the case of $k = \#J = 2$, the matrix $\hat{A}$ has the 2×2 block structure

$$\hat{A} = \begin{bmatrix} A_1 & A_{12} \\ A_{12}^H & A_2 \end{bmatrix} \quad \text{with } A_\varkappa \text{ from (5) and } A_{12} := p_1^H A p_2. \tag{11.2.16c}$$

11.2.5 Classical Schwarz Iteration

A situation not comparable with the Gauß-Seidel or Jacobi iteration arises in case (9b). The classical Schwarz iteration corresponds to the choice

$$J = \{1,2\},\ I_1 = \{1,\ldots,n_1\},\ I_2 = \{n - n_2 + 1,\ldots,n\},\ p_\varkappa \text{ from (10a)}, \tag{11.2.18}$$

where $n_1 \geqslant n - n_2 + 1$ ensures the overlapping of index sets $I_\varkappa$.

As an example, we choose the one-dimensional problem $-u'' = f$ $(0 \leqslant x \leqslant 1)$ discretised as in (10.3.1). The index sets (18) with

$$n_1 = b/h - 1, \qquad n - n_2 = a/h \qquad (a, b \text{ multiples of } h)$$

correspond to the subdomains $\Omega_1 = (0, b)$ and $\Omega_2 = (a, 1)$ with $0 < a < b < 1$. The iteration step Φ_k $(k = 1, 2)$ from (11) corresponds to the solution in Ω_k, where for $k = 1$, the right boundary value x_{n_1+1} and for $k = 2$, the left boundary value x_{n-n_2} are used. Convergence with respect to the maximum norm can be verified as follows. Let x^0 and e^0 be the starting iterates and the initial error. The error $\bar{e}$ of $\bar{x} := \Phi_1(x^0, b)$ satisfies $(A\bar{e})_i = 0$ for all $1 \leqslant i \leqslant n_1$. For the given example, this yields the explicit error representation $\bar{e}_i = ie^0_{n_1+1}/(n_1 + 1)$ for $1 \leqslant i \leqslant n_1$ and $\bar{e}_i = e^0_i$ otherwise. Correspondingly, the second partial step $x^1 := \Phi_2(\bar{x}, b)$ yields $e^1_i = \bar{e}_i = ie^0_{n_1+1}/(n_1 + 1)$ for $i \leqslant n - n_2$ and

$$e^1_i = (n - i)\bar{e}_{n-n_2}/n_2 = (n - n_2)(n - i)/[n_2(n_1 + 1)]e^0_{n_1+1} \quad \text{for } i > n - n_2.$$

Exercise 11.2.10. Prove for the example discussed above that $\|e^1\|_\infty \leqslant \frac{a}{b}\|e^0\|_\infty$ and find an even more favourable convergence rate than $\frac{a}{b}$.

11.2.6 Approximate Solution of the Subproblems

In the case of blockwise Jacobi methods, one chooses the block structure (and thus the block-diagonal $D = \text{blockdiag}\{D_\varkappa\colon \varkappa \in J\}$) such that equations of the form $D_\varkappa y^\varkappa = c^\varkappa$ are easy to solve. Even if the additive Schwarz iteration can partially be interpreted as the block-Jacobi method, this does not mean that the subproblems (7): $A_\varkappa y^\varkappa = c^\varkappa$ are also easily solved. The reason is that, in general, (7) also discretises the same differential equation (in a subdomain).

If no direct solver for (7): $A_\varkappa y^\varkappa = c^\varkappa$ is available, this subproblem must be solved again by some iteration $\Phi^{(\varkappa)}$. Then a composed iteration arises, which differs from the one studied in §8.4 only by the fact that k subproblems are iteratively approximated during each outer iteration step. Denote the matrix of the third normal form of $\Phi^{(\varkappa)}$ by $W_\varkappa$. We only consider symmetric iterations with

$$\delta W_\varkappa \leqslant A_\varkappa := r_\varkappa A p_\varkappa \leqslant \Delta W_\varkappa \qquad (\varkappa \in J, 0 < \delta \leqslant \Delta). \tag{11.2.19}$$

Often, $\Delta < 2$ is required, i.e., $\Phi^{(\varkappa)}$ is convergent.

As in §8.4 we can construct the iteration $\Phi_{(m)}$ for solving $Ax = b$ by the additive or multiplicative Schwarz iteration, where the exact solution of the subproblems (7): $A_\varkappa y^\varkappa = c^\varkappa$ is replaced by $m > 0$ applications of $\Phi^{(\varkappa)}$. Since the m-fold (semi-)iteration $\Phi^{(\varkappa)m}$ can be regarded as one new iteration, formally, we may fix the number m of secondary iterations by $m = 1$ (the case $m > 1$ is mentioned in Exercise 3.7). Then the projection $P_\varkappa = p_\varkappa A_\varkappa^{-1} r_\varkappa A$ from

(8a) becomes

$$\Pi_\varkappa = p_\varkappa W_\varkappa^{-1} r_\varkappa A \quad (\varkappa \in J). \tag{11.2.20a}$$

$I - \Pi_\varkappa$ is the iteration matrix of the iteration $\Phi_\varkappa$ replacing (11):

$$\Phi_\varkappa(x, b) := x - p_\varkappa W_\varkappa^{-1} r_\varkappa(Ax - b) \qquad (\varkappa \in J). \tag{11.2.20b}$$

Note that $\Pi_\varkappa$ is no longer a projection. Some of its properties are gathered in

Remark 11.2.11. (a) The mapping $\Pi_\varkappa$ is A-adjoint, i.e., $\Pi_\varkappa^H = A\Pi_\varkappa A^{-1}$ and $\langle \Pi_\varkappa x, y\rangle_A = \langle x, \Pi_\varkappa y\rangle_A$ holds (cf. (8b)).
(b) Its spectrum satisfies $\sigma(\Pi_\varkappa) \subset \sigma(W_\varkappa^{-1} A_\varkappa) \cup \{0\}$, where the equality sign "$=$" holds instead of "$\subset$", if $n_\varkappa = \dim X_\varkappa < n = \dim X$.
(c) $\varrho(\Pi_\varkappa) \leqslant \Delta$ is equivalent to the last inequality in (19):

$$A_\varkappa \leqslant \Delta W_\varkappa \qquad (\varkappa \in J) \tag{11.2.21}$$

and gives rise to the later estimate (23) instead of (15b).

The iteration matrices of the multiplicative and additive Schwarz iteration with approximate subspace solvers (20b) are

$$M^{\mathrm{multSI}} = (I - \Pi_k)(I - \Pi_{k-1})\cdot \ldots \cdot (I - \Pi_1), \tag{11.2.22a}$$

$$M_\Theta^{\mathrm{addSI}} = I - \Theta\left(\sum_{\varkappa \in J} p_\varkappa W_\varkappa^{-1} r_\varkappa\right) A = I - \Theta \sum_{\varkappa \in J} \Pi_\varkappa. \tag{11.2.22b}$$

In the following, we assume always that approximate subspace solvers are applied. The case of exact subspace solutions considered in §11.2.3 can be regarded as the special choice $W_\varkappa = A_\varkappa$ leading to $\delta = \Delta = 1$ in (19).

Exercise 11.2.12. Assume (19) and prove the following. (a) The additive Schwarz iteration based on (20b) is a symmetric iteration.
(b) Define $N^{\mathrm{addSI}} = \sum_{\varkappa \in J} p_\varkappa W_\varkappa^{-1} r_\varkappa$ and $W^{\mathrm{addSI}} := (N^{\mathrm{addSI}})^{-1}$ according to (15a).
The estimate (15b) is now to be replaced by

$$A \leqslant k\Delta W^{\mathrm{addSI}}. \tag{11.2.23}$$

11.2.7 Strengthened Estimate $A \leqslant \Gamma W$

We say that two indices $\varkappa, \lambda \in J$ (or the respective subdomains) are *connected* if

$$\langle A p_\varkappa x^\varkappa, p_\lambda y^\lambda\rangle \neq 0 \quad \text{for suitable } x^\varkappa \in X_\varkappa,\ y^\lambda \in X_\lambda. \tag{11.2.24}$$

Exercise 11.2.13. Prove that if there are subsets $I_\varkappa, I_\lambda \subset I$, such that $p_\varkappa$ and p_λ satisfy (10a), then (24) holds if and only if the graph $G(A)$ contains an edge between some knots $\alpha \in I_\varkappa$ and $\beta \in I_\lambda$.

In Figs. 1.1a, b the subdomains indexed by 1 and 3 were not connected because of the sparsity of A. If, as in the block-tridiagonal case, the block-index i is connected only to $i^{\pm}1$, $J = \{1, 2, \ldots, k\}$ can be split into $J_1 = \{1, 3, \ldots\}$ and $J_2 = \{2, 4, \ldots\}$ and satisfies the assumptions of the following lemma with $K = 2$.

Lemma 11.2.14. *Let $A > 0$ and assume that J can be decomposed into K subsets $J_1, \ldots, J_K$, so that property* (24) *applies only to indices $\varkappa \neq \lambda$ from different sets J_i, J_j $(i \neq j)$. Then* (23) *holds in the strengthened form*

$$A \leqslant K\Delta W^{\text{addSI}}. \tag{11.2.25}$$

Proof. Write $N := N^{\text{addSI}}$ from (15a) as a sum $N_1 + \cdots + N_K$ with

$$N_i := \sum_{\varkappa \in J_i} p_\varkappa W_\varkappa^{-1} r_\varkappa \leqslant \Delta N_i', \quad \text{where } N_i' := \sum_{\varkappa \in J_i} p_\varkappa A_\varkappa^{-1} r_\varkappa.$$

By definition, (24) does not hold for indices $\varkappa$, $\lambda \in J_I$ with $\varkappa \neq \lambda$; hence, $\text{range}(p_\varkappa) \perp_A \text{range}(p_\lambda)$. This proves that N_i' is an A-orthogonal projection and therefore, as in the proof of Lemma 7, satisfies $N_i' \leqslant A^{-1}$. Summation of $N_i \leqslant \Delta N_i' \leqslant \Delta A^{-1}$ yields $N \leqslant K\Delta A^{-1}$, implying (25). $\square$

Another bound Γ in $A \leqslant \Gamma W^{\text{addSI}}$ can be derived from the matrix

$$E = (\varepsilon_{\varkappa\lambda})_{\varkappa, \lambda \in J} \in \mathbb{R}^{J \times J}, \tag{11.2.26a}$$

whose entries are the smallest bounds in

$$|\langle p_\varkappa x^\varkappa, p_\lambda y^\lambda \rangle_A| \leqslant \varepsilon_{\varkappa\lambda} \|x^\varkappa\|_W \|y^\lambda\|_W \quad \text{for all } x^\varkappa \in X_\varkappa,\ y^\lambda \in X_\lambda, \tag{11.2.26b}$$

where $\|\cdot\|_W$ are the following norms on $X_\varkappa$:

$$\|x^\varkappa\|_W := \|W_\varkappa^{1/2} x^\varkappa\|_2 = \langle W_\varkappa x^\varkappa, x^\varkappa \rangle^{1/2} \quad \text{for } x^\varkappa \in X_\varkappa. \tag{11.2.26c}$$

Lemma 11.2.15. (a) *For the case $W_\varkappa = A_\varkappa$ (exact subspace solution), the W-norm coincides with the energy norm: $\|x^\varkappa\|_W = \|p_\varkappa x^\varkappa\|_A$.*
(b) *Under assumption* (21), *$\|p_\varkappa x^\varkappa\|_A \leqslant \Delta \|x^\varkappa\|_W$ holds.*

Proof. Part (a) is shown by $\|x^\varkappa\|_W^2 = \langle W_\varkappa x^\varkappa, x^\varkappa \rangle = \langle A_\varkappa x^\varkappa, x^\varkappa \rangle = \langle p_\varkappa^H A p_\varkappa x^\varkappa, x^\varkappa \rangle = \langle A p_\varkappa x^\varkappa, p_\varkappa x^\varkappa \rangle = \langle p_\varkappa x^\varkappa, p_\varkappa x^\varkappa \rangle_A = \|p_\varkappa x^\varkappa\|_A^2$. $\square$

Because of Lemma 15a, we may call (26b) inequalities of the Cauchy-Schwarz type. If $\varepsilon_{\varkappa\lambda}$ is sufficiently small, (26b) is a strengthened Cauchy-Schwarz inequality (cf. Theorem 3.6c).

Remark 11.2.16. The Cauchy-Schwarz inequality implies $0 \leqslant \varepsilon_{\varkappa\lambda} \leqslant \Delta$. In the case $W_\varkappa = A_\varkappa$, $\varepsilon_{\varkappa\lambda} = 1$ holds if and only if the ranges of $p_\varkappa$ and p_λ have a nontrivial intersection. In particular, $\varepsilon_{\varkappa\varkappa} = 1$ holds for all $\varkappa \in J$. $\varepsilon_{\varkappa\lambda}$ vanishes if and only if the respective ranges are A-orthogonal.

Abbreviate $N := N^{\text{addSI}}$ and estimate NAx by

$$\begin{aligned}\|NAx\|_A^2 &= \left\langle \sum_\varkappa p_\varkappa W_\varkappa^{-1} r_\varkappa Ax, \sum_\lambda p_\lambda W_\lambda^{-1} r_\lambda Ax \right\rangle_A \\ &\leqslant \sum_{\varkappa,\lambda} \left| \langle p_\varkappa W_\varkappa^{-1} r_\varkappa Ax, p_\lambda W_\lambda^{-1} r_\lambda Ax \rangle_A \right| \\ &\leqslant \sum_{\varkappa,\lambda} \varepsilon_{\varkappa\lambda} \|W_\varkappa^{-1} r_\varkappa Ax\|_W \|W_\lambda^{-1} r_\lambda Ax\|_W \\ &\leqslant \varrho(E) \sum_\varkappa \|W_\varkappa^{-1} r_\varkappa Ax\|_W^2.\end{aligned}$$

$\sum_\varkappa \|W_\varkappa^{-1} r_\varkappa Ax\|^2 = \sum_\varkappa \langle r_\varkappa Ax, W_\varkappa^{-1} r_\varkappa Ax\rangle = \sum_\varkappa \langle Ax, p_\varkappa W_\varkappa^{-1} r_\varkappa Ax\rangle = \langle Ax, NAx\rangle = \langle x, NAx\rangle_A \leqslant \|x\|_A \|NAx\|_A$ yields $\|NAx\|_A^2 \leqslant \varrho(E)\|NAx\|_A \|x\|_A$; hence, $\|NAx\|_A \leqslant \varrho(E)\|x\|_A$. This inequality is equivalent to $\|A^{1/2}NA^{1/2}\|_2 \leqslant \varrho(E)$, $A^{1/2}NA^{1/2} \leqslant \varrho(E)I$, $N \leqslant \varrho(E)A^{-1}$, and finally $A \leqslant \varrho(E)W^{\text{addSI}}$. It proves

Theorem 11.2.17. *$A \leqslant \Gamma W^{\text{addSI}}$ holds with $\Gamma = \varrho(E)$ and E from* (26a, b):

$$A \leqslant \varrho(E) W^{\text{addSI}}. \tag{11.2.27}$$

The trivial estimate $\varrho(E) \leqslant \|E\|_\infty \leqslant \#J\Delta$ leads us back to (23).

11.3 Properties of the Additive Schwarz Iteration

11.3.1 Parallelism

Actual interest in the additive Schwarz iteration is due to its parallelism, which makes the method well-suited for parallel computers. Therefore, we consider the single steps of the algorithm in detail.

(i) After we compute the partitioned defect $d^\varkappa := p_\varkappa^H(Ax^m - b)$ for all $\varkappa \in J$, the steps

$$d^\varkappa \mapsto W_\varkappa^{-1} d^\varkappa = W_\varkappa^{-1} r_\varkappa (Ax^m - b) \mapsto \delta x_\varkappa := p_\varkappa W_\varkappa^{-1} d^\varkappa = p_\varkappa W_\varkappa^{-1} r_\varkappa (Ax^m - b)$$

are completely independent of each other and can be computed by different processors without any communication.

(ii) Even the correction step $x^{m+1} := x^m - \Theta \sum_{\varkappa \in J} \delta x_\varkappa$ ($\delta x_\varkappa$ as defined above) can be executed in parallel if (2.9a) and (2.10a) hold. In this case, one uses the processors of index $\varkappa \in J$ for storing the block of index $\varkappa \in J$. Then the correction simplifies to $(x^{m+1})^\varkappa = (x^m)^\varkappa - \Theta(\delta x_\varkappa)^\varkappa$ and therefore requires only local quantities. In the overlapping case (2.9b), an additional communication is necessary.

(iii) If according to (ii), the blocks $\{(x^m)^\varkappa : \varkappa \in J\}$ are distributed over the local storage of the processors, the computation of $d^\varkappa := r_\varkappa(Ax^m - b)$ requires communication with all processors of indices $\lambda \in J$ that are connected to $\varkappa$ (cf. property (2.24)).

11.3.2 Condition Estimates

A general assumption for the following considerations is $A > 0$. Let the matrix $W = W^{\text{addSI}} = \Theta W_\Theta^{\text{addSI}}$ be defined as in Lemma 2.7b. As known from (2.10.9) or Lemma 7.3.11, the condition number $\sigma(W^{-1}A)$ is the ratio Γ/γ of the optimal bounds in

$$\gamma W^{\text{addSI}} \leqslant A \leqslant \Gamma W^{\text{addSI}}. \tag{11.3.1}$$

Note that the condition number is independent of the choice of Θ. In (2.23), (2.25), and (2.27) we have determined $\Gamma = k\Delta$, $K\Delta$, and $\varrho(E)$, respectively, as upper bounds. The following theorem enables us to arrive at an explicit description of a lower bound γ.

Theorem 11.3.1 (Widlund [3]). *Assume that $A > 0$ and $W_\varkappa = A_\varkappa$. Let C be a constant such that for each $x \in X$, a representation $x = \sum_{\varkappa \in J} p_\varkappa x^\varkappa$ $(x^\varkappa \in X_\varkappa)$ exists with*

$$\sum_{\varkappa \in J} \langle A p_\varkappa x^\varkappa, p_\varkappa x^\varkappa \rangle \leqslant C \langle Ax, x \rangle. \tag{11.3.2}$$

Then the first inequality $\gamma W^{\text{addSI}} \leqslant A$ in (1) holds with $\gamma = 1/C$.

Often, Theorem 1 is referred to as the «Lemma of P. Lions», although his contribution in Glowinski–Golub–Meurant–Périaux [1] contains this statement only indirectly. However, it can be found in the thesis of Nepomnyashich [1] from 1986. In case (2.9a) («disjoint subdomains»), the decomposition $x = \sum p_\varkappa x^\varkappa$ is unique (cf. Remark 2.3); in the opposite case (2.9b) («overlapping subdomains»), one can choose an appropriate representation $x = \sum p_\varkappa x^\varkappa$ for (2) from infinitely many ones.

The proof of Theorem 1 can be omitted, since it is the special case $W_\varkappa = A_\varkappa$ of the later Theorem 4 (cf. Lemma 2.15a).

The estimate from Theorem 1 is strict, as shown by the following result of Xu [2].

Corollary 11.3.2. *If C is the smallest possible constant in (2), then $\gamma = 1/C$ is the largest constant in $\gamma W^{\text{addSI}} \leqslant A$.*

Proof. Given x, define $y := A^{-1} W^{\text{addSI}} x$. We may choose the decomposition $x^\varkappa := A_\varkappa^{-1} r_\varkappa A y \in X^\varkappa$, because $\sum p_\varkappa x^\varkappa = \sum p_\varkappa A_\varkappa^{-1} r_\varkappa A y = N^{\text{addSI}} A y = x$.

$$\begin{aligned}
\sum \langle A p_\varkappa x^\varkappa, p_\varkappa x^\varkappa \rangle &\underset{(2.8b)}{=} \sum \langle A P_\varkappa y, P_\varkappa y \rangle \\
&\underset{(2.8c)}{=} \sum \langle A P_\varkappa^2 y, y \rangle \\
&= \left\langle A \sum P_\varkappa y, y \right\rangle \\
&= \langle A N^{\text{addSI}} A y, y \rangle
\end{aligned}$$

$$= \langle N^{\text{addSI}} Ay, Ay \rangle$$
$$= \langle x, W^{\text{addSI}} x \rangle$$
$$\underset{A \geqslant \gamma W}{\leqslant} \frac{1}{\gamma} \langle x, Ax \rangle$$

proves (2) with $C = 1/\gamma$. □

A corresponding result due to Björstad–Mandel [1] exists for the reverse direction.

Remark 11.3.3. Let $A > 0$ and $W_\varkappa = A_\varkappa$. If for all $x \in X$ and *all* decompositions $x = \sum_{\varkappa \in J} p_\varkappa x^\varkappa$ $(x^\varkappa \in X_\varkappa)$, the inequality $\sum_{\varkappa \in J} \langle A p_\varkappa x^\varkappa, p_\varkappa x^\varkappa \rangle \geqslant C \langle Ax, x \rangle$ holds with $C > 0$, (1) is satisfied with $\Gamma = 1/C$.

Inequality (2) can also be written as $\sum \|p_\varkappa x^\varkappa\|_A^2 \leqslant C \|x\|_A^2$. Replacing the energy norm $\|\cdot\|_A$ by the W norm introduced in (2.26c), we obtain a generalisation of Theorem 1 to $W_\varkappa \neq A_\varkappa$.

Theorem 11.3.4. *Assume $A > 0$ and $W_\varkappa > 0$ $(\varkappa \in J)$. Let C' be a constant such that for any $x \in X$, a decomposition $x = \sum_{\varkappa \in J} p_\varkappa x^\varkappa$ $(x^\varkappa \in X_\varkappa)$ exists with*

$$\sum_{\varkappa \in J} \|x^\varkappa\|_W^2 \leqslant C' \|x\|_A^2. \tag{11.3.3}$$

Then the first inequality $\gamma W^{\text{addSI}} \leqslant A$ in (1) holds with $\gamma = 1/C'$.

Proof. Squaring the inequality

$$\begin{aligned}
\|x\|_A^2 &= \langle x, x \rangle_A \\
&= \left\langle x, \sum p_\varkappa x^\varkappa \right\rangle_A \\
&= \sum \langle Ax, p_\varkappa x^\varkappa \rangle \\
&= \sum \langle r_\varkappa Ax, x^\varkappa \rangle \\
&= \sum \langle W_\varkappa^{-1/2} r_\varkappa Ax, W_\varkappa^{1/2} x^\varkappa \rangle \\
&\leqslant \sum \|W_\varkappa^{-1/2} r_\varkappa Ax\|_2 \|W_\varkappa^{1/2} x^\varkappa\|_2 \\
&\leqslant \left(\sum \|W_\varkappa^{-1/2} r_\varkappa Ax\|_2^2 \right)^{1/2} \left(\sum \|W_\varkappa^{1/2} x^\varkappa\|_2^2 \right)^{1/2} \\
&= \left(\sum \|W_\varkappa^{-1/2} r_\varkappa Ax\|_2^2 \right)^{1/2} \left(\sum \|x^\varkappa\|_W^2 \right)^{1/2} \underset{(3)}{\leqslant} \\
&= \left(\sum \|W_\varkappa^{-1/2} r_\varkappa Ax\|_2^2 \right)^{1/2} (C' \|x\|_A^2)^{1/2}
\end{aligned}$$

and cancelling the factor $\|x\|_A^2$ yields $\langle Ax, x\rangle \leqslant C' \sum \|W_\varkappa^{-1/2} r_\varkappa A x\|_2^2$. Since $\sum \|W_\varkappa^{-1/2} r_\varkappa A x\|_2^2 = \sum \langle A(p_\varkappa W_\varkappa^{-1} r_\varkappa) A x, x\rangle = \langle ANAx, x\rangle$, we arrive at the inequality

$$A \leqslant C'ANA \qquad (N = N^{\text{addSI}}),$$

which is equivalent to $A^{-1} \leqslant C'N = C'(W^{\text{addSI}})^{-1}$ and $A \geqslant \frac{1}{C'} W^{\text{addSI}}$. □

If inequality (2) can be verified more easily than (3), the use of Theorem 4 can be avoided as shown in

Exercise 11.3.5. Assume that (2) is valid with the constant C. Prove that (3) holds with $C' := C/\delta$, where δ is the lower bound in (2.9): $\delta W_\varkappa \leqslant A_\varkappa$.

This would be desirable if the bounds γ and Γ from (1) were h-independent. Even if the number k of subdomains is independent of h, k might be a large number (depending on the number of available parallel processors), so that the k-independence of γ and Γ also seems to be desirable. Therefore, the bound $\Gamma = k$ from (2.15b) is not optimal. However, Lemma 2.14 already yields a criterion for $\Gamma = K$, where K does not depend on the number k of the subdomains, but only on the degree of their mutual connectivity. Moreover, Theorem 2.17 may help, if $\varrho(E)$ is independent of the parameters. An h- or k-independent lower bound γ can be obtained from Theorem 1 [Theorem 4], if the constant C [C'] used there is h- or k-independent.

11.3.3 Convergence Statements

The additive Schwarz iteration $\Phi_\Theta^{\text{addSI}}$ yields a convergent iteration, provided that suitable damping is applied. According to (4.4.5), the optimal damping factor is $\Theta = 2/(\gamma + \Gamma)$ with γ, Γ from (1). This leads us to the contraction number $\varrho(M_\Theta^{\text{addSI}}) = \|M_\Theta^{\text{addSI}}\|_A \leqslant (\Gamma - \gamma)/(\Gamma + \gamma)$. The same rate holds for the gradient method with $\Phi_\Theta^{\text{addSI}}$ as the basic iteration. The best convergence rate $(\sqrt{\Gamma} - \sqrt{\gamma})/(\sqrt{\Gamma} + \sqrt{\gamma})$ is given by the cg method applied to $\Phi_\Theta^{\text{addSI}}$. In any case, a small ratio Γ/γ is favourable. In the latter cases, the value of Θ does not matter. The choice $\Theta = 1$ leads to $W_\Theta^{\text{addSI}} = W^{\text{addSI}}$ and $N_\Theta^{\text{addSI}} = N^{\text{addSI}}$.

A simply analysed situation is the case of two disjoint subdomains (the weakly 2-cyclic case).

Theorem 11.3.6. *Assume* $A > 0$ *and* (2.9a) *with* $k = 2$ *(2 disjoint domains).*
(a) *Then the optimal bounds* γ, Γ *in* (1) *have the form*

$$\gamma = 1 - \delta, \qquad \Gamma = 1 + \delta \quad \text{with} \tag{11.3.4a}$$

$$\delta := \|A_1^{-1/2} p_1^H A p_2 A_2^{-1/2}\|_2 < 1 \qquad (A_\varkappa \text{ from } (2.5)). \tag{11.3.4b}$$

(b) $\Theta = 1$ *is the optimal damping factor of the additive Schwarz iteration and yields the convergence rate* $\varrho(M_1^{\mathrm{addSI}}) = \|M_1^{\mathrm{addSI}}\|_A = \delta$. *The cg method applied to* $\Phi_\Theta^{\mathrm{addSI}}$ *has the asymptotical rate* $\delta/(1 + \sqrt{1 - \delta^2})$.
(c) *The number* δ *from* (4b) *is also the best bound in the* strengthened Cauchy-Schwarz inequality

$$|\langle x, y\rangle_A| \leqslant \delta \|x\|_A \|y\|_A \quad \text{for all } x \in \mathrm{range}(p_1),\ y \in \mathrm{range}(p_2). \qquad (11.3.4c)$$

Proof. (i) Inserting $x = p_1 x^1$ and $y = p_2 x^2$ in (4c) with $x^\varkappa \in X_\varkappa$ and exploiting $\|p_\varkappa x^\varkappa\|_A = \langle A p_\varkappa x^\varkappa, p_\varkappa x^\varkappa\rangle^{1/2} = \langle A_\varkappa x^\varkappa, x^\varkappa\rangle^{1/2} = \|A_\varkappa^{1/2} x^\varkappa\|_2$, one arrives at the coincidence of the optimal δ from (4c) with the norm (4b).
(ii) The coincidence of the additive Schwarz iteration with the block-Jacobi iteration applied to $\hat{A}\hat{x} = \hat{b}$ ($\hat{A}$ from (2.16c)) leads to the equivalence of inequality (1) to $\gamma D \leqslant \hat{A} \leqslant \Gamma D$ with $D := \mathrm{blockdiag}\{A_1, A_2\}$. The latter can be rewritten as $(\gamma - 1)D \leqslant \hat{A} - D \leqslant (\Gamma - 1)D$. Lemma 5.2.1 can be applied to $B := D^{-1/2}(\hat{A} - D)D^{-1/2}$ and yields $\lambda_{\max}(B) = \delta$, $\lambda_{\min}(B) = -\delta$. The remaining statements follow from $\gamma - 1 = -\delta$ and $\Gamma - 1 = \delta$. □

The constant δ from (4c) coincides with ε_{12} from (2.26a, b).

A corresponding analysis for $k = 2$ *overlapping* subdomains is given by Björstad–Mandel [1]. Different from the nonoverlapping case in Theorem 6, $\Gamma = 2$ cannot be improved. For the proof, one chooses $0 \neq x \in \mathrm{range}(p_1) \cap \mathrm{range}(p_2)$, which is possible because of (2.9b). $p_\varkappa A_\varkappa^{-1} r_\varkappa A x = x$ for $\varkappa = 1, 2$ leads to $NAx = 2x$ or $Ax = 2Wx$ and implies $\Gamma \geqslant 2$. On the other hand, (2.15b) ensures that $\Gamma \leqslant k = 2$. Since all eigenvectors $x \in \mathrm{range}(p_1) \cap \mathrm{range}(p_2)$ are associated with the eigenvalue 2, this hardly troubles the cg method, but it influences Φ^{addSI} as well as the gradient method based on Φ^{addSI}.

Let $\Phi_{(m)}$ be the additive Schwarz iteration combined with an m-fold application of the solver $\Phi^{(\varkappa)}$ for the subproblem $A_\varkappa y^\varkappa = c^\varkappa$. Above we have analysed the case $m = 1$. The case $m \geqslant 1$ is discussed in

Exercise 11.3.7. For any $m \geqslant 1$, $\Phi_{(m)}$ is again a symmetric iteration. If (2.1.9) holds with $\Delta < 2$, $\Phi_{(m)}$ converges monotonically with respect to the energy norm. The corresponding matrix $N_{(m)}$ of the second normal form has the representation

$$N_{(m)} = \sum_{\varkappa \in J} p_\varkappa (I - (I - W_\varkappa^{-1} A_\varkappa)^m) A_\varkappa^{-1} r_\varkappa. \qquad (11.3.5a)$$

For *even* m it satisfies the estimates

$$\gamma(1 - \max\{(1 - \delta)^m, (1 - \Delta)^m\}) A^{-1} \leqslant N_{(m)} \leqslant \Gamma A^{-1}. \qquad (11.3.5b)$$

(5b) implies (5c) for the matrix $W_{(m)}$ of the third normal form:

$$\gamma(1 - \max\{(1 - \delta)^m, (1 - \Delta)^m\}) W_{(m)} \leqslant A \leqslant \Gamma W_{(m)}. \qquad (11.3.5c)$$

Concerning the corresponding statement for odd M, compare §8.4.3. Instead of the iterative solution of $A_\varkappa y^\varkappa = c^\varkappa$, one could also take into consideration a semi-iterative treatment.

According to Exercise 7, the condition of the additive iteration deteriorates from Γ/γ for the exact solution of the subspace problems to $\Gamma/(\gamma[1 - \max\{(1-\delta)^m, (1-\Delta)^m\}])$ for the approximate one. Concerning a suitable choice of m, the considerations from §8.4.4 are to be repeated.

11.4 Analysis of the Multiplicative Schwarz Iteration

11.4.1 Convergence Statements

The case of $k = 2$ nonoverlapping domains is very easy to analyse.

Exercise 11.4.1. Prove that under the assumptions of Theorem 3.6, the *multiplicative* Schwarz iteration has the convergence rate δ^2. *Hint*: Use Remark 5.5.2.

The following general analysis for $k > 1$ is based on the presentation of Xu [2] (cf. also Bramble–Pasciak–Wang–Xu [1]) and private communications with H. Yserentant. From the beginning we consider the case that the subproblems are solved approximately using W_i $(i \in J = \{1, \ldots, k\})$ (cf. §11.2.6). The choice $A_i = W_i$ corresponds to the exact solution.

The convergence will be based on two inequalities similar to (3.3) and (2.26b). As discussed in Remark 2.16, the constants $\varepsilon_{\varkappa\lambda}$ in (2.26b) are smaller the smaller the subspaces $X_\varkappa$ are. Therefore, we introduce the set $\{Y_j: 1 \leqslant j \leqslant k\}$ of subspaces with the following properties:

$$Y_j \subset X_j, \qquad \sum_{j=1}^{k} p_j Y_j = X, \tag{11.4.1}$$

where $p_j Y_j = \{p_j y^j : y^j \in Y_j\}$ is the range of p_j restricted to Y_j. The second property ensures that every $x \in X$ has a representation $x = \sum p_j y^j$ with $y^j \in Y_j$. In the overlapping case (2.9b) one may, e.g., choose Y_j such that $\{Y_j: 1 \leqslant j \leqslant k\}$ is nonoverlapping.

Inequality (3.3) is now required in the following form. There is a bound C_1 such that for each $x \in X$ we have

$$\sum_{j=1}^{k} \|y^j\|_W^2 \leqslant C_1 \|x\|_A^2 \quad \text{for a suitable decomposition } x = \sum p_j y^j \text{ with } y^j \in Y_j. \tag{11.4.2}$$

If $Y_j = X_j$, (2) and condition (3.3) in Theorem 3.4 are identical; otherwise, condition (2) is stronger, since there may be fewer decompositions $x = \sum p_j y^j$ with $y^j \in Y_j$ than $x = \sum p_j x^j$ with $x^j \in X_j$ as in (3.3).

Besides (2), we need estimates similar to strengthened Cauchy inequalities. Let ε_{ij}^{XY} be the smallest numbers with

$$|\langle p_i x^i, p_j y^j \rangle_A| \leqslant \varepsilon_{ij}^{XY} \|x^i\|_W \|y^j\|_W \quad \text{for all } x^i \in X_i,\ y^j \in Y_j, \text{ and } i < j. \tag{11.4.3a}$$

For $i \geqslant j$, we define $\varepsilon_{ij}^{XY} := 0$. If $X_j = Y_j$ and $i < j$, (3a) is identical with (2.26b), i.e., $\varepsilon_{ij}^{XY} = \varepsilon_{ij}$; otherwise, $\varepsilon_{ij}^{XY} \leqslant \varepsilon_{ij}$ may become a strict inequality. We form the strictly upper triangular matrix

$$E_{XY} := (\varepsilon_{ij}^{XY})_{i,j=1,\ldots,k} \tag{11.4.3b}$$

and denote its spectral norm by

$$C_2 := \|E_{XY}\|_2. \tag{11.4.3c}$$

An estimate of this $k \times k$ matrix is given by

Exercise 11.4.2. Prove that $\varepsilon_{ij}^{XY} \leqslant \Delta$ with Δ from (2.21) for all $i < j$ and $C_2 \leqslant \Delta\sqrt{k(k-1)/2} < k\Delta$. *Hint*: $\|E_{XY}\|_2^2 \leqslant \|E_{XY}^T E_{XY}\|_\infty$.

The first convergence result is

Theorem 11.4.3. *Assume* (2.21): $A_j \leqslant \Delta W_j$ *with* $\Delta < 2$. *Let* C_1 *and* C_2 *be the numbers defined in* (2) *and* (3c). *Then the multiplicative Schwarz iteration converges monotonically with respect to the energy norm. The contraction number can be estimated by*

$$\|M^{\text{multSI}}\|_A \leqslant \sqrt{1 - (2-\Delta)/[C_1(1+C_2)^2]}. \tag{11.4.4}$$

Inserting the bound from Exercise 3, one obtains the k-dependent rate $\|M^{\text{multSI}}\|_A \leqslant 1 - O(k^{-2})$. If, however, the bounds C_1, C_2 are k-independent, the convergence is also. If the subspace problems are solved exactly ($A_j = W_j$), the factor $2 - \Delta$ in (4) becomes 1.

The second convergence statement replaces the estimates (3a–c) by

$$\|x\|_A^2 \leqslant C_3 \sum_{j=2}^{k} \frac{1}{j-1} \|y^j\|_W^2 \quad \text{for any } x = \sum_{j=2}^{k} p_j y^j \text{ with } y^j \in Y_j. \tag{11.4.5}$$

The relation to (3a–c) can be seen from the sufficient condition stated in

Lemma 11.4.4. *Let* δ_{ij} $(1 \leqslant i,j \leqslant k)$ *be the smallest numbers satisfying*

$$\langle p_i y^i, p_j y^j \rangle_A \leqslant \delta_{ij} \|y^i\|_W \|y^j\|_W \quad \textit{for all } y^i \in Y_i,\ y^j \in Y_j. \tag{11.4.6a}$$

Define the symmetric $k \times k$ *matrix*

$$E_{YY} := (\sqrt{i-1}\sqrt{j-1}\,\delta_{ij})_{i,j=1,\ldots,k}. \tag{11.4.6b}$$

Then estimate (5) follows with $C_3 := \varrho(E_{YY})$.

Proof. The norm of $x = \sum_{j=2}^{k} p_j y^j$ can be estimated by

$$\begin{aligned}
\|x\|_A^2 &= \sum_{i,j=2}^{k} \langle p_i y^i, p_j y^j \rangle_A \\
&\leqslant \sum_{i,j=2}^{k} \delta_{ij} \|y^i\|_W \|y^j\|_W \\
&= \sum_{i,j=2}^{k} (E_{YY})_{ij} [(i-1)^{-1/2} \|y^i\|_W][(j-1)^{-1/2} \|y^j\|_W] \\
&\leqslant \varrho(E_{YY}) \left[\sum_{j=2}^{k} (j-1)^{-1} \|y^i\|_W^2 \right].
\end{aligned}$$

□

Note that $\delta_{ij} \leqslant \varepsilon_{ij}^{XY}$ with ε_{ij}^{XY} from (3a) for $i < j$ and $A_i = W_i$, since both arguments y^i and y^j are from the possibly smaller subspaces. The weights $\sqrt{i-1}$ in (6b) demonstrate that the ordering of the substeps in the multiplicative approach is essential. Unfortunately, C_3 is not k-independent as demonstrated in part (b) of

Exercise 11.4.5. (a) Prove that (6a) is always true for the choice $\delta_{ij} = \Delta$, which leads to $C_3 = \Delta k(k-1)/2$. Furthermore, one may use the estimate $C_3 \leqslant (k-1)\varrho((\delta_{ij})_{i,j=1,\dots,k})$.
(b) Let $W_i = A_i$ and prove that $C_3 \geqslant k - 1$.

The convergence result involving inequality (5) is the following.

Theorem 11.4.6. *Assume* (2.21): $A_j \leqslant \Delta W_j$ *with* $\Delta < 2$. *Let* C_1 *and* C_3 *be the numbers defined in* (2) *and* (5). *Then the multiplicative Schwarz iteration converges monotonically with respect to the energy norm. The contraction number can be estimated by*

$$\|M^{\text{multSI}}\|_A \leqslant \sqrt{1 - (2-\Delta)/[C_1(1 + \sqrt{\Delta C_3})^2]}. \tag{11.4.7}$$

It may happen that C_2 from (3c) is larger than $O(1)$ because of the first k_0 rows in the matrix E_{XY}. Then

$$C_{2,k_0} := \varrho(E_{XY,k_0}), \qquad E_{XY,k_0} := (\varepsilon_{ij}^{XY})_{i,j=k_0,\dots,k} \tag{11.4.8}$$

might be smaller than $C_2 = C_{2,0}$. The following result is due to Dryja–Widlund [3]:

Corollary 11.4.7. *Assume* (2.21): $A_j \leqslant \Delta W_j$ *with* $\Delta < 2$. *Let* C_1 *be defined by* (2) *and* C_{2,k_0} by (8) *for some* $k_0 \in \{1, \dots, k\}$. *Then*

$$\|M^{\text{multSI}}\|_A \leqslant \sqrt{1 - (2-\Delta) \Big/ \left\{ C_1 \left[1 + \sqrt{C_{2,k_0}^2 + \Delta k_0 \left(\frac{2}{C_1} + \Delta(k_0+1) \right)} \right]^2 \right.}.$$

11.4.2 Proofs of the Convergence Theorems

The products

$$M_i := (I - \Pi_i)(I - \Pi_{i-1})\cdot \ldots \cdot (I - \Pi_1) \qquad (1 \leqslant i \leqslant k) \qquad (11.4.9a)$$

with $\Pi_i := p_i W_i^{-1} p_i^H A$ from (2.20a) are the iteration matrices corresponding to the products $\Phi_i \circ \ldots \circ \Phi_1$ of the first i substeps. As usual, the empty product is

$$M_0 := I. \qquad (11.4.9b)$$

According to (2.22a), we have $M^{\text{multSI}} = M_k$. Summing the identities

$$M_{i-1} - M_i = \Pi_i M_{i-1} \qquad (1 \leqslant i \leqslant k), \qquad (11.4.10a)$$

we obtain

$$I - M_i = \sum_{j=1}^{i} \Pi_j M_{j-1} \qquad (1 \leqslant i \leqslant k). \qquad (11.4.10b)$$

Lemma 11.4.8. $\|\Pi_j x\|_A^2 \leqslant \Delta\langle \Pi_j x, x\rangle_A$ *with* Δ *from* (2.21) *holds for all* $x \in X$ *and* $j \in J$.

Proof. $A_j \leqslant \Delta W_j$ and

$$\begin{aligned}\|\Pi_j x\|_A^2 &= \langle A p_j W_j^{-1} r_j A x, p_j W_j^{-1} r_j A x\rangle \\ &= \langle A p_j W_j^{-1} r_j A p_j W_j^{-1} r_j A x, x\rangle \\ &= \langle A p_j W_j^{-1} A_j W_j^{-1} r_j A x, x\rangle\end{aligned}$$

show $\|\Pi_j x\|_A^2 \leqslant \Delta\langle A p_j W_j^{-1} r_j A x, x\rangle = \Delta\langle \Pi_j x, x\rangle_A$. □

Lemma 11.4.9. *For all* $x \in X$ *we have*

$$\|x\|_A^2 - \|M^{\text{multSI}} x\|_A^2 \geqslant (2 - \Delta) \sum_{i=1}^{k} \langle \Pi_i M_{i-1} x, M_{i-1} x\rangle_A. \qquad (11.4.11)$$

Proof. (10a) and Lemma 8 show

$$\begin{aligned}\|M_{i-1}x\|_A^2 - \|M_i x\|_A^2 &= \|M_{i-1}x\|_A^2 - \|(M_{i-1} - \Pi_i M_{i-1})x\|_A^2 \\ &= 2\langle M_{i-1}x, \Pi_i M_{i-1}x\rangle_A - \|\Pi_i M_{i-1}x\|_A^2 \\ &\geqslant (2 - \Delta)\langle \Pi_i M_{i-1}x, \Pi_i M_{i-1}x\rangle_A.\end{aligned}$$

Summation over $1 \leqslant i \leqslant k$ yields (11), since $M_0 = I$ and $M_k = M^{\text{multSI}}$. □

Lemma 11.4.10. *Let* C_1 *be the bound from* (5). *Then, for all* x, $x_j \in X$, *the decomposition* $x = \sum p_j y^j$ *from* (2) *with* $y^j \in Y_j$ *is such that*

$$\sum_{j=1}^{k} \langle p_j y^j, x_j\rangle_A \leqslant \sqrt{C_1}\,\|x\|_A \sqrt{\sum_{j=1}^{k} \langle \Pi_j x_j, x_j\rangle_A}. \qquad (11.4.12)$$

Proof. Choose the decomposition according to (2). Summation of

$$\begin{aligned}\langle p_j y^j, x_j\rangle_A &= \langle y^j, r_j A x_j\rangle \\ &= \langle W_j^{1/2} y^j, W_j^{-1/2} r_j A x_j\rangle \\ &\leqslant \|y^j\|_W \|W_j^{-1/2} r_j A x_j\|_2\end{aligned}$$

together with $\|W_j^{-1/2} r_j A x_j\|_2^2 = \langle A p_j W_j^{-1} r_j A x_j, x_j\rangle = \langle \Pi_j x_j, x_j\rangle_A$ yields $\sum\langle p_j y^j, x_j\rangle_A \leqslant \sum \|y^j\|_W \langle \Pi_j x_j, x_j\rangle_A^{1/2} \leqslant [\sum \|y^j\|_W^2]^{1/2} [\sum \langle \Pi_j x_j, x_j\rangle_A]^{1/2}$. Applying (2) to the first square root, we arrive at (12). □

Proof of Theorem 4. For a given $x \in X$, we choose a decomposition $x = \sum p_j y^j$ with $y^j \in Y_j$ satisfying the estimate (2). We write $\|x\|_A^2$ as

$$\begin{aligned}\|x\|_A^2 &= \sum_j \langle x, p_j y^j\rangle_A \\ &= \sum_j \langle M_{j-1} x, p_j y^j\rangle_A + \sum_j \langle (I - M_{j-1}) x, p_j y^j\rangle_A. \qquad (11.4.13a)\end{aligned}$$

To estimate the first term on the right-hand side, apply Lemma 10 with $x_j := M_{j-1} x \in X$:

$$\sum_j \langle M_{j-1} x, p_j y^j\rangle_A \leqslant \sqrt{C_1}\,\|x\|_A \sqrt{\sum_{j=1}^{k} \langle \Pi_j M_{j-1} x, M_{j-1} x\rangle_A}. \qquad (11.4.13b)$$

For the second term in (13a) use (10b):

$$\langle (I - M_{j-1}) x, p_j y^j\rangle_A = \sum_{i=1}^{j-1} \langle \Pi_i M_{i-1} x, p_j y^j\rangle_A.$$

Since $\Pi_i M_{i-1} x = p_i W_i^{-1} r_i A M_{i-1} x \in p_i X_i$, (3a) can be applied and yields

$$\begin{aligned}\sum_{j=1}^{k} \langle (I - M_{j-1}) x, p_j y^j\rangle_A &= \sum\sum_{1 \leqslant i < j \leqslant k} \langle \Pi_i M_{i-1} x, p_j y^j\rangle_A \\ &\leqslant \sum\sum_{1 \leqslant i < j \leqslant k} \varepsilon_{ij}^{XY} \|W_i^{-1} r_i A M_{i-1} x\|_W \|y^j\|_W \\ &= \langle E_{XY}\alpha, \beta\rangle, \qquad (11.4.13c)\end{aligned}$$

where the vectors $\alpha, \beta \in \mathbb{R}^k$ have the components $\alpha_i = \|W_i^{-1} r_i A M_{i-1} x\|_W$ and $\beta_j = \|y^j\|_W$. From $\langle E_{XY}\alpha, \beta\rangle \leqslant \|E_{XY}\|_2 \|\alpha\|_2 \|\beta\|_2 = C_2 \|\alpha\|_2 \|\beta\|_2$ and

$$\begin{aligned}\|\alpha\|_2^2 &= \sum_i \|W_i^{-1} r_i A M_{i-1} x\|_W^2 \\ &= \sum_i \langle r_i A M_{i-1} x, W_i^{-1} r_i A M_{i-1} x\rangle \\ &= \sum_i \langle A p_i W_i^{-1} r_i A M_{i-1} x, M_{i-1} x\rangle \\ &= \sum_i \langle \Pi_i M_{i-1} x, M_{i-1} x\rangle_A,\end{aligned}$$

we conclude that

$$\sum_{j=1}^{k} \langle (I - M_{j-1}) x, p_j y^j\rangle_A \leqslant C_2 \sqrt{\sum_j \|y^j\|_W^2} \sqrt{\sum_i \langle \Pi_i M_{i-1} x, M_{i-1} x\rangle_A}.$$

Inequality (2) proves that

$$\sum_{j=1}^{k} \langle (I - M_{j-1})x, p_j y^j \rangle_A \leqslant C_2 \sqrt{C_1} \|x\|_A \sqrt{\sum_i \langle \Pi_i M_{i-1} x, M_{i-1} x \rangle_A}. \tag{11.4.13d}$$

Together, (13a, b, d) yield

$$\|x\|_A^2 \leqslant \sqrt{C_1}(1 + C_2)\|x\|_A \sqrt{\sum_i \langle \Pi_i M_{i-1} x, M_{i-1} x \rangle_A}.$$

Therefore, $\sum_i \langle \Pi_i M_{i-1} x, M_{i-1} x \rangle_A$ is bounded from below by

$$\sum_i \langle \Pi_i M_{i-1} x, M_{i-1} x \rangle_A \geqslant \|x\|_A^2 / [C_1(1 + C_2)^2].$$

This estimate and Lemma 9 show that

$$\begin{aligned} \|M^{\mathrm{multSI}} x\|_A^2 &\leqslant \|x\|_A^2 - (2 - \Delta) \sum_i \langle \Pi_i M_{i-1} x, M_{i-1} x \rangle_A \\ &\leqslant \|x\|_A^2 - (2 - \Delta)\|x\|_A^2 / [C_1(1 + C_2)^2]. \end{aligned}$$

Since $x \in X$ is arbitrary, the last inequality implies the estimate (4) and ends the proof of Theorem 4. □

Proof of Theorem 6. The proof differs only with respect to the estimation of the left-hand side of (13c). Using (10b), we conclude as follows:

$$\begin{aligned} \sum_{j=1}^{k} \langle (I - M_{j-1})x, p_j y^j \rangle_A &= \sum\sum_{1 \leqslant i < j \leqslant k} \langle \Pi_i M_{i-1} x, p_j y^j \rangle_A \\ &= \sum_{i=1}^{k-1} \sum_{j=i+1}^{k} \langle \Pi_i M_{i-1} x, p_j y^j \rangle_A \\ &= \sum_{i=1}^{k-1} \left\langle \Pi_i M_{i-1} x, \sum_{j=i+1}^{k} p_j y^j \right\rangle_A \\ &\leqslant \sum_{i=1}^{k-1} \|\Pi_i M_{i-1} x\|_A \left\| \sum_{j=i+1}^{k} p_j y^j \right\|_A \\ &\leqslant \sqrt{\sum_{i=1}^{k-1} \|\Pi_i M_{i-1} x\|_A^2} \sqrt{\sum_{i=1}^{k-1} \left\| \sum_{j=i+1}^{k} p_j y^j \right\|_A^2} \end{aligned}$$

Applying (5) to $x = \sum_{j=i+1}^{k} p_j y^j$ (i.e., $y^2 = y^3 = \cdots = y^i = 0$), we conclude

$$\begin{aligned} \sum_{i=1}^{k-1} \left\| \sum_{j=i+1}^{k} p_j y^j \right\|_A^2 &\leqslant \sum_{i=1}^{k-1} C_3 \sum_{j=i+1}^{k} \frac{1}{j-1} \|y^j\|_W^2 \\ &= C_3 \sum_{j=2}^{k} \sum_{i=1}^{j-1} \frac{1}{j-1} \|y^j\|_W^2 \\ &= C_3 \sum_{j=2}^{k} \|y^j\|_W^2. \end{aligned}$$

Lemma 8 shows $\|\Pi_i M_{i-1}x\|_A^2 \leqslant \Delta\langle\Pi_i M_{i-1}x, M_{i-1}x\rangle_A$. Together, we obtain

$$\sum_{j=1}^{k} \langle (I - M_{j-1})x, p_j y^j\rangle_A \leqslant \sqrt{\Delta C_3}\sqrt{\sum_{j=2}^{k} \|y^j\|_W^2}\sqrt{\sum_{i=1}^{k-1} \langle\Pi_i M_{i-1}x, M_{i-1}x\rangle_A}$$

$$\underset{(2)}{\leqslant} \sqrt{\Delta C_1 C_3}\|x\|_A \sqrt{\sum_{i=1}^{k-1} \langle\Pi_i M_{i-1}x, M_{i-1}x\rangle_A}.$$

A comparison with (13d) shows that C_2 from (13d) is replaced by $\sqrt{\Delta C_3}$. Substituting C_2 in (4) by $\sqrt{\Delta C_3}$, we have shown (7). □

Proof of Corollary 7. The sum in (13c) over $1 \leqslant i < j \leqslant k$ can be split into a first part with $i \leqslant k_0$ and a second with $k_0 < i < j \leqslant k$. The estimate of the latter sum is obtained by replacing the lower index bound 1 by $k_0 + 1$:

$$\sum_I := \sum_{j=k_0+1}^{k} \langle (I - M_{j-1})x, p_j y^j\rangle_A$$

$$\leqslant C_{2,k_0}\sqrt{C_1}\|x\|_A \left\{\sum_{j=k_0+1}^{k} \langle\Pi_i M_{i-1}x, M_{i-1}x\rangle_A\right\}^{1/2}.$$

The first sum with $i \leqslant k_0$ reads

$$\sum_{II} = \sum_{j=1}^{k} \sum_{i=1}^{\min(j-1,k_0)} \langle\Pi_i M_{i-1}x, p_j y^j\rangle_A$$

$$= \sum_{i=1}^{k_0} \left\langle \Pi_i M_{i-1}x, \sum_{j=i+1}^{k} p_j y^j \right\rangle_A$$

$$= \sum_{i=1}^{k_0} \left\langle \Pi_i M_{i-1}x, x - \sum_{j=1}^{i} p_j y^j \right\rangle_A.$$

Applying the Cauchy-Schwarz inequality and using $\|p_j y^j\|_A \leqslant \sqrt{\Delta}\|y^j\|_W$, (2), and Lemma 8, we arrive at

$$\sum_{II}^{2} \leqslant \Delta k_0(2 + \Delta C_1(k_0 + 1))\|x\|_A^2 \sum_{i=1}^{k_0} \langle\Pi_i M_{i-1}x, M_{i-1}x\rangle_A.$$

Following the lines of the proof of Theorem 4, we can show the statement of Corollary 7. □

11.5 Examples

11.5.1 Schwarz Methods with Proper Domain Decomposition

Let the square $\Omega \subset (0,1) \times (0,1)$ be partitioned as in Fig. 1 into disjoint smaller squares $\Omega_\varkappa$ ($\varkappa \in J$) of the side length H, so that $\overline{\Omega} = \bigcup \overline{\Omega}_\varkappa$. In order to obtain overlapping subdomains, which are typical for the Schwarz iteration,

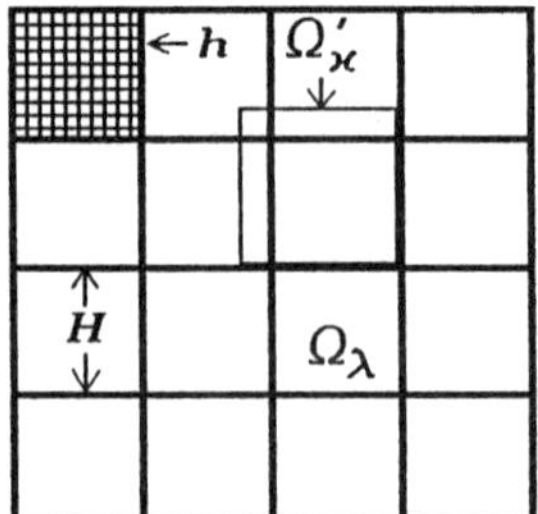

Fig. 11.5.1 $\Omega_\varkappa, \Omega'_\varkappa$

we enlarge $\Omega_\varkappa$ to the square $\Omega'_\varkappa$ by elongating the side in the left and upper direction by the factor αH (cf. Fig. 1). Close to the boundary, $\Omega'_\varkappa$ is to be restricted to Ω. The grid Ω_h of the Poisson-model problem with step size h has to be compatible with the domain decomposition: i.e., H/h an $\alpha H/h$ must be integers.

The entire index set is $I = \Omega_h$. The subsets $I_\varkappa$ are $I_\varkappa := I \cap \Omega'_\varkappa$. Vectors $x^\varkappa \in X_\varkappa$ can be regarded as grid functions on $\Omega'_\varkappa$. For $p_\varkappa$ we make the trivial choice (2.10a). The matrices $A_\varkappa = p_\varkappa^H A p_\varkappa$ are again the matrices of the Poisson-model case (but on a smaller square).

The arising multiplicative variant of the Schwarz iteration is the classical Schwarz method, only the number of subdomains $\Omega'_\varkappa$ generalises from 2 to H^{-2}. For the convergence analysis of the additive variant, one has to determine the quantities γ, Γ from (3.1). Lemma 2.14 can be applied as follows. Decompose the set of squares $\Omega'_\varkappa$ analogously to four-colour numbering (4.2.12) into four classes $J_1, \ldots, J_4$. Two squares $\Omega'_\varkappa, \Omega'_\lambda$ with $\varkappa \neq \lambda$, $\varkappa \in J_i$, $\lambda \in J_j$, $i \neq j$ have the distance $(1-\alpha)H > h$, if $\alpha < 1 - H/h$. Therefore, $\varkappa$ and λ are not connected (cf. (2.24)); hence, Lemma 2.14 proves $\Gamma = 4$.

The bound γ does prove to be h- but not H-independent: $\gamma = O(H^2)$ (cf. Widlund [2]). The reason can be understood by means of Theorem 3.1. A smooth grid functon like $x = \sin(\xi\pi)\sin(\eta\pi)$ $((\xi,\eta) \in \Omega)$ cannot be decomposed into $x = \sum p_\varkappa x^\varkappa$ with smooth subgrid functions $p_\varkappa x^\varkappa$, since the supports of $p_\varkappa x^\varkappa$ are overlapping and the sum $\sum p_\varkappa x^\varkappa$ would be nonsmooth. However, for smooth x and nonsmooth $p_\varkappa x^\varkappa$, the product $\langle Ax, x\rangle$ would be small compared with $\langle Ap_\varkappa x^\varkappa, p_\varkappa x^\varkappa\rangle$, so that $C = 1/\gamma$ becomes large in (3.2).

The deterioration of condition Γ/γ for decreasing H can be recognised as unavoidable by taking the limit $H = h$. Then the square $\Omega'_\varkappa$ (being open) contains exactly one grid point. Therefore, the additive Schwarz iteration is the classical *pointwise* (eventually damped) Jacobi method, for which $\Gamma/\gamma = O(h^{-2}) = O(H^{-2})$ holds. The same convergence order holds for the multiplicative variant (proper Schwarz iteration), which coincides with the classical *pointwise* Gauß-Seidel iteration.

11.5.2 Additive Schwarz Iteration with Coarse-Grid Correction

To overcome the unfavourable convergence results of §11.5.1, a coarse-grid correction is added (cf. Dryja [4], Dryja–Widlund [1,2]). The set $J = \{1,\dots,H^{-2}\}$ from §11.5.1 associated with the subsquares $\Omega'_\varkappa$ ($\varkappa \in J$) is enlarged by the index 0. The new index set $I_0 = \Omega_H$ consists of all (interior) grid points corresponding to the coarser step size H. The prolongation p_0: $X_0 = \mathbb{K}^{I_0} \to X = \mathbb{K}^I$ belonging to $0 \in J$ is defined differently from (2.10a). It is sufficient to explain the application of p_0 to unit vectors. Let $e_{\xi,\eta} \in X_0$ be the vector with the value 1 at the grid point $(\xi,\eta) \in \Omega_H = I_0$ and 0 elsewhere. A first approach to p_0 is the piecewise bilinear interpolation in $I = \Omega_h$:

$$(p_0 e_{\xi,\eta})(x,y) = (1 - |\xi - x|/H)(1 - |\eta - y|/H) \quad \text{for } (x,y) \in \Omega_h \text{ with } |\xi - x|, |\eta - y| < H, \tag{11.5.1a}$$

$$(p_0 e_{\xi,\eta})(x,y) = 0 \quad \text{otherwise.} \tag{11.5.1b}$$

By adding $0 \in J$, according to Lemma 2.14, the previous bound $\Gamma = 4$ can increase at most to $\Gamma = 5$. However, concerning γ, now better estimates can be expected, since instead of the grid function x, only the remainder $x - p_0 x^0$ of the interpolation [we choose $x^0(\xi,\eta) := x(\xi,\eta)$ for $(\xi,\eta) \in \Omega_H$] has to be represented by $\sum p_\varkappa x^\varkappa$. The condition number Γ/γ proves to be $O(1 + \log(H/h))$. It becomes h- and H-independent, if p_0 describes the orthogonal projection (with respect to $\|\cdot\|_2$) (cf. Dryja–Widlund [1,2], Dryja [4]).

A similar idea with nonoverlapping subdomains $\Omega_\varkappa = \Omega'_\varkappa$ is also the basis of the method of Bramble–Pasciak–Schatz [2].

11.5.3 Formulation in the Case of a Galerkin Discretisation

Let the boundary value problem be described in the form (10.6.13): «$v \in V$, $a(v,w) = (f,w)_U$ for all $w \in V$» and discretised by (10.6.14b):

$$v_h \in V_h, \qquad a(v_h, w) = (f,w)_U \quad \text{for all } w \in V_h, \tag{11.5.2}$$

where V_h is a finite dimensional subspace of V. The bijective relation $v_h = P_h x = \sum_{\alpha\in I} x_\alpha b_\alpha$ (b_α: basis functions of V_h; cf. (10.6.15)) connects the functions $v_h \in V_h$ and the coefficient vectors $x \in X$.

The sub*domain* $\Omega_\varkappa \subset \Omega$ corresponds to the sub*space*

$$V_{h,\varkappa} := \{v_h \in V_h \colon v_h(\xi) = 0 \text{ for } \xi \in \Omega\backslash\Omega_\varkappa\}, \tag{11.5.3}$$

i.e., $v_h \in V_{h,\varkappa}$ has its support in $\overline{\Omega}_\varkappa$. For the usual finite elements each component x_α of $x \in X$ represents the function value of v_h in a corresponding «nodal point» $\xi_\alpha \in \Omega$ (cf. Hackbusch [15]). Choose $I_\varkappa := \{\alpha \in I \colon \xi_\alpha \in \Omega_\varkappa\}$. For a suitable choice of the subdomain $\Omega_\varkappa$, this definition coincides with $I_\varkappa := \{\alpha \in I$:

support$(b_\alpha) \subset \overline{\Omega}_\varkappa\}$. Then we have

$$V_{h,\varkappa} = \{P_h x \colon x \in \operatorname{range}(p_\varkappa)\} = \{P_h p_\varkappa x^\varkappa \colon x^\varkappa \in X_\varkappa\}, \quad \text{where } X_\varkappa := \mathbb{K}^{I_\varkappa}. \tag{11.5.4}$$

Since the matrix A is defined by $\langle Ax, y\rangle = a(P_h x, P_h y)$ $(x, y \in X)$, the following representations are equivalent:

$$\begin{aligned}
\langle A p_\varkappa x^\varkappa, p_\lambda x^\lambda\rangle &= a(P_h p_\varkappa x^\varkappa, P_h p_\lambda x^\lambda) && \text{with } x^\varkappa \in X_\varkappa \\
&= a(P_h x^{(\varkappa)}, P_h p_\lambda x^{(\lambda)}) && \text{with } x^{(\varkappa)} := p_\varkappa x^\varkappa \in \operatorname{range}(p_\varkappa) \\
&= a(v_h^{(\varkappa)}, v_h^{(\lambda)}) && \text{with } v_h^{(\varkappa)} := P_h x^{(\varkappa)} \in V_{h,\varkappa}.
\end{aligned} \tag{11.5.5}$$

$P_h p_\varkappa \colon X_\varkappa \to V_{h,\varkappa}$ is the bijective mapping that allows us to transfer formulations in the vector spaces $X_\varkappa$ into those in the Galerkin subspaces $V_{h,\varkappa}$ and vice versa. Relation (5) allows us, e.g., to formulate the strengthened Cauchy–Schwarz inequality (3.4c) in the following equivalent form:

$$|a(v_h, w_h)| \leqslant \delta \sqrt{a(v_h, v_h) a(w_h, w_h)} \quad \text{for all } \begin{cases} v_h \in V_{h,\varkappa}, \\ w_h \in V_{h,\lambda}. \end{cases} \tag{11.5.6}$$

In general, $a(v, w)$ is an integral $\int_\Omega \dots dx$ over the domain Ω (cf. §10.6.3.2 and Hackbusch [15]; occasionally, additional boundary integrals may also be involved). Let $a_\tau(v, w) = \int_\tau \dots dx$ be the respective integral over $\tau \subset \Omega$. For a disjoint decomposition of $\Omega = \bigcup \tau_i$ into subsets τ_i, one obtains

$$\sum_i a_{\tau_i}(v, w) = a(v, w) \quad \text{for all } v, w \in V. \tag{11.5.7}$$

The following simple lemma is an important tool for many proofs, since it requires the demonstration of (6) only over the subsets τ_i (e.g., over the triangles of a finite element triangulation) instead of Ω.

Lemma 11.5.1. *If for all i, inequality* (6) *holds with* a_{τ_i} *instead of a, then* (6) *is also satisfied for a with the same constant* δ.

Proof. The Schwarz inequality $(\sum \alpha_i \beta_i)^2 \leqslant \sum \alpha_i^2 \sum \beta_i^2$ yields

$$\begin{aligned}
|a(v_h, w_h)| &= \left| \sum a_{\tau_i}(v_h, w_h) \right| \\
&\leqslant \sum |a_{\tau_i}(v_h, w_h)| \\
&\leqslant \delta \sum |a_{\tau_i}(v_h, v_h) a_{\tau_i}(w_h, w_h)|^{1/2} \\
&\leqslant \delta \left(\sum a_{\tau_i}(v_h, v_h) \right)^{1/2} \left(\sum a_{\tau_i}(w_h, w_h) \right)^{1/2} \\
&= \delta \sqrt{a(v_h, v_h) a(w_h, w_h)}.
\end{aligned}$$

□

11.6 Multi-Grid Methods as Subspace Decomposition Method

The variant described in §11.5.2 has already shown its similarity to multi-grid methods. On the other hand, multi-grid variants have been described and analysed with respect to convergence in such a way that they can immediately be interpreted as a multiplicative Schwarz iteration.

11.6.1 The Analysis of Braess

The first proof of the monotone two-grid convergence $\|M_\ell^{\mathrm{TGM}}\|_A \leqslant \zeta < 1$ without need of regularity assumptions was due to Braess [1] (cf. §10.6.3.5, §10.6.6). Approaches to this proof can be found in Bank–Dupont [2]. Different from the previous choice, we define the coarse grid of the Poisson-model problem as a grid rotated by 45° with step size $h_{\ell-1} = \sqrt{2}h_\ell$. Figure 1 shows the grid for a more general domain than the unit square. View the grid Ω_ℓ as a triangular grid. It is well-known that finite-element discretisation with linear triangular elements leads again to the five-point star (1.2.4a). The coarse-grid equation is the finite-element discretisation corresponding to the thick-lined triangles in Fig. 1. The *canonical prolongation* p of the multi-grid method is the linear interpolation along the hypotenuses of the coarse-grid triangles.

As *smoothing*, we choose the chequer-board Gauß-Seidel iteration $\mathscr{S}_\ell$. The «red» points from Ω_h coincide with the coarse-grid points: $\Omega_\ell^r = \Omega_{\ell-1}$, while $\Omega_\ell^b = \Omega_\ell \backslash \Omega_{\ell-1}$ contains the «black» points. Correspondingly, $\mathscr{S}_\ell$ is the product $\mathscr{S}_\ell^b \circ \mathscr{S}_\ell^r$ of one Gauß-Seidel half-step on the red points, followed by a half-step on the black ones. For the following considerations, it suffices to reduce the smoothing step to the half Gauß-Seidel iteration $\mathscr{S}_\ell^b$. The two-grid method then reads as $\Phi_\ell^{\mathrm{TGM}} := \Phi_\ell^{\mathrm{CGC}} \circ \mathscr{S}_\ell^b$, where Φ_ℓ^{CGC} denotes the coarse-grid correction.

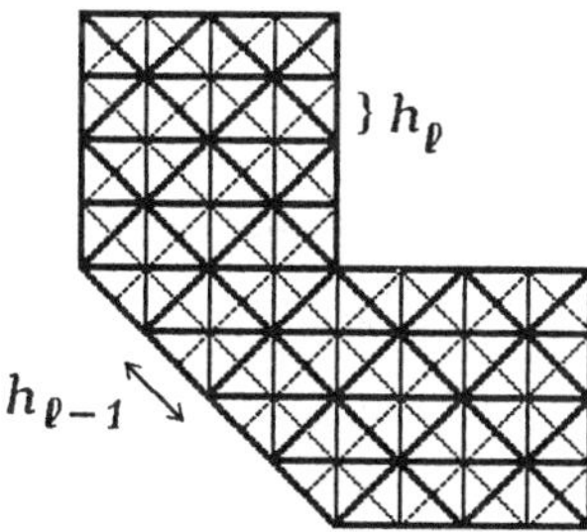

Fig. 11.6.1 Coarse and fine grid

For analysis, we need the subspaces

$$V_{\ell,1} := \{v \in V_\ell : v_\ell(\xi,\eta) = 0 \text{ for all } (\xi,\eta) \in \Omega_\ell^r = \Omega_{\ell-1}\}, \tag{11.6.1a}$$

$$V_{\ell,2} := V_{\ell-1}, \tag{11.6.1b}$$

where V_ℓ is the (finite-element) space of the continuous functions being linear over all triangles of the grids Ω_ℓ. Analogously, functions from $V_{\ell-1} \subset V_\ell$ are linear on the larger triangles of $\Omega_{\ell-1}$. All $v \in V_\ell$ satisfy $v(\xi,\eta) = 0$ for boundary points $(\xi,\eta) \in \partial\Omega$.

Decompose the complete index set $I = \Omega_\ell$ into

$$I_1 := \Omega_\ell \backslash \Omega_{\ell-1}, \qquad I_2 := \Omega_{\ell-1}. \tag{11.6.2}$$

The vector spaces

$$X_{\ell,1} := \mathbb{K}^{I_1}, \qquad X_{\ell,2} := \mathbb{K}^{I_2} \tag{11.6.3}$$

correspond to I_1 and I_2. The second one coincides with the vector space denoted in (10.1.9) by $X_{\ell-1}$. The prolongations (in the sense of the domain decomposition method) are chosen as

$$p_1 \colon X_{\ell,1} \to X_\ell \quad \text{according to (2.10a)}, \tag{11.6.4a}$$

$$p_2 = p \colon X_{\ell,2} = X_{\ell-1} \to X_\ell \quad \text{is the canonical prolongation.} \tag{11.6.4b}$$

Exercise 11.6.1. (a) Prove identity (5) for $P = P_\ell$ from (10.6.15):

$$V_{\ell,\varkappa} = \operatorname{range}(Pp_\varkappa) \quad \text{for } \varkappa = 1, 2. \tag{11.6.5}$$

(b) Let $A = A_\ell$ be an arbitrary five-point formula. Prove that the chequerboard Gauß-Seidel half-steps $\mathcal{S}_\ell^b$ and $\mathcal{S}_\ell^r$ are projections. If $A > 0$, $\mathcal{S}_\ell^b$ and $\mathcal{S}_\ell^r$ are symmetric iterations (cf. (5.8.1/2)).

Lemma 11.6.2. *Let* $A > 0$. Φ_ℓ^{CGC} *and* $\mathcal{S}_\ell^b$ *are A-orthogonal projections onto* $\operatorname{range}(p_\varkappa) \subset X$. (2.3) *holds*: $\operatorname{range}(p_1) + \operatorname{range}(p_2) = X_\ell$ *and* (2.9a): $n_1 + n_2 = n$ *for the dimensions* $n_\varkappa := \dim(\operatorname{range}(p_\varkappa)) = \# I_\varkappa$, $n := \dim X_\ell$.

Proof. $\mathcal{S}_\ell^b$ and Φ_ℓ^{CGC} are projections, as can be concluded from Exercise 1b and Lemma 10.1.6 (the assumption (10.1.26) is satisfied for Galerkin discretisations by Exercise 10.6.15b). By Exercise 1b and Lemma 10.7.1 with $v = 0$, $\mathcal{S}_\ell^b$ and Φ_ℓ^{CGC} are symmetric. From Exercise 2.2c we conclude that $\mathcal{S}_\ell^b$ and Φ_ℓ^{CGC} are A-orthogonal projections. □

The results of Lemma 2 and the identity $\Phi_\ell^{\mathrm{TGM}} := \Phi_\ell^{\mathrm{CGC}} \circ \mathcal{S}_\ell^b$ prove

Remark 11.6.3. The two-grid method Φ_ℓ^{TGM} described above is the multiplicative Schwarz iteration characterised by the prolongations (4a, b). It corresponds to the case (2.9a) of two disjoint domains.

For convergence analysis, Theorem 3.6 is applicable. The quantity δ of the strengthened Cauchy–Schwarz inequality (3.4c) can be determined by Lemma 5.1. For this purpose, the triangles of the grid $\Omega_{\ell-1}$ are used as subsets τ_i. $v \in V_{\ell,1}$ is a linear function on τ_i, whereas $w \in V_{\ell,2}$ is piecewise linear on the smaller triangles of the grid Ω_ℓ and vanishes at all corners of τ_i. The estimation of the bilinear form $a_{\tau_i}(v,w) = \int_{\tau_i} \langle \nabla v, \nabla w \rangle \, dx$ yields the bound $\delta[a\tau_j(v,v)a_{\tau_i}(w,w)]^{1/2}$ with the constant $\delta = 1/\sqrt{2}$. Lemma 5.1 proves

Theorem 11.6.4. *The two-grid method Φ_ℓ^{TGM} described above converges monotonically with respect to the energy norm with the contraction number $\|M_\ell^{\mathrm{TGM}}\|_A \leqslant \frac{1}{2}$. The same bound holds for the case of several smoothing steps with $\mathscr{L}_\ell = \mathscr{S}_\ell^b \circ \mathscr{S}_\ell^r$.*

Proof of second part. The power $\mathscr{S}_\ell^\nu$ has the form $\mathscr{S}_\ell^b \circ (\mathscr{S}_\ell^r \circ \cdots \circ \mathscr{S}_\ell^b \circ \mathscr{S}_\ell^r)$. $\|M_\ell^{\mathrm{TGM}}(\nu,0)\|_A \leqslant \|M_\ell^{\mathrm{TGM}}\|_A \|S_\ell^r\|_A \cdot \ldots \cdot \|S_\ell^b\|_A \|S_\ell^r\|_A = \|M_\ell^{\mathrm{TGM}}\|_A \leqslant \frac{1}{2}$ follows, since $\|\mathscr{S}_\ell^r\|_A = \|\mathscr{S}_\ell^b\|_A = 1$ holds for A-orthogonal projections. □

The given two-grid convergence proof requires no regularity assumption. Often, it is viewed as an advantage when convergence can be shown for multi-gridlike methods without regularity requirements. On the other hand, one sacrifices a possible increase of efficiency that can be achieved by means of more smoothing steps. The variant $\Phi_\ell^{\mathrm{CGC}} \circ \mathscr{S}_\ell^b$ discussed here is a typical example. As explained in Braess [2], an improved form of the Cauchy–Schwarz inequality (6) (thanks to an implicit regularity assumption!) leads to quantitative convergence statements for $\Phi_\ell^{\mathrm{CGC}} \circ (\mathscr{S}_\ell^b \circ \mathscr{S}_\ell^r)^\nu$, demonstrating that the half-smoothing step $\mathscr{S}_\ell^b$ is not optimal with respect to efficiency.

11.6.2 V-Cycle Interpreted as Multiplicative Schwarz Iteration

Let ℓ be the maximal level, for which $A = A_\ell$ and $X = X_\ell$ are identified. In the following, we study the V-cycle $\Phi_\ell^{\mathrm{V}}(\nu, 0)$ with ν pre- and no post-smoothing (cf. §10.7.5). The spaces X_i $(0 \leqslant i \leqslant \ell)$ of dimension n_i introduced in (10.1.9) for the multi-grid method are also taken as subspaces for the domain decomposition. The index set $J = J_\ell$ is $J = \{0, 1, \ldots, \ell\}$ so that $k = \ell + 1$ is the number of subspaces.

Let p: $X_{i-1} \to X_i$ be the multi-grid prolongation (10.1.10). To indicate the levels involved, we call this mapping $p_{i,i-1}$. Their products define

$$p_{i,j} := p_{i,i-1} \cdot \ldots \cdot p_{j+1,j} \quad \text{for } 0 \leqslant j < i \leqslant \ell \text{ and } p_{i,i} = I \text{ for } i = j. \qquad (11.6.6\text{a})$$

The prolongations needed for domain decomposition are defined as

$$p_i := p_{\ell,i}\colon X_i \to X = X_\ell \qquad (0 \leqslant i \leqslant \ell). \qquad (11.6.6\text{b})$$

In contrast to the previous examples, the ranges of p_i are not disjoint or partially overlapping, but monotonically increasing: $\mathrm{range}(p_0) \subset \mathrm{range}(p_1) \subset \cdots \subset \mathrm{range}(p_\ell) = X$.

Let the coarse-grid matrices be defined by the Galerkin product (10.1.26). Multiple application of the identity (10.1.26) yields

$$A_i = p_i^H A_\ell p_i \qquad (0 \leqslant i \leqslant \ell; A = A_\ell) \tag{11.6.7}$$

according to (2.5).

Wc introduce the auxiliary iteration Ψ_i on $X = X_\ell$ that corresponds to the solution of the ith subproblem by one V-cycle step: $\Psi_i(x, b) = x - p_i W_i^{-1} p_i^H(Ax - b)$. Here, the matrix W_i^{-1} corresponds to the V-cycle $\Phi_i^{\mathrm{V}}(v, 0)$. Using $M_i^{\mathrm{V}} = I - W_i^{-1} A_i$, we obtain the representation

$$\Psi_i(x, b) = x - p_i(I - M_i^{\mathrm{V}}) A_i^{-1} p_i^H(A_\ell x - b). \tag{11.6.8a}$$

For $i = \ell$ we regain the V-cycle at the level ℓ because of $p_i = I$:

$$\Psi_\ell = \Phi_\ell^{\mathrm{V}}(v, 0). \tag{11.6.8b}$$

The following presentation simplifies if we do not solve exactly at level $i = 0$ but apply the v-fold pre-smoothing: $M_0^{\mathrm{V}} = S_0^v$. This leads us to

$$\Psi_0(x, b) = x - p_0(I - S_0^v) A_0^{-1} p_0^H(A_\ell x - b). \tag{11.6.8c}$$

Essential for interpreting the V-cycle as a multiplicative Schwarz iteration is the following

Lemma 11.6.5. *Assume* (6a, b), (7), *and* $M_0^{\mathrm{V}} = S_0^v$. *Then*

$$\Phi_\ell^{\mathrm{V}}(v, 0) = \tilde{\Phi}_0 \circ \tilde{\Phi}_1 \circ \cdots \circ \tilde{\Phi}_\ell, \quad \textit{where} \tag{11.6.9a}$$

$$\tilde{\Phi}_i(x, b) := x - p_i(I - S_i^v) A_i^{-1} p_i^H(A_\ell x - b) \tag{11.6.9b}$$

represents the approximative solution of the ith subproblem $A_i x^i = c^i := p_i^H(A_\ell x - b)$ *by* v *smoothing steps.*

Proof. Because of (8b, c), it is sufficient to prove

$$\Psi_i = \Psi_{i-1} \circ \tilde{\Phi}_i \qquad (1 \leqslant i \leqslant \ell). \tag{11.6.9c}$$

The iteration matrix of Ψ_i equals

$$M_{\Psi,i} = I - p_i(I - M_i^{\mathrm{V}}) A_i^{-1} p_i^H A_\ell = I - p_i A_i^{-1} p_i^H A_\ell + p_i M_i^{\mathrm{V}} A_i^{-1} p_i^H A_\ell.$$

In the recursion formula (4.13b): $M_i^{\mathrm{V}} = [I - p(I - M_{i-1}^{\mathrm{V}}) A_{i-1}^{-1} r A_i] S_i^v$, we now have to write $p = p_{i,i-1}$ and $r = p_{i,i-1}^H$. Its insertion into $M_{\Psi,i}$ yields

$$M_{\Psi,i} = I - p_i A_i^{-1} p_i^H A_\ell + p_i[I - p(I - M_{i-1}^{\mathrm{V}}) A_{i-1}^{-1} r A_i S_i^v] A_i^{-1} p_i^H A_\ell.$$

Noting $p_i p = p_{\ell,i} p_{i,i-1} = p_{i-1}$ and $r A_i = p_{i,i-1}^H A_i = p_{i,i-1}^H p_i^H A_\ell p_i = p_{i-1}^H A_\ell p_i$, we may write the last term as

$$[I - p_{i-1}(I - M_{i-1}^{\mathrm{V}}) A_{i-1}^{-1} p_{i-1}^H A_\ell] p_i S_i^v A_i^{-1} p_i^H A_\ell = M_{\Psi,i-1} p_i S_i^v A_i^{-1} p_i^H A_\ell.$$

The projection $I - P_i = I - p_i A_i^{-1} p_i^H A_\ell$ (cf. (2.8a)) satisfies $p_{i-1}^H A_\ell (I - P_i) = 0$, so that $I - P_i = M_{\Psi,i-1}(I - P_i)$. This enables us to formulate the representation

$$
\begin{aligned}
M_{\Psi,i} &= I - P_i + M_{\Psi,i-1} p_i S_i^\nu A_i^{-1} p_i^H A_\ell \\
&= M_{\Psi,i-1}(I - P_i + p_i S_i^\nu A_i^{-1} p_i^H A_\ell) \\
&= M_{\Psi,i-1}(I - p_i A_i^{-1} p_i^H A_\ell + p_i S_i^\nu A_i^{-1} p_i^H A_\ell) \\
&= M_{\Psi,i-1}[I - p_i(I - S_i^\nu) A_i^{-1} p_i^H A_\ell] \\
&= M_{\Psi,i-1} M_{\tilde{\Phi},i},
\end{aligned}
$$

where $M_{\tilde{\Phi},i} = I - p_i(I - S_i^\nu) A_i^{-1} p_i^H A_\ell$ is the iteration matrix of $\tilde{\Phi}_i$. Hence, the product form (9c) is proved. □

11.6.3 Proof of the V-Cycle Convergence

The interpretation of the V-cycle as a Schwarz iteration is less interesting for the purpose of algorithmic performance than for convergence analysis. We are investigating convergence by means of Theorem 4.3. First we discuss estimates of W_i.

In the case of the model problem, we know that $\|A_i\|_2 \leqslant C h_i^{-2}$ holds for the uniform grid size h_i at level i. Assuming the coarsest mesh to be fixed, we have $\|A_i\|_2 \leqslant C 4^i$ if $h_i = h_0/2^i$ (cf. (10.1.7b)). The same bound holds for a general finite element discretisation provided that in the refinement process from level $i - 1$ to i the element size is halved at most. Without loss of generality, we may assume that only one step of the smoothing procedure is performed (otherwise, redefine S_i^ν by S_i); however, S_i must be symmetric and convergent. The latter property implies (2.21): $A_i \leqslant \Delta W_i$ with $\Delta < 2$ for the matrix W_i of the third normal form of the smoothing iteration. The upper bound of W_i should be of the same order as the upper bound of A_i discussed above:

$$W_i \leqslant 4^i C_W p_i^H p_i. \tag{11.6.10}$$

To obtain the constants C_1 and C_2 from (4.2) and (4.3c), one has to select a suitable decomposition $x = \sum p_i x^i$ with $x^i \in Y_i$ for appropriate subspaces $Y_i \subset X_i$. The orthogonal projection (with respect to the Euclidean scalar product) onto range(p_i) is

$$Q_i := p_i (p_i^H p_i)^{-1} p_i^H \qquad (0 \leqslant i \leqslant \ell). \tag{11.6.11a}$$

Since $p_\ell = I$, we also have $Q_\ell = I$. Following Bramble–Pasciak–Xu [1], we decompose x into

$$x = Q_\ell x = (Q_\ell - Q_{\ell-1})x + (Q_{\ell-1} - Q_{\ell-2})x + \cdots + (Q_1 - Q_0)x + Q_0 x.$$

Note that $(Q_i - Q_{i-1})x \in \mathrm{range}(p_i)$. Introducing $Q_{-1} := 0$, we write

$$x = \sum_{i=0}^{\ell} p_i x^i \quad \text{with } p_i x^i := (Q_i - Q_{i-1})x, \tag{11.6.11b}$$

i.e., $x^i = (p_i^H p_i)^{-1} p_i^H (Q_i - Q_{i-1})x$. These x^i belong to the subspaces

$$Y_i := \mathrm{range}\{(p_i^H p_i)^{-1} p_i^H (Q_i - Q_{i-1})\} = \{x^i \in X_i \colon Q_{i-1} p_i x^i = 0\}. \tag{11.6.11c}$$

Oswald [1] proves that the energy norm $\|\cdot\|_A$ is equivalent to the norm defined by

$$|||x|||^2 = \|Q_0 x\|_A^2 + \sum_{i=0}^{\ell} 4^i \|(Q_i - Q_{i-1})x\|_2^2. \tag{11.6.12a}$$

A compact proof can be found in Bornemann–Yserentant [1]. Applying this statement in particular to $x = p_i x^i$ with $x^i \in Y_i$, we obtain

$$4^i \|p_i x^i\|_2^2 \leqslant C_E \|x\|_A^2 = C_E \|x^i\|_A^2, \tag{11.6.12b}$$

where C_E is the equivalence constant: $|||x|||^2 \leqslant C_E \|x\|_A^2$. Let $x = \sum p_i x^i$ with $x^i \in Y_i$. Then, inequality (10) implies that

$$\begin{aligned} \sum \|x^i\|_W^2 &= \sum \langle W_i x^i, x^i \rangle \\ &\leqslant C_W \sum 4^i \|p_i x^i\|_2^2 \\ &\leqslant C_W |||x|||^2 \\ &\leqslant C_E C_W \|x\|_A^2. \end{aligned}$$

Hence, condition (4.2) holds with $C_1 = C_E C_W$.

As in Lemma 3.3 of Bornemann–Yserentant [1], one can prove the strengthened Cauchy–Schwarz inequality

$$|\langle x^i, x^j \rangle_A| \leqslant C 2^{(i-j)/2} \|x^i\|_A 2^j \|p_j x^j\|_2 \quad \text{for } j > i,\ x^i \in X_i,\ x^j \in X_j \tag{11.6.13a}$$

with respect to the subspaces X_i, X_j. Thanks to (12b), we conclude that

$$|\langle x^i, y^j \rangle_A| \leqslant C' 2^{(i-j)/2} \|x^i\|_A \|y^j\|_A \quad \text{for } j > i,\ x^i \in X_i,\ y^j \in Y_j \tag{11.6.13b}$$

with $C' := C C_E^{1/2}$. Since $\|\cdot\|_A \leqslant \Delta \|\cdot\|_W$ (cf. (2.21)), the estimates (4.3a) hold with $\varepsilon_{ij}^{XY} \leqslant 2^{(i-j)/2}$ for $j > i$. Therefore, the matrix E_{XY} from (3.4b) has the row-sum norm $\|E_{XY}\|_\infty < \sum_{\nu=1}^{\infty} 2^{-\nu/2} = 2 + \sqrt{2}$. Since the same estimate holds for E_{XY}^T, the constant from (3.4c) is bounded by

$$C_2 \leqslant C'(2 + \sqrt{2}) \quad \text{with } C' \text{ from (13b)} \tag{11.6.13c}$$

(cf. Exercise 2.9.6a). Both constants, C_1 and C_2, are independent of the dimension and the number of levels. Hence, Theorem 4.3 proves h-independent convergence of the V-cycle.

The V-cycle $\Phi_\ell^{\mathrm{V}}(\nu, 0)$ $(\nu > 0)$ is unsymmetric. Symmetry, however, holds for $\Phi_\ell^{\mathrm{V}}(\nu, \nu)$ (cf. Lemma 10.7.1). By Exercise 10.7.16a, this iteration can be interpreted as product $\Phi_\ell^{\mathrm{V}}(\nu, \nu) = \Phi_\ell^{\mathrm{V}}(0, \nu) \circ \Phi_\ell^{\mathrm{V}}(\nu, 0)$. Since $\Phi_\ell^{\mathrm{V}}(0, \nu)$ is the adjoint of $\Phi_\ell^{\mathrm{V}}(v,0)$ (cf. (10.7.3a)), we arrive at the product representation

$$\Phi_\ell^{\mathrm{V}}(v,v) = \tilde{\Phi}_\ell \circ \cdots \circ \tilde{\Phi}_1 \circ \tilde{\Phi}_0 \circ \tilde{\Phi}_0 \circ \tilde{\Phi}_1 \circ \cdots \circ \tilde{\Phi}_\ell$$

(symmetry of the $\tilde{\Phi}_i$ and thereby of the smoothing iteration is assumed), which corresponds to the symmetrisation described in Exercise 2.8. Hence, the symmetric V-cycle $\Phi_\ell^{\mathrm{V}}(v,v)$ can also be interpreted within the framework of multiplicative Schwarz iterations.

11.6.4 Method of the Hierarchical Basis

In §11.6.2 we used subspaces which overlap completely with the foregoing ones: $p_\ell V_\ell \supset p_{\ell-1} V_{\ell-1}$. The method of the hierarchical basis is a nonoverlapping subspace decomposition.

Figures 2a–c show a sequence of refining triangulations for a Galerkin discretisation with piecewise linear functions. Let V^h be the space of the piecewise linear functions on the grid Ω_h from Fig. 2c. Similarly, V^{2h} corresponds to Ω_{2h} and V^{4h} to Ω_{4h}. The inclusion $V^{4h} \subset V^{2h} \subset V^h$ holds. The Galerkin subspace V^h can be written as the sum of V^{2h} and

$$V_2 := \{v \in V^h \colon v = 0 \text{ in all nodal points of } \Omega_{2h}\}. \tag{11.6.14a}$$

Hence, $V^h = V_2 + V^{2h}$. The prescribed zeros of $v \in V_2$ are marked in Fig. 2c by «**o**». Correspondingly, $V^{2h} = V_1 + V^{4h}$ holds with

$$V_1 := \{v \in V^{2h} \colon v = 0 \text{ in all nodal points of } \Omega_{4h}\}. \tag{11.6.14c}$$

With the notation $V_0 := V^{4h}$, one obtains the decomposition

$$V^h = V_0 + V_1 + V_2. \tag{11.6.14c}$$

In all spaces $V_\varkappa$ $(0 \leqslant \varkappa \leqslant 2)$, one may choose the usual (nodal) basis functions (corresponding to the differently large triangles). According to (14c), the union of these bases yields a basis for V^h, the «*hierarchical basis*» (cf. Yserentant [2]). Of course, more general (e.g., irregular) triangulations than in Fig. 2a–c and a larger number of grid levels may be used. In the latter case, (14c) becomes $V^h = V_0 + \cdots + V_\ell$. We introduce $V^i := V^{h_i}$ as auxiliary subspaces:

$$V^i = V_0 + V_1 + \cdots + V_i \quad \text{for } 0 \leqslant i \leqslant \ell. \tag{11.6.14d}$$

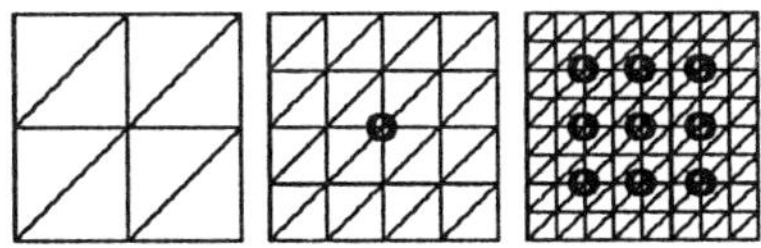

Fig. 11.6.2 Refining grids of size (a) $4h$ (b) $2h$ (c) h

We have to distinguish between *three different representations*. (a) V^i is the finite-element subspace, which can be written as the sum of the subspaces V_j $(0 \leqslant j \leqslant i)$ according to (14d). (b) The standard nodal basis representation of the functions from V^i yields the coefficients forming the vector space X^i. (c) The coefficients of the hierarchical basis belong to $X_0 \times X_1 \times \cdots \times X_i$. This product corresponds to the decomposition (14d).

The dimension of V^i is the number of nodal points in $I_i := \Omega_i \backslash \Omega_{i-1}$, where $\Omega_i := \Omega_{h_i}$ and $\Omega_{-1} := \varnothing$. These nodal points also serve as indices of the vector $x^i \in X_i$. The coefficient vector x^i represents the finite element function $u \in V_i \subset V^i \subset V^\ell$ defined by

$$u = \sum_{Q \in I_i} x_Q^i b_Q^i$$

where $b_Q^i \in V_i \subset V^i$ is the basic function of level i characterised by $b_Q^i(R) = \delta_{QR}$ for all $R \in I_i$. Therefore, the coefficients of x^i are the nodal values of u in the subset I_i: $x_Q^i = u(Q)$ for $Q \in I_i$. The prolongation p_i: $X_i \to X$ is defined by

$$(p_i x^i)_Q := \sum_{R \in I_i} x_R^i b_R^i(Q) \quad \text{for all } Q \in I_i.$$

The isomorphism P_h: $X \to V^h = V^\ell$ from §11.5.3 is given by

$$P_h x = \sum_{Q \in Q_\ell} x_Q^i b_Q^\ell,$$

where $b_Q^\ell \in V^\ell$ is the basic function of level ℓ characterised by $b_Q^\ell(R) = \delta_{QR}$ for all $R \in \Omega_\ell$. The interpolation of $u = P_h x$ with $x = \sum_{j=0}^{\ell} p_j x^j$ $(x^j \in X_j)$ in the points of Ω_i is the partial sum $u^i = P_h \sum_{j=0}^{i} p_j x^j \in V^i$. An important estimate of this interpolant is due to Yserentant [3]:

$$\left\| \sum_{j=0}^{i} p_j x^j \right\|_A^2 \leqslant C_Y(\ell - i + 1) \|x\|_A^2 \quad \text{for all } x = \sum_{j=0}^{\ell} p_j x^j,\ x^j \in X_j,\ 0 \leqslant i \leqslant \ell. \tag{11.6.15}$$

Let $s^i := \sum_{j=0}^{i} p_j x^j$. (15) implies

$$\|p_i x^i\|_A^2 = \|s_i - s_{i-1}\|_A^2 \leqslant 2\|s_i\|_A^2 + 2\|s_{i-1}\|_A^2 \leqslant 2C_Y(2\ell - 2i + 3)\|x\|_A^2$$

for $i > 0$, whereas $\|p_0 x^0\|_A^2 \leqslant C_Y(\ell + 1)\|x\|_A^2$ is identical to (14) for $i = 0$. Summing up all inequalities, we obtain

$$\sum_{j=0}^{\ell} \|p_j x^j\|_A^2 \leqslant C_Y' \ell^2 \quad \text{with } C_Y' := 5C_Y.$$

Hence, in the case of an exact solution of the subproblems (2.7), we have proved the inequality (3.2) with $C = C_Y' \ell^2$. The exact solution of $A_i y^i = c^i$ will be maintained only at the level $i = 0$ (this corresponds to the exact solution on the coarsest grid in (10.4.2a)). For $i > 0$, we apply the Jacobi iteration: $W_i := D_i := \operatorname{diag}\{A_i\}$. The matrices A_i and D_i turn out to be spectrally equivalent independently of h_i. In particular, $W_i \leqslant \text{const}\, A_i$ holds. Hence, (3.2) also implies inequality (3.3) with $C = O(\ell^2)$ and proves $\gamma = O(\ell^{-2})$ in (3.1) for the additive Schwarz variant. Concerning the estimate from above, one can determine Γ in (3.1) by means of Theorem 2.17. The technique described in Lemma 5.1 leads us to

$$\langle p_i x^i, p_j x^j \rangle_A \leqslant \varepsilon_{ij} \| p_i x^i \|_A \| p_j x^j \|_A \quad \text{with } \varepsilon_{ij} \leqslant C_E 2^{-|i-j|/2}. \qquad (11.6.16)$$

Hence, $\varrho(E) \leqslant \|E\|_\infty$ is bounded by a constant and (2.27) proves $\Gamma = O(1)$. In the end, we obtain convergence of the additive Schwarz iteration with the rate $1 - O(1/\ell^2)$.

The additive iteration (cf. Yserentant [3]) described above can be viewed as the Jacobi iteration for the system associated with the hierarchical basis, where the block-diagonal $D = \mathrm{blockdiag}\{A_0, D_1, \ldots, D_\ell\}$ is defined pointwise for the block indices $1 \leqslant i \leqslant \ell$. The respective multiplicative variant corresponds to the Gauß-Seidel method for the hierarchical basis system. The multiplicative Schwarz iteration (cf. Bank–Dupont–Yserentant [1]) also converges with the rate $1 - O(1/\ell^2)$, as one easily derives from (4.4) with $C_1 = O(\ell^2)$ and $C_2 = O(1)$ using $Y_j = X_j$.

Concerning algorithmic implementation, in particular, the fast transformation between the hierarchical representation $(x^0, x^1, \ldots, x^\ell) \in X_0 \times X_1 \times \cdots \times X_\ell$ and the nodal basis representation $x = \sum p_i x^i \in X$, we refer the interested reader to Yserentant [3].

Remark 11.6.8. The multiplicative Schwarz iteration defined by means of the hierarchical bases can be viewed as a special multi-grid method (V-cycle). The solution of the ith subproblem $A_i y^i = c^i$ by a secondary iteration step with $W_i := D_i := \mathrm{diag}\{A_i\}$ describes smoothing at the ith level by a (pointwise) Jacobi step. However, there is a remarkable difference. The smoothing is not performed at all fine grid points of Ω_i, but only at points $(x, y) \in \Omega_i \backslash \Omega_{i-1}$, which do not belong to the coarse grid.

The amount of work per iteration step equals $O(n)$ even when condition (10.4.7): $n_{i-1} \leqslant n_i / C_h$ is violated. This allows local refinements that insert only *few* additional fine grid points $\Omega_i \backslash \Omega_{i-1}$.

The presented convergence results are restricted to the boundary value problem in less than three spatial variables. Otherwise, the interpolation is to be replaced by the $L^2(\Omega)$-orthogonal projection projections onto V^i. The latter method is due to Bramble–Pasciak–Xu [2]. The relation of both methods is discussed by Yserentant [5] (cf. also Dryja–Widlund [2]).

11.6.5 Multi-Level Schwarz Iteration

A characteristic of the Schwarz iteration from §11.5.2 is the coarse-grid correction connected to $I_0 = \Omega_H$. The two-grid situation $\{h, H\}$ can be generalised to the multi-grid case $\{h = h_\ell < h_{\ell-1} < \cdots < h_0 = H\}$. For this purpose we rewrite the previous decomposition as $\{I_{0,\ell}, I_{1,\ell}, \ldots, I_{k_\ell,\ell}\}$, where $I_{\varkappa,\ell}$ $(1 \leqslant \varkappa \leqslant k_\ell)$ corresponds to the overlapping subdomains $\Omega'_\varkappa$ and $I_{0,\ell}$ is related to the coarse-grid $\Omega_{h_{\ell-1}}$. The analogous domain decomposition can be repeated for solving the coarse-grid equation: $I_{0,\ell}$ is replaced by $\{I_{0,\ell-1}, I_{1,\ell-1}, \ldots, I_{k_{\ell-1},\ell-1}\}$, where $I_{\varkappa,\ell-1}$ $(1 \leqslant \varkappa \leqslant k_{\ell-1})$ represents the overlapping subdomains in the coarse grid and $I_{0,\ell-1} = \Omega_{h_{\ell-2}}$ is related to the next coarser grid. Recursive replacement of $I_{0,\ell-1}, \ldots, I_{0,1}$ leads to $\{I_{0,0}, I_{\varkappa,\lambda}: 1 \leqslant \varkappa \leqslant k_\lambda, 1 \leqslant \lambda \leqslant \ell\}$ with corresponding prolongations $p_{\varkappa,\lambda}$.

In contrast to the usual multi-grid method, the multi-level additive Schwarz iteration works in parallel at all levels. For this variant, Dryja–Widlund [2, Theorem 3.2]) proved the conditions number $\Gamma/\gamma = O(\ell^2)$, which weakly deteriorates at an increasing number of levels.

11.6.6 Further Approaches for Decompositions into Subspaces

Thus far, subspaces (subdomains) have been constructed as proper domain decompositions or decompositions with respect to different grid sizes. A further possibility is the deomposition of a function space by means of symmetries (cf. Allgower–Böhmer–Zhen [1]). The prolongations appearing in the frequency decomposition variant of the multi-grid method (cf. Hackbusch [11]) can also be used directly as prolongations (2.1) of a domain decomposition method.

11.6.7 Indefinite and Unsymmetric Systems

Domain decomposition methods for more general (i.e., not positive definite) problems are discussed by Cai–Widlund [1]. A very simple but elegant approach is due to Xu [3]. He uses a product iteration, where the first factor is a coarse-grid correction (10.1.24) corresponding to the coarsest grid ($\ell = 0$), while the second factor has the iteration matrix $M = I - W^{-1}A$, where W is taken from a (fast) iteration applied to $A_0 x = b_0$, where $A_0 > 0$ is the positive definite part of A.

11.7 Schur Complement Methods

Several methods place the treatment of the Schur complement (cf. (6.4.12)) in the foreground.

11.7.1 Nonoverlapping Domain Decomposition with Interior Boundary

Matrices A representing (at most) nine-point formulae have the following property. The grid points I_1 and I_2 from Fig. 1a, which we use as index sets,

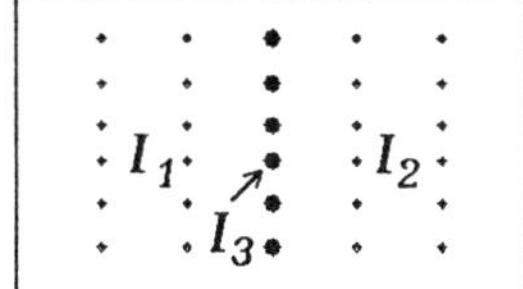

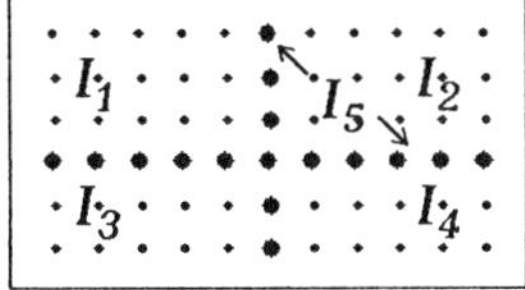

Fig. 11.7.1a $k = 3$ **Fig. 11.7.1b** $k = 5$
$k - 1$ subgrids with interior boundary I_k

are separated by a grid line I_3. Let the prolongations $p_\varkappa$ be defined as in (2.10a). Then the block indices 1 and 3 as well as 2 and 3 are connected, but 1 and 2 are not. The grid line I_3 is the minimal set that is able to separate I_1 and I_2 in this sense (if A is not a 9- but 25-point formula, one has to choose I_3 as a double line). In general, $k - 1$ subdomains $I_\varkappa$ $(1 \leqslant \varkappa \leqslant k - 1)$ can be separated by an interior boundary line I_k (illustrated in Fig. 1b for $k = 5$). The arising index subsets $I_\varkappa$ are disjoint (case (2.9a)).

For a suitable numbering of the indices (first those from I_1, then those from I_2, etc.), A has the block structure

$$A = \begin{bmatrix} A_{11} & & 0 & A_{1,k} \\ & \ddots & & \vdots \\ 0 & & A_{k-1,k-1} & A_{k-1,k} \\ A_{1,k}^H & \cdots & A_{k-1,k}^H & A_{k,k} \end{bmatrix}. \tag{11.7.1}$$

We define $I_I := \bigcup_{\varkappa=1}^{k-1} I_\varkappa$ (interior indices)and $I_R := I_k$ (boundary indices) and obtain a 2×2 block form of A:

$$A = \begin{bmatrix} A_{II} & A_{IR} \\ A_{IR}^H & A_{RR} \end{bmatrix}, \qquad A_{II} := \mathrm{blockdiag}\{A_{\varkappa\varkappa}: 1 \leqslant \varkappa \leqslant k - 1\}. \tag{11.7.2}$$

11.7.2 Direct Solution

Blockwise Gauß elimination of the entries $A_{1,k}^H, \ldots, A_{k-1,k}^H$ in (1) yields

$$\begin{bmatrix} A_{11} & & & A_{1,k} \\ & \ddots & & \vdots \\ & & A_{k-1,k-1} & A_{k-1,k} \\ 0 & \ldots & 0 & S \end{bmatrix} x = \begin{bmatrix} b^1 \\ \vdots \\ b^{k-1} \\ \hat{b}^k \end{bmatrix}, \quad \text{where} \tag{11.7.3}$$

$$S := A_{kk} - \sum_{\varkappa=1}^{k-1} A_{\varkappa,k}^H A_{\varkappa\varkappa}^{-1} A_{\varkappa,k} \tag{11.7.4}$$

is the *Schur complement*. The dimension of the matrix S is $n_k \times n_k$, where $n_k = O(h^{-1}k)$ is a typical size. If one tolerates the expensive computation of the Schur complement and with direct methods for solving $Sx^k = c^k$, one obtains a direct and largely parallelisable method.

11.7.3 Capacitance Matrix Method

The Schur complement S is also known under the name *capacitance matrix*. Although it is expensive to compute the matrix entries of S, the *matrix-vector product* $x^k \mapsto Sx^k$ for some $x^k \in X_k$ can be performed relatively easily via the (in parallel executable) partial steps (i) $x^k \mapsto c^k := A_{\varkappa,k}x^k$, (ii) solve $A_{\varkappa\varkappa}y^\varkappa = c^\varkappa$, and (iii) $Sx^k := A_{kk}x^k - \sum_\varkappa A^H_{\varkappa,k}y^\varkappa$, provided that direct solvers for $A_{\varkappa\varkappa}y^\varkappa = c^\varkappa$ are available.

The positive definiteness of A carries over to S, hence, the cg method is applicable to $Sx^k = c^k$. Depending on the size of the condition number $\varkappa(S)$, this proposal is recommendable or not. As soon as it is possible to solve the last block equation $Sx^k = \hat{b}^k$ in (3) (at least approximately), the other blocks are obtainable from $A_{\varkappa\varkappa}x^\varkappa = b^\varkappa - A_{\varkappa,k}x^k$.

In the case of the Poisson-model problem, the condition number $\varkappa(S) = O(h^{-1})$ results. Hence, the cg method (applied to the Richardson iteration) yields the unfavourable convergence rate $1 - O(h^{1/2})$. One can try to find suitable preconditioning for the matrix S. Dryjia [2] proposed the square root of the second differences on the grid line I_3, which is computable by means of the fast Fourier transform (cf. also Mróz [1]).

Concerning the preconditioning of a Hermitian 2×2 block matrix $A = \begin{bmatrix} A_{11} & A_{12} \\ A_{21} & A_{22} \end{bmatrix}$ by block-diagonal matrices, two alternatives can be considered. The first one is the block-Jacobi method with $W = \text{blockdiag}\{A_{11}, D\}$, which would be the block-diagonal Jacobi method if $D = A_{22}$. The second approach is the preconditioning of the Schur complement by the same matrix D. Mandel [3] proved that under suitable conditions the latter alternative is more favourable.

A variant of the capacitance matrix method termed the «method of fictitious domains» in the Russian literature is based on the embedding of domain Ω into a larger domain Ω'. For example, one may choose Ω' as a rectangle in order to obtain an easily solvable discretisation of the differential equation in Ω'. For this subject, we refer the reader to Proskurowski–Widlund [1], Astrachancev [2], and Börgers–Widlund [1].

11.7.4 Domain Decomposition Method with Nonoverlapping Domains

Instead of looking directly for a preconditioning of S, one may try to precondition S by means of the Schur complement of a similar problem. The following representation is taken from the papers of Widlund [2] and Björstad–Widlund [1]. There further references to earlier literature on this subject can also be found.

Let the domain Ω be decomposed into disjoint subdomains $\Omega_\varkappa$, as in Fig. 1.1a. The interior grid points (nodal points) are the sets that in Fig. 1a, b are denoted by $I_\varkappa$ $(1 \leqslant \varkappa \leqslant k-1)$. The inner boundaries $\partial\Omega_\varkappa \cap \partial\Omega_\lambda$ $(1 \leqslant \varkappa, \lambda \leqslant k-1)$ correspond to the set I_k from Fig. 1a, b. For the sake of simplicity, we assume $k = 3$ (i.e., two subdomains; cf. Fig. 1a). The use of the classical Schwarz iteration with the index sets $I_1 \cup I_3$ and I_2 (or I_1 and $I_2 \cup I_3$) cannot be recommended because of its slowness. Instead, the preconditioning of the Schur complement S can be constructed by means of the following Galerkin discretisation on Ω_1:

$$v \in V^h \quad \text{with } a_{\Omega_1}(v, w) = (f, w)_U \text{ for all } w \in V^h. \tag{11.7.5}$$

Here, a_{Ω_1} is the bilinear form a with the integration over Ω_1 instead of Ω (cf. §11.5.3). (5) yields the (discrete) solution of the differential equation on Ω_1 with the natural boundary condition on $\partial\Omega_1 \cap \partial\Omega_2$ (cf. Hackbusch [15, §7.5]). Problem (5) involves only the index sets I_1 and I_3. The arising system of equations has the block structure

$$\begin{bmatrix} A_{11} & A_{13} \\ A_{13}^H & A'_{33} \end{bmatrix} \begin{bmatrix} x^1 \\ x^3 \end{bmatrix} = \begin{bmatrix} c^1 \\ c^3 \end{bmatrix} \tag{11.7.6}$$

with a different matrix $A'_{33} \neq A_{33}$. The corresponding Schur complement

$$S' := A'_{33} - A_{13}^H A_{11}^{-1} A_{13} \tag{11.7.7}$$

is used as a preconditioning matrix for S. The proof of $\varkappa(S'^{-1}S) = O(1)$ requires tools from functional analysis (continuation theorems; cf. Widlund [2]). The solution of a system with the matrix

$$A' := \begin{bmatrix} A_{11} & 0 & A_{13} \\ 0 & A_{22} & A_{23} \\ A_{13}^H & A_{23}^H & A'_{33} \end{bmatrix}$$

starts with the subproblem of blocks x^1, x^3 using the matrix of (6) and is completed by solving for x^2. If S' is a good preconditioner for S, then A' is also a good preconditioner for A as stated in

Exercise 11.7.1. (a) Show that $S' \leqslant S$ for S, S' from (4) and (7).
(b) Let $\gamma \leqslant 1 \leqslant \Gamma$. Prove the equivalence of $\gamma S' \leqslant S \leqslant \Gamma S'$ and $\gamma A' \leqslant A \leqslant \Gamma A'$.

The solution of $A'x = c$ can be obtained from the exact solution of the subproblems (6) and $A_{22}x^2 = c^2$. The exact solution can be approximated by replacing A' by A''. According to Lemma 8.3.10, one estimates $\varkappa(A''^{-1}A)$ by $\varkappa(A''^{-1}A')\varkappa(A'^{-1}A) = \varkappa(A''^{-1}A')\varkappa(S'^{-1}S) \leqslant C\varkappa(A''^{-1}A')$.

11.7.5 Multi-Gridlike Domain Decomposition Methods

Domain decomposition can be performed in the following recursive manner: First, the index set I is decomposed into disjoint sets: $I = I_\prime \cup I'_\prime$, then we decompose $I'_\ell = I_{\ell-1} \cup I'_{\ell-1}, I'_{\ell-1} = I_{\ell-2} \cup I'_{\ell-2}, \ldots, I'_1 = I_1 \cup I_0$, and end up with the disjoint splitting $I = I_\ell \cup I_{\ell-1} \cup \cdots \cup I_1 \cup I_0$. For instance, the indices I_μ may correspond to the grid-point set $\Omega_\mu \backslash \Omega_{\mu-1}$ ($\Omega_{-1} := \varnothing$), where $\Omega_0 \subset \Omega_1 \subset \cdots \subset \Omega_\ell$ is a grid hierarchy associated with the step size sequence $h_\mu = 2^{-\mu} h_0$. The index decomposition $I = I_\ell \cup I'_\ell$ corresponds to a partitioning of the matrix $A = A^{(\ell)}$. This block-matrix can be written as the product

$$A^{(\ell)} = \begin{bmatrix} A_{11}^{(\ell)} & A_{12}^{(\ell)} \\ A_{12}^{(\ell)H} & A_{22}^{(\ell)} \end{bmatrix} = L^{(\ell)} \begin{bmatrix} A_{11}^{(\ell)} & 0 \\ 0 & S^{(\ell)} \end{bmatrix} L^{(\ell)H}, \qquad L^{(\ell)} = \begin{bmatrix} I & 0 \\ A_{12}^{(\ell)H} A_{11}^{(\ell)-1} & I \end{bmatrix}, \tag{11.7.8}$$

where $S^{(\ell)}$ denotes again the Schur complement. Following Exercise 1, for the preconditioning of A one needs a good preconditioner of $S^{(\ell)}$ by $S'^{(\ell)}$. Since $S^{(\ell)}$ corresponds to a coarse-grid correction, $S'^{(\ell)}$ can be defined recursively on ℓ. Such approaches can be found in Axelsson (in Monegato [1]), Axelsson–Vassilevski [1], Vassilevski [1], Kuznetsov [1–3]. They lead to methods that are similar to the hierarchical basis multi-grid method described in §11.6.4. Occasionally, the term «*algebraic multi-grid method*» is applied to these methods. On this point, we remark that this name is already given to a really algebraic method (for which the information contained in the algebraic system $Ax = b$ is sufficient; see the end of §10.9.3), whereas an algebraic multi-grid method in the new sense involves discretisation hierarchies that are to be prescribed explicitly, as in the case of standard multi-grid methods.

11.7.6 Further Remarks

Instead of constructing multi-grid methods by means of domain decompositions, one can adapt the multi-grid iteration to a given decomposition of the domain Ω into overlapping subdomains. The method described in Hackbusch [14, §15.3.3] uses an in parallel executable smoothing procedure, which can be regarded as an approximation of the additive version of the classical Schwarz method. The slowness of the convergence of the Schwarz method presented in §11.4.1 is harmless, since it corresponds to the smooth error components. For a combination of the Schwarz iteration concept with multi-grid methods of the second kind, compare Hackbusch [5].

Bibliography

[Prog] *Source codes of the Pascal programs in this book*; the disk can be ordered from the author

Alefeld, G.: [1] Zur Konvergenz des Peaceman-Rachford-Verfahrens. *Numer. Math.* **26** (1976) 409–419

Alefeld, G.: [2] On the convergence of the symmetric SOR method for matrices with red-black ordering. *Numer. Math.* **39** (1982) 113–117

Alefeld, G. and R. S. Varga: [1] Zur Konvergenz des symmetrischen Relaxationsverfahrens. *Numer. Math.* **25** (1976) 291–295

Allgower, E., K. Böhmer, and M. Zhen: [1] On a problem decomposition for semilinear nearly symmetric elliptic equations. In: Hackbusch [19] 1–17

Ashby, S. F., Th. A. Manteuffel, P. E. Saylor: [1] Adaptive polynomial preconditioning for hermitian indefinite linear systems. *BIT* **29** (1989) 583–609

Astrachancev, G. P.: [1] An iterative method of solving elliptic net problems. *USSR Comput. Math. and Math. Phys.* **11**, 2 (1971) 171–182

Astrachancev, G. P.: [2] Methods of fictitious domains for a second-order elliptic equation with natural boundary conditions. *USSR Comput. Math. and Math. Phys.* **18** (1978) 114–121

Axelsson, O.: [1] Solution of linear systems of equations: iterative methods. In: Barker [1] 1–51

Axelsson, O.: [2] A survey of preconditioned iterative methods for linear systems of algebraic equations. *BIT* **25** (1985) 166–187

Axelsson, O.: [3] A restarted version of a generalized preconditioned conjugate gradient method. *Comm. in Appl. Numer. Methods* **4** (1988) 521–530

Axelsson, O. and V. A. Barker: [1] *Finite element solution of boundary value problems.* Academic Press, Orlando, 1984

Axelsson, O., S. Brinkkemper, and V. P. Il'in: [1] On some versions of incomplete block-matrix factorization iterative methods. *LAA* **58** (1984) 3–15

Axelsson, O., V. Eijkhout, B. Polman, and P. Vassilevski: [1] Incomplete block-matrix factorization iterative methods for convection-diffusion problems. *BIT* **29** (1989) 867–889

Axelsson, O. and B. Polman: [1] A robust preconditioner based on algebraic substructuring and two-level grids. In: Hackbusch [18] 1–26

Axelsson, O. and P. S. Vassilevski: [1] Algebraic multilevel preconditioning methods, Part I: *Numer. Math.* **56** (1989) 157–177; Part II: *SIAM J.Numer. Anal.* **27** (1990) 1569–1590

Axelsson, O. and P. S. Vassilevski: [2] *Construction of variable-step preconditioners for inner-outer iteration methods.* Report 9107, Universiteit Nijmegen, 1991

Bachvalov, N. S.: [1] On the convergence of a relaxation method with natural constraints on the elliptic operator. *USSR Comput. Math. and Math. Phys.* **6**, 5 (1966) 101–135

Bank, R. E.: [1] A comparison of two multilevel iterative methods for nonsymmetric and indefinite elliptic finite element equations. *SIAM J. Numer. Anal.* **18** (1981) 724–743

Bank, R. E.: [2] PLTMG: *A software package for solving elliptic partial differential equations. Users' guide 6.0.* SIAM, Philadelphia, 1990

Bank, R.E. and T. F. Chan: [1] An analysis of the composite step biconjugate gradient method. To appear (Manuscript 1992)

Bank, R. E. and C. C. Douglas: [1] Sharp estimates for multigrid rates of convergence with general smoothing and acceleration. *SIAM J. Numer. Anal.* **22** (1985) 617–633

Bank, R. E. and T. F. Dupont: [1] *Analysis of a two-level scheme for solving finite element equations.* Rep. CNA-159, Univ. Texas, Austin, 1980

Bank, R.E. and T. F. Dupont: [2] An optimal order process for solving elliptic finite element equations. *Math. Comp.* **36** (1981) 35–51

Bank, R. E., T. F. Dupont, and H. Yserentant: [1] The hierarchical basis multigrid method. *Numer. Math.* **52** (1988) 427–458

Bank, R. E., B. D. Welfert, and H. Yserentant: [1] A class of iterative methods for solving saddle point problems. *Numer. Math.* **56** (1990) 645–666

Bank, R. E. and H. Yserentant: [1] Some remarks on the hierarchical basis multigrid method. In: Chan–Glowinski–Périaux–Widlund [1] 140–146

Barker, V. A. (ed.): [1] *Spare matrix techniques.* Proceedings, Copenhagen, Aug. 1976. Lecture Notes in Mathematics **572**. Springer-Verlag, Berlin, 1977

Beauwens, R.: [1] Approximate factorizations with s/p consistently ordered M-factors. *BIT* **29** (1989) 658–681

Berg, L.: [1] *Lineare Gleichungssysteme mit Bandstruktur.* Deutscher Verlag der Wissenschaften, Berlin, 1986

Berman, A. and R. J. Plemmons: [1] *Nonnegative matrices in the mathematical sciences.* Academic Press, New York, 1979

Björstad (Bjørstad), P. E.: [1] The direct solution of a generalized biharmonic equation on a disk. In: Hackbusch [17] 1–9

Björstad, P. E.: [2] Multiplicative and additive Schwarz methods: Convergence in the two-domain case. In: Chan–Glowinski–Périaux–Widlund [1] 147–159

Björstad, P. E. and J. Mandel: [1] On the spectra of sums of orthogonal projections with applications to parallel computing. *BIT* **31** (1991) 76–88

Björstad, P. E., R. Moe, and M. Skogen: [1] Parallel decomposition and iterative refinement algorithms. In: Hackbusch [19] 28–46

Björstad, P. E. and O. B. Widlund: [1] Iterative methods for the solution of elliptic problems on regions partitioned into substructures. *SIAM J. Numer. Anal.* **23** (1986) 1097–1120

Böhmer, K.: see Allgower, E. et al.

Börgers, Ch. and O. B. Widlund: [1] On finite element domain embedding methods. *SIAM J. Numer. Anal.* **27** (1990) 963–978

Bornemann, F. and Yserentant, H.: [1] A basic norm equivalence for the theory of multilevel methods. To appear in *Numer. Math.*

Braess, D.: [1] The contraction number of a multigrid method for solving the Poisson equation. *Numer. Math.* **37** (1981) 387–404

Braess, D.: [2] The convergence rate of a multigrid method with Gauß-Seidel relaxation for the Poisson equation. *Math. Comp.* **42** (1984) 505–519
Braess, D.: [3] A multigrid method for the membrane problem. *Computational Mechanics* **3** (1985) 617–633
Braess, D.: [4] On the combination of the multigrid method and conjugate gradients. In: Hackbusch–Trottenberg [2] 52–64
Braess, D.: [5] *Finite Elemente.* Springer-Verlag, Berlin, 1992
Braess, D.: see Peisker, P. et al.
Braess, D. and W. Hackbusch: [1] A new convergence proof for the multigrid method including the V-cycle. *SIAM J. Numer. Anal.* **20** (1983) 967–975
Braess, D., W. Hackbusch, and U. Trottenberg (eds.): [1] *Advances in multi-grid methods.* Proceedings, Oberwolfach, December 1984. Notes on Numerical Fluid Mechanics **11**, Vieweg, Braunschweig, 1985
Brakhage, H.: [1] Über die numerische Behandlung von Integralgleichungen nach der Quadraturformelmethode. *Numer. Math.* **2** (1960) 183–196
Bramble, J. H. and J. E. Pasciak: [1] New convergence estimates for multigrid algorithms. *Math. Comp.* **49** (1987) 311–329
Bramble, J. H., J. E. Pasciak, and A. H. Schatz: [1] An iterative method for elliptic problems on regions partitioned into substructures. *Math. Comp.* **46** (1986) 361–369
Bramble, J. H., J. E. Pasciak, and A. H. Schatz: [2] The construction of preconditioners for elliptic problems by substructuring. I: *Math. Comp.* **47** (1986) 103–134; II: *Math. Comp.* **49** (1987) 1–16; III: *Math. Comp.* **51** (1988) 415–430; IV: *Math. Comp.* **53** (1989) 1–24
Bramble, J. H., J. E. Pasciak, and J. Xu: [1] The analysis of multigrid algorithms for nonsymmetric and indefinite elliptic problems. *Math. Comp.* **51** (1988) 389–414
Bramble, J. H., J. E. Pasciak, and J. Xu: [2] Parallel multilevel preconditioners. *Math. Comp.* **55** (1990) 1–22
Bramble, J. H., J. E. Pasciak, J. Wang, and J. Xu: [1] Convergence estimates for multigrid algorithms without regularity assumptions. *Math. Comp.* **57** (1991) 23–45
Brandt, A.: [1] Multi-level adaptive solutions to boundary-value problems. *Math. Comp.* **31** (1977) 333–390
Brandt, A.: [2] Guide to multigrid development. In: Hackbusch–Trottenberg [1] 220–312
Buleev, N. I. (Булеев, Н. И.): [1] Численный метод решения двумерных и трехмерных уравнеий диффузии (A numerical method for solving two- and three-dimensional diffusion equations). *Math. Sb.* **51** (1960) 227–238
Bulirsch, R., R. D. Grigorieff, and J. Schröder (eds.): [1] *Numerical treatment of differential equations.* Proceedings, Oberwolfach, July 1976. Lecture Notes in Mathematics **631**. Springer-Verlag, Berlin, 1978
Bulirsch, R.: see Stoer, J.
Buneman, O.: [1] *A compact non-iterative Poisson solver.* SUIPR Report 294, Stanford, 1969
Bunse, W. and A. Bunse-Gerstner: [1] *Numerische lineare Algebra.* Teubner, Stuttgart, 1985
Cai, Xiao-Chuan and O. B. Widlund: [1] Domain decomposition algorithms for indefinite elliptic problems. To appear in *SIAM J. Sci. Stat. Comput.*
Chan, T. F., R. Glowinski, J. Périaux and O. Widlund (eds.): [1] *Domain decomposition methods.* Proceedings, January 1988, Los Angeles. SIAM, Philadelphia, 1989
Chan, T. F., R. Glowinski, J. Périaux, and O. Widlund (eds.): [2] *Domain decomposition methods.* Proceddings, March 1989, Houston. SIAM, Philadelphia, 1990
Chan, T. F., D. E. Keyes, G. Meurant, J. S. Scroggs, R. G. Vogt (eds.): [1] *Domain decomposition methods.* Proceedings, May 1991. SIAM, Philadelphia (to appear)
Concus, P. and G. H. Golub: [1] A generalized conjugate gradient method for nonsymmetric systems of linear equations. In: Glowinski-Lions [1] 56–65

Concus, P., G. H. Golub, and G. Meurant: [1] Block preconditioning for the conjugate gradient method. *SIAM J. Sci. Stat. Comput.* **6** (1985) 220–252

Concus, P., G. H. Golub, and D. P. O'Learyy: [1] Numerical solution of nonlinear elliptic partial differential equations by a generalized conjugate gradient method. *Computing* **19** (1978) 321–339

Dahlquist, G.: see Kronsjö, L. et al.

de Boor, C. and J. R. Rice: [1] Extremal polynomials with applications to Richardson iteration for indefinite linear systems. *SIAM J. Sci. Stat. Comput.* **3** (1982) 47–57

de Zeeuw, P. M.: [1] Matrix-dependent prolongations and restrictions in a blackbox multigrid solver. *J. of Comput. and Appl. Math.* **33** (1990) 1–27

de Zeeuw, P. M.: see Sonneveld, P. et al., Hemker, P. W. et al.

Deulhard, P., R. Freund, and A. Walter: [1] Fast secant methods for the iterative solution of large nonsymmetric linear systems. *Impact of Comput. in Sc. and Eng.* **2** (1990) 244–276

Douglas, C. C.: see Bank, R. E. et al.

Dryja, M.: [1] A capacitance matrix method for Dirichlet problem on polygon region. *Numer. Math.* **39** (1982) 51–64

Dryja, M.: [2] A finite element—capacitance matrix method for elliptic problems on regions partitioned into subregions. *Numer. Math.* **44** (1984) 153–168

Dryja, M.: [3] A method of domain decomposition for 3-dimensional elliptic problems. In: Glowinski–Golub–Meurant–Périaux [1] 43–61

Dryja, M.: [4] An additive Schwarz algorithm for two- and three-dimensional finite element elliptic problems. In: Chan–Glowinski–Périaux–Widlund [1] 168–172

Dryja, M. and O. B. Widlund: [1] Towards a unified theory of domain decomposition algorithms for elliptic problems. In: Chan–Glowinski–Périaux–Widlund [2] 3–21

Dryja, M. and O. B. Widlund: [2] Multilevel additive methods for elliptic finite element problems. In: Hackbusch [19] 58–69

Dryja, M. and O. B. Widlund: [3] Additive Schwarz methods for elliptic finite element problems in three dimensions. In: Chan–Keyes–Meurant–Scroggs–Vogt [1]

Duff, I. S., A. M. Erisman, and J. K. Reid: [1] *Direct methods for sparse matrices.* Clarendon Press, Oxford, 1989

Duff, I. S. and G. A. Meurant: [1] The effect of ordering on preconditioned conjugate gradients. *BIT* **29** (1989) 635–657

Duff, I. S. and G. W. Stewart (eds.): [1] *Sparse matrix proceedings 1978.* Proceedings, Knoxville, November 1978. SIAM, Philadelphia, 1979

Dupont, T. F.: see Bank, R. E. et al.

D'Yakonov, E. G. (Дьяхонов, Е, Г.): [1] The construction of iterative methods based on the use of spectrally equivalent operators. *USSR Comput. Math. and Math. Phys.* **6**, 1 (1966) 14–46

D'Yakonov, E. G.: [2] О шодимости одногa итерационного процесса (On the convergence of an iterative process). *Usp. Mat. Nauk* **21** (1966) 179–182

D'Yakonov, E. G.: [3] Минимизация вычислительной работы—Асимптотичесхи оптимальные алгоритмы для эллиптчесхих задач (*Minimisation of the computational work—asymptotically optimal algorithms for elliptic equations*). Nauka, Moscow, 1989

Eiermann, M., I. Marek, and W. Niethammer: [1] On the solution of singular systems of algebraic equations by semiiterative methods. *Numer. Math.* **53** (1988) 265–283

Eiermann, M., W. Niethammer, and A. Ruttan: [1] Optimal successive overrelaxation iterative methods for p-cyclic matrices. *Numer. Math.* **57** (1990) 593–606

Eiermann, M., W. Niethammer, and R. S. Varga: [1] A study of semiiterative methods for nonsymmetric systems of linear equations. *Numer. Math.* **47** (1985) 505–533

Eijkhout, V.: see Axelsson, O. et al.

Elman, H. C.: [1] Relaxed and stabilized incomplete factorizations for non-self-adjoint linear systems. *BIT* **29** (1989) 890–915

Erisman, A. M.: see Duff, I. S. et al.
Ewing, R. E.: [1] Preconditioned conjugate gradient methods for large-scale fluid flow applications. *BIT* **29** (1989) 850–866
Fedorenko, R. P.: [1] A relaxation method for solving elliptic difference equations. *USSR Comput. Math. and Phys.* **1**, 5 (1961) 1092–1096
Fedorenko, R. P.: [2] The speed of convergence of one iterative process. *USSR Comput. Math. and Math. Phys.* **4**, 3 (1964) 227–235
Finogenov, S. A.: see Lebedev, V. I. et al.
Fischer, B. and R. Freund: [1] Chebyshev polynomials are not always optimal. *J. Approx. Theory* **65** (1991) 261–272
Fischer, B. and R. Freund: [2] On the constrained Chebyshev approximation problem on ellipses. *J. Approx. Theory* **62** (1990) 297–315
Forsythe, G. E. and E. G. Strauss: [1] On best conditioned matrices. *Proc. Amer. Soc.* **6** (1955) 340–345
Freund, R.: [1] On conjugate gradient type methods and polynomial preconditioners for a class of complex non-Hermitian matrices. *Numer. Math.* **57** (1990) 285–312
Freund, R.: see Deulhard, P. et al., Fischer, B. et al.
Fridman, V. M.: [1] The method of minimum iterations with minimum errors for a system of linear algebraic equations with a symmetrical matrix. *USSR Comput. Math. and Math. Phys.* **2** (1963) 362–363
Frobenius, G.: [1] Über Matrizen aus positiven Elementen. *Sitzungsbericht Akad. Wiss. Phys.-math. Klasse Berlin* 417–476 (1908) and 514–518 (1909)
Gantmacher, F. R.: [1] *The theory of matrices*, Vol. I and II. Chelsea, New York, 1959
George, J. A.: [1] Solution of linear systems of equations: Direct methods for finite element problems. In: Barker [1] 52–101
Glowinski, R.: see Chan, T. F. et al.
Glowinski, R., G.H. Golub, G. A. Meurant, and J. Périaux (eds.): [1] *First international symposium on domain decomposition methods for partial differential equations.* Proceedings, January 1987, Paris. SIAM, Philadelphia, 1988
Glowinski, R. and J. L. Lions (eds.): [1] *Computing methods in applied sciences and engineering.* Lecture Notes in Economics and Math. Systems **134**, Springer-Verlag, New York, 1976
Golub, G.: [1] Direct methods for solving elliptic difference equations. In: Morris [1] 1–19
Golub, G. H. and D. O'Leary: [1] Some history of the conjugate gradient and Lanczos algorithms: 1948–1976. *SIAM Rev.* **31** (1989) 50–102
Golub, G. H. and M. L. Overton: [1] The convergence of inexact Chebyshev and Richardson iterative methods for solving linear systems. *Numer. Math.* **53** (1988) 571–593
Golub, G. H. and Ch. F. van Loan: [1] *Matrix computations.* North Oxford Academic, Oxford, 1983
Golub, G. H.: see Concus, P. et al., Glowinski, R. et al.
Grigorieff, R. D.: see Bulirsch, R. et al.
Gunn, J. E.: [1] The solution of elliptic difference equations by semiexplicit iterative techniques. *SIAM J. Numer. Anal.* **2** (1964) 24–45
Gustafsson, I.: [1] A class of first order factorization methods. *BIT* **18** (1978) 142–156
Gutknecht, M. H.: [1] A completed theory of the unsymmetric Lanczos process and related algorithms. Part I: *SIAM J. Matrix Anal. Appl.* **13** (1992) 594–639–Part II: to appear in *SIAM J. Matrix Anal. Appl.*
Gutknecht, M. H.: [2] The unsymmetric Lanczos algorithms and their relations to Padé approximation, continued fractions, the QD algorithm, biconjugate gradient squared methods, and fast Hankel solvers. Publication planned in *SIAM Review*
Haase, G. and U. Langer: [1] On the use of multigrid preconditioners in the domain decomposition method. In: Hackbusch [19] 101–110

Hackbusch, W.: [1] A fast iterative method solving Poisson's equation in a general region. In: Bulirsch–Grigorieff–Schröder [1] 51–62

Hackbusch, W.: [2] *On the convergence of a multi-grid iteration applied to finite element equations.* Report 77–8, Universität zu Köln, 1977

Hackbusch, W.: [3] On the multi-grid method applied to difference equations. *Computing* **20** (1978) 291–306

Hackbusch, W.: [4] Convergence of multi-grid iterations applied to difference equations. *Math. Comp.* **34** (1980) 425–440

Hackbusch, W.: [5] The fast numerical solution of very large elliptic difference schemes. *J. Inst. Maths Applics* **26** (1980) 119–132

Hackbusch, W.: [6] Die schnelle Auflösung der Fredholmschen Integralgleichung zweiter Art. *Beiträge Numer. Math.* **9** (1981) 47–62

Hackbusch, W.: [7] On the convergence of multi-grid iterations. *Beiträge Numer. Math.* **9** (1981) 213–239

Hackbusch, W.: [8] On the regularity of difference schemes. *Ark. Mat.* **19** (1981) 71–95

Hackbusch, W.: [9] On the regularity of difference schemes–part II: regularity estimates for linear and nonlinear problems. *Ark. Mat.* **21** (1983) 3–28

Hackbusch, W.: [10] Multi-grid convergence theory. In: Hackbusch–Trottenberg [1] 177–219

Hackbusch, W.: [11] The frequency decomposition multi-grid method. Part I: Application to anisotropic equations. *Numer. Math.* **56** (1989) 229–245

Hackbusch, W.: [12] A parallel conjugate gradient method. *J. Numer. Linear Algebra with Appl.* **1** (1992) 133–147

Hackbusch, W.: [13] The solution of large systems of BEM equations by the multi-grid and panel clustering method technique. In: Monegato [1] 133–160

Hackbusch, W.: [14] *Multi-grid methods and applications.* Springer-Verlag, Berlin, 1985

Hackbusch, W.: [15] *Theory and numerical treatment of elliptic differential equations.* Springer-Verlag, Berlin, 1992.—*Theorie und Numerik elliptischer Differentialgleichungen.* Teubner, Stuttgart, 1986

Hackbusch, W.: [16] *Integralgleichungen—Theorie und Numerik.* Teubner, Stuttgart, 1989

Hackbusch, W. (ed.): [17] *Efficient solvers for elliptic systems.* Notes on Numerical Fluid Mechanics **10**. Vieweg, Braunschweig, 1984

Hackbusch, W. (ed.): [18] *Robust Multi-Grid Methods.* Proceedings, Kiel, January 1988. Notes on Numerical Fluid Mechanics **23**. Vieweg, Braunschweig, 1988

Hackbusch, W. (ed.): [19] *Parallel Algorithms for PDEs.* Proceedings, Kiel, January 1990. Notes on Numerical Fluid Mechanics **31**. Vieweg, Braunschweig, 1991

Hackbusch, W.: [20] Comparison of different multi-grid variants for nonlinear equations. *ZAMM* **72** (1992) 148–151

Hackbusch, W.: see Braess, D. et al.

Hackbusch, W. and A. Reusken: [1] Analysis of a damped nonlinear multilevel method. *Numer. Math.* **55** (1989) 225–246

Hackbusch, W. and U. Trottenberg (eds.): [1] *Multi-grid methods.* Proceedings, Köln-Porz, November 1981. Lecture Notes in Mathematics **960**. Springer-Verlag, Berlin, 1982

Hackbusch, W. and U. Trottenberg (eds.): [2] *Multi-grid methods II.* Proceedings, Köln, October 1985. Lecture Notes in Mathematics **1228**. Springer-Verlag, Berlin, 1986

Hackbusch, W. and U. Trottenberg (eds.): [3] *Multi-grid methods III.* Proceedings, Bonn, October 1990. ISNM **98**, Birkhäuser, Basel, 1991

Hageman, L. A. and D. M. Young: [1] *Applied Iterative Methods.* Academic Press, New York, 1981

Hanke, M., M. Neumann, and W. Niethammer: [1] On the SOR method for symmetric positive definite systems. *LAA* **154–156** (1991) 457–472

Hemker, P. W.: [1] The incomplete LU-decomposition as a relaxation method in multi-grid algorithms. In: Miller [1] 306–311

Hemker, P. W.: [2] Mixed defect correction iteration for the accurate solution of the convection diffusion equation. In: Hackbusch–Trottenberg [2] 485–501

Hemker, P. W.: [3] On the comparison of line-Gauss-Seidel and ILU relaxation in multigrid algorithms. In: Miller [2] 269–277

Hemker, P., R. Kettler, P. Wesseling, and P. M. de Zeeuw: [1] Multigrid methods: Development of fast solvers. *Appl. Math. Comput.* **13** (1983) 311–326

Hemker, P. W. and H. Schippers: [1] Multiple grid methods for the solution of Fredholm integral equations of the second kind. *Math. Comp.* **36** (1981) 215–232

Hestenes, M. R.: [1] *Conjugate direction methods in optimization.* Springer-Verlag, New York, 1980

Hestenes, M. R. and E. Stiefel: [1] Methods of conjugate gradients for solving linear systems. *J. Res. Nat. Bur. Standards* **49** (1952) 409–436

Holstein, H.: see Paddon, D. J. et al.

Il'in, V. P.: [1] Some estimates for conjugate gradient methods. *USSR Comput. Math. and Math. Phys.* **16**, 4 (1976) 22–30

Il'in, V. P.: see Axelsson, O. et al.

Jennings, A. and G. M. Malik: [1] Partial elimination. *J. IMA* **20** (1977) 307–316

Jensen, K. and N. Wirth: [1] *PASCAL user manual and report.* 2nd ed. Springer-Verlag, New York, 1978

Kaczmarz, S.: [1] Angenäherte Auflösung von Systemen linearer Gleichungen. *Bulletin de l'Academie Polonaise des Sciences et Lettres* **A35** (1937) 355–357

Kershaw, D. S.: [1] The incomplete Cholesky-conjugate gradient method for the iterative solution of systems of linear equations. *J. Comput. Phys.* **26** (1978) 43–65

Kettler, R.: [1] Analysis and comparison of relaxation schemes in robust multigrid and preconditioned conjugate gradient methods. In: Hackbusch–Trottenberg [1] 502–534

Kettler, R.: see Hemker, P. W. et al.

Kosmol, P.: [1] *Methoden zur numerischen Behandlung nichtlinearer Gleichungen und Optimierungsaufgaben.* Teubner, Stuttgart, 1989

Kosmol, P. and Xinlong Zhou: [1] The limit points of affine iterations. *Numer. Funct. Analysis and Optim.* **11** (1990) 403–409

Kronsjö, L. and G. Dahlquist: [1] On the design of nested iterations for elliptic difference equtions. *BIT* **11** (1971) 63–71

Kuznetsov, Ju. A.: [1] Algebraic multigrid domain decomposition methods, I. *Soviet J. Numer. Anal. and Math. Model.* **4** (1989) 361–392

Kuznetsov, Ju. A.: [2] Multigrid domain decomposition methods for elliptic problems. *Comp. Meth. in Appl. Mech. and Eng.* **75** (1989) 185–193

Kuznetsov, Ju. A.: [3] Multigrid domain decomposition methods. In: Chan–Glowinski–Périaux–Widlund [2] 290–313

Langer, U.: see Haase, G. et al.

Lebedev, V. I.: [1] On a Zolotarev problem in the method of alternating directions. *USSR Comput. Math. and Math. Phys.* **17**, 2 (1977) 58–76

Lebedev, V. I. and S. A. Finogenov: [1] On the order of choice of the iteration parameters in the Chebyshev cyclic iteration method. *USSR Comput. Math. and Math. Phys.* **11**, 2 (1971) 155–170 and **13**, 1 (1973) 21–41

Liebau, F.: see Wittum, G. et al.

Lions, J. L.: see Glowinski, R. et al.

Loan, Ch. F. van: see Golub, G. H. et al.

Luenberger, D. G.: [1] *Introduction to linear and nonlinear programming.* Addison-Wesley, Reading, Mass., 1973

Maess, G.: [1] *Vorlesungen über numerische Mathematik. I. Lineare Algebra.* Birkhäuser, Basel, 1985
Maess, G.: [2] Projection methods solving rectangular systems of linear equations. *J. of Comput. and Appl. Math.* **24** (1988) 107–119
Maitre, J.-F. and F. Musy: [1] Multigrid methods: convergence theory in a variational framework. *SIAM J. Numer. Anal.* **21** (1984) 657–671
Maitre, J.-F., F. Musy, and P. Nigon: [1] A fast solver for the Stokes equations using multigrid with a Uzawa smoother. In: Braess–Hackbusch–Trottenberg [1]
Malik, G. M.: see Jennings, A. et al.
Mandel, J.: [1] A multilevel iterative method for symmetric, positive definite problems. *Appl. Math. Optim.* **11** (1984) 77–95
Mandel, J.: [2] Multigrid convergence for nonsymmetric, indefinite variational problems and one smoothing step. *Appl. Math. Optim.* **19** (1986) 201–216
Mandel, J.: [3] On block diagonal and Schur complement preconditioning. *Numer. Math.* **58** (1990) 79–93
Mandel, J. and S. McCormick: [1] Iterative solution of elliptic equations with refinement. In: Chan–Glowinski–Périaux–Widlund [1] 81–102
Mandel, J.: see Björstad, P. E. et al.
Manteuffel, T. A.: [1] Adaptive procedure for estimating parameters for the nonsymmetric Tchebychev iteration. *Numer. Math.* **31** (1978) 183–208
Manteuffel, T. A.: [2] An incomplete factorization technique for positive definite linear systems. *Math. Comp.* **34** (1980) 473–497
Manteuffel, Th. A.: see Ashby, S. F. et al.
Marcowitz, U.: see Meis, Th. et al.
Marek, I.: [1] *Iterative methods of solving linear systems with a rectangular matrix.* Report 8132, Universiteit Nijmegen, 1981
Marek, I.: see Eiermann, M. et al.
McCormick, S. (ed.): *Multigrid methods.* SIAM, Philadelphia, 1987
McCormick, S.: see Mandel, J. et al.
Mehrmann, V.: see Varga, R. S. et al.
Meijerink, J. A.: [1] *Iterative methods for the solution of linear equations based on incomplete factorisation of the matrix.* Shell Publ. 643, Rijswijk, 1983
Meijerink, J. A. and H. A. van der Vorst: [1] An iterative solution method for linear systems of which the coefficient matrix is a symmetric M-matrix. *Math. Comp.* **31** (1977) 148–162
Meis, Th. and U. Marcowitz: [1] *Numerical solution of partial differential equations.* Springer-Verlag, New York, 1981. German version: 1978
Meurant, G.: see Concus, P. et al., Duff, I. S. et al., Glowinski, R. et al.
Miller, J. J. H. (ed.): [1] *Boundary and interior layers—computational and asymptotic methods.* Proceedings, Dublin, June 1980. Boole Press, Dublin, 1980
Miller, J. J. H. (ed.): [2] *Computational and asyymptotic methods for boundary and interior layers.* Proceedings, Dublin, June 1982. Boole Press, Dublin, 1982
Moe, R.: see Björstad, P. E. et al.
Moncgato, G. (ed.): [1] *Numerical Methods.* Proceedings, Turin, June 1990. Rend. Semin. Mat., Univ. Pol. Torino. Fasc. Speciale, 1991
Morris, J. Ll. (ed.): [1] *Symposium on the theory of numerical analysis.* Proceedings, Dundee, Sept. 1970. Lecture Notes in Mathematics 193. Springer-Verlag, Berlin, 1971
Mróz, M.: Domain decomposition method for elliptic mixed boundary value problems. *Computing* **42** (1989) 45–59
Munksgaard, N.: [1] Solving sparse symmetric sets of linear equations by preconditioned conjugate gradients. *ACM Trans. Math. Software* **6** (1980) 206–219
Musy, F.: see Maitre, J.-F. et al.
Natterer, F.: [1] *The mathematics of computerized tomography.* Wiley and Teubner, Stuttgart, 1986

Nepomnyashich, S. V. [1] *Domain decomposition and Schwarz methods in a subspace for the approximate solution of elliptic boundary value problems.* Thesis, Novosibirsk, 1986 (in Russian)
Neumaier, A. and R. S. Varga: [1] Exact convergence and divergence domains for the symmetric successive overrelaxation iterative (SSOR) method applied to H-matrices. *LAA* **58** (1984) 261–272
Neumann, M.: see Hanke, M. et al.
Nicolaides, R. A.: [1] On multiple grid and related techniques for solving discrete elliptic systems. *J. Comput. Phys.* **19** (1975) 418–431
Nicolaides, R. A.: [2] On the ℓ^2 convergence of an algorithm for solving finite element equations. *Math. Comp.* **31** (1977) 892–906
Niethammer, W.: [1] Relaxation bei nichtsymmetrischen Matrizen. *Math. Zeitschr.* **85** (1964) 319–327
Niethammer, W.: [2] Relaxation bei komplexen Matrizen. *Math. Zeitschr.* **86** (1964) 34–40
Niethammer, W.: [3] The SOR method on parallel computers. *Numer. Math.* **56** (1989) 247–254
Niethammer, W. and R. S. Varga: [1] The analysis of k-step iterative methods for linear systems from summability theory. *Numer. Math.* **41** (1983) 177–206
Niethammer, W.: see Eiermann, M. et al., Hanke, M. et al., Starke, G. et al.
Nigon, P.: see Maitre, J.-F. et al.
Nikolaev, E. S.: see Samarskii, A. A. et al.
O'Leary, D. P.: [1] The block conjugate gradient algorithm and related methods. *LAA* **29** (1980) 293–322
O'Leary, D. P.: [2] Parallel implementation of the block conjugate gradient algorithm. *Parallel Computing* **5** (1987) 127–139
O'Leary, D. P.: see Concus, P. et al., Golub, G. H. et al.
Opfer, G. and G. Schober: [1] Richardson's iteration for nonsymmetric matrices. *LAA* **58** (1984) 343–361
Ortega, J. M.: [1] *Introduction to parallel vector solution of linear systems.* Plenum Press, New York, 1988
Oswald, P.: [1] On function spaces related to finite element approximation theory. *Zeitschrift für Analysis und ihre Anwendungen* **9** (1990) 43–64
Ostrowski, A. M.: [1] Über die Determinanten mit überwiegender Hauptdiagonale. *Commentarii Mathematici Helvetici* **10** (1937) 69–96
Ostrowski, A. M.: [2] On the linear iteration procedures for symmetric matrices. *Rend. Math. e. Appl.* **14** (1954) 140–163
Overton, M. L.: see Golub, G. H. et al.
Paddon, D. J. and H. Holstein (eds.): [1] *Multigrid methods for integral and differential equations.* Proceedings, Bristol, September 1983. Clarendon Press, Oxford, 1985
Paige, C. C. and M. A. Saunders: [1] Solution of sparse indefinite systems of linear equations. *SIAM J. Numer. Anal.* **12** (1975) 617–629
Parlett, B. N.: [1] *The symmetric eigenvalue problem.* Prentice-Hall, Englewood Cliffs, 1980
Pasciak, J. E.: see Bramble, J. H. et al.
Peaceman, D. W. and Rachford, H. H.: [1] The numerical solution of parabolic and elliptic differential equations. *J. SIAM* **3** (1955) 28–41
Pearcy, C.: [1] An elementary proof of the power inequality for the numerical radius. *Michigan Math. J.* **13** (1966) 289–291
Peisker, P. and D. Braess: [1] A conjugate gradient method and a multigrid algorithm for Morley's finite element approximation of the biharmonic equations. *Numer. Math.* **55** (1987) 567–586
Périaux, J.: see Chan, T. F. et al., Glowinski, R. et al.
Perron, O.: [1] Zur Theorie der Matrices. *Math. Ann.* **64** (1907) 248–263

Plemmons, R. J.: see Berman, A. et al.
Polman, B.: see Axelsson, O. et al.
Proskurowski, W. and O. Widlund: [1] On the numerical solution of Helmholtz's equation by the capacitance matrix method. *Math. Comp.* **30** (1976) 433–468
Reid, J. K.: [1] On the method of conjugate gradients for the solution of large sparse systems of linear equations. In: Reid [2] 231–254
Reid, J. K. (ed.): [2] *Large sparse sets of linear equations.* Proceedings, Oxford, April 1970. Academic Press, New York, 1971
Reid, J. K.: [3] Solution of linear systems of equations: Direct methods (general). In: Barker [1] 102–129
Reid, J. K.: see Duff, I. S. et al.
Reusken, A.: [1] Steplength optimization and linear multigrid methods. *Numer. Math.* **58** (1991) 819–838
Reusken, A.: [2] *A new lemma in multigrid convergence theory.* Report RANA 91–07, Eindhoven, 1991
Reusken, A.: see Hackbusch, W. et al.
Rice, J. R.: see de Boor, C. et al.
Rosenberg, R. L.: see Stein, P. et al.
Ruttan, A.: see Eiermann, M. et al.
Saad, Y. and M. H. Schultz: [1] GMRES: A generalized minimal residual method for solving nonsymmetric linear systems. *SIAM J. Sci. Statist. Comput.* **7** (1986) 856–869
Saff, E.B.: see Varga, R. S. et al.
Samarskij, A. A. and E. S. Nikolaev: [1] *Numerical methods for grid equations. Vol II: Iterative methods.* Birkhäuser, Basel, 1989
Saunders, M. A.: see Paige, C. C. et al.
Saylor, P. E.: see Ashby, S. F. et al.
Schatz, A. H.: see Bramble, J. H. et al.
Schippers, H.: see Hemker, P. W. et al.
Schober, G.: see Opfer, G. et al.
Schröder, J. and U. Trottenberg: [1] Reduktionsverfahren für Differenzengleichungen bei Randwertaufgaben I. *Numer. Math.* **22** (1973) 37–68
Schultz, M. H.: see Saad, Y. et al.
Sheldon, J.: [1] On the numerical solution of elliptic difference equations. *Math. Tables Aids Comput.* **9** (1955) 101–112
Skogen, M.: see Björstad, P. E. et al.
Sluis, A. van der: see van der Sluis, A.
Sonneveld, P., P. Wesseling, and P. M. de Zeeuw: [1] Multigrid and conjugate gradient methods as convergence acceleration techniques. In: Paddon–Holstein [1] 117–167
Southwell, R.V.: [1] Stress-calculation in frameworks by the method of "systematic relaxation of constraints"–parts I, II: *Proc. Roy. Soc. (A)* **151** (1935) 56–95; part III: *Proc. Roy. Soc. (A)* **153** (1935) 41–76
Southwell, R. V.: [2] *Relaxation methods in engineering science—a treatise on approximate computation.* Oxford Univ. Press, London, 1940
Southwell, R. V.: [3] *Relaxation methods in theoretical physics.* Clarendon Press, Oxford, 1946
Spedicato, E. (ed.): [1] Computer algorithms for solving linear algebraic equations. The state of art. Proceedings, September 1990. NATO ASI Series F, Vol. **77**. Springer-Verlag, Berlin, 1991
Starke, G.: [1] Optimal ADI parameter for nonsymmetric systems of linear equations. *SIAM J. Numer. Anal.* **28** (1991) 1431–1445
Starke, G. and W. Niethammer: [1] SOR for $AX - XB = C$. *LAA* **154–156** (1991) 355–375

Stein, P. and R. L. Rosenberg: [1] On the solution of linear simultaneous equations by iteration. *J. London Math. Soc.* **23** (1948) 111–118

Stiefel, E.: [1] Über einige Methoden der Relaxationsrechnung. *Z. Angew. Math. Phys.* **3** (1952) 1–33

Stiefel, E.: [2] Relaxationsmethoden bester Strategie zur Lösung linearer Gleichungssysteme. *Comment. Math. Helv.* **29** (1955) 157–179

Stiefel, E.: see Hestenes, M. R. et al.

Stoer, J.: [1] *Einführung in die Numerische Mathematik I*. 5th ed., Springer-Verlag, Berlin, 1989

Stoer, J.: [2] Solution of large systems of linear equations by conjugate gradient type methods. In: *Mathematical Programming, the state of art* (A. Bachem, M. Grötschel, B. Korte, eds.). Springer-Verlag, Berlin, 1983

Stoer, J. and R. Bulirsch: [1] *Einführung in die Numerische Mathematik II*. 3. Auflage, Springer-Verlag, Berlin, 1990

Stoer, J. and R. Bulirsch: [2] *Introduction to numerical analysis*. Springer-Verlag, Berlin, 1980

Stone, H. L.: Iterative solution of implicit approximations of multidimensional partial differential equations. *SIAM J. Numer. Anal.* **5** (1968) 530–558

Strakoš, Z.: [1] On the real convergence rate of the conjugate gradient method. *LAA* **154** (1991) 535–549

Strauss, E. G.: see Forsythe, G. E. et al.

Stüben, K.: [1] Algebraic multigrid (AMG): Experiences and comparisons. *Appl. Math. Comput.* **13** (1983) 419–451

Tanabe, K.: [1] Projection method for solving a singular system of linear equations and its applications. *Numer. Math.* **17** (1971) 203–214

Todd, J.: [1] Applications of transformation theory: A legacy from Zolotarev (1847–1878). In: *Approximation theory and spline functions* (S. P. Singh et al., eds.), Reidel Publ., Dordrecht, 1984, pages 207–245

Trottenberg, U.: see Braess, D. et al., Frehse, J. et al., Hackbusch, W. et al., Schröder, J. et al.

Van der Sluis, A.: [1] Condition numbers and equilibrium matrices. *Numer. Math.* **14** (1969) 14–23

van der Sluis, A. and H. A. van der Vorst: [1] The rate of convergence of conjugate gradients. *Numer. Math.* **48** (1986) 543–560

van der Sluis, A.: see Axelsson, O. et al.

van der Vorst, H. A.: [1] Iterative solution mthods for certain sparse linear systems with a non-symmetric matrix arising from pde-problems. *J. Comput. Phys.* **44** (1981) 1–19

van der Vorst, H. A.: [2] A vectorizable variant of some ICCG methods. *SIAM J. Sci. Statist. Comput.* **3** (1982) 350–356

van der Vorst, H. A.: [3] Bi-CGSTAB: A fast and smoothly converging variant of Bi-CG for the solution of nonsymmetric linear systems. *SIAM J. Sci. Statist. Comput.* **13** (1992) 631–644

van der Vorst, H. A.: see Meijerink, J. A. et al., van der Sluis, A. et al.

van Loan, Ch. F.: see Golub, G. H. et al.

Varga, R. S.: [1] Factorization and normalized iterative methods. In: Boundary problems in differential equations (R. E. Langer, ed.), University of Wisconsin Press, Madison, 1960, pages 121–142

Varga, R. S.: [2] *Matrix iterative analysis*. Prentice-Hall, Englewood Cliffs, 1962

Varga, R. S.: [3] On recurring theorems on diagonal dominance. *Lin. Alg. Appl.* **13** (1976) 1–9

Varga, R. S., E. B. Saff and V. Mehrmann: [1] Incomplete factorizations of matrices and connections with H-matrices. *SIAM J. Numer. Anal.* **17** (1980) 787–793

Varga, R. S.: see Alefeld, G. et al., Eiermann, M. et al., Neumaier, A. et al., Niethammer, W. et al.

Vassilevski, P.: [1] Multilevel preconditioning matrices and multigrid V-cycle methods. In: Hackbusch [18] 200–208

Vassilevski, P.: see Axelsson, O. et al.

Vorst, H. A. van der: see van der Vorst, H. A.

Wachspress, E. L.: [1] *Iterative solution of elliptic systems and applications to the neutron diffusion equations of reactor physics.* Prentice-Hall, Englewood Cliffs, 1966

Walker, H. F.: [1] Implementation of the GMRES method using Householder transformations. *SIAM J. Sci. Statist. Comput.* **9** (1988) 152–163

Welfert, B. D.: see Bank, R. E. et al.

Wesseling, P.: [1] Theoretical and practical aspects of a multigrid method. *SIAM J. Sci. Statist. Comput.* **3** (1982) 387–407

Wesseling, P.: [2] A robust and efficient multigrid method. In: Hackbusch–Trottenberg [1] 614–630

Wesseling, P.: [3] *An introduction to multigrid methods.* Wiley, Chichester, 1991

Wesseling, P.: see Sonneveld, P. et al., Hemker, P. W. et al.

Widlund, O.: [1] A Lanczos method for a class of nonsymmetric systems of linear equations. *SIAM J. Numer. Anal.* **15** (1978) 801–812

Widlund, O.: [2] Iterative substructuring methods: Algorithms and theory for elliptic problems in the plane. In: Glowinski–Golub–Meurant–Périaux [1] 113–128

Widlund, O.: [3] Optimal iterative refinement methods. In: Chan–Glowinski–Périaux–Widlund [1] 114–125

Widlund, O.: see Björstad, P. E. et al., Börgers, Ch. et al., Cai et al., Chan, T. F. et al., Dryja, M. et al., Proskurowski, W. et al.

Windisch, G.: [1] *M-matrices in numerical analysis.* Teubner-Texte zur Mathematik, Vol. 115. Teubner, Leipzig, 1989

Wirth, N.: see Jensen, K. et al.

Wittum, G.: [1] *Distributive Iterationen für indefinite Systeme als Glätter im Mehrgitterverfahren am Beispiel der Stokes- und Navier-Stokes-Gleichungen mit Schwerpunkt auf unvollständige Zerlegungen.* Dissertation, Kiel, 1986

Wittum, G.: [2] Multi-grid methods for Stokes and Navier-Stokes equations. Transforming smoothers: Algorithms and numerical results. *Numer. Math.* **54** (1989) 543–563

Wittum, G.: [3] On the convergence of multi-grid methods with transforming smoothers. Theory with applications to the Navier-Stokes equations. *Numer. Math.* **57** (1989) 15–38

Wittum, G.: [4] Linear iterations as smoothers in multigrid methods: Theory with applications to incomplete decompositions. *Impact of Comput. in Sc. and Eng.* **1** (1989) 180–215

Wittum, G.: [5] On the robustness of ILU smoothing. *SIAM J. Sci. Stat. Comput.* **10** (1989) 699–717

Wittum, G.: [6] An ILU-based smoothing correction scheme. In: Hackbusch [19] 228–240

Wittum, G.: [7] *Filternde Zerlegungen: Schnelle Löser für große Gleichungssysteme* (Filtering decomposition, Fast solvers for large systems of equations). Teubner Skripten zur Numerik, Teubner, Stuttgart, 1992

Wittum, G. and F. Liebau: [1] On truncated incomplete decompositions. *BIT* **29** (1989) 719–740

Wozniakowski, H.: [1] Round-off error analysis of iterations for large linear systems. *Numer. Math.* **30** (1978) 301–314

Xu, Jinchao: [1] *Theory of multilevel methods.* Thesis. Penn. State University, 1989

Xu, Jinchao: [2] Iterative methods by space decomposition and subspace correction: a unifying approach. *SIAM Review*, to appear

Xu, Jinchao: [3] A new class of iterative methods for nonself-adjoint or indefinite problems. *SIAM J. Numer. Anal.* **29** (1992) 303–319

Xu, J.: see Bramble, J. H. et al.

Young, D. M.: [1] *Iterative methods for solving partial differential equations of elliptic type.* Thesis, Harvard University, 1950

Young, D. M.: [2] *Iterative solution of large linear systems.* Academic Press, New York, 1971

Young, D. M.: [3] On the accelerated SSOR method for solving large linear systems. *Adv. in Math.* **23** (1977) 215–271

Young, D. M.: see Hageman, L. A. et al.

Yserentant, H.: [1] On the convergence of multi-level methods for strongly nonuniform families of grids and any number of smoothing steps per level. *Computing* **30** (1983) 305–313

Yserentant, H.: [2] Hierarchical bases of finite element spaces in the discretization of nonsymmetric elliptic boundary value problems. *Computing* **35** (1985) 39–49

Yserentant, H.: [3] On the multi-level splitting of finite element spaces. *Numer. Math.* **49** (1986) 379–412

Yserentant, H.: [4] Preconditioning indefinite discretization matrices. *Numer. Math.* **54** (1989) 719–734

Yserentant, H.: [5] Two preconditioners based on the multi-level splitting of finite element spaces. *Numer. Math.* **58** (1990) 163–184

Yserentant, H.: [6] Hierarchical bases. Proceedings of ICIAM91, SIAM, Philadelphia (to appear)

Yserentant, H.: see Bank, R. E. et al., Bornemann, F. et al.

Zeeuw, P. M. de: see de Zeeuw, P. M.

Subject Index

Page numbers in bold denote pages where definitions or headings appear. Related or more specific terms are given parenthetically, preceeded by an arrow.

D

E

Index of PASCAL Terms (Procedures, Functions, and Types)

Page numbers in bold face indicate that the program text of the procedure or function can be found at that page.

D

E

F

G

H

I

J

Applied Mathematical Sciences

(continued from page ii)

52. *Chipot:* Variational Inequalities and Flow in Porous Media.
53. *Majda:* Compressible Fluid Flow and System of Conservation Laws in Several Space Variables.
54. *Wasow:* Linear Turning Point Theory.
55. *Yosida:* Operational Calculus: A Theory of Hyperfunctions.
56. *Chang/Howes:* Nonlinear Singular Perturbation Phenomena: Theory and Applications.
57. *Reinhardt:* Analysis of Approximation Methods for Differential and Integral Equations.
58. *Dwoyer/Hussaini/Voigt (eds):* Theoretical Approaches to Turbulence.
59. *Sanders/Verhulst:* Averaging Methods in Nonlinear Dynamical Systems.
60. *Ghil/Childress:* Topics in Geophysical Dynamics: Atmospheric Dynamics, Dynamo Theory and Climate Dynamics.
61. *Sattinger/Weaver:* Lie Groups and Algebras with Applications to Physics, Geometry, and Mechanics.
62. *LaSalle:* The Stability and Control of Discrete Processes.
63. *Grasman:* Asymptotic Methods of Relaxation Oscillations and Applications.
64. *Hsu:* Cell-to-Cell Mapping: A Method of Global Analysis for Nonlinear Systems.
65. *Rand/Armbruster:* Perturbation Methods, Bifurcation Theory and Computer Algebra.
66. *Hlavácek/Haslinger/Necasl/Lovísek:* Solution of Variational Inequalities in Mechanics.
67. *Cercignani:* The Boltzmann Equation and Its Applications.
68. *Temam:* Infinite Dimensional Dynamical Systems in Mechanics and Physics.
69. *Golubitsky/Stewart/Schaeffer:* Singularities and Groups in Bifurcation Theory, Vol. II.
70. *Constantin/Foias/Nicolaenko/Temam:* Integral Manifolds and Inertial Manifolds for Dissipative Partial Differential Equations.
71. *Catlin:* Estimation, Control, and the Discrete Kalman Filter.
72. *Lochak/Meunier:* Multiphase Averaging for Classical Systems.
73. *Wiggins:* Global Bifurcations and Chaos.
74. *Mawhin/Willem:* Critical Point Theory and Hamiltonian Systems.
75. *Abraham/Marsden/Ratiu:* Manifolds, Tensor Analysis, and Applications, 2nd ed.
76. *Lagerstrom:* Matched Asymptotic Expansions: Ideas and Techniques.
77. *Aldous:* Probability Approximations via the Poisson Clumping Heuristic.
78. *Dacorogna:* Direct Methods in the Calculus of Variations.
79. *Hernández-Lerma:* Adaptive Markov Processes.
80. *Lawden:* Elliptic Functions and Applications.
81. *Bluman/Kumei:* Symmetries and Differential Equations.
82. *Kress:* Linear Integral Equations.
83. *Bebernes/Eberly:* Mathematical Problems from Combustion Theory.
84. *Joseph:* Fluid Dynamics of Viscoelastic Fluids.
85. *Yang:* Wave Packets and Their Bifurcations in Geophysical Fluid Dynamics.
86. *Dendrinos/Sonis:* Chaos and Socio-Spatial Dynamics.
87. *Weder:* Spectral and Scattering Theory for Wave Propagation in Perturbed Stratified Media.
88. *Bogaevski/Povzner:* Algebraic Methods in Nonlinear Perturbation Theory.
89. *O'Malley:* Singular Perturbation Methods for Ordinary Differential Equations.
90. *Meyer/Hall:* Introduction to Hamiltonian Dynamical Systems and the N-body Problem.
91. *Straughan:* The Energy Method, Stability, and Nonlinear Convection.
92. *Naber:* The Geometry of Minkowski Spacetime.
93. *Colton/Kress:* Inverse Acoustic and Electromagnetic Scattering Theory.
94. *Hoppensteadt:* Analysis and Simulation of Chaotic Systems.
95. *Hackbusch:* Iterative Solution of Large Sparse Systems of Equations.
96. *Marchioro/Pulvirenti:* Mathematical Theory of Incompressible Nonviscous Fluids.
97. *Lasota/Mackey:* Chaos, Fractals, and Noise: Stochastic Aspects of Dynamics, 2nd ed.
98. *de Boor/Höllig/Riemenschneider:* Box Splines.
99. *Hale/Lunel:* Introduction to Functional Differential Equations.
100. *Sirovich:* Trends and Perspectives in Applied Mathematics.
101. *Nusse/Yorke:* Dynamics: Numerical Explorations.
102. *Chossat/Iooss:* The Couette-Taylor Problem.